An Introduction to Quantum Optics (Second Edition)

An open systems approach

Online at: https://doi.org/10.1088/978-0-7503-6014-2

Series Editor

R Barry Johnson, a Senior Research Professor at Alabama A&M University, has been involved for over 50 years in lens design, optical systems design, electro-optical systems engineering, and photonics. He has been a faculty member at three academic institutions engaged in optics education and research, has been employed by a number of companies, and has provided consulting services.

Dr Johnson is an IOP Fellow, an SPIE Fellow and Life Member, an OSA Fellow, and was the 1987 President of SPIE. He serves on the editorial board of *Infrared Physics & Technology* and *Advances in Optical Technologies*. Dr Johnson has been awarded many patents, has published numerous papers and several books and book chapters, and was awarded the 2012 OSA/SPIE Joseph W Goodman Book Writing Award for *Lens Design Fundamentals* (second edition). He is a perennial co-chair of the annual SPIE Current Developments in Lens Design and Optical Engineering Conference.

Foreword

Until the 1960s the field of optics was primarily concentrated in the classical areas of photography, cameras, binoculars, telescopes, spectrometers, colorimeters, radiometers, etc. In the late 1960s optics began to blossom with the advent of new types of infrared detector, liquid crystal display (LCDs), light emitting diode (LEDs), charge coupled device (CCDs), laser, holography, and fiber optics along with new optical materials, advances in optical and mechanical fabrication, new optical design programs, and many more technologies. With the development of the LED, LCD, CCD, and other electro-optical devices, the term 'photonics' came into vogue in the 1980s to describe the science of using light in the development of new technologies and the operation of a myriad of applications. Today optics and photonics are truly pervasive throughout society and new technologies are continuing to emerge. The objective of this series is to provide students, researchers, and those who enjoy self-education with a wide-ranging collection of books, each of which focuses on a topic relevant to the technologies and applications of optics and photonics. These books will provide knowledge to prepare the reader to be better able to participate in these exciting areas now and in the future. The title of this series is *Emerging Technologies in Optics and Photonics*, in which 'emerging' is taken to mean 'coming into existence', 'coming into maturity', and 'coming into prominence'. IOP Publishing and I hope that you will find this series of significant value to you and your career.

A list of titles published in this series can be found here: https://iopscience.iop.org/bookListInfo/emerging-technologies-in-optics-and-photonics.

An Introduction to Quantum Optics (Second Edition)

An open systems approach

Perry Rice
Department of Physics, Miami University, Oxford, United States of America

IOP Publishing, Bristol, UK

ISBN 978-0-7503-6014-2 (ebook)
ISBN 978-0-7503-6012-8 (print)
ISBN 978-0-7503-6015-9 (myPrint)
ISBN 978-0-7503-6013-5 (mobi)

DOI 10.1088/978-0-7503-6014-2

Version: 20250901

IOP ebooks

British Library Cataloguing-in-Publication Data: A catalogue record for this book is available from the British Library.

Published by IOP Publishing, wholly owned by The Institute of Physics, London

IOP Publishing, No.2 The Distillery, Glassfields, Avon Street, Bristol, BS2 0GR, UK

US Office: IOP Publishing, Inc., 190 North Independence Mall West, Suite 601, Philadelphia, PA 19106, USA

This book is dedicated to the memory of my dad, Paul Richard Rice. A kind and loving man with a great fondness for music.

Contents

Preface

Preface for the first edition

In this volume, we hope to introduce some mathematical fundamentals of quantum optics from a statistical physics point of view. The material is meant to be understandable and usable by upperclassmen and beginning graduate students. Chapter 1 is a brief review of what is meant, what I mean, by quantum optics. The next two chapters, 2 and 3, are reviews of classical electromagnetism and quantum mechanics. The reviews are not comprehensive, but they provide the tools we shall need to proceed. Next, in chapter 4, we turn to the interaction of a simple two-level system with an electromagnetic wave to begin our studies of emission and absorption. The basics of the quantum treatment of the electromagnetic field are presented in chapter 5. In chapter 6, we revisit the simple two-level system, but let it interact with a quantized electromagnetic field. Measurement plays a key role in quantum optics and in quantum mechanics in general. Various detection schemes are discussed in chapter 7. The essence the matter is in chapters 8 and 9. Chapter 8 introduces the density matrix formulation of quantum mechanics for an open quantum system—one that exchanges energy and information with its environment. Chapter 9 discusses quantum trajectory theory, a method of unraveling the density matrix to allow for ease of computation as well as intuition. This is particularly useful for undergraduates who understand the evolution of quantum wave functions and do not shy away from the need for quantum jumps. Finally, we conclude in chapter 10 with the use of quasi-probability functions to describe quantum optical systems.

This book is the result of teaching quantum optics to undergraduates and master's students for 30 years at Miami University. The purpose was to introduce students to the mathematical formalism of quantum optics in the most physical way I knew how. There are far too many students to thank for suffering through the development of this book over that time, but I must thank my MS advisees Xiaosong Yin, Tim Burt, Bobby Jones, Pranaw Rungta, James Clemens, Puneet Swarup, Elliot Strimbu, Joe Leach, Mambwe Mamba, Dyan Jones, Jeff Hyde, Habtom Woldekristos, Todd vanWoerkom, Charlie Baldwin, Thomas Jenkins, Jared Goettemoeller, Robert McCutcheon, Tyler Thurtell, Arkan Hassan, Daniel King, and Mitch Mazzei. Also, Matt Terraciano and Joe Reiner from other institutions. Undergraduate research students who were helpful in the development of this book were James Walden, Chris Baird, Shohini Ghose, Scott Secrest, Jeffrey Hyde, Luke Keltner, Bill Konyk, Robert Rogers, and Jason Rook. Most of these students produced research articles using the techniques described within.

I must also thank my undergraduate mentors at Wright State University, Merrill Andrews, Gust Bambikidis, Joe Hemsky, Mitchell Simpson, and David Wood. The attention given to their students was the beginning of my development as a teacher. In graduate school at the University of Arkansas, I had the pleasure of interacting with Howard Carmichael, James Cresser, William Harter, Peter Milonni, Surendra

Singh, and Reeta Vyas. One could not have asked for a better set of teachers and role models. Fellow travelers of note were Paul Alsing, Greg Holliday, Steve Montgomery, John Shultz, Jan Yarrison, David Weeks, and Murray Wolinsky. I have been fortunate to have met Luis Orozco at this time and have begun a lifetime of collaboration and friendship.

My interest in quantum optics began when my colleagues Leno Pedrotti went to graduate school working with Marlon Scully, and Bob Brecha worked with H Jeffrey Kimble. Leno and Bob have remained great friends and quantum colleagues over the years, and I have treasured that. I was working for the USAF wondering what to do when I discovered Richard Cook, who patiently worked through Glauber's initial papers, and directed me to the University of Arkansas and Peter Milonni. Before Peter moved to Los Alamos, I was able to learn much about the nature of the quantum vacuum, how beautiful quantum electrodynamics is, and quantum field theory in general. My research advisor was Howard Carmichael. He perhaps has had the most impact on how I view the quantum world. One is biased of course, but I think he is the finest advisor and friend one can have.

Finally, my loving family who started me out deserve notice. My dad Paul, mom Lucy, grandma Gertrude, aunt Fern, and uncle Bud. At present, my wife Jan and son Patrick are at my side. I feel very lucky to have walked the path I have, with the company I had.

Furthermore, as they say, all mistakes in the book are solely attributable to me. I hope this is of value to people starting to learn about the mysteries of quantum optics.

Perry Rice, Xenia, 14 March 2020

Preface for the second edition

This edition is expanded and updated. There are five new chapters, and two of the old chapters have been split into two. Chapter 1 has a discussion of light passing through three polarizers at $\pi/4$ axis angles with respect to each other, a simple system that exemplifies much of 'quantum weirdness'. Chapter 2 is basically unchanged, and chapter 3 has added to the review of quantum mechanics a discussion of the Feynman path integral as it is now used later in the book. Chapter 4 now includes a discussion on collective emission from two atoms and a brief discussion of superradiance. The next chapter, chapter 5, is new and deals with the semiclassical theory of multilevel atoms, with emphasis on three-level systems. There are sections on quantum beats (semiclassical versus quantum), driven systems and electromagnetically induced transparency (EIT), classical EIT, slow and stationary light, quantum memories for photons, cavity induced transparency, and two-photon absorption. The chapter of quantum fields has been split into two and expanded. Chapter 6 is the old chapter for the most part, and chapter 7 discusses beam splitters, single- and two-photon interference, and gravitational wave detection using interferometers. In chapter 8, we have added a discussion of the relation between the Jaynes–Cummings model and supersymmetry. A discussion of out-of-time correlation functions and one on detailed balance have been added to chapter 10. Chapter 11 on quantum trajectory theory has a discussion of schemes to observe quantum jumps in three-level atoms, single-atom optical bistability, weak field cavity quantum electrodynamics, the quantum Zeno effect, and a description of a recent experiment to catch and reverse a quantum jump with high efficiency. The chapter on quasiprobabilty distributions has been split into two, chapter 12 is the bulk of that and chapter 13 has an expanded discussion of Langevin equations, input–output theory and its applications to an optical parametric oscillator, and a discussion of cascaded quantum systems. Chapter 14 is new, and has a discussion of the Schwinger–Keldysh formalism for quantum mechanical statistical mechanics, widely used in condensed matter physics. Chapter 15 is also new, and discusses double-sided Feynman diagrams and their application in multidimensional Fourier spectroscopy of nonlinear systems. A brief introduction to quantum entanglement, its uses, and ways to quantify it are in the new chapter 16. Finally chapter 17 is essentially a list of topics for further study. I hope this makes the volume more useful for readers and researchers. I would like to thank Mike Raymer, Andy Marcus, and Brian Smith for hosting a visit to the Oregon Optical, Molecular, and Quantum Sciences, as well as Dylan Heusseman, Justin Kittell, Jack Maurer, and Tiemo Landes.

Perry Rice, Xenia, 31 March 2025

Author biography

Dr Perry Rice

Dr Perry Rice earned a BS in Physics from Wright State University in 1981, and a PhD in Physics from the University of Arkansas in 1988. He then spent 30 years as a professor at Miami University where he is an Emeritus Professor, followed by a year at the Oregon Center for Optical, Molecular, and Atomic Physics. His research focuses on the interaction of quantized light with matter and the properties of quantum noise, including resonance fluorescence, cavity quantum electrodynamics, and nonlinear optics. Currently he uses neural networks to work on waveguide quantum electrodynamics, quantum phase transitions, error correction codes, and the distribution of multipartite entanglement. In his spare time, he enjoys playing and listening to jazz, blues, folk, and rock music.

Chapter 1

Introduction

1.1 What is quantum optics?

The simplest 'definition' of quantum optics is the discipline that examines 'what is a photon?'. A more detailed description would be a 'non-relativistic quantum field theory of atoms interacting with light'. The field grew out of the invention of the laser and the question of 'what is the difference between laser light and a light bulb?'. The acronym LASER stands for 'light amplification by stimulated emission of radiation'. Often in textbooks, it is stated that laser light is *directional, intense, and monochromatic.* I can make the light directional via a fiber optic, I can increase the power until the thing melts, and I can filter the output to limit the bandwidth. But what about the stimulated emission? When a photon interacts with a detector, it does not announce its origin as stimulated or spontaneous; it is just an excitation of an electromagnetic field mode. My definition of a photon is a click in an absorptive detector, as we will detail later. It is not a particle or a wave, but an excitation of a bosonic field. So how *do* you tell whether the light is coming from an incandescent bulb or a HeNe laser? Roy Glauber [1] taught us that the difference is in *photon correlations.* One must look past expectation values and examine two-time correlations at the least. The fluctuations in the light field provide information that cannot be accessed by just examining the averages. It will turn out that for laser light the statistics of the photo detection events are Poissonian, each one being independent of the other. For light from an incandescent bulb, thermal light, we find that the photons are more likely to come in pairs than they would be for a random source. This is the origin of the Hanbury–Brown–Twiss effect [2], whereby correlations in intensity can be detected from two spatially separated detectors. We will find that light sources exist where the photons are less likely to come in pairs than from a laser, which is called photon anti-bunching. This type of light is quite interesting as it *cannot* be modeled by a deterministic random process. A cartoon of this concept is shown in figure 1.1.

doi:10.1088/978-0-7503-6014-2ch1

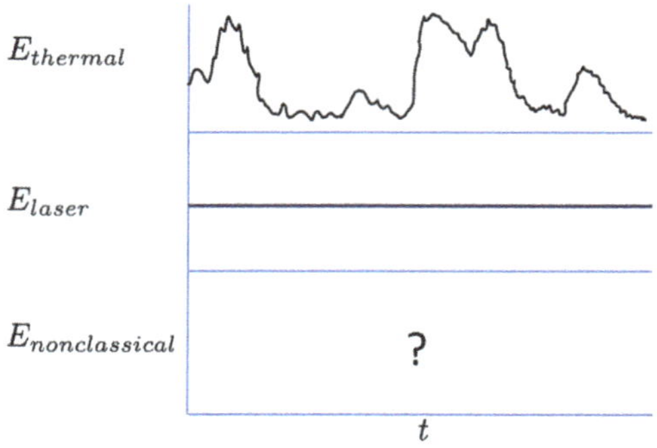

Figure 1.1. Electric field amplitude time evolution for a noisy, thermal source, a laser source, and a nonclassical or manifestly quantum light source.

In thermal light there are large fluctuations and during periods of higher intensity you are more likely to detect a photon. You are also more likely to get a second photon, simply because then there are more photons. For a laser, the output is described by a Poisson stochastic process; after the detection of a photon, you are no more or less likely to detect a second photon. A fully quantum theory of the laser based on Scully and Lamb [3] showed the Poissonian nature of the photon statistics and pointed out the key role of spontaneous emission in determining the non-zero linewidth of the laser output. What about the anti-correlations? The simplest example of such a source is a single two-level atomic transition driven by a laser, and examining the fluorescence. To spontaneously emit, the atom must be excited. Upon emission, the atom goes to the ground state and hence cannot emit a second photon. The photon statistics and spectra of this system of resonance fluorescence were the focus of intense work in the mid 1970s, and a fully quantum model of the electromagnetic field was found to be necessary to understand and predict the results of the experiments.

How can we differentiate between different field states if we cannot measure the field directly? Let us consider a field inside a box and ask how many photons are in the box. We will open the box and let all the photons leak out and count how many we detect. Of course, we assume perfect photo detection for now. Then let us do that over and over, as our answer will be a random number generally. Different field states will have different statistics and different probability distributions for the number of photons we detect. Examples of such distributions are shown in figure 1.2. There are states of the field with a definite number of photons, called number or Fock states. The coherent state that is output from a laser has a Poissonian distribution, and the thermal field has the decaying exponential distribution characteristic of a Bose–Einstein distribution. More interesting distributions result from nonclassical states such as a displaced squeezed vacuum or squeezed state. How do we prepare such states in a box? Good question. What we are generally more interested in is a slightly leaky box, where we also inject photons. The result is often a steady state that is not in thermal equilibrium, a non-equilibrium steady state. The leaky box can be two mirrors forming an optical cavity, such as in a

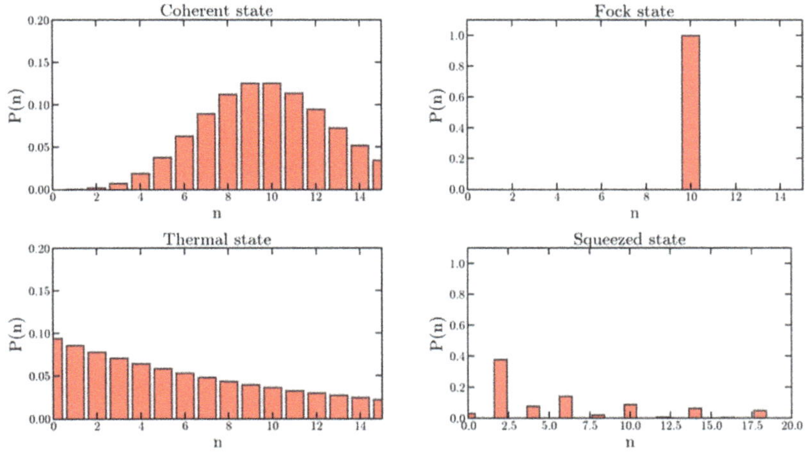

Figure 1.2. Probability distribution of photons $P(n)$ for the given field states with $\langle n \rangle = 10$.

Fabry–Perot, or a semiconductor wave guide with polished ends. We can use an essentially classical source, the output of a laser. We can put single atoms, a gas of many atoms, or crystals inside this slightly leaky box and create the field states of our choice. Some are more difficult than others, of course. We can infer what is inside the box by the statistical properties of what leaks out. This could be through a non-perfect mirror or fluorescence out through the sides of the cavity, perhaps via spontaneous emission. The leaky part is important because we would like to create a field source and the leak is essentially our output. A laser with two perfect mirrors would generate a large field inside the cavity, but if no light comes out, we cannot make laser pointers, blow holes in things, or perform laser spectroscopy.

The focus of this book is to examine how to calculate the statistics of photo counts, as well as various spectral quantities in various systems. We will mainly look at two-time correlations, but of course the probability distribution is determined in full by the entire set of n-time correlation functions, just as in quantum electrodynamics and other field theories.

1.2 Open quantum systems

In this example above, resonance fluorescence, the atom is not an isolated system. Energy is put into the system via the driving laser and energy leaves the system via interaction with the field modes (all directions, frequencies, and polarizations) that are in the ground or vacuum state. It is via the latter interaction that quantum fluctuations, or vacuum fluctuations, enter the picture. Such a system that interacts with its environment is known as an *open quantum system*. Such a system is not described by a wave function, unless one considers the wave function of the atom and all of the field modes as an infinite number. Therefore, such an approach is not tenable, and we must resort to approximations. If we want a description of our atom and its interaction with its environment, we must use a density operator or density matrix. In the simplest models, the density operator evolves via the Schrödinger

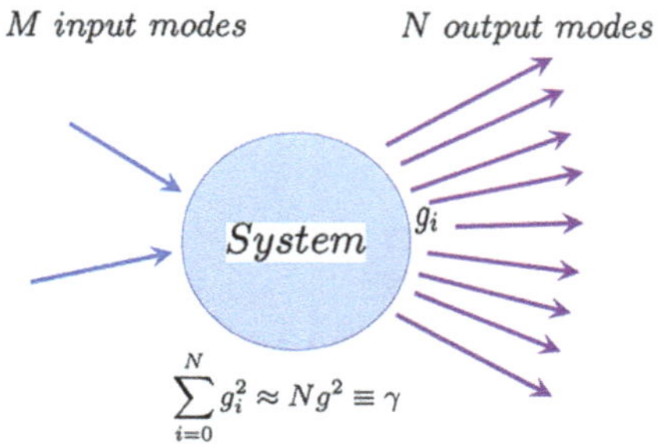

Figure 1.3. Sketch of an open quantum system with a (typically small) M number of input modes and a large number N of output modes. The coupling of the system to a given mode is g_i.

supplemented by corrections due to the environmental interactions. These corrections will depend only on the gross properties of the environment, such as temperature, pressure, etc. In this procedure, we reduce the number of equations by omitting the details of the environment, thereby losing information about the environment. A good example might be dropping an ice cube into the ocean. The ice cube will eventually melt, and the temperature of the ocean will not change very much, nor will any other properties of the ocean. The ocean can be considered a heat bath at a constant temperature. This is mainly due to the large volume of the ocean; an ice cube dropped into a glass of water will lower the temperature of the resulting water by a measurable amount, and as such, the thermodynamic properties of water can be measured via calorimetry. In the case of the quantum electromagnetic vacuum, we have an infinite number of modes, an infinite number of frequencies, each with an infinite number of propagation directions, and two polarizations. Therefore, it seems reasonable that an approach based on the interaction of a closed quantum system with a type of heat bath will be useful.

We consider the coupling of the atom to a given field mode (g) as very small, that it scales with the inverse square root of the volume of the field mode (which is infinite). However, the number of modes N is infinite. We find that the spontaneous emission rate of the atom depends on the product Ng^2 which can take on a finite value, as shown in figure 1.3. The fancy title for what we are doing is quantum statistical mechanics.

1.3 The mystery of the polaroids

Perhaps you are familiar with a polarizing filter (figure 1.4). This device allows light polarized in one direction to pass through, while the orthogonal polarization is not passed through. With these, we can explore the essence of quantum mechanics, following an idea attributed to Dirac [4]. We will assume linear polarization for now, but the concepts apply to other orthogonal pairs of polarizations as well. You may have these in a pair of sunglasses. Light that reflects off a surface such as a lake becomes predominantly horizontally polarized. Using a lens that only passes vertically polarized light, eliminates those reflections, or glare.

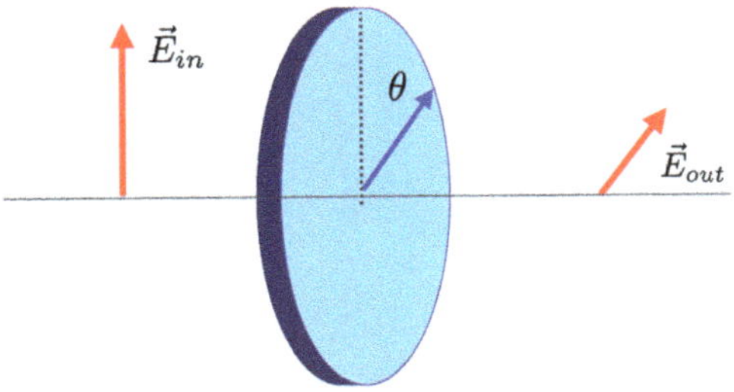

$$|E_{out}| = |E_{in}| cos\ \theta$$

$$I_{out} = I_{in}\ cos^2\theta$$

Figure 1.4. Linear polarization filter.

These filters are usually made of some type of plastic or other material with long 'stringy' molecules, which are predominantly oriented along an axis perpendicular to the transmission axis. Light polarized along the direction of the molecules excites the vibrational levels of the molecule, taking energy from the beam. The amount of energy taken out is defined by an extinction ratio, and we will assume our filter is all or nothing. Light polarized perpendicular to the molecules does not excite vibrations and passes through the filter. If we have a vertically polarized light incident on a filter with the transmission axis at an angle of θ, the transmitted intensity is given by Malus's law:

$$I_{\text{out}} = I_{\text{in}}\cos^2\theta. \tag{1.1}$$

This law has been known from ancient times. This is due to the partial excitation of the vibrational modes of the molecules. The key thing to notice is that the light passing through the polarizer is polarized along the transmission axis. The state of the light changes, not just the intensity. If we send light through a polarizer, the output light is polarized along the transmission axis. If we add a second polarizer with a transmission axis perpendicular to the first, no light passes through the pair (see figure 1.5).

Now let us put a third filter in between those two, oriented at $\pi/4$ with respect to them (figure 1.6). Light passing through the first polarizer is vertically polarized. Light passing through the second polarizer (the new one) is reduced in intensity by a factor of 1/2 but, crucially, the light is polarized at an angle of $\pi/4$ with the transmission axis of the first. Passing through the final polarizer, we obtain another reduction in the intensity of 1/2, and the resultant light is polarized horizontally. It is evident that a polarizer does more than block light.

It essentially makes a measurement of the polarization and changes the axis of the resultant light. In terms of the quantum mechanics, the polarizer is akin to a Hermitian operator, and the transmission axis is one of the eigenvectors of that operator with eigenvalue 1. The perpendicular axis is also an eigenvector with eigenvalue 0. If we define a polarization vector

$$|P\rangle = \begin{pmatrix} \psi_H \\ \psi_V \end{pmatrix}, \tag{1.2}$$

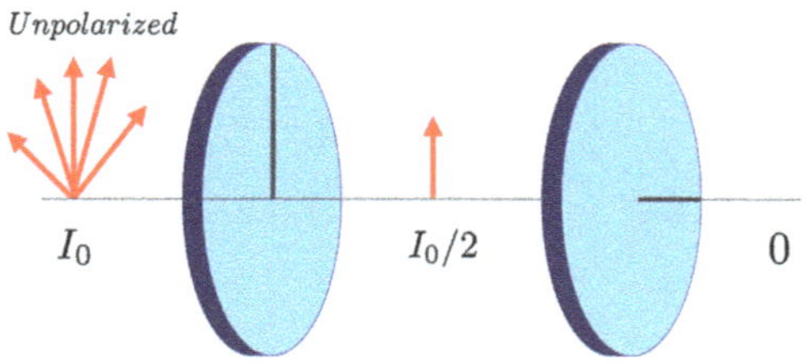

Figure 1.5. A pair of filters orthogonal to each other.

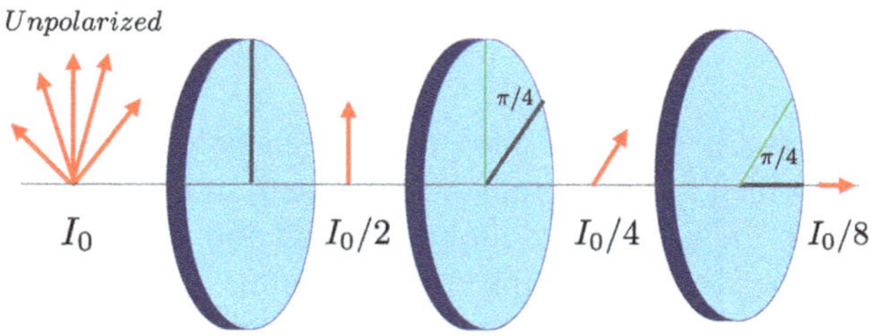

Figure 1.6. A trio of polarization filters.

the operator representing a filter with a horizontal transmission axis is

$$H = \begin{pmatrix} 1 & 0 \\ 0 & 0 \end{pmatrix}. \tag{1.3}$$

1.4 This book

This book resulted from notes developed for teaching quantum optics to upper-level undergraduates, graduate students starting out in the field, or anyone with a good grasp of quantum mechanics who wishes to know more. Quantum optics is one of the seminal fields for the revolution underway in quantum information processing, the detection of gravitational waves, many-body quantum systems, nanophotonics, and others. As such, an audience beyond specialists will want/need to understand the basics of quantum optics. This book seeks to present the simplest non-trivial models that illustrate the principles and is not meant to be exhaustive. In the 1980s whilst in graduate school, the key books (at least for me) were those of Loudon [5], Allen and Eberly [6], and Sargent, Scully, and Lamb [7]. I would suggest that the modern trifecta would include the books by Orszag [8], Scully and Zubairy [9], and Mandel and Wolf [10]. I would also recommend the trilogy written by Carmichael [11–13], and the work of Milonni [14–17] for more detail. My interactions with the latter two largely form the basis of my approach. A similar approach is that of Bruer and Pettrucione [18]. Other books I am familiar enough to recommend highly are those of Walls and Milburn [19], Gardiner and Zoller [20], Gerry and Knight [21], Meystre and Sargent [22], and Agarwai [23]. I would be remiss not to direct the readers' attention to the set of notes on quantum and atom optics notes available online, provided by Dan Steck [24]. There are many good texts on quantum optics these days, so why not one more? The novelty of this book is the accessibility of the material to a newcomer to the field, and it is more pedagogical than exhaustive.

The great Sidney Coleman once said: 'The career of a young theoretical physicist consists of treating the harmonic oscillator in ever-increasing levels of abstraction.' I say add the two-level atom to get the fermions involved, and you have the Jaynes–Cummings model and quantum optics. For me, the joy of this field is the ability to study quantized fields and matter deeply, and their interactions, *and* have access to clever experimentalists that can actually measure what we theorists calculate. Furthermore, most know theory well, and then have the magic of making it real. I think this is an amazing field for theorists or experimentalists just starting out, or for folks who have not picked a 'camp' yet. It prepares you well for fascinating physics in any field that uses quantum principles, and more and more each day, non-academic positions exploring the frontier of quantum information.

Many of the figures in this book were made using the Quantum Toolbox in Python 5.1 (QuTiP) [25–27].

References

[1] Glauber R 1963 Coherent and incoherent states of the radiation field *Phys. Rev.* **131** 2766–88

[2] Hanbury-Brown R and Rwiss R Q 1956 Correlation between photons in two coherent beams of light *Nature* **177** 27–9

[3] Scully M O and Lamb W 1967 Quantum theory of an optical maser. I *General theory Phys. Rev.* **159** 208–26

[4] Dirac P A M 2001 *The Principles of Quantum Mechanics* 4th edn (Oxford: Oxford University Press)

[5] Loudon R 2000 *The Quantum Theory of Light* (Oxford: Oxford University Press)

[6] Allen L and Eberly J 1975 *Optical Resonance and Two-Level Atoms* (New York: Wiley)

[7] Sargent M, Scully M and Lamb W 1975 *Laser Physics* (Reading, MA: Addison-Wesley)

[8] Orszag M 2000 *Quantum Optics* Advanced Texts in Physics (Berlin: Springer)

[9] Scully M O and Zubairy M S 1997 *Quantum Optics* (Cambridge: Cambridge University Press)

[10] Mandel L and Wolf E 1995 *Optical Coherence and Quantum Optics* (Cambridge: Cambridge University Press)

[11] Carmichael H J 1993 *An Open Systems Approach to Quantum Optics* (Berlin: Springer)

[12] Carmichael H J 1999 *Statistical Methods in Quantum Optics 1: Master Equations and Fokker–Planck Equations* (Berlin: Springer)

[13] Carmichael H J 2007 *Statistical Methods in Quantum Optics 2: Non-Classical Fields* (Berlin: Springer)

[14] Milonni P W and Eberly J H 1994 *Laser Physics* (New York: Wiley)

[15] Milonni P W 1994 *The Quantum Vacuum: An Introduction to Quantum Electrodynamics* (New York: Academic)

[16] Milonni P 2004 *Fast Light, Slow Light, and Left-Handed Light* (Boca Raton, FL: CRC Press)

[17] Milonni P 2019 *An Introduction to Quantum Optics and Quantum Fluctuations* (Oxford: Oxford University Press)

[18] Breuer H and Petruccione F 2002 *The Theory of Open Quantum Systems* (Oxford: Oxford University Press)
[19] Walls D and Milburn G 2008 *Quantum Optics* (Berlin: Springer)
[20] Gardiner C and Zoller P 2004 *Quantum Noise* (Berlin: Springer)
[21] Gerry C and Knight P 2004 *Introductory Quantum Optics* (Cambridge: Cambridge University Press)
[22] Meystre P and Sargent M 2007 *Elements of Quantum Optics* (Berlin: Springer)
[23] Agarwal G 2012 *Quantum Optics* (Cambridge: Cambridge University Press)
[24] Steck D 2007 *Quantum and atom optics notes* The Quantum Well https://publish.obsidian.md/myquantumwell/Literature/Steck%2C+D.+A.%2CQuantum+and+Atom+Optics%2C+2007+(last+revised+September+2021)
[25] Johansson J R, Nation P D and Nori F 2012 QuTiP: an open-source Python framework for the dynamics of open quantum systems *Comput. Phys. Commun.* **183** 1760–72
[26] Johansson J R, Nation P D and Nori F 2013 QuTiP 2: a Python framework for the dynamics of open quantum systems *Comput. Phys. Commun.* **184** 1234–40
[27] Lambert N *et al* 2024 QuTiP 5: the quantum toolbox in Python arXiv: 2412.04705

IOP Publishing

An Introduction to Quantum Optics (Second Edition)

An open systems approach

Perry Rice

Chapter 2

Classical electromagnetism and linear optics

2.1 Maxwell's equations and electromagnetic waves

Here we review the basics of electromagnetism and applications to wave propagation. Standard references would include [1–6]. We begin with a discussion of Maxwell's equations. Recall that these are

Gauss's law	$\vec{\nabla}\cdot\vec{E} = \rho/\epsilon_0$
The law with no name	$\vec{\nabla}\cdot\vec{B} = 0$
Faraday's law	$\vec{\nabla}\times\vec{E} = -\partial\vec{B}/\partial t$
Ampère's law	$\vec{\nabla}\times\vec{B} = \mu_0\vec{J}+\frac{1}{c^2}\partial\vec{E}/\partial t$

In these we have the total electric charge density ρ and the total current density $\vec{J}$ as sources for the electric and magnetic fields, $\vec{E}$ and $\vec{B}$. Also we have defined $c^2 = 1/\epsilon_0\mu_0$. The second equation, the law with no name (TLWNN) is automatically satisfied if we have

$$\vec{B} = \vec{\nabla}\times\vec{A} \tag{2.1}$$

as we have the identity that $\vec{\nabla}\cdot(\vec{\nabla}\times\overrightarrow{\text{anything}}) = 0$; the divergence has no curl. This allows us to introduce $\vec{A}$, the vector potential. The usual electric potential emerges by using equation (2.1) in Faraday's law,

$$\begin{aligned}\vec{\nabla}\times\vec{E} &= \frac{-\partial\vec{B}}{\partial t}\\ &= -\vec{\nabla}\times\frac{\partial\vec{A}}{\partial t},\end{aligned} \tag{2.2}$$

doi:10.1088/978-0-7503-6014-2ch2

which allows us to write

$$\vec{E} = -\frac{\partial \vec{A}}{\partial t} + \vec{\eta}, \tag{2.3}$$

where we must have $\vec{\nabla} \times \vec{\eta} = 0$. We can now make use of the identity $\vec{\nabla} \times \vec{\nabla}$ (any scalar function) $= 0$, or the gradient has no divergence, to write

$$\vec{\eta} = -\vec{\nabla}\phi, \tag{2.4}$$

where ϕ is the usual electric potential. We now have a relation between the potentials and the fields

$$\vec{E} = -\frac{\partial \vec{A}}{\partial t} - \vec{\nabla}\phi \tag{2.5}$$

$$\vec{B} = \vec{\nabla} \times \vec{A}. \tag{2.6}$$

Let us note that we now have four variables (functions really); the three components of the vector potential and the scalar potential. How many equations do we have? Well, TLWNN and Faraday's law are automatically satisfied by the potentials, so they can be viewed as constraints akin to initial or boundary conditions. Thus we have four first order equations (one vector equation in Ampere's law and one scalar equation, Gauss's law) and four constraints, which allows us to have a solution. The potentials are *not* unique as we shall see. Consider the following transformation, known as a gauge transformation:

$$\vec{A}' = \vec{A} - \vec{\nabla}\lambda \tag{2.7}$$

$$\phi' = \phi + \frac{\partial \lambda}{\partial t}. \tag{2.8}$$

What fields do these potentials produce?

$$\begin{aligned} \vec{B}' &= \vec{\nabla} \times \vec{A}' \\ &= \vec{\nabla} \times \vec{A} \\ &= \vec{B} \end{aligned} \tag{2.9}$$

$$\begin{aligned} \vec{E}' &= -\vec{\nabla}\phi' - \frac{\partial \vec{A}'}{\partial t} \\ &= -\vec{\nabla}\phi - \frac{\partial}{\partial t}\vec{\nabla}\lambda - \frac{\partial \vec{A}}{\partial t} + \frac{\partial}{\partial t}\vec{\nabla}\lambda \\ &= -\frac{\partial \vec{A}}{\partial t} - \vec{\nabla}\phi \\ &= \vec{E}. \end{aligned} \tag{2.10}$$

So we see that our electric and magnetic fields will be unchanged by the gauge transformation, thus it would seem that the potentials are not unique. What about the fields themselves? They are unique, a fact guaranteed by Helmholtz's theorem which states that the curl and divergence of a vector field uniquely define that field.

So we can use the potentials instead of the fields; in introductory physics the electric potential is introduced early on as it is often easier to solve for a scalar function and then takes its gradient to obtain the electric field. There is no such utility with the vector potential, hence it appears later! But what equations do these potentials satisfy? Let us start down a well trodden path by examining Faraday's equation in terms of potentials. We now need to find out more about the vector and scalar potentials. We do this by examining what equations they need to satisfy. We have

$$\vec{\nabla}\times\vec{B} = \vec{\nabla}\times(\vec{\nabla}\times\vec{A}) = \frac{1}{c^2}\left(\vec{\nabla}\frac{\partial\phi}{\partial t} - \frac{\partial^2\vec{A}}{\partial t^2}\right) + \mu_0\vec{J} \tag{2.11}$$

and

$$\vec{\nabla}(\vec{\nabla}\cdot\vec{A}) - \nabla^2\vec{A} + \frac{1}{c^2}\vec{\nabla}\frac{\partial\phi}{\partial t} + \frac{1}{c^2}\frac{\partial^2\vec{A}}{\partial t^2} = \mu_0\vec{J} \tag{2.12}$$

$$\vec{\nabla}\cdot\vec{E} = -\vec{\nabla}\cdot(\vec{\nabla}\phi) - \frac{\partial}{\partial t}(\vec{\nabla}\cdot\vec{A}) = \frac{\rho}{\epsilon_0}. \tag{2.13}$$

So that when we combine all of these, the two resultant equations are given as

$$\vec{\nabla}(\vec{\nabla}\cdot\vec{A}) - \nabla^2\vec{A} + \frac{1}{c^2}\vec{\nabla}\frac{\partial\phi}{\partial t} + \frac{1}{c^2}\frac{\partial^2\vec{A}}{\partial t^2} = \mu_0\vec{J} \tag{2.14}$$

$$-\nabla^2\phi - \frac{\partial}{\partial t}(\vec{\nabla}\cdot\vec{A}) = \frac{\rho}{\epsilon_0}. \tag{2.15}$$

Now, we want to work in the Coulomb gauge, $\vec{\nabla}\cdot\vec{A} = 0$. This can be explicitly shown by letting $\vec{A} \longrightarrow \vec{A_0} - \vec{\nabla}\eta$ and $\phi \longrightarrow \phi_0 + \partial\eta/\partial t$ for any function η. The Coulomb gauge is defined by $\vec{\nabla}\cdot\vec{A} = 0$. This gauge makes our equations a bit simpler, because since

$$\nabla^2\eta = \vec{\nabla}\cdot\vec{A}, \tag{2.16}$$

we can write our two equations for the potentials as

$$\nabla^2\vec{A} - \frac{1}{c^2}\frac{\partial}{\partial t}\vec{\nabla}\phi = -\mu_0\vec{J} \tag{2.17}$$

$$\nabla^2\phi = -\frac{\rho}{\epsilon_0}. \tag{2.18}$$

If $\vec{\nabla}\cdot\vec{A}\neq 0$, we plug $\vec{A}$ into (2.16) and solve for the $\vec{\eta}$ needed to make it so. The scalar potential satisfies Poisson's equation with a static solution

$$\phi(\vec{r}) = \frac{1}{4\pi\epsilon_0}\int\frac{\rho(\vec{r}')}{|\vec{r}-\vec{r}'|}d^3\vec{r}'. \tag{2.19}$$

We next decompose the current $\vec{J}$ into transverse and longitudinal terms $\vec{J} = \vec{J_T} + \vec{J_L}$ so that

$$\vec{\nabla}\times\vec{J_L} = 0 \tag{2.20}$$

$$\vec{\nabla}\cdot\vec{J_T} = 0. \tag{2.21}$$

Now, since the curl of $\vec{J_L} = 0$, we can write $\vec{J_L} = \vec{\nabla}\,\psi$. We then have

$$\vec{\nabla}\cdot\vec{J_L} = -\frac{\partial\rho}{\partial t} \tag{2.22}$$

and

$$\vec{\nabla}\cdot(\vec{\nabla}\times\vec{B}) = \epsilon_0\,\vec{\nabla}\cdot\frac{\partial\vec{E}}{\partial t} + \vec{\nabla}\cdot\vec{J}. \tag{2.23}$$

The longitudinal current density satisfies

$$\begin{aligned}\vec{\nabla}\cdot\vec{J_L} &= \vec{\nabla}\cdot(\vec{\nabla}\psi)\\ &= \nabla^2\psi\\ &= -\frac{\partial\rho}{\partial t}.\end{aligned} \tag{2.24}$$

This is nonzero if there are charges are moving transverse to the direction of propagation for some reason. We then have

$$\psi = \epsilon_0\frac{\partial\phi}{\partial t} \tag{2.25}$$

$$\vec{J_L} = \epsilon_0\,\vec{\nabla}\,\frac{\partial\phi}{\partial t}. \tag{2.26}$$

Finally then, we have

$$-\nabla^2\vec{A} + \frac{1}{c^2}\frac{\partial^2\vec{A}}{\partial t} = \mu_0\vec{J_T} \tag{2.27}$$

with the form of the solution being

$$\vec{A} = \frac{\mu_0}{4\pi} \int \frac{\vec{J_T}(\vec{r}', t)}{|\vec{r} - \vec{r}'|} d^3\vec{r}'. \tag{2.28}$$

The equations that determine the scalar potential are

$$\nabla^2 \phi = \frac{\rho}{\epsilon_0} \tag{2.29}$$

$$\epsilon_0 \vec{\nabla} \frac{\partial \phi}{\partial t} = \vec{J_L}. \tag{2.30}$$

In relativistic notation we have the four-vectors

$$J^\nu = (c\rho, \vec{J}) \tag{2.31}$$

$$A^\nu = (\phi/c, \vec{A}) \tag{2.32}$$

with Maxwell's equations written as

$$\Box A^\nu = \mu_0 J^\nu, \tag{2.33}$$

with the D'Alembertion operator

$$\begin{aligned} \Box &= \partial^\nu \partial_\nu \\ &= (\partial^2/c^2\partial t^2, -\nabla^2). \end{aligned} \tag{2.34}$$

The equation of continuity, exhibiting global charge conservation is

$$\frac{\partial J^\alpha}{\partial x^\alpha} = \frac{\partial \rho}{\partial t} + \vec{\nabla} \cdot \vec{J} = 0. \tag{2.35}$$

This is seen from

$$\frac{dQ_{\text{tot}}}{dt} = \frac{d}{dt} \int dV \rho = -\int \vec{J} \cdot dS = I\infty = 0, \tag{2.36}$$

where the volume integral is over all space, and the surface integral is over the surface at infinity, and we assume there are no currents on that surface. Finally in the language of differential forms, which we will not go into, the equations take the form

$$dF = 0 \tag{2.37}$$

$$d^*F = J \tag{2.38}$$

or in terms of potentials,

$$F = dA \tag{2.39}$$

$$d^*dA = J. \tag{2.40}$$

2.2 Wave equation for fields in a medium

Let us now consider generalized Maxwell's equations in a medium. These would be

Gauss's law	$\vec{\nabla}\cdot\vec{D} = \rho_{\text{free}}/\epsilon_0$
The law with no name	$\vec{\nabla}\cdot\vec{B} = 0$
Faraday's law	$\vec{\nabla}\times\vec{E} = -\partial\vec{B}/\partial t$
Ampere's law	$\vec{\nabla}\times\vec{H} = \mu_0 \vec{J}_{\text{free}} + \partial\vec{D}/\partial t$

where we also have the constitutive relations

$$\vec{D} = \epsilon_0 \vec{E} + \vec{P} \tag{2.41}$$

$$\vec{B} = \mu_0 \vec{H} + \vec{M} \tag{2.42}$$

with the polarization of the medium $\vec{P}$ and magnetization $\vec{M}$, and the sources are the free charges and free currents. These relations are heuristic in nature and not fundamental. In this book, we will consider nonmagnetic materials and set $\vec{M} = 0$. To see how a polarization may arise, consider an electrically neutral atom or molecule. If an external field is applied, the atomic nucleus will want to move along the direction of the field as it is positively charged, and the electrons will want to move in the opposite direction. This leads to a separation of charge, with a resultant field that opposes the applied field. This is a dipole field, with $\vec{P} = N\vec{\mu} = Ne\vec{x}$ for the case of N atoms, with $\vec{x}$ the separation between the nucleus and the electron, for an alkali element for example. In a linear, isotropic, homogeneous material, the polarization is given by $\vec{P} = \epsilon_0 \chi \vec{E}$. The quantity χ is the susceptibility of the material. Generally the situation is more complicated than that. Consider a material with a different response along different directions, an anisotropic material.

If we apply a field E along the x axis, as shown in figure 2.1, we find a polarization $P_x = \epsilon_0 \chi_x E$. If instead, we apply the field along the y axis, we find a polarization $P_y = \epsilon_0 \chi_y E$. In both cases, the polarization is in the same direction as the applied field. But what if we apply a field E aligned at a 45° angle? What is the resulting polarization? We can find it simply by vector addition, but the result is that the polarization is no longer along the direction of the applied field. To handle such situations, we must promote the susceptibility χ from a scalar to a matrix (technically a tensor, more on that later). We can then write the relationship between the polarization and the applied electric field

$$\vec{P} = \epsilon_0 \boldsymbol{\chi} \vec{E}. \tag{2.43}$$

Here we see that $\boldsymbol{\chi}$ transforms, or maps the vector $\epsilon_0 \vec{E}$ into a new vector $\vec{P}$, and the two vectors need not be collinear. In the case of a homogeneous material (i.e. no directional dependence), we have $\boldsymbol{\chi} = \chi \mathbf{I}$, where the scalar χ is the usual

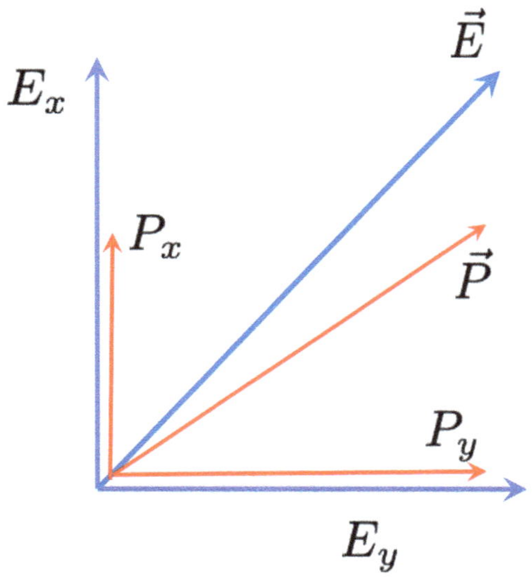

Figure 2.1. Polarization induced with fields along the x and y axes, and the resultant.

susceptibility, and **I** is the identity matrix. In this case $\vec{E}$ and $\vec{P}$ are collinear. In the case illustrated in figure 2.1,

$$\chi = \begin{pmatrix} \chi_x & 0 \\ 0 & \chi_y \end{pmatrix} \tag{2.44}$$

and we have in component form

$$P_x = \epsilon_0\, \chi_x\, E_x \tag{2.45}$$

$$P_y = \epsilon_0\, \chi_y\, E_y. \tag{2.46}$$

More generally, the susceptibility matrix (soon to be promoted to tensor status) has nonzero off diagonal matrix elements,

$$\chi = \begin{pmatrix} \chi_{xx} & \chi_{xy} \\ \chi_{yx} & \chi_{yy} \end{pmatrix} \tag{2.47}$$

or in component form

$$P_x = \epsilon_0(\chi_{xx}E_x + \chi_{xy}E_y) \tag{2.48}$$

$$P_y = \epsilon_0(\chi_{yx}E_y + \chi_{yy}E_y). \tag{2.49}$$

There are special directions in space where the polarization *will* be collinear with the applied field. If the field is applied along one of these special directions, we can replace the susceptibility matrix with a scalar. These special directions will be determined by the eigenvectors of χ, and the value of the susceptibility for that direction will be the corresponding eigenvalue. The situation may be even more complicated; if the material is non-homogeneous, the susceptibility matrix elements may depend on position. Furthermore, for a nonlinear material, the susceptibility matrix elements may be dependent on field strength

$$\chi_{ij} \equiv \chi_{ij}(\vec{r}, \vec{E}). \tag{2.50}$$

In fact, the simplest quantum system, a two-level atom, will have a nonlinear susceptibility. All materials will respond nonlinearly at some applied field level; if nothing else an atom will ionize at some point. It is often useful to express the polarization as a Taylor series in applied field strength,

$$P(E) = \epsilon_0 \left(\chi^{(0)} + \chi^{(1)}E + \chi^{(2)}E^2 + \chi^{(3)}E^3 + \dots\right). \tag{2.51}$$

Here we consider a component of P and E for an isotropic, but nonlinear, material. $\chi^{(0)}$ is related to a permanent dipole moment, $\chi^{(1)}$ is the usual linear susceptibility, and terms such as $\chi^{(2)}$ and $\chi^{(3)}$ represent nonlinear terms. If we have a material with inversion symmetry, reversing the direction of the applied field should also reverse the direction of the polarization. We would then have

$$\begin{aligned} P(-E) &= \epsilon_0 \left(\chi^{(0)} + \chi^{(1)}(-E) + \chi^{(2)}(-E)^2 + \chi^{(3)}(-E)^3 + \dots\right) \\ &= \epsilon_0\left(\chi^{(0)} - \chi^{(1)}E + \chi^{(2)}E^2 - \chi^{(3)}E^3 + \dots\right) \\ &= -P(E). \end{aligned} \tag{2.52}$$

For this to hold true we must have $\chi^{(0)} = \chi^{(2)} = 0$ as well as all even terms. So for an isotropic material, we cannot have a permanent dipole moment, and the nonlinearity is cubic at best.

Let us now consider the wave equation for electric fields in a material. We start by taking the curl of Faraday's law,

$$\begin{aligned} \vec{\nabla} \times (\vec{\nabla} \times \vec{D}) &= -\vec{\nabla} \times \frac{\partial \vec{B}}{\partial t} \\ &= -\nabla^2 \vec{D} + \vec{\nabla}(\vec{\nabla} \cdot \vec{D}) \\ &= -\mu_0 \vec{J} + \mu_0 \frac{\partial \vec{D}}{\partial t}. \end{aligned} \tag{2.53}$$

Utilizing the constitutive relation $\vec{D} = \epsilon_0 \vec{E} + \vec{P}$, we find

$$\nabla^2 \vec{E} - \frac{1}{c^2}\frac{\partial^2 \vec{E}}{\partial t^2} = \mu_0 \frac{\partial^2 \vec{P}}{\partial t^2} + \mu_0 \frac{\partial \vec{J}}{\partial t}, \tag{2.54}$$

where we have taken $\nabla \cdot \vec{P} = 0$. In free space, we have

$$\nabla^2 \vec{E} - \frac{1}{c^2}\frac{\partial^2 \vec{E}}{\partial t^2} = 0 \tag{2.55}$$

and any vector with components of the form $f(k(x - ct)) = f(kx - \omega t)$ is a solution. This is a wave that propagates at speed $c = \sqrt{1/\epsilon_0 \mu_0}$, with wavelength $\lambda = 2\pi/k$ and angular frequency $\omega = ck$. The source terms on the right-hand side generate electromagnetic waves. As $\vec{P} = Ne\vec{x}$ and $\vec{J} \sim \vec{v}$, both of these terms are proportional to the acceleration of a charge, hence as you have been long told, accelerating charges radiate electromagnetic radiation. Sometimes, we wish to take into account some loss mechanism for the field, and this can be accomplished mathematically by taking the current density to be that of an Ohmic material, $\vec{J} = \sigma \vec{E}$ where σ is the electrical conductivity. This leads to a wave equation of the form

$$\nabla^2 \vec{E} - \frac{1}{c^2} \frac{\partial^2 \vec{E}}{\partial t^2} - \mu_0 \sigma \frac{\partial \vec{E}}{\partial t} = \mu_0 \frac{\partial^2 \vec{P}}{\partial t^2}. \tag{2.56}$$

In the case of an electromagnetic field confined to a cavity, this Ohmic loss will in reality stem from light leaking out an imperfect mirror. Let us ignore losses at this point, and consider a linear material with $\vec{P} = \epsilon_0 \chi \vec{E}$, where χ is the linear susceptibility. We then have

$$\nabla^2 \vec{E} - \frac{1 + \chi}{c^2} \frac{\partial^2 \vec{E}}{\partial t^2} = 0. \tag{2.57}$$

This is just like the free space case, only now the speed of the wave is $v = c/n$, where we have defined an index of refraction $n = \sqrt{1 + \chi}$. Waves would be of the form $E = \exp(i(Kx - \omega t))$, where $K = nk$. We consider the medium to be linear, and so it responds at the frequency it is driven at. Since $v = c/n = f\lambda$, if f does not change, then the wavelength must decrease, $\lambda \to \lambda/n$ or $k \to nk$. This applies in the linear regime where the oscillator responds at the frequency it is driven. This is the Abraham, or kinetic momentum of the photon; it differs from the Minkowski, or canonical momentum [7, 8]. Measurements of the mechanical force of photons will involve the kinetic momentum. Let us consider what would happen if the index of refraction were complex, $n = n_R + in_I$. In this case, wave solutions would take the form

$$E = \exp(i(n_R kx - \omega t))\exp(-n_I kx), \tag{2.58}$$

where k is the free space wave number. We see that the real part of the index of refraction is what we are used to having, something that slows the wave down and shortens its wavelength. If there is an imaginary part, we have loss as the wave propagates if $n_I > 0$, this is a way to model absorption. If the imaginary part if the index is negative, we would have gain, as in a laser. One must have a population inversion to have a negative imaginary index in the steady state, and so one must have at least three levels for a laser system.

2.3 Slowly varying envelope approximation

For optical frequencies, it is often very convenient to separate the electric field into a rapidly oscillating part, with a slowly varying envelope function, of the form

$$E = A(t)e^{i(kx-\omega t+\phi)}. \tag{2.59}$$

In what follows, we assume that the frequencies involved in the Fourier expansion of the slowly varying amplitude $A(t)$ are small compared to the center, or carrier, frequency ω. Formally this means we take $\omega A \gg \dot{A}$. We ignore any spatial dependence of the envelope for now. Let us substitute our form for the field back into the wave equation. To do this, we need the following results,

$$\frac{\partial E}{\partial z} = ikEe^{i(kx-\omega t+\phi)} \tag{2.60}$$

$$\frac{\partial^2 E}{\partial z^2} = -k^2 Ee^{i(kx-\omega t+\phi)} \tag{2.61}$$

$$\frac{\partial E}{\partial t} = (\dot{A} - i(\omega + \dot{\phi})A)e^{i(kx-\omega t+\phi)} \tag{2.62}$$

$$\frac{\partial^2 E}{\partial t^2} = (\ddot{A} - 2i(\omega + \dot{\phi})\dot{A} - (\omega + \dot{\phi})^2 A - i\ddot{\phi}A)e^{i(kx-\omega t+\phi)}. \tag{2.63}$$

We must also decompose the polarization into rapidly and slowly varying components

$$P = \mathcal{P}(t)e^{i(kx-\omega t+\phi)}. \tag{2.64}$$

One then has

$$\frac{\partial^2 P}{\partial t^2} = (\ddot{\mathcal{P}} - 2i(\omega + \dot{\phi})\dot{\mathcal{P}} - (\omega + \dot{\phi})^2\mathcal{P} - i\ddot{\phi}\mathcal{P})e^{i(kx-\omega t+\phi)}. \tag{2.65}$$

The wave equation then becomes

$$(\Omega^2 - (\omega + \dot{\phi})^2)A - i\omega\frac{\sigma}{\epsilon_0} - 2i\omega\dot{A} = \frac{\omega^2}{\epsilon_0}\mathcal{P}, \tag{2.66}$$

where we have used the slowly varying envelope approximation, $\ddot{O} \ll \omega\dot{O}$ and $\dot{O} \ll \omega O$, for $O = \phi$, A, or $\mathcal{P}$ and used $\Omega = ck$. Another approximation is to ignore $\dot{\phi}$ compared to $\omega + \Omega$ and take $\Omega \sim \omega$, which allows us to write

$$\begin{aligned} \Omega^2 - (\omega + \dot{\phi})^2 &= (\Omega - (\omega + \dot{\phi}))(\Omega + \omega + \dot{\phi}) \\ &\sim 2\omega(\Omega - \omega - \dot{\phi}). \end{aligned} \tag{2.67}$$

On equating real and imaginary parts then, we find

$$\dot{A} = -\kappa A - \frac{1}{2}\frac{\omega}{\epsilon_0}\mathrm{Im}(\mathcal{P}) \tag{2.68}$$

$$\dot{\phi} = \omega_c - \omega - \frac{1}{2A}\frac{\omega}{\epsilon_0}\mathrm{Re}(\mathcal{P}), \tag{2.69}$$

where in terms of our fictional conductivity we have $\kappa = \omega/2Q$, where Q is the quality factor $\epsilon_0\,\omega/\sigma$. Usually this loss rate results from a lossy mirror in a cavity. A photon has a probability of $T = 1 - R$ of leaving the mirror when it hits, where R is the reflectivity and T is the transmission of the mirror. The rate at which the photon interacts with the mirror is $1/\tau$, with $c\tau = 2L$. Overall this leads to $\kappa = (1 - R)c/2L$. For a linear material, with an imaginary susceptibility $\chi = \chi_R + i\chi_I$, we have

$$\dot{A} = -\kappa A - \frac{1}{2}\omega\chi_I A \tag{2.70}$$

$$\Delta\omega = \dot{\phi}\omega_c = -\omega - \frac{1}{2}\omega\chi_R, \tag{2.71}$$

with ω_c the resonant frequency of the empty cavity, and $\Delta\omega$ is the frequency of the field due to mode pulling. Here we see that the sign of the imaginary part of the susceptibility determines whether the interaction with the medium results in gain or loss, and that there is a frequency shift of the field that depends on the real part of the susceptibility. Recall that the relation to the index of refraction is $n = \sqrt{1 + \chi} \sim 1 + \chi_R/2 + i\chi_I/2$, where we assume a dilute medium ($n \sim 1$).

It is sometimes more convenient to to consider a complex field variable

$$X = Ae^{i\phi} \tag{2.72}$$

which satisfies the equation

$$\dot{X} = -\kappa X - \frac{\omega}{\epsilon_0}P. \tag{2.73}$$

2.4 Lorentz oscillator model of the atom

The simplest model of the atom, hydrogen say, is that of an electron and proton connected by a spring. At first this seems ridiculous until we realize that due to the Coulomb interaction the proton and electron form a bound state. For any bound state, there must be a minimum in the potential.

Near that minimum, we can expand the potential in a Taylor series,

$$V(x) = V(x_0) + \frac{dV}{dx}\mid_{x=x_0}(x - x_0) + \frac{1}{2}\frac{d^2V}{dx^2}\mid_{x=x_0}(x - x_0)^2 + \dots. \tag{2.74}$$

We are always free to define $V(x_0) = 0$ and, as it is a minimum, we have $\frac{dV}{dx}\mid_{x=x_0}(x - x_0) = 0$, which leaves us with

$$V(x) \sim \frac{1}{2}\frac{d^2V}{dx^2}\mid_{x=x_0}(x - x_0)^2 = \frac{1}{2}k(x - x_0)^2, \tag{2.75}$$

where k is an effective spring constant, not to be confused with the wave number $k = 2\pi/\lambda$, Coulomb's constant $k = 1/4\pi\epsilon_0$, or Boltzmann's constant. Can you imagine the turmoil if we are to study potassium (K) ions (interacting with a Coulomb force) with light of wave number k, at thermal equilibrium at temperature T. Anyway, I digress. The equation of motion of the oscillator driven by a cosinusoidal electric field with amplitude E and angular frequency ω would be given by

$$\ddot{x} = -\omega_0^2 x - \Gamma\dot{x} + \frac{eE}{m}\cos(\omega t), \tag{2.76}$$

where we have introduced a frictional force that is proportional to the velocity. Later we will relate that damping force to spontaneous emission. As we have a linear oscillator, it will respond in the steady state at the frequency it is driven at, with some amplitude and phase. We then assume a steady state solution of the form

$$x = X_0\cos(\omega t + \theta). \tag{2.77}$$

Substituting this into the equation of motion we find

$$(\omega_0^2 - \omega^2)X\cos(\omega t + \theta) - \omega\Gamma X\sin(\omega t + \theta) = -\frac{eE}{m}\cos(\omega t). \tag{2.78}$$

We use trig identities of the form $\cos(\omega t + \theta) = \cos(\omega t)\cos(\theta) - \sin(\omega t)\sin(\theta)$ and a similar one for $\sin(\omega t + \theta)$, and equate the sin and cos terms, and we find

$$\theta = \tan^{-1}\frac{\omega\Gamma}{\omega_0^2 - \omega^2} \tag{2.79}$$

and

$$X_0 = \frac{-eE/m}{\sqrt{(\omega_0^2 - \omega^2)^2 + \omega^2\Gamma^2}}. \tag{2.80}$$

Here we see the familiar results that the amplitude of the response is maximized on resonance ($\omega = \omega_0$), and that on resonance the response is $\pi/2$ out of phase with the driving field. We can write the response of the oscillator as

$$\begin{aligned} x(t) &= X_0\cos(\omega t)\cos(\phi) - X_0\sin(\omega t)\sin(\phi) \\ &= X_0(U\cos(\omega t) - V\sin(\omega t)), \end{aligned} \tag{2.81}$$

with $U = \cos(\phi)$ the relative amplitude in phase with the driving field, and $V = \sin\phi$ the relative amplitude in quadrature ($\pi/2$ out of phase) with the driving field. There is a slightly easier method to obtain these results, where we use a complex representation. The driving field has a time dependence $\cos(\omega t) = \mathrm{Re}(e^{i\omega t})$. As this is a linear system, we may take the driving field to be complex, find the response, and then take the real part at the end. Thus we need to solve

$$\ddot{x} = -\omega_0^2 x - \Gamma\dot{x} + \frac{eE}{m}e^{i\omega t} \tag{2.82}$$

with steady state response

$$x(t) = \tilde{X}_0 e^{i\omega t}. \tag{2.83}$$

This leads to

$$\tilde{X}_0 = \frac{-eE/m}{(\omega_0^2 - \omega^2) - i\omega\Gamma} \tag{2.84}$$

which leads to

$$x(t) = X_0 \cos(\omega t + \theta) \tag{2.85}$$

with the same X_0 and θ we found earlier. This gives us an expression for the steady state polarization of the medium via

$$\begin{aligned} P &= Nex(t) \\ &= NeX_0 \cos(\omega t + \theta) \\ &= \epsilon_0 \chi E(t), \end{aligned} \tag{2.86}$$

which leads to an expression for the susceptibility

$$\chi = \frac{Ne^2/m\epsilon_0}{(\omega_0^2 - \omega^2) - i\omega\Gamma}, \tag{2.87}$$

which is a complex expression with real and imaginary parts

$$\chi_R = \frac{Ne^2/m\epsilon_0}{(\omega_0^2 - \omega^2)^2 + \Gamma^2\omega^2}(\omega_0^2 - \omega^2) \tag{2.88}$$

$$\chi_I = \frac{Ne^2/m\epsilon_0}{(\omega_0^2 - \omega^2)^2 + \Gamma^2\omega^2}\omega\Gamma. \tag{2.89}$$

For a dilute medium, we then have $n_R - 1 = \chi_R/2$ and $n_I = \chi_I/2$.

The power dissipated by the oscillator is Power $= Fv = F\dot{x}$ which leads to

$$\text{Power} = \frac{e^2E^2\omega/m}{\sqrt{(\omega_0^2 - \omega^2)^2 + \omega^2\Gamma^2}} \cos(\omega t)\cos(\omega t + \beta), \tag{2.90}$$

where we have taken $\beta = \theta - \pi/2$ which is zero on resonance. A time average of this expression yields

$$\begin{aligned} \text{Power} &= \frac{e^2E^2\omega/2m}{\sqrt{(\omega_0^2 - \omega^2)^2 + \omega^2\Gamma^2}} \cos(\beta) \\ &= \frac{e^2E^2\omega/2m}{(\omega_0^2 - \omega^2)^2 + \omega^2\Gamma^2} \\ &= \omega\epsilon_0 E^2 n_I. \end{aligned} \tag{2.91}$$

This is consistent with our earlier observation that the imaginary part of the index of refraction is responsible for changes in the amplitude of a wave as it propagates. Here we are putting energy into the oscillator via the applied field, which is then dissipated as shown above. What is the physical mechanism for this? Well the electron is not connected to the proton by a spring, which has some velocity dependent friction associated with it! How would a real atom dissipate energy? The only mechanism we know of in a real, quantum mechanical theory is spontaneous emission. When we examine the quantum theory of an atom interacting with an applied field, we will find that for weak excitation, the results above are correct from a quantum mechanical point of view if we identify Γ as half the spontaneous emission rate.

How about the dynamical response of the oscillator? The time dependence of the approach to the steady state is given by equation (2.76). We will make the decomposition

$$x(t) = X_0(U\cos(\omega t) - V\sin(\omega t)) \tag{2.92}$$

and substitute this ansatz into equation (2.76), which upon equating coefficients of sin and cos, leads to equations

$$\dot{U} = -\Delta V - \Gamma U \tag{2.93}$$

$$\dot{V} = \Delta U - \Gamma V + \Omega_{\text{cl}}, \tag{2.94}$$

where we define $\Delta = \omega_0 - \omega$, $\Omega_{\text{cl}} = eE/2m\omega X_0$ we will refer to as a classical Rabi frequency, and we have made liberal use of the slowly varying amplitude approximation, for example $\omega U \gg \dot{U}$ and $\omega V \gg \dot{V}$. In terms of these quantities, the polarization is

$$\begin{aligned} P &= NeX_0(U\cos(\omega t) - V\sin(\omega t)) \\ &= \frac{NeX_0}{2}((U + iV)e^{i\omega t} + (U - iV)e^{-i\omega t}), \end{aligned} \tag{2.95}$$

which in a complex representation leads to

$$\chi_R = \frac{NeX_0}{2}\frac{U}{E} \tag{2.96}$$

$$\chi_I = \frac{NeX_0}{2}\frac{V}{E}. \tag{2.97}$$

Then we can write down the full set of coupled equations for the field and the polarization, ignoring the mode pulling term

$$\dot{U} = -\Delta V - \Gamma U \tag{2.98}$$

$$\dot{V} = \Delta U - \Gamma V + A \tag{2.99}$$

$$\dot{A} = -\kappa A - \frac{1}{2}\omega V \tag{2.100}$$

$$\dot{\phi} = \omega_c - \omega - \frac{1}{2}\omega U. \tag{2.101}$$

References

[1] Jackson J D 1975 *Classical Electrodynamics* (New York: Wiley)

[2] Cohen-Tannoudji C, Dupont-Roc J and Grynberg G 1989 *Photons and Atoms: Introduction to Quantum Electrodynamics* (New York: Wiley)

[3] Cohen-Tannoudji C, Dupont-Roc J and Grynberg G 1998 *Atom-Photon Interactions: Basic Processes and Applications* (New York: Wiley)

[4] Landau L D, Bell J S, Kearsley M J, Pitaevskii L P, Lifshitz E M and Sykes J B 2013 *Electrodynamics of Continuous Media* (Amsterdam: Elsevier Science)

[5] Griffiths D J 1999 *Introduction to Electrodynamics* (Englewood Cliffs, NJ: Prentice Hall)

[6] Bromley D A and Greiner W 1998 *Classical Electrodynamics Classical Theoretical Physics* (New York: Springer)

[7] Hinds E A and Barnett S M 2009 Momentum exchange between light and a single atom: Abraham or Minkowski? *Phys. Rev. Lett.* **102** 050403

[8] Barnett S M 2010 Resolution of the Abraham–Minkowski dilemma *Phys. Rev. Lett.* **104** 070401

IOP Publishing

An Introduction to Quantum Optics (Second Edition)

An open systems approach

Perry Rice

Chapter 3

Quantum mechanics review

We provide a brief review of quantum mechanics, with attention to the usage of Hilbert spaces and Dirac notation. Standard references include [1–7]. In this chapter we denote quantum operators with a caret ($\hat{x}$) and classical variables without. In subsequent chapters we will not do this, the meaning should be quite clear in the context.

3.1 Wave mechanics

We provide a brief review of quantum mechanics, with attention to the usage of Hilbert spaces and Dirac notation. Standard references include [1–7]. In all of your previous quantum mechanics, you have dealt with the wave function $\Psi(\vec{x}, t)$. This function describes the state of the quantum system. The time evolution of this function is given by the Schrödinger equation,

$$i\hbar\frac{\partial\Psi(\vec{x}, t)}{\partial t} = H\Psi(\vec{x}, t). \tag{3.1}$$

The Hamiltonian is generally the sum of the kinetic energy and the potential energy operators

$$H = p^2/2m + V(\vec{x}, t). \tag{3.2}$$

In the position representation, the momentum operator is given by $p_x = -\imath\hbar\partial/\partial x$ in one dimension, or $\vec{p} = -\imath\hbar\vec{\nabla}$ in three spatial dimensions, which makes the Schrödinger equation take the familiar form

$$\imath\hbar\frac{\partial\Psi(\vec{x}, t)}{\partial t} = -\frac{\hbar^2}{2m}\nabla^2\Psi(\vec{x}, t) + V(\vec{x}, t)\Psi(\vec{x}, t). \tag{3.3}$$

In the special case where the potential energy depends only on position but not time, we may use the separation of variables to reduce the Schrödinger equation, a partial

doi:10.1088/978-0-7503-6014-2ch3

differential equation in time and space, to two equations involving spatial dependence and temporal dependence. We use the ansatz

$$\Psi(\vec{x}, t) = \psi(\vec{x})F(t). \tag{3.4}$$

Substituting this into the Schrödinger equation and dividing it by $\Psi(\vec{x}, t)$ yields

$$\frac{\imath\hbar}{F(t)}\frac{\partial F(t)}{\partial t} = -\frac{\hbar^2}{2m\psi(\vec{x})}\nabla^2\psi(\vec{x}) + V(\vec{x}) = E, \tag{3.5}$$

where E is a constant of separation. The function $F(t)$ is then given by

$$F(t) = e^{-\imath Et/\hbar}. \tag{3.6}$$

We have ignored an overall constant factor because the full wave function $\Psi(\vec{x}, t)$ must be normalized, and we choose to normalize the spatial and temporal parts separately. As written, $F^*(t)F(t) = 1$, and ultimately we normalize the spatial part of the wave function. The function $\psi(\vec{x})$ then satisfies the time-independent Schrödinger equation

$$-\frac{\hbar^2}{2m}\nabla^2\psi(\vec{x}) + V(\vec{x}, t)\psi(\vec{x}) = H\psi(\vec{x}) = E\psi(\vec{x}). \tag{3.7}$$

As the left-hand side of the equation is the sum of the potential and kinetic energy operators, we associate the constant E with the total energy of the system.

Most introductory work on quantum mechanics concerns specifying a potential energy $V(\vec{x})$, and solving the above equation for the allowed energies E_n and wave functions $\psi_n(\vec{x})$. The subscript n is used to indicate a set of possible solutions, and not just one. It is important to realize that the time-independent Schrödinger equation $H\psi_n(\vec{x}) = E_n\psi_n(\vec{x})$ is an eigenvalue equation for *both* $\psi_n(\vec{x})$ as well as E_n. Generally in quantum mechanics, we deal with Hamiltonians that are Hermitian operators, so the eigenenergies E_n are real, and the eigenfunctions $\psi_n(\vec{x})$ are orthogonal, and can be made orthonormal,

$$\int_{\infty}^{\infty}\psi_n^*(\vec{x})\psi_m(\vec{x})d\vec{x} = \delta_{n,m}, \tag{3.8}$$

where we have normalized the individual wave functions via

$$\int_{\infty}^{\infty}\psi_n^*(\vec{x})\psi_n(\vec{x})d\vec{x} = 1. \tag{3.9}$$

Recall that the concept of a Hermitian operator in infinite dimensions, as we have here (an infinite number of possible positions $\vec{x}$) means that the operator is self-adjoint and that the solutions must satisfy certain boundary conditions. As one choice of appropriate boundary conditions is that the wave functions and their derivatives vanish at $\pm\infty$, this is automatically satisfied for any wave function that can be normalized. One can look at a matrix and quickly determine whether it is Hermitian or not; for a differential operator, we can only tell whether it is self-adjoint or not. However, with the appropriate boundary conditions that we require in quantum mechanics, the two concepts are essentially one and the same.

The individual energy eigenstates are 'stationary' as $|\Psi(\vec{x}, t)|^2 = |\psi(\vec{x})|^2$ is independent of time. We recall the Born interpretation that the probability of finding the particle between x and $x + dx$ is given by $P(x)dx = |\psi(\vec{x})|^2 dx$, so the stationarity is that of the spatial probability distribution. As the Schrödinger equation is linear, the principle of superposition yields the most general form of the solution

$$\Psi(\vec{x}, t) = \sum_{n=1}^{\infty} C_n \psi_n(\vec{x}) e^{-\imath\omega_n t}, \tag{3.10}$$

where we used the Bohr condition $E_n = \hbar\omega_n$. It can be shown that

$$\sum_{n=1}^{\infty} |C_n|^2 = 1. \tag{3.11}$$

From this follows the interpretation of $|C_n|^2 = C_n C_n^*$ as a probability of finding E_n because of measuring the energy of the system. After such a measurement, the wave function is in the state $\psi_n(\vec{x})$, the 'collapse' of the wave function. We will have more to talk about regarding this 'collapse' later when we talk about decoherence in chapter 8.

Note that the most general state is *not* stationary! The probability of obtaining a certain value of $E = E_n$ is independent of time, but other probabilities, such as the probability of finding the particle at a certain position x, are not. In fact, for a simple superposition such as $\psi(\vec{x}) = \sqrt{1/2}\,(\psi_1 + \psi_3)$, we have (in 1D)

$$P(x)dx = |\Psi|^2 dx = (1/2)\Big(|\psi_1|^2 + |\psi_3|^2 + 2\,\mathrm{Re}\Big(\psi_1^*\psi_3 e^{-\imath(\omega_3 - \omega_1)t}\Big)\Big)dx. \tag{3.12}$$

In a very real sense, the system is simultaneously in all eigenstates with a nonzero probability amplitude C_n. Quantum mechanics is certainly statistical in nature, in that we cannot tell before a measurement of energy which eigenvalue E_n we will obtain, but only the probability of each possible result. But it is more than statistical. We cannot say that the system has a definite energy, but we have a certain lack of knowledge about which one it is. The work of Bell and others has shown us that the implications of quantum mechanics are deeper than that.

We recall that once the wave function has been found, the information about a system in that state, or more precisely the average value of a measurement of some observable O on an ensemble of identically prepared systems, is given by

$$\langle O \rangle = \int_{-\infty}^{\infty} \Psi^*(\vec{x}, t) O \Psi(\vec{x}, t) d\vec{x}. \tag{3.13}$$

We have again assumed that the individual eigenstates have been normalized. It is also convenient to consider the spread, or 'uncertainty' of a set of measurements,

$$(\Delta O)^2 = \langle (O - \langle O \rangle)^2 \rangle = \langle O^2 \rangle - \langle O \rangle^2. \tag{3.14}$$

At this point, we also note that representing physical observables by Hermitian differential operators preserves the feature that one notices in the lab in the quantum realm; that is, the order of measurement matters! If I make a position measurement and then a momentum measurement, I will in general obtain different results than if I make a momentum measurement first and then a position measurement. Consider the commutator of the operators x and p, acting on the wave function ψ,

$$\begin{aligned}[x, p]\psi &\equiv (xp - px)\psi \\ &= -\imath\hbar\left(x\frac{\partial}{\partial x} - \frac{\partial}{\partial x}x\right)\psi \\ &= -\imath\hbar x\frac{\partial\psi}{\partial x} + \imath\hbar\frac{\partial(x\psi)}{\partial x} \\ &= \imath\hbar\psi.\end{aligned} \tag{3.15}$$

Here we see that differential operators can consider that 'order matters', as the operators representing physical quantities do not commute. There are other choices of mathematical objects that do not generally commute, such as matrices, which we shall also use later on. If x and p were numbers, then obviously $[x, p] = 0$, but in the quantum case, we find that $[x, p] = \imath\hbar$ which is *not* zero, unless we take the classical limit that $\hbar \to 0$. This non-commutativity leads directly to the uncertainty principle for the two Hermitian operators A and B,

$$\Delta A\Delta B \geqslant \frac{1}{-}|\langle -i[A, B]\rangle|^2, \tag{3.16}$$

which is generally state dependent. For position and momentum this becomes

$$\Delta x\Delta p \geqslant \hbar/2. \tag{3.17}$$

3.2 Dirac notation

We now move to a more abstract view of the state of a quantum system. We will represent the state of a quantum system by a ray in a linear vector space. For a review of linear vector spaces and their application to quantum mechanics, we refer you to Shankar's texts [8], and Arfken and Weber's texts [9]. Typically, the ray representing the quantum state is notated as $|\Psi(\vec{x}, t)\rangle$. This is known as a 'ket'. This is also known as the state vector and may have a representation as a column of numbers in a given basis (or in simpler terms an orthonormal set of vectors that forms a coordinate system). However, it is an abstract property of a quantum system, independent of the basis. Think about the electric field at some point in space, it exists independent of our description of it; our representation of it (the three components) may vary in different coordinate systems. There is also a corresponding vector referred to as the 'bra', $\langle\Psi(\vec{x}, t)|$, which usually has a representation as a row vector, whose elements are the complex conjugates of the components of the column vector 'ket'. The bra is given by

$$\langle\psi(x)| = (\psi(x_1)\psi(x_2)\psi(x_3)\ldots). \tag{3.18}$$

How do we connect these objects? What about $\psi(x)$, the wave function. Is there any way to represent a function as a vector? The answer is yes. We merely list the value of the function at all points x_i. This vector is infinite in length of course, and so it is less than useful other than in the abstract:

$$|\psi(x)\rangle = \begin{pmatrix} \psi(x_1) \\ \psi(x_2) \\ \psi(x_3) \\ \vdots \end{pmatrix}. \tag{3.19}$$

If we were just to sample the wave function at certain equally spaced points, we would have a function 'vector' such as you have created in Excel or another spreadsheet to make a plot of the function. In the abstract, however, we wish to consider sampling at every point, which does make for a long vector. The bra is given by

$$\langle\psi(x)| = (\psi^*(x_1), \psi^*(x_2), \psi^*(x_3)\ldots). \tag{3.20}$$

The inner product $\langle\psi(x)|\phi(x)\rangle$ is an infinite sum that in general does not converge. The sum is generally infinite due to the infinite length of the vectors, independent of the nature of the functions. If we consider sampling the wave function over an interval L at N points separated by a distance dx, the size of the inner product scales as N. If we multiply the usual inner product by $dx = L/N$, and taking the limit of the sum over into an integral, we find

$$\langle\psi(x)|\phi(x)\rangle = \int \psi(x)^*\phi(x)dx. \tag{3.21}$$

This defines the inner product on our vector space, called a Hilbert space. It is linear and has other nice properties. We note that we choose *this* inner product over others because of its utility in quantum mechanics. The probability of finding the particle between x and $x + dx$ is given by $P(x)dx = \psi(x)^*\psi(x)dx$. The requirement that the probability of finding the particle anywhere is one results in

$$\langle\psi(x)|\psi(x)\rangle = \int_{-\infty}^{\infty} \psi(x)^*\psi(x)dx = 1, \tag{3.22}$$

which is referred to as the normalization condition. As $A\psi(x)$ is a solution to the Schrödinger equation if $\psi(x)$ is, the wave function must be normalized after finding the form of the solution. But why do we need to consider $\psi(x)^*$? Wave functions are in general complex of course, but why is the inner product defined in this way?

We could have defined the square of the 'length' of the quantum state vector in the usual sense of the dot product of the vector with itself. This would not always generate a real number, or even one with a positive real part. If for example

$$|\psi\rangle = \begin{pmatrix} i \\ 1 \\ i \end{pmatrix}, \tag{3.23}$$

then the 'length' of the quantum state would be $-1 + 1 - 1 = -1$. It is more convenient to generalize the dotted product into an inner product. To do this, we define a bra

$$\langle\psi| = \begin{pmatrix} C_1^* & C_2^* & C_3^* \end{pmatrix}. \tag{3.24}$$

We then define the inner product as the matrix multiplication of the bra and ket

$$\langle\psi||\psi\rangle \equiv \langle\psi|\psi\rangle \tag{3.25}$$

$$= \begin{pmatrix} C_1^* & C_2^* & C_3^* \end{pmatrix} \begin{pmatrix} C_1 \\ C_2 \\ C_3 \end{pmatrix} \tag{3.26}$$

$$= |C_1|^2 + |C_2|^2 + |C_3|^2. \tag{3.27}$$

This general definition of an inner product always results in a positive definite quantity. For a normalized quantum state, where the total probability of being in any state is one, this inner product yields unity, as $|C_1|^2 + |C_2|^2 + |C_3|^2 = P_1 + P_2 + P_3 = 1$. One can take the inner product $\langle\psi|\phi\rangle$ which will not generally yield a real positive number. The names bra and ket stem from the fact that the inner product $\langle\psi|\phi\rangle$ is formed inside a 'bracket'. This notation is due to Paul Dirac, and is called the Dirac notation, and generalizes to a state with N levels:

$$|\psi\rangle = \begin{pmatrix} C_1 \\ C_2 \\ C_3 \\ \vdots \end{pmatrix}. \tag{3.28}$$

3.3 Representations

So far, we have only dealt with what is called the position representation, where we use

$$\begin{aligned} \hat{p}_x &= -i\hbar\frac{\partial}{\partial x} \\ \hat{x} &= x. \end{aligned} \tag{3.29}$$

This representation is obtained by considering the matrix representation of the position and momentum operators on the basis of the eigenstates of the position operator. We are interested in states that satisfy $\hat{x}|x\rangle = x|x\rangle$. These states form a complete set because they are eigenvectors of a Hermitian operator,

$$\int dx|x\rangle\langle x| = 1. \tag{3.30}$$

This leads us to the normalization of the position eigenstates as

$$\begin{aligned}\langle x|x'\rangle &= \int dx'' \langle x|x''\rangle\langle x|x'\rangle \\ &= \int dx' \delta(x - x'')\langle x''|x'\rangle\end{aligned} \tag{3.31}$$

and we see that we have $\langle x|x'\rangle = \delta(x - x')$ via the definition of the Dirac delta function. This is zero for $x \neq x'$, but infinite for $x = x'$. Recall that the Dirac delta function is really a distribution and makes much more sense when it lives inside an integral, but we shall have occasion to use it outside of integration. At this point, we can make a definite connection between the ket $|\psi\rangle$ and our old friend the wave function $\psi(x)$. Let us use the completeness relationship to understand the connection:

$$\begin{aligned}|\psi\rangle &= \int dx|x\rangle\langle x|\psi\rangle \\ &= \int dx\langle x|\psi\rangle|x\rangle \\ &= \int dx\ C(x)|x\rangle.\end{aligned} \tag{3.32}$$

This is merely a way to express that we can write the state vector as a linear superposition of the position eigenstates, with the probability amplitude $C(x) = \langle x|\psi\rangle$. Recall from wave mechanics the requirement of the normalization of the wave function

$$\int_{-\infty}^{\infty} \psi^*(x)\psi(x)dx = \int_{-\infty}^{\infty} P(x)dx = 1. \tag{3.33}$$

This allows us to interpret $P(x)dx = \psi^*(x)\psi(x)dx$ as a probability density for finding the particle in a region of space $x \to x + dx$. In the Dirac formalism we have, using the completeness relationship,

$$\begin{aligned}\langle\psi|\psi\rangle &= \int dx\langle\psi|x\rangle\langle x|\psi\rangle \\ &= \int C^*(x)C(x)dx \\ &= 1,\end{aligned} \tag{3.34}$$

where we have normalized the 'length' of the state vector to unity. Requiring this normalization for the expansion equation (3.32) leads us to $P(x) = |C(x)|^2$, or that the expansion coefficient is just $C(x) = \psi(x)$.

Using the eigenstates of the position operator yields the position representation. The matrix elements of the position operator are

$$\begin{aligned}\langle x'|\hat{x}|x\rangle &= \int dx'' \langle x'|x''\rangle\langle x''|\hat{x}|x\rangle \\ &= \int dx'' x \langle x'|x''\rangle\langle x''|x\rangle \\ &= \int dx'' \delta(x' - x'')\delta(x - x'') \\ &= x\delta(x - x').\end{aligned} \tag{3.35}$$

Now consider the expectation value of an operator defined by

$$
\begin{aligned}
\langle \hat{O} \rangle &\equiv \langle \psi | \hat{O} | \psi \rangle \\
&= \int dx \int dx' \langle \psi | x' \rangle \langle x' | \hat{O} | x \rangle \langle x | \psi \rangle \\
&= \int dx \int dx' \psi^*(x') O_{x,x'} \psi(x)
\end{aligned}
\tag{3.36}
$$

with the matrix element of the operator $\hat{O}$ between two position eigenstates defined as $\langle x' | \hat{O} | x \rangle \equiv O_{x,x'}$. This is just the inner product of a column vector and a row vector, with a matrix for the operator sandwiched between them. For $\hat{O} = \hat{x}^n$ we have

$$
\begin{aligned}
\langle \hat{x}^n \rangle &= \int dx \int dx' \psi^*(x') x^n \delta(x - x') \psi(x) \\
&= \int dx \psi^*(x) x^n \delta(x - x') \psi(x) \\
&= \int dx x^n P(x),
\end{aligned}
\tag{3.37}
$$

which again tells us that $P(x) = |\psi(x)|^2$ is a probability distribution with respect to x. Furthermore, in this representation based on the eigenstates of the position operator, we can always replace the operator $\hat{x}$ with the scalar x. The general recipe equation (3.36) holds for all operators.

The matrix elements of $\hat{K} = -i$ times the derivative operator in the position representation are

$$
\langle x | \hat{K} | x' \rangle = -\imath \delta'(x - x'),
\tag{3.38}
$$

where $\delta'(x - x')$ is the derivative of the delta function with respect to x'. This operator is Hermitian; if we multiply it by $\hbar$ we find the momentum operator $\hat{p} = \hbar \hat{K}$. Calculating the expectation values of the momentum operator using equation (3.36) we find

$$
\begin{aligned}
\langle \hat{p}^n \rangle &= \int dx \int dx' \psi^*(x') (-\imath\hbar)^n \delta'^n(x - x') \psi(x) \\
&= (-\imath\hbar)^n \int dx \psi^*(x) \delta'^n(x - x') \psi(x) \\
&= (-\imath\hbar)^n \int dx \psi^*(x) \frac{d^n \psi(x)}{dx^n}.
\end{aligned}
\tag{3.39}
$$

Thus, as long as we remember to put the momentum operator *between* the wave function and its complex conjugate in the integral, we may always use $\hat{p} \to -\imath\hbar d/dx$.

Why should position eigenstates occupy such a special place in our theory? We should be able to use any complete set of states to expand our state vector, and the eigenstates of any Hermitian operator should yield a complete set of normalizable eigenstates. What about using the position eigenstates defined by

$$
\hat{p} | p \rangle = p | p \rangle,
\tag{3.40}
$$

where the momentum eigenstates form a complete basis satisfying

$$\int dp|p\rangle\langle p| = 1. \tag{3.41}$$

Going through a similar analysis, we find that for these states, we have

$$\langle \hat{p}^n \rangle = \int_{-\infty}^{\infty} dpp^n \Phi^*(p)\Phi(p) \tag{3.42}$$

$$\langle \hat{x}^n \rangle = (\imath\hbar)^n \int_{-\infty}^{\infty} dp\Phi^*(p)\frac{d^n\Phi(p)}{dp^n}, \tag{3.43}$$

where we defined the momentum space wave function $\Phi(p) = \langle p|\psi\rangle$.

The connection between these two representations is found by considering the character of the momentum eigenstates in the position representation, $\langle x|p\rangle$. Inserting the appropriate completeness relationships, we find that

$$\psi_p(x) = \langle x|p\rangle = \frac{1}{\sqrt{2\pi}}\exp(ipx/\hbar). \tag{3.44}$$

This is just a plane wave, a particle with a definite momentum $p = \hbar k$, definite energy (for a free particle) of $p^2/2m$, and a completely indeterminate position, $P(x) =$ constant. Using this, we find

$$\begin{aligned}\Phi(p) &= \langle p|\psi\rangle \\ &= \int dx\,\langle p|x\rangle\langle x|\psi \\ &= \frac{1}{\sqrt{2\pi}}\int dx e^{ipx/\hbar}\psi(x)\end{aligned} \tag{3.45}$$

and so the position and momentum space wave functions form a Fourier pair. As Fourier pairs satisfy the relationship $\Delta x\Delta k \geqslant 1/2$, we have the uncertainty principle showing up. Multiply by $\hbar$ to obtain $\Delta x\Delta p \geqslant \hbar/2$, where we have $p = \hbar k$. Let us pause and examine the cause of the 2π, we have played a little fast and loose with it so far. Recall that the plane wave function cannot be normalized in the ordinary sense. It has a mod squared of unity, and the integral over all x would lead to infinity. It is just as problematic as a Dirac delta function, as well it should be, as they form a Fourier pair! We require the orthonormality of the momentum space wave functions to be given in terms of a Dirac delta function, as we did for the position eigenstates, and we have

$$\langle k|k'\rangle = \delta(k - k') = \int_x \langle k|x\rangle\langle x|k'\rangle = \frac{1}{2\pi}\int dx e^{\imath(k-k')x}, \tag{3.46}$$

where the 2π factor comes from our use of an integral representation of the Dirac delta function.

Why have you not seen the momentum space representation before? It is just as valid. The main reason is that we usually deal with potentials that are functions of position rather than momentum, and so we have an easier time solving differential equations in the position representation. Let us first consider the harmonic oscillator. We have $\hat{H} = \hat{p}^2/2m + k\hat{x}^2/2$. In the position representation, the time-independent Schrödinger equation becomes

$$\frac{-\hbar^2}{2m}\frac{d^2\psi_n}{dx^2} + \frac{kx^2\psi_n}{2} = E_n\psi_n, \tag{3.47}$$

which can be solved to find the energy levels given by $E_n = (n + 1/2)\hbar\omega$ with $\omega^2 = k/m$; and wave functions that are a Gaussian multiplied by a Hermite polynomial of order n. How would this look in the momentum space representation?

$$\frac{p^2\Phi_l}{2m} + \frac{k}{2}\frac{d^2\Phi_l}{dp^2} = E_l\Phi_l. \tag{3.48}$$

They are the same form, a term quadratic in the variable natural to the representation, and a second derivative with respect to that same variable. Obviously, something is very special about the simple harmonic oscillator! We obtain the same type of wave function independent of our choice of special variable.

What about other systems? Consider a simple 1D Coulomb-type potential, $V(x) = \eta/x$. In the position representation, we have

$$\frac{-\hbar^2}{2m}\frac{d^2\psi_n}{dx^2} + \frac{\eta\psi_n}{x} = E_n\psi_n. \tag{3.49}$$

Now this may or may not have pretty solutions, but it does not seem too bad for now. What about the momentum representation?

$$\frac{p^2\Phi_l}{2m} + \frac{\eta}{\partial/\partial p}\Phi = E_l\Phi_l. \tag{3.50}$$

It is a very different form in this representation, and it is not quite clear how to interpret the derivative in the denominator acting on Φ. Obviously, there is something very special about the harmonic oscillator, which we will discuss in more detail later.

3.4 Pictures

In quantum mechanics, we also have pictures. What is a picture? It is a choice of where to put the time dependence in our equations. Usually, one uses the Schrödinger picture where the wave function evolves in time via

$$\imath\hbar\frac{\partial|\psi(x, t)\rangle}{\partial t} = \hat{H}\psi(x, t) \tag{3.51}$$

and operators that are independent of time such as

$$\hat{H} = \hat{p}^2/2m + V(\hat{x}) \tag{3.52}$$

$$\hat{x} \rightarrow x \tag{3.53}$$

$$\hat{p} \rightarrow -\imath\hbar\partial/\partial x. \tag{3.54}$$

We can write down a formal solution to the Schrödinger equation,

$$|\psi(x, t)\rangle = \exp(-\imath\hat{H}t/\hbar)|\psi(0, t\rangle). \tag{3.55}$$

To this point in the discussion, as we are using the time-independent Schrödinger equation, remember that $\hat{H}$ is independent of time, at least for now. We refer to the operator $\exp(-\imath\hat{H}t/\hbar) = U(t)$ as the propagator, as it propagates our state vector in time.

Maybe you have not run across the exponential of an operator before. It is generally well defined, consider

$$e^x = \sum_{n=0}^{\infty}\frac{x^n}{n!} = 1 + x + x^2/2 + x^3/6 + \ldots. \tag{3.56}$$

So as long as the multiplication of an operator is well defined, as the multiple action of the operator on a state vector, the functions of the operators are also well defined. For example,

$$\hat{X}^2|\psi(x, t)\rangle = x^2|\psi(x, t)\rangle. \tag{3.57}$$

The propagator yields unitary evolution if $\hat{H}$ is Hermitian, as we generally require; that is, the norm of the state function does not change under this temporal evolution.

The propagator can be used to solve the Schrödinger equation numerically, essentially using Simpson's rule for small time step Δt,

$$|\psi(t + \Delta t)\rangle = \exp(-\imath\hat{H}t/\hbar)|\psi(t\rangle \approx (1 - \imath\hat{H}\Delta t/\hbar))|\psi(t)\rangle. \tag{3.58}$$

Fancier integration schemes may be more efficient, stable, or both. Using the propagator, we can write down how the expectation value of an operator changes with time:

$$\begin{aligned}\langle\hat{O}(t)\rangle &= \langle\psi(t)|\hat{O}|\psi(t)\rangle \\ &= \langle\psi(0)|e^{\imath\hat{H}t/\hbar}\hat{O}e^{-\imath\hat{H}t/\hbar}|\psi(0)\rangle.\end{aligned} \tag{3.59}$$

We can use this as a starting point to define the Heisenberg picture, where we take all the time dependence to be in the operators, and none of the time dependence is in the state vector. The two pictures correspond at some initial time $t = 0$. We define the Heisenberg state vector as $|\psi_H(t)\rangle = |\psi_H(0)\rangle = |\psi_S(0)\rangle$. To preserve our expression for the time evolution of expectation values, the things we actually measure, we must define the Heisenberg operator. The Heisenberg operator is time dependent and given by

$$\hat{O}_H(t) = e^{\imath\hat{H}t/\hbar}\hat{O}e^{-\imath\hat{H}t/\hbar}. \tag{3.60}$$

If we take a time derivative of the above, we find the equation of motion for the operators

$$\begin{aligned}\frac{d\hat{O}_H}{dt} &= \frac{\imath}{\hbar}\hat{H}e^{\imath\hat{H}t/\hbar}\hat{O}_H e^{-\imath\hat{H}t/\hbar} \\ &+ e^{\imath\hat{H}t/\hbar}\frac{\partial\hat{O}_H}{\partial t}e^{-\imath\hat{H}t/\hbar} \\ &- \frac{\imath}{\hbar}e^{\imath\hat{H}t/\hbar}\hat{O}_H e^{-\imath\hat{H}t/\hbar}.\end{aligned} \tag{3.61}$$

Using the definition of the Heisenberg operator and the fact that the Hamiltonian commutes with itself (but not in general $\hat{O}$), we have

$$\frac{d\hat{O}_H}{dt} = \frac{\imath}{\hbar}[\hat{H}, \hat{O}_H] + \frac{\partial\hat{O}}{\partial t}. \tag{3.62}$$

Taking the expectation values of both sides, one finds Ehrenfest's theorem

$$\left\langle \frac{d\hat{O}_H}{dt} \right\rangle = \frac{\imath}{\hbar}[\langle \hat{H}, \hat{O}_H \rangle] + \left\langle \frac{\partial\hat{O}}{\partial t} \right\rangle. \tag{3.63}$$

Usually, the term involving the partial derivative of the operator with time is zero; after all operators such as $\hat{x}$ and $\hat{p}$ have no explicit time dependence. The first thing we notice is that if an operator has no explicit time dependence (time dependence even in the Schrödinger picture), if it commutes with the Hamiltonian, its expectation value does not change with time. This is the basis of the conservation laws in quantum mechanics.

To further illustrate the utility of the Heisenberg picture, consider the equations for the operators $\hat{x}$ and $\hat{p}$. For $\hat{x}$ we have

$$\begin{aligned}\frac{d\hat{x}}{dt} &= \frac{\imath}{\hbar}\left[\frac{\hat{p}^2}{2m}, \hat{x}\right] + \frac{\imath}{\hbar}[V(\hat{x}), \hat{x}] \\ &= \frac{\imath}{2m\hbar}(\hat{p}\hat{p}\hat{x} - \hat{x}\hat{p}\hat{p}) \\ &= \frac{\imath}{2m\hbar}(\hat{p}\hat{p}\hat{x} - \hat{p}\hat{x}\hat{p} + \hat{p}\hat{x}\hat{p} - \hat{x}\hat{p}\hat{p}) \\ &= \frac{\imath}{2m\hbar}(\hat{p}[\hat{p}, \hat{x}] - [\hat{p}, \hat{x}]\hat{p}) \\ &= \frac{\imath}{2m\hbar}(-2\imath\hbar\hat{p}) \\ &= \hat{p}/m,\end{aligned} \tag{3.64}$$

where we have used the fact that $\hat{x}$ commutes with itself and $[\hat{x}, \hat{p}] = \imath\hbar$. This is reminiscent of the Newtonian relation $p = mv$. Let us now consider

$$\begin{aligned}\frac{d\hat{p}}{dt} &= \frac{\imath}{\hbar}\left[\frac{\hat{p}^2}{2m}, \hat{p}\right] + \frac{\imath}{\hbar}[V(\hat{x}), \hat{p}] \\ &= \frac{\imath}{\hbar}\left[V(\hat{x}), \frac{\partial}{\partial x}\right].\end{aligned} \tag{3.65}$$

To calculate this last commutator, we use a test function

$$\begin{aligned}\left[V(\hat{x}), \frac{\partial}{\partial x}\right]|\psi\rangle &= \left(V(\hat{x})\frac{\partial|\psi\rangle}{\partial x} - V(\hat{x})\frac{\partial|\psi\rangle}{\partial x} - \frac{\partial V(\hat{x})}{\partial x}|\psi\rangle\right) \\ &= -\frac{\partial V(\hat{x})}{\partial x}|\psi\rangle.\end{aligned} \tag{3.66}$$

As this holds for any test function that is well behaved, we have

$$\frac{d\hat{p}}{dt} = -\frac{\partial V(\hat{x})}{\partial x}. \tag{3.67}$$

This is all that is left of Newton's second equation in quantum mechanics. Recall that this does not give us Newtonian equations as averages over quantum uncertainty. Newton says that $dp/dt = -dV/dx$. If we take an ensemble average over both sides, that would result in

$$\frac{d\langle p_{\text{Newton}}\rangle}{dt} = -\frac{\partial\langle V(x)\rangle}{dx}, \tag{3.68}$$

where Ehrenfest's theorem tells us that

$$\frac{d\langle\hat{p}\rangle}{dt} = -\left\langle\frac{\partial V(\hat{x})}{\partial x}\right\rangle. \tag{3.69}$$

This is in general different, as taking a derivative and an ensemble average do not commute.

Finally, there is one final picture we would like to consider, the interaction picture. It is used often in quantum optics as well as particle physics. Here, there is time dependence in *both* the state vector and the operators. At first sight, this seems to be more complicated, letting both things change. But we will see that there is a utility for this, otherwise we would not consider it! In many cases of interest, we will have a quantum system with a bare Hamiltonian $\hat{H}_0$ which will have a set of known eigenstates and energies. It may be a simple harmonic oscillator, a hydrogen atom, or some other nice system. We will then take this system and apply a time-dependent interaction, which we will see causes transitions, $\hat{H}_I$. We will put the time dependence arising from the bare Hamiltonian into the evolution of the operators, and the time dependence arising from the interaction into the state vectors evolution. Using the expression for the evolution of expectation values

$$\begin{aligned}\langle \hat{O}(t) \rangle &= \langle \psi(t)|\hat{O}|\psi(t) \rangle \\ &= \langle \psi(0)|e^{\imath\hat{H}t/\hbar}\hat{O}e^{-\imath\hat{H}t/\hbar}|\psi(0) \rangle \\ &= \langle \psi(0)|e^{\imath\hat{H}_I t/\hbar}e^{\imath\hat{H}_0 t/\hbar}\hat{O}e^{-\imath\hat{H}_0 t/\hbar}e^{\imath\hat{H}_I t/\hbar}|\psi(0) \rangle\end{aligned} \tag{3.70}$$

we can define an interaction picture state vector via

$$e^{\imath\hat{H}_I t/\hbar}|\psi(0), \tag{3.71}$$

which satisfies the time-dependent Schrödinger equation with *just* the interaction Hamiltonian

$$\imath\hbar\frac{\partial\psi_I(t)}{\partial t} = H_I\psi(t). \tag{3.72}$$

and the interaction picture operators

$$\hat{O}_I(t) = e^{\imath\hat{H}_0 t/\hbar}\hat{O}_I e^{-\imath\hat{H}_0 t/\hbar}, \tag{3.73}$$

which satisfy equations of the form

$$\frac{d\hat{O}_I}{dt} = \frac{\imath}{\hbar}[\hat{H}_0, \hat{O}_I] + \frac{\partial\hat{O}_S}{\partial t}. \tag{3.74}$$

We have cheated just a bit here, we have assumed that we can write

$$\exp(-\imath(\hat{H}_I + \hat{H}_0)t/\hbar) = \exp(-\imath\hat{H}_I t/\hbar)\exp(-\imath\hat{H}_0 t/\hbar). \tag{3.75}$$

We cannot always do this. Recall

$$\begin{aligned}\exp(A + B) &= 1 + (A + B) + (1/2)(A + B)^2 + \ldots \\ &= 1 + (A + B) + (1/2)(A^2 + B^2 + AB + BA) + \ldots.\end{aligned} \tag{3.76}$$

On the other hand, we have

$$\begin{aligned}\exp(A)\exp(B) &= (1 + A + (1/2)A^2 + \ldots)(1 + B + (1/2)B^2 + \ldots) \\ &= 1 + (A + B) + (1/2)(A^2 + B^2 + 2AB) + \ldots.\end{aligned} \tag{3.77}$$

These two forms are only equivalent if $[A, B] = 0$, which is true for real and complex numbers, but not generally for operators. The implication for the interaction picture is that our above formulae are only correct if we have $[\hat{H}_I, \hat{H}_0] = 0$, which will be true in the cases that interest us the most. We realize that often

$$[H(t), H(t')] \neq 0 \tag{3.78}$$

and in that case, the propagator takes the form

$$U(t) = \int exp(-iH(t)/\hbar). \tag{3.79}$$

3.5 The path integral method

The Feynman path integral formulation provides an alternative way of understanding quantum mechanics by expressing the probability amplitude of a particle's behavior as a sum over all possible paths it could take between two points [10]. The idea is to describe a quantum system not in terms of a single trajectory (like classical mechanics) but as a superposition of all possible trajectories that connect an initial and final state. In the standard formulation of quantum mechanics, the probability amplitude $\psi(x, t)$ is governed by the Schrödinger equation:

$$i\hbar\frac{\partial}{\partial t}\psi(x, t) = \hat{H}\psi(x, t),$$

where $\hat{H}$ is the Hamiltonian operator of the system. The transition amplitude $K(x_f, t_f; x_i, t_i)$ describes the probability amplitude for a particle to go from position x_i at time t_i to position x_f at time t_f. In the standard formalism, the transition amplitude can be written in terms of the time evolution operator $e^{-\frac{i}{\hbar}\hat{H}(t_f - t_i)}$, so

$$K(x_f, t_f; x_i, t_i) = \langle x_f|e^{-\frac{i}{\hbar}\hat{H}(t_f - t_i)}|x_i\rangle .$$

At this point, we break the time evolution into N steps, and each time step includes an identity of the form

$$I = \int dx|x\rangle\langle x|$$

and find

$$\begin{aligned} K(x_f, t_f; x_i, t_i) &= \langle x_f|e^{-\frac{i}{\hbar}\hat{H}(t_f - t_i)}|x_i\rangle \\ &= \int dx_1\langle x_f|e^{-\frac{i}{\hbar}\hat{H}\Delta t}|x_{N-1}\rangle\langle x_{N-1}|e^{-\frac{i}{\hbar}\hat{H}\Delta t}|x_{N-2}\rangle \ldots \langle x_1|e^{-\frac{i}{\hbar}\hat{H}\Delta t}|x_i\rangle , \end{aligned}$$

where Δt is the time interval between insertions. For small time steps Δt, we can approximate the matrix elements $\langle x_{n+1}|e^{-\frac{i}{\hbar}\hat{H}\Delta t}|x_n\rangle$ by considering the Trotter approximation for the time evolution operator. For small time intervals Δt, we approximate the exponential of the Hamiltonian by splitting the operator into its kinetic and potential parts. This is known as the Trotter approximation:

$$e^{-\frac{i}{\hbar}\hat{H}\Delta t} \approx e^{-\frac{i}{\hbar}\frac{\hat{p}^2}{2m}\Delta t}e^{-\frac{i}{\hbar}V(\hat{x})\Delta t}.$$

This means that we approximate the full evolution operator as a product of two separate operators, the kinetic part, $e^{-\frac{i}{\hbar}\frac{\hat{p}^2}{2m}\Delta t}$, and the potential part, $e^{-\frac{i}{\hbar}V(\hat{x})\Delta t}$. We now combine the kinetic and potential parts:

$$\langle x_{n+1}|e^{-\frac{i}{\hbar}\hat{H}\Delta t}|x_n\rangle \approx \sqrt{\frac{m}{2\pi i\hbar\Delta t}}\exp\left(\frac{im(x_{n+1} - x_n)^2}{2\hbar\Delta t} - \frac{i}{\hbar}V(x_n)\Delta t\right).$$

The expression for $\langle x_{n+1}|e^{-\frac{i}{\hbar}\hat{H}\Delta t}|x_n\rangle$ contains both the kinetic and potential terms, and the combined terms are related to the classical Lagrangian for the system. The Lagrangian $L(x_{n+1}, x_n)$ is given by the classical formula

$$L(x_{n+1}, x_n) = \frac{m}{2}\left(\frac{x_{n+1} - x_n}{\Delta t}\right)^2 - V(x_n),$$

where $\frac{m}{2}\left(\frac{x_{n+1} - x_n}{\Delta t}\right)^2$ represents the kinetic energy term, and $V(x_n)$ represents the potential energy at position x_n. Thus, we can rewrite the matrix element as

$$\langle x_{n+1}|e^{-\frac{i}{\hbar}\hat{H}\Delta t}|x_n\rangle \approx \exp\left(\frac{i}{\hbar}L(x_{n+1}, x_n)\Delta t\right).$$

To obtain the total propagator, we multiply this matrix element for each time step and integrate over all intermediate positions $x_1, x_2, \ldots, x_{N-1}$

$$\begin{aligned} K(x_f, t_f; x_i, t_i)- &= \lim_{N\to\infty} \int dx_1 \int dx_2 \cdots \int dx_{N-1} \Pi_{n=1}^{N-1} \\ &\times \sqrt{\frac{m}{2\pi i\hbar\Delta t}} \exp\left(\frac{im(x_{n+1} - x_n)^2}{2\hbar\Delta t} - \frac{i}{\hbar}V(x_n)\Delta t\right). \end{aligned} \tag{3.80}$$

In the limit $N \to \infty$, we replace the sum over separate positions with an integral over continuous paths. The result is the path integral formulation of the propagator

$$K(x_f, t_f; x_i, t_i) = \int_{x(t_i)=x_i}^{x(t_f)=x_f} \mathcal{D}[x(t)]\exp\left(\frac{i}{\hbar}S[x(t)]\right), \tag{3.81}$$

where the action $S[x(t)]$ is given by

$$S[x(t)] = \int_{t_i}^{t_f} \left(\frac{m}{2}\dot{x}^2 - V(x)\right)dt. \tag{3.82}$$

This can be visualized as shown in Zee [11]. First, consider the paths available in a double-slit experiment (figure 3.1). We know this can yield interference on the screen if we do not have information on which path was taken.

Now add a second set of slits, three of them (figure 3.2). Now we see there are more paths to go from the source to a given point on the screen.

Now go to the limit of an infinite number of slits on an infinite set of slits (figure 3.3). At this point, many paths are possible. We associate from the math

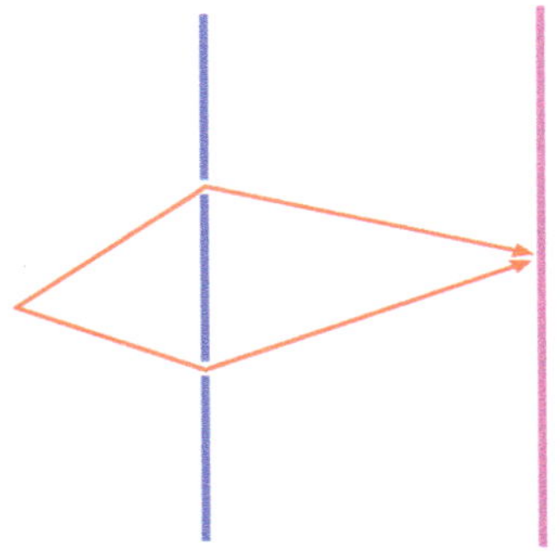

Figure 3.1. Possible paths for the double slit.

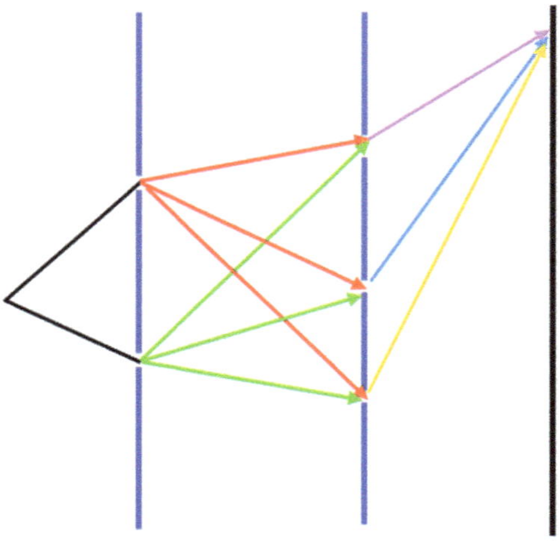

Figure 3.2. More possible paths for more slits.

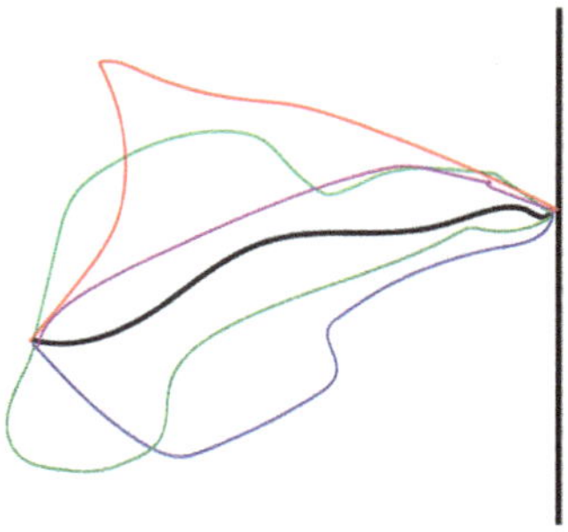

Figure 3.3. Infinite sets of infinite slits.

above a phase with each path of e^{iS}. The path in black is the classical path, which has a zero action, and hence makes the largest contribution to the path integral.

3.6 Density matrix

The density operator is defined as

$$\rho = |\psi\rangle\langle\psi|, \tag{3.83}$$

where you will recall that $|\psi\rangle$ is the quantum state ket, and $\langle\psi|$ is the corresponding bra. For now, the length of the quantum state vector will be some finite N, corresponding perhaps to some internal states of the operator such as energy or angular momentum. The expectation value of an operator $\hat{O}$ is given in the Dirac notation by

$$\langle\hat{O}\rangle = \langle\psi|\hat{O}|\psi\rangle. \tag{3.84}$$

This is just the multiplication of three matrices, a $1 \times N(\langle\psi|)$, an $N \times N$ (a matrix representation of $\hat{O}$), and finally an $N \times 1$, the ket $|\psi\rangle$. The result is a 1×1 matrix, or scalar, representing the average result of the measurements of the observable corresponding to $\hat{O}$ on many identical systems. We can also write this as

$$\begin{aligned}\langle \hat{O} \rangle &= \langle \psi | \hat{O} | \psi \rangle \\ &= \mathrm{Tr}\left(\langle \psi | \hat{O} | \psi \rangle\right) \\ &= \mathrm{Tr}(|\psi\rangle\langle\psi|\hat{O}) \\ &= \mathrm{Tr}(\rho\hat{O}),\end{aligned} \tag{3.85}$$

where we have used the cyclic property of the trace. We can also derive how the density matrix evolves in time using the Schrödinger equation $i\hbar d|\psi\rangle/dt = \hat{H}|\psi\rangle$,

$$\begin{aligned}\frac{d\rho}{dt} &= \frac{d}{dt}|\psi\rangle\langle\psi| \\ &= \frac{-i}{\hbar}(\hat{H}|\psi\rangle\langle\psi| - |\psi\rangle\langle\psi|\hat{H}) \\ &= \frac{-i}{\hbar}[\hat{H}, \rho].\end{aligned} \tag{3.86}$$

Combining the above two results, we find

$$\begin{aligned}\frac{d\langle \hat{O} \rangle}{dt} &= \frac{d\,\mathrm{Tr}(\rho\hat{O})}{dt} \\ &= \mathrm{Tr}\left(\hat{O}\frac{d(\rho)}{dt}\right) \\ &= \frac{-i}{\hbar}\langle[\hat{H}, \hat{O}]\rangle.\end{aligned} \tag{3.87}$$

We will have more to talk about regarding the density matrix when we deal with dissipation in quantum mechanics in chapter 10.

3.7 Choice of basis and measurement

How do we represent measurement in quantum mechanics? We have described how to obtain expectation values and standard deviations and the like from the state vector/wave function. The most common measurement made in the laboratory and discussed theoretically is the projective measurement. Let us say we are interested in some physical observable O, which in 1D quantum mechanics is some function, $O(x, p)$. If O is physically measurable, it must be represented by a Hermitian operator. For the rest of this section, the term measurement is considered a projective measurement. The allowed values of a measurement are the eigenvalues of the operator that correspond to the physical observable. Our meter always gives us a real number, we never get $3 + i$ for an energy measurement, so the fact that we want real values for measurement means we should only consider Hermitian matrices as they are guaranteed to have real eigenvalues. Operators such as $\hat{x}$ and $\hat{p}$ are Hermitian. Other operators we shall encounter are not, such as the creation and annihilation operators $\hat{a} = \hat{x} + i\hat{p}$. This means that we cannot measure a physical observable that corresponds to $\hat{a}$.

Another nice property of the Hermitian operators is that their eigenvectors form a complete orthogonal set in the case of non-degenerate eigenvalues. We can always find such a set in the case of degenerate eigenvalues as well, it just makes more work. This means that we can use the eigenvectors of O to expand our state vector,

$$|\psi\rangle = \sum C_n |o_n\rangle , \tag{3.88}$$

where we have defined

$$O|o_n\rangle = o_n|o_n\rangle \tag{3.89}$$

and $\langle o_n|o_m\rangle = \delta_{n,m}$. We can easily show that $|C_n|^2$ can be interpreted as the probability of finding o_n as a result of the measurement of O, and that

$$\langle O\rangle = \sum o_n |C_n|^2 . \tag{3.90}$$

We now see that we can use *any* Hermitian operator as a preferred variable, not just position and momentum!

OK, now what is the state of the system after the measurement? If the result of the measurement is the value o_n, the state after the measurement that yielded that value is

$$|\psi_{\text{after}}\rangle = |o_n\rangle . \tag{3.91}$$

We can formalize this a bit by introducing a projection operator,

$$P_n = |o_n\rangle \langle o_n| . \tag{3.92}$$

Then, the state after the measurement is

$$|\psi_{\text{after}}\rangle = \frac{P_n|\psi\rangle}{\sqrt{\langle\psi|P^\dagger P|\psi\rangle}} . \tag{3.93}$$

3.8 Executive summary

1. Our quantum system is described by the state vector $|\psi\rangle$ in a Hilbert space.
2. The state vector evolves via the Schrödinger equation. The Hermitian nature of the Hamiltonian implies that only unitary evolution is allowed. Any unitary operator has a Hamiltonian operator underlying it, even if it is very complicated. Simple unitarity can have complicated Hamiltonians underlying them.
3. Operators must be represented mathematically in a way that respects the experimental fact that order matters. This is why differential operators and matrices are common ways to represent quantum operators; their operation on a state is not necessarily commutative.
4. Projective measurements of an operator yield results that are eigenvalues of the Hermitian operator that correspond to the measurable. Recall that the

Hermitian operators have real eigenvalues (and hence can be related to a single measurement), as well as orthogonal eigenvectors (that can be used as a basis for expanding the wave function).

5. After a projective measurement, the state is in the corresponding eigenvector. Mathematically, this is expressed as a projection operator.
6. Other types of measurements, weak measurements and positive operator value measurements, are also possible and will be discussed later, in the context of balanced homodyne detection in chapter 9.

References

[1] Griffiths D J 2017 *Introduction to Quantum Mechanics* (Cambridge: Cambridge University Press)
[2] Shankar R 1994 *Principles of Quantum Mechanics* (Berlin: Springer)
[3] Bellac M L and de Forcrand-Millard P 2011 *Quantum Physics* (Cambridge: Cambridge University Press)
[4] Ballentine L E 1998 *Quantum Mechanics: A Modern Development* (Singapore: World Scientific)
[5] Cohen-Tannoudji C, Diu B and Laloe F 1991 *Quantum Mechanics* vol 1 (New York: Wiley)
[6] Beck M 2012 *Quantum Mechanics: Theory and Experiment* (Oxford: Oxford University Press)
[7] Sakurai J J and Napolitino J 2006 *Modern Quantum Mechanics* (Pearson Education)
[8] Shankar R 1003 *Principles of Quantum Mechanics* 2nd edn (Plenum Press)
[9] Arfken G B, Weber H J and Harris F E 2012 *Mathematical Methods for Physicists: A Comprehensive Guide* 7th edn (Academic Press)
[10] Feynman P R and Hibbs A R 2010 *Quantum Mechanics and Path Integrals: Emended Edition* (New York: Dover)
[11] Zee A 2012 *Quantum Field Theory in a Nutshell* (Levant)

IOP Publishing

Chapter 4

Two-level dynamics

4.1 Two-level atoms

We consider the quantum dynamics of two-level systems now. It is commonly discussed in introductory classes that shining light (photons/electromagnetic waves) on an atom, we can cause the atom to change state, as shown in figure 4.1.

We could also use molecules, solids, liquids, whatever. This occurs via absorption, in which a photon disappears from the field and the atom goes to a higher energy state. This can occur when the angular frequency of the light, ω, satisfies the Bohr condition $\Delta E = \hbar\omega$, where ΔE is the energy difference between the two states of interest. The atom can also go to a *lower* state (or energy level) by emitting a photon, in a process called stimulated emission. An atom in an excited state will also spontaneously emit to a lower state, which is called spontaneous emission. We will examine these phenomena using the Schrödinger equation and a simple two-level model. No real atom has only two energy levels, but if the light we shine is near resonance with a transition between two levels, we can ignore all other levels as a superb approximation. By 'light' we mean electromagnetic radiation of a well-defined frequency, whether it is in the visible, infrared, ultraviolet, or even gamma rays. Our two-level system will be taken to be two levels of an atom, probably a member of the alkali family in practice (cesium, rubidium, and sodium are common in quantum optics experiments), and we will illuminate this atom with monochromatic laser light. Initially, we treat the field as a classical quantity but, in later chapters, we more correctly treat it as a quantum mechanical quantity. Many of the results we derive apply to other physical systems, such as a nucleus with spin 1/2 in a time-varying magnetic field as in nuclear magnetic resonance, but we stick to lasers and atoms as a picture in our heads.

One can more realistically prepare a two-level atom by using optical pumping. Consider the states for $J = 2$ and $J = 1$ as shown in figure 4.2. Driving the system with right circular polarized light will drive the atom up and over one magnetic sublevel. Spontaneous emission can bring you straight down ($\Delta m = 0$) with linearly

doi:10.1088/978-0-7503-6014-2ch4

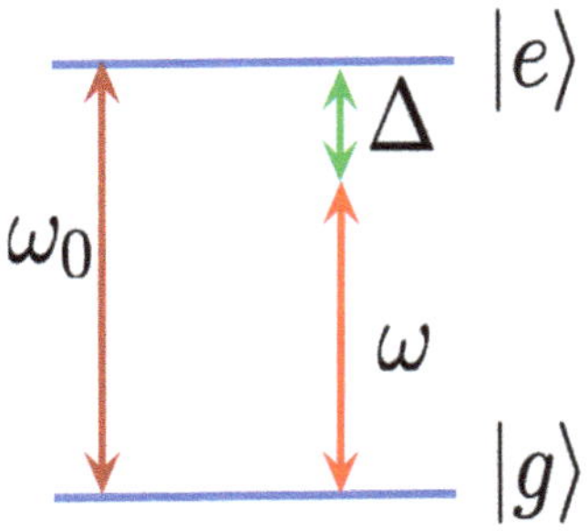

Figure 4.1. Schematic of a two-level system with resonance frequency ω_0, and a driving field at ω, detuned by $\Delta = \omega - \omega_0$.

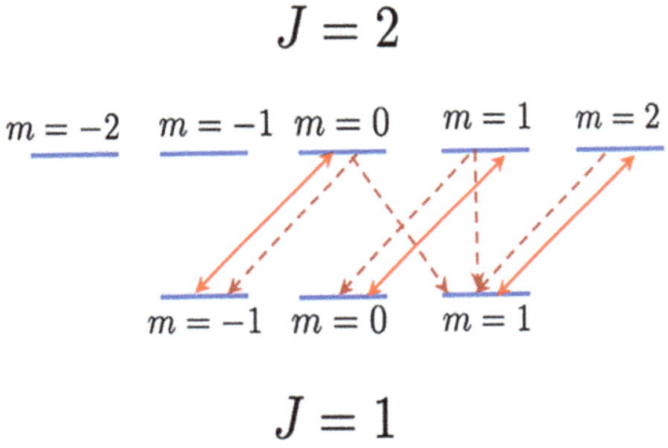

Figure 4.2. Optical pumping scheme.

polarized light, or with right/left circular light ($\Delta m = \pm 1$). Eventually, the atom is restricted to the two-state manifold at the far right of the diagram.

The general form for the quantum state of a two-level atom in the absence of a time-dependent interaction potential is [1, 2]

$$|\Psi\rangle = C_e e^{-i\omega_e t}|e\rangle + C_g e^{-i\omega_g t}|g\rangle\,, \tag{4.1}$$

where the subscripts e and g refer to the excited and ground states, respectively. Here, we have expanded the state vector in the energy eigenbasis and kept only the two relevant terms. Quantities C_e and C_g are the probability amplitudes for the excited and ground states. The probability of finding the atom in the excited state (if we actually look!) is $P_e = |C_e|^2$, similarly the probability of finding the atom in the ground state (again, if we actually look!) is $P_g = |C_g|^2$. It is important to recall that in quantum mechanics, we know nothing until we make a measurement, and then we only know what we have measured and no more. In the general state equation (4.1), it is *not* as if the atom is sometimes in the ground state and sometimes in the excited state, and hops back and forth. If that were the case, we would merely have statistical physics. But realizing this is a loose language, an atom in state equation (4.1) is simultaneously in the ground *and* excited states. This may seem meaningless, and really it is, as we do not know what state the atom is in until we make a measurement of the energy, then we find that atom in either the ground or excited state, with the probabilities given above. What state is the atom really in, in a superposition? In quantum mechanics, we know nothing until we (or someone or something) make a measurement. As Asher Peres says, 'an unperformed measurement has no outcome'. As we proceed, we will assume that the quantum statics has been done; analytically, numerically, or experimentally, that is, we assume that the eigenstates and eigenvalues of the atom are known to us:

$$H_0|e\rangle = \hbar\omega_e|e\rangle \tag{4.2}$$

$$H_0|g\rangle = \hbar\omega_g|g\rangle. \tag{4.3}$$

We assume that H_0 is independent of time and that the resulting eigenstates are stationary. The general superposition is *not,* we recall! However, if the system starts out in the ground or excited state, with no admixture from the other, it will sit there in that state for all eternity. In reality, the quantum vacuum *always* provides a time-dependent interaction, which is inescapable. Here, we take the eigenstates $|e\rangle$ and $|g\rangle$ as stationary, and proceed with the introduction of a time-dependent interaction Hamiltonian due to an external laser field in our later detailed calculations. We take the Hamiltonian to now be

$$H = H_0 + H_I. \tag{4.4}$$

We have two avenues open to us. We can take this new Hamiltonian and find the eigenstates and eigenvalues (or eigenenergies). These will be time dependent in general. This leads to the so-called dressed states and quasi-energies, and we will examine this path later on. For now, we realize that regardless of the state of the system, the states $|e\rangle$ and $|g\rangle$ still form an orthonormal basis, and our state can be expanded in terms of them via

$$|\Psi\rangle = C_e(t)e^{-i\omega_e t}|e\rangle + C_g(t)e^{-i\omega_g t}|g\rangle, \tag{4.5}$$

where we now treat the probability amplitudes C_e and C_g as functions of time; this stems from the fact that the Hamiltonian is dependent on time and will lead to transitions between states. If the system starts in a state such that $C_g(0) = 1$ and $C_e(0) = 0$, i.e. that the atom is in the ground state initially, we will later find that $C_e(t) \neq 0$. The problem will be to find equations for how these probability amplitudes change with time. The probabilities for funding the atom in a given state are also time dependent, for example, the excited-state probability $P_e(t) = |C_e(t)|^2$. To find the equations that the probability amplitudes satisfy, we substitute our Hamiltonian and state vector into the Schrödinger equation. The time derivative of the state vector equation (4.5) is

$$i\hbar\frac{\partial|\Psi\rangle}{\partial t} = i\hbar(\dot{C}_e - i\omega_e C_e)e^{-i\omega_e t}|g\rangle + \left(\dot{C}_g - i\omega_g C_g\right)e^{-i\omega_g t}|g\rangle. \tag{4.6}$$

The Hamiltonian acting on the state vector is given by

$$\begin{aligned} H|\Psi\rangle &= (H_0 + H_I)|\Psi\rangle \\ &= (\hbar\omega_e + H_I)C_e(t)e^{-i\omega_e t}|e\rangle + \left(\hbar\omega_g + H_I\right)C_g(t)e^{-i\omega_g t}|g\rangle, \end{aligned} \tag{4.7}$$

where we have used the fact that the states $|e\rangle$ and $|g\rangle$ are eigenstates of the bare Hamiltonian H_0. Using the above results, we obtain

$$\begin{aligned} &i\hbar(\dot{C}_e - i\omega_e C_e)e^{-i\omega_e t}|g\rangle + \left(\dot{C}_g - i\omega_g C_g\right)e^{-i\omega_g t}|g\rangle = \\ &(\hbar\omega_e + H_I)C_e(t)e^{-i\omega_e t}|e\rangle + \left(\hbar\omega_g + H_I\right)C_g(t)e^{-i\omega_g t}|g\rangle \end{aligned} \tag{4.8}$$

and then using the Bohr condition, we find

$$i\hbar(\dot{C}_e e^{-i\omega_e t}|e\rangle + \dot{C}_g e^{-i\omega_g t}|g\rangle) = (C_e(t)e^{-i\omega_e t}H_I|e\rangle + C_g(t)e^{-i\omega_g t}H_I|g\rangle). \tag{4.9}$$

To obtain an equation for $\dot{C}_g$, we act on the right of the above equation with the bra $\langle g|$. Using the orthonormality of the excited and ground states, $\langle i|j\rangle = \delta_{ij}$, with i and j either e or g, we have

$$i\hbar\dot{C}_g e^{-i\omega_g t} = (C_e(t)e^{-i\omega_e t}\langle g|H_I|e\rangle + C_g(t)e^{-i\omega_g t}\langle g|H_I|g\rangle) \tag{4.10}$$

or

$$\dot{C}_g = \frac{-i}{\hbar}(C_e(t)e^{-i\omega_0 t}\langle g|H_I|e\rangle + C_g(t)\langle g|H_I|g\rangle), \tag{4.11}$$

where we have defined the resonant frequency of the transition as $\omega_0 \equiv \omega_e - \omega_g$. In a similar fashion, we obtain

$$\dot{C}_e = \frac{-i}{\hbar}(C_g(t)e^{i\omega_0 t}\langle e|H_I|g\rangle + C_e(t)\langle e|H_I|e\rangle). \tag{4.12}$$

We cannot solve these equations until we are told what the interaction Hamiltonian is, and we can compute the matrix elements of the Hamiltonian, $\langle i|H_I|j\rangle$. It is convenient to use a vector representation of the state vector, sometimes as

$$|\Psi\rangle = \begin{pmatrix} C_e \\ C_g \end{pmatrix}. \tag{4.13}$$

The equations for the probability amplitudes may then be written as

$$\begin{pmatrix} \dot{C}_e \\ \dot{C}_g \end{pmatrix} = -\frac{i}{\hbar}\begin{pmatrix} H_I^{ee} & H_I^{eg}e^{i\omega_0 t} \\ H_I^{eg}e^{-i\omega_0 t} & H_I^{gg} \end{pmatrix}\begin{pmatrix} C_e \\ C_g \end{pmatrix}, \tag{4.14}$$

where we have defined $H_I^{ij} \equiv \langle i|H_I|j\rangle$, the matrix elements of the interaction Hamiltonian. For this to be Hermitian we must have $\langle i|H_I|j\rangle = \langle j|H_I|i\rangle^*$.

4.2 Atom–field interaction in the electric dipole approximation

At this point, we wish to specify an interaction Hamiltonian appropriate for the interaction of an atom with laser light, the electric dipole interaction

$$H_I = -\vec{\mu}\cdot\vec{E}. \tag{4.15}$$

The quantity $\vec{\mu} = -e\vec{x}$ is the dipole moment operator. Recall that for two opposite charges separated by a distance d, that the two monopole fields cancel to order $1/r^2$, leaving a residual dipole field proportional to ed/r^3, and $ed = \mu$ is the dipole moment. The total energy of the field is proportional to the square of the total field, $E_{\text{tot}} = E_{\text{app}} + E_{\text{dip}}$. The interaction energy stems from the cross term and goes as $E_{\text{app}}E_{\text{dip}} \propto \mu E_{\text{app}}$. We take $E \equiv E_{\text{app}}$ from now on. The dipole moment is a vector that points from the positive charge to the negative charge and has magnitude ed. In the case of a hydrogen atom, the two charges are the proton and the neutron. For an alkali element, we have a nucleus surrounded by closed orbitals, with a net positive charge, and a single valence electron. These are hydrogen-like systems, particularly for circular states where the quantum number m_l is maximized. Thus, knowing the wave functions of the two states of interest, we could calculate

$$\langle i|H_I|j\rangle = -e\int \psi_i^*\vec{x}\cdot\vec{E}\,\psi_j dV. \tag{4.16}$$

However, in any event, we will assume that the energies of the ground and excited states are known, or really just the difference of the two energies; as well as the interaction matrix elements. This knowledge can be obtained by knowing the wave functions of a tractable system, such as a hydrogen or a simple harmonic oscillator, numerical solutions of the Schrödinger equation, or measured by spectroscopic means. We will take them to be known and proceed.

The field has contributions from the electric quadrapole and magnetic dipole, as well as the electric octopole and magnetic quadrapole terms The electric dipole term is generally the largest, and each term in the multipole expansion generally decreases as $1/c$. The key physical assumption is that the electric field does not change much over the size of the atom, and one can approximate $\cos(kx) \simeq 1$. As the size of an atom is around $0.2 \to 10.0$ nm, if we consider radiation in the near UV to infrared ($200 \to 1500$ nm), this condition is easily met. The electric field will not vary much over the location of the electron relative to the (assumed stationary) nucleus. We take the perturbing field to be of the form $\vec{E} = E_0\cos(\omega t)i$, that is polarized along the x axis. We shall take it to be propagating along the z axis, with an amplitude of E_0 and an angular frequency ω that is in general it different than the resonant frequency of the atomic transition, ω_0.

The latter will allow us to consider what happens when the laser is not on resonance, but of course it must be nearer this resonance than any others, or else our approximation of a two-level system is in jeopardy. Hence, our interaction Hamiltonian is

$$H_I = -exE_0\cos(\omega t). \tag{4.17}$$

We will discuss later what happens if the laser is polarized along a different direction, or if one has a collection of atoms whose dipole moments are oriented randomly with respect to the polarization of the laser, as well as circular polarization.

Now let us consider the calculation of the matrix elements of this interaction Hamiltonian. We assume in what follows that the atomic states of interest have a definite parity, that is $\psi(\vec{x}) = \pm\psi(-\vec{x})$, or rather that $|\psi(\vec{x})|^2$ is an even function. This will be true as long as the probability of the electron to be at a position $\vec{x}$ with respect to the nucleus is the same as the probability for it to be found at $-\vec{x}$ or, in more physical terms, the atom has no permanent dipole moment. All of our assumptions to this point can be relaxed, but we wish to look at the simplest case first. With this in hand, we first examine a diagonal matrix element:

$$\langle e|H_I|e\rangle = -eE_0\cos(\omega t)\int \psi_i^* x\psi_j dx = -eE_0\int x|\psi(\vec{x})|^2 dx = 0. \tag{4.18}$$

The same occurs for the other diagonal matrix element, $\langle g|H_I|g\rangle = 0$. There are some systems, particularly molecules, that do have a permanent dipole moment, but we ignore that the possibility for now.

Now consider the off-diagonal matrix elements

$$\langle e|H_I|g\rangle = \langle g|H_I|e\rangle^* = -eE_0\cos(\omega t)\int \psi_e^* x\psi_g dx. \tag{4.19}$$

This is in general nonzero, and we will assume that the quantity

$$\langle e|H_I|g\rangle = -eE_0\cos(\omega t)\int \psi_e^* x\psi_g dx \equiv \mu_{eg} \tag{4.20}$$

is known somehow. Then, our interaction Hamiltonian is given in matrix form by

$$H_I = -i\hbar E_0\cos(\omega t)\begin{pmatrix} 0 & \mu_{eg} \\ \mu_{eg}^* & 0 \end{pmatrix}. \tag{4.21}$$

There will be times when $\mu_{eg} = 0$, for example, when the states have the same parity. For example if ψ_e and ψ_g are both odd (or both even), then we will have $\mu_{eg} = 0$. In this case, we are left with $\dot{C}_g = \dot{C}_e = 0$ and there are no transitions. The probabilities $P_e = |C_e(t)|^2$ and $P_g = |C_g(t)|^2$ are stationary. This is the source of the selection rules, which tell us which transitions are allowed. For example, the parity of a hydrogen state is given by $P = (-1)^l$, and hence the quantum number l must change in a transition leading to the rule $\Delta l = \pm 1$. A photon is a spin 1 particle and can only carry away one unit of angular momentum. Another selection rule for hydrogen is $\Delta m_l = 0, \pm 1$. These correspond to the absorption and emission of circularly polarized photons ($\Delta m_l = \pm 1$) or linearly polarized photons ($\Delta m_l = 0$). Recall that linearly polarized light is a linear combination of the right and left circular polarized light. However, remember that transitions forbidden for electric dipole interactions can still proceed by higher order terms (for example electric quadrapole and magnetic dipole), which will result in a smaller amplitude for that component of the spectrum.

If $\Delta m_l = 0$, then for the hydrogen states, μ_{eg} is real, as the only complex nature of the matrix element is the ϕ dependence, $\exp\big(i(m_f - m_i)\big)$, and for convenience we specialize in that case here. The resulting equations are

$$\dot{C}_g = -i\Omega\cos(\omega t)e^{-i\omega_0 t}C_e \tag{4.22}$$

$$\dot{C}_e = -i\Omega\cos(\omega t)e^{i\omega_0 t}C_g, \tag{4.23}$$

where we have defined $\Omega = \mu_{eg}E_0/\hbar$, and we refer to it as the Rabi frequency. By writing the cos in terms of complex exponentials, we find

$$\dot{C}_g = -i\Omega/2(e^{i\Delta t} + e^{-i(\omega_0 + \omega)t})C_e \tag{4.24}$$

$$\dot{C}_e = -i\Omega/2(e^{-i\Delta t} + e^{i(\omega_0 + \omega)t})C_g. \tag{4.25}$$

With $\Delta = \omega - \omega_0$ as the laser detuning from the resonance. We will use the rotating wave approximation, which involves dropping the rapidly oscillating terms $e^{(i(\omega+\omega_0)t)}$; this is known as the rotating wave approximation (RWA). We will explore this approximation more in this chapter and in chapters 6 and 7 when we explore quantum fields. We are then left with

$$\dot{C}_g = \frac{-i\Omega}{2}e^{i\Delta t}C_e \tag{4.26}$$

$$\dot{C}_e = \frac{-i\Omega}{2}e^{-i\Delta t}C_g. \tag{4.27}$$

So far, the only approximations made are the dipole approximation and the rotating wave approximation. We will solve these equations using the perturbation theory first, and then look at the full solutions, on and off resonance. Let us assume that these probability amplitudes have a Taylor series expansion in the parameter Ω,

$$C_e(t) = C_e^{(0)}(t) + C_e^{(1)}(t) + C_e^{(2)}(t) + \ldots \tag{4.28}$$

$$C_e(g) = C_g^{(0)}(t) + C_g^{(1)}(t) + C_g^{(2)}(t) + \ldots, \tag{4.29}$$

where (n) indicates the power of Ω for each term. The initial conditions would be

$$C_e^{(0)} = C_e(0) \tag{4.30}$$

$$C_g^{(0)} = C_g(0) \tag{4.31}$$

and for all other values of (n), we have $C_e^{(n)} = C_g^{(n)} = 0$. Plugging this back into equations (4.26) and (4.27), and equating terms in equal powers of Ω we find

$$\dot{C}_g^{(n)} = \frac{-i\Omega}{2}e^{i\Delta t}C_e^{(n-1)} \tag{4.32}$$

$$\dot{C}_e^{(n)} = \frac{-i\Omega}{2} e^{-i\Delta t} C_g^{(n-1)}. \tag{4.33}$$

Another way to develop this formalism is to start with the formal solutions

$$C_e(t) = C_e(0) - \frac{i\Omega}{2} \int_0^t dt' e^{i\Delta t'} C_g(t') \tag{4.34}$$

$$C_g(t) = C_g(0) - \frac{i\Omega}{2} \int_0^t dt' e^{-i\Delta t'} C_e(t') \tag{4.35}$$

and then iterate it by substituting the solutions on the left-hand side for $C_e(t)$ and $C_g(t)$ back into the formal solutions:

$$\begin{aligned} C_e(t) = C_e(0) &- \frac{i\Omega}{2} \int_0^t dt' e^{i\Delta t'} C_g(0) \\ &+ \left(\frac{i\Omega}{2}\right)^2 \int dt' \int dt'' e^{i\Delta(t'-t'')} C_e(t'') \end{aligned} \tag{4.36}$$

$$\begin{aligned} C_g(t) = C_g(0) &- \frac{i\Omega}{2} \int_0^t dt' e^{-i\Delta t'} C_e(0) \\ &+ \left(\frac{-i\Omega}{2}\right)^2 \int dt' \int dt'' e^{-i\Delta(t'-t'')} C_e(t''). \end{aligned} \tag{4.37}$$

One can continue and, in this manner, generate a power series for the coefficients in powers of Ω. We will now truncate the series after the first integral, assuming that Ω is small. Recall that this will occur for weak driving fields *or* short interaction times. We also assume that the atom starts in the ground state, so that $C_g(0) = 1$ and $C_e(0) = 0$. This yields

$$C_e(t) = -\frac{i\Omega}{2} \frac{(e^{i\Delta t} - 1)}{i\Delta}. \tag{4.38}$$

We then use the following trick,

$$e^{ix} - 1 = e^{-ix/2}(e^{-ix/2} - e^{ix/2}) = -2ie^{ix/2} \sin(x/2), \tag{4.39}$$

to write

$$C_e(t) = \frac{i\Omega}{\Delta} e^{i\Delta t/2} \sin(\Delta t/2). \tag{4.40}$$

It will sometimes be convenient to express this in the form

$$C_e(t) = \frac{i\Omega t}{2} e^{i\Delta t/2} \frac{\sin(\Delta t/2)}{\Delta t/2}. \tag{4.41}$$

This is convenient to use in the resonant case when $\Delta = 0$, and we find

$$C_e(t) = \frac{i\Omega t}{2} e^{i\Delta t/2}, \tag{4.42}$$

recalling that $\lim_{x\to 0}\sin(x)/x = 1$. On resonance then, the probability of the atom being found in the excited state at t, given that it was in the ground state initially when the laser was turned on, is

$$P_e(t) = |C_e(t)|^2 = \frac{\Omega^2 t^2}{4}. \tag{4.43}$$

Obviously, if we let t become much larger, the probability of being in the excited state exceeds one! Well, all this means is that we can no longer ignore the higher order terms in the perturbation series above, which goes as powers of Ωt. In fact, the above result defines what we mean by a 'short ' time or 'weak' field, it is when

$$t \ll 2/\Omega. \tag{4.44}$$

When we are off resonance, the result is

$$P_e(t) = |C_e(t)|^2 = \frac{\Omega^2 \sin^2(\Delta t/2)}{\Delta^2}. \tag{4.45}$$

Plots of this probability for various detunings are shown in figure 4.3.

Perhaps the surprising result here is that even if one is detuned from the resonance, at a given time it is possible to find the atom in the excited state! Well, that seems to be at odds with energy conservation; we are putting in photons with energy $\hbar\omega$, which is larger or smaller than the energy difference between the two levels $\hbar\omega_0$. This is a problem for you to figure out! Think about how that atom knows what frequency is illuminating it, and when.

If we had kept the non-RWA terms in equation (4.25) that oscillated at $\omega + \omega_0$, the resulting integrals would have yielded terms such as

$$\frac{e^{i(\omega+\omega_0)t} - 1}{i(\omega + \omega_0)}. \tag{4.46}$$

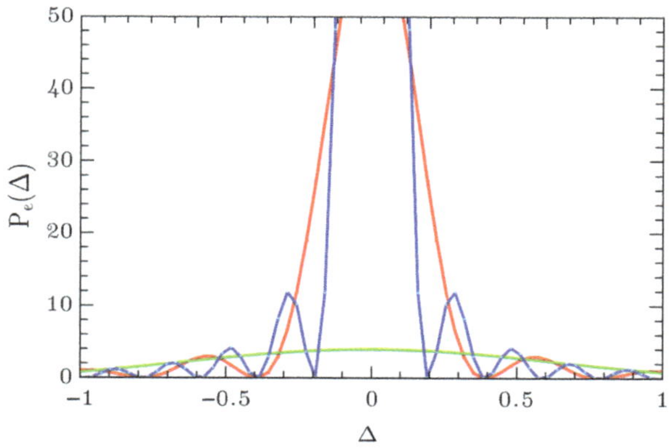

Figure 4.3. Excited state population as a function of detuning at various times.

The numerator of this term would oscillate between 0 and 2 in amplitude, while the denominator is of order 10^{15} for visible light. Obviously, in the perturbation theory, these terms will have little effect. Later, when we quantize the field, we find that they make no contributions to the transition rates in the dipole approximation. They do, however, result in energy shifts of the energies, known as the Bloch–Siegert shifts. Let us now go beyond the perturbation theory, beginning with resonance. The relevant equations are

$$\dot{C}_g = \frac{-i\Omega}{2} C_e \tag{4.47}$$

$$\dot{C}_e = \frac{-i\Omega}{2} C_g. \tag{4.48}$$

We take the derivative of the first equation, which yields

$$\ddot{C}_e = \frac{-i\Omega}{2} \dot{C}_g = \frac{-\Omega^2}{4} C_e. \tag{4.49}$$

This has the solution

$$C_e(t) = A\cos(\Omega t/2) + B\sin(\Omega t/2). \tag{4.50}$$

Knowing C_e, we can use the differential equation for $\dot{C}_e$ to find C_g,

$$C_g(t) = iA\sin(\Omega t/2) - iB\cos(\Omega t/2). \tag{4.51}$$

We have the normalization condition $P_e + P_g = |A|^2 + |B|^2$, where A and B are generally complex and found from the initial conditions.

With the atom initially in the ground state, we have $C_e(0) = A = 0$, and $C_g(0) = 1 = iB$, or $B = -i$. Thus, we have

$$C_e(t) = i\sin(\Omega t/2) \tag{4.52}$$

$$C_g(t) = \cos(\Omega t/2). \tag{4.53}$$

The associated probabilities are

$$P_e(t) = \sin^2(\Omega t/2) = \frac{1}{2}(1 - \cos(\Omega t)) \tag{4.54}$$

$$P_g(t) = \cos^2(\Omega t/2) = \frac{1}{2}(1 + \cos(\Omega t)). \tag{4.55}$$

This is shown in figure 4.4.

Again, we see that the sum of the probabilities is unity at all times, which must occur due to the unitary evolution of the quantum state. The probabilities oscillate at the frequency Ω, known as the Rabi frequency. Notice that this exact result reduces to our perturbation result for $\Omega t \ll 1$, where we have previously determined that the perturbation theory would hold.

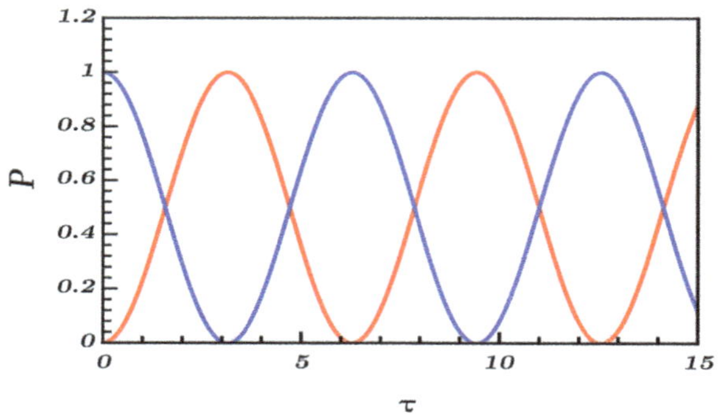

Figure 4.4. Excited (red) and ground (blue) state populations as functions of time when the system is driven on resonance.

Notice that by turning the laser off at a time $\Omega T = \pi$, the atom is in the excited state, with no admixture from the ground state. This is referred to as a π pulse. Another special case is when the laser is turned off at $T = \pi/2\Omega$ or $\Omega T = \pi/2$. Here the atom is left in the superposition state

$$|\Psi\rangle = \frac{1}{\sqrt{2}}(i|e\rangle + |g\rangle). \tag{4.56}$$

The atom is in a state where it is equally likely to be found in the ground or excited states if an energy measurement is made, but it is neither in the ground or excited state *unless* such a measurement is made. A 2π pulse would return the atom to the ground state. Such pulses can be useful for communication applications as no energy is deposited in the atoms of, say, the fiber. This can only occur in realistic systems for certain pulse shapes, which results in the phenomena of solitons and self-induced transparency. If the laser is not turned on and off abruptly (is it ever?) the relevant parameter is the pulse 'area' $A = \int_0^T \Omega(t)dt$.

What about off resonance? Our trick of differentiating an equation and substituting that into the other will still work, but it is a little more complicated, resulting in

$$\ddot{C}_e + i\Delta \dot{C}_e + \frac{\Omega^2}{4}C_e = 0 \tag{4.57}$$

$$\ddot{C}_g - i\Delta \dot{C}_e + \frac{\Omega^2}{4}C_e = 0. \tag{4.58}$$

These equations can be solved using the ansatz $C_e(t) = e^{i\lambda t}$. The first equation then yields

$$\left[-\lambda^2 - \lambda\Delta + \frac{\Omega^2}{4}\right] = 0. \tag{4.59}$$

The two roots are

$$\lambda_{1,2} = -\frac{1}{2}(\Delta \pm \Omega'^2), \tag{4.60}$$

where the generalized Rabi frequency is $\Omega' = \sqrt{\Omega^2 + \Delta^2}$. The general form of the solutions is

$$C_e(t) = Ae^{i\lambda_1 t} + Be^{i\lambda_2 t} \tag{4.61}$$

$$C_g(t) = \frac{-2i}{\Omega} e^{i\Delta t/2}[\lambda_1 A e^{i\lambda_1 t} + \lambda_2 B e^{i\lambda_2 t}]. \tag{4.62}$$

With the atom initially in the ground state, we have

$$C_e(t) = -i\frac{\Omega}{\Omega'} \sin(\Omega' t/2) e^{i\Delta t/2} \tag{4.63}$$

$$C_g(t) = \cos(\Omega' t/2) - i\frac{\Delta}{\Omega'} \sin(\Omega' t/2) e^{-i\Delta t/2}. \tag{4.64}$$

A plot of the excited-state probability is shown in figure 4.5. As we increase the detuning, we find that the maximum excited-state probability decreases, and the oscillations become faster, oscillating at the generalized Rabi frequency.

Notice that the probability oscillates at a larger frequency $\Omega' \gg \Omega$ than on resonance, and note that the atom never reaches the excited state.

Notice that we have never explicitly used the fact that the excited state had a greater energy than the ground state, so our results also work for stimulated emission as well. Actually, there are some phase changes for emission, but the probabilities remain the same. Mathematically, one would make the substitution $\omega_0 = -|\omega_0|$, and the terms dropped in the RWA would be kept. However, the terms we kept would then involve sums of frequencies and would be dropped.

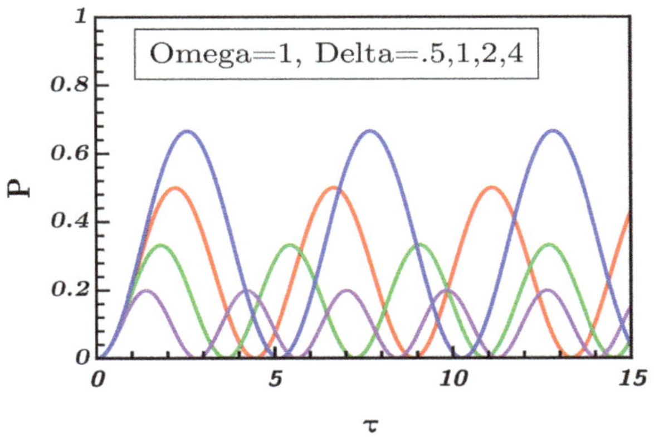

Figure 4.5. Excited (red) and ground (blue) state populations as functions of time when the system is driven off resonance.

4.3 Introduction to the dressed states

Until now, we have used the excited and ground states as basis states to expand the general quantum state of the two-level atom; let us now consider a different tack and talk about the eigenstates and eigenvalues of the Hamiltonian $H = H_0 + H_I$. These eigenstates are time dependent as the Hamiltonian is, they are *not* stationary. However, if we make a time-dependent transformation to our probability equations, we can find some useful results. This transformation is

$$D_g(t) = C_g(t) \tag{4.65}$$

$$D_e(t) = C_e(t)e^{-i\Delta t}. \tag{4.66}$$

The new probability amplitudes satisfy the equations

$$\dot{D}_g = \frac{-i\Omega}{2}D_e \tag{4.67}$$

$$\dot{D}_e = \frac{-i\Omega}{2}D_g - i\Delta D_e. \tag{4.68}$$

These equations can be written in the form

$$\begin{pmatrix} \dot{D}_e \\ \dot{D}_g \end{pmatrix} = -\frac{i}{\hbar}\begin{pmatrix} -\hbar\Delta & \hbar\Omega/2 \\ \hbar\Omega/2 & 0 \end{pmatrix}\begin{pmatrix} D_e \\ D_g \end{pmatrix}. \tag{4.69}$$

In this new frame, our new Hamiltonian has no time dependence, and the eigenvalues or quasi-energies are

$$E_{1,2} = \frac{\hbar}{2}(-\Delta \pm \Omega'). \tag{4.70}$$

The new (unnormalized) eigenstates are

$$|+\rangle = \begin{pmatrix} \Omega' - \Delta \\ \Omega \end{pmatrix} \tag{4.71}$$

$$|-\rangle = \begin{pmatrix} \Omega' + \Delta \\ -\Omega \end{pmatrix}. \tag{4.72}$$

On resonance, we have

$$|+\rangle = \frac{1}{\sqrt{2}}\begin{pmatrix} 1 \end{pmatrix} \tag{4.73}$$

$$|-\rangle = \frac{1}{\sqrt{2}}\begin{pmatrix} 1 \end{pmatrix}. \tag{4.74}$$

It is often convenient to express the eigenstates in the form

$$|+\rangle = \cos(\Theta/2)||e\rangle + \sin(\Theta/2)|g\rangle \tag{4.75}$$

$$|-\rangle = -\sin(\Theta/2)|e\rangle + \cos(\Theta/2)|g\rangle \tag{4.76}$$

with

$$\tan(\Theta) = \frac{\Omega}{\Delta}. \tag{4.77}$$

Again, on resonance, we have $\Theta = \pi$, with $\Theta/2 = \pi/2$, with

$$|+\rangle = \frac{1}{\sqrt{2}}(|e\rangle + |g\rangle) \tag{4.78}$$

$$|-\rangle = \frac{1}{\sqrt{2}}(|g\rangle - |e\rangle). \tag{4.79}$$

The two states effectively split into two, separated by 2Ω. The details of this doubling are a hallmark of Floquet states, of which the dressed states are an example. It is obvious that we will get spectral peaks at ω_0 as well as sidebands at $\omega_0 \pm \Omega$. The details of the widths and relative heights are different in a fully quantum theory such as that considered in chapter 11 than the predictions of the semiclassical theory. We delay this discussion until later.

Very far off resonance, the energies are

$$E_1 = E_g + \frac{\hbar\Omega^2}{4\Delta} \tag{4.80}$$

$$E_2 = E_e - \frac{\hbar\Omega^2}{4\Delta}. \tag{4.81}$$

The energy shifts quadratic in the driving field are known as 'light shifts', or AC Stark shifts, as shown in figure 4.6. In some situations for the cooling and trapping of

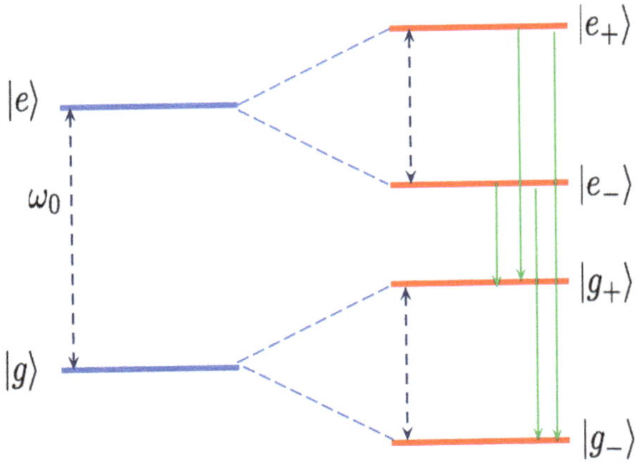

Figure 4.6. Dressed states.

atoms, the field (and hence Ω and the light shifts) is position dependent, leading to an energy gradient or force!

4.4 Floquet theory

We have just introduced the dressed states, where it seems that two energy levels turn into four! This can be formalized using Floquet theory, which is applicable for periodic interactions. Here,

$$H_m(t + T_L) = H_m(t),$$

where $T_L \equiv 2\pi/\omega_L$ and ω_L denotes the laser frequency. The Schrödinger equation has time-periodic coefficients. Via Floquet's theorem, there exists a complete set of solutions labeled by quantum numbers r of the form

$$\psi^r_m(t) = u_r(t)\exp\{-i\varepsilon_r t/\hbar\}.$$

The time-independent quantities ε_r are called quasi-energies or Floquet indices, and Floquet states $u_r(t)$ are periodic in time,

$$u_r(t + T_L) = u_r(t).$$

The Floquet states form a complete basis, so $\psi_m(t)$ of the time-dependent Schrödinger equation can be written as

$$\psi_m(t) = \sum_r a_r u_r(t)\exp\{-i\varepsilon_r t/\hbar\}$$

with time-independent amplitudes a_r. The Floquet wave functions $u_r(t)$ may be obtained from the Floquet eigenvalue equation

$$H_F u_r(t) \equiv \left\{H_m(t) - i\hbar\frac{\partial}{\partial_t}\right\}u_r(t) = \varepsilon_r u_r(t)$$

with periodic boundary conditions. In this equation, the Floquet Hamiltonian H has to be regarded as an operator on the extended Hilbert space consisting of all time-periodic wave functions, and $-i\hbar\partial/\partial_t$ is an operator in that space that is canonically conjugated to the time coordinate t. These operators are Hermitian with respect to the scalar product

$$\langle\langle u|v\rangle\rangle \equiv \int_0^{T_L} \frac{dt}{T_L}\langle u(t)|v(t)\rangle$$

In the extended Hilbert space. This is the same type of extended Hilbert space that one encounters in the studies of non-Hermitian quantum mechanics. In chapter 9, we find such a formalism useful in the context of quantum trajectory theory. Here, the non-Hermitian evolution is supplemented by stochastic 'quantum jumps' to fully simulate an open quantum system. Here we have bra and ket states that are *not* evolved unitarily, resulting in the need to consider a doubled Hilbert space. This is the source of the doubled number of states. There is a lot of work on non-Hermitian systems without the jumps, in the context of parity–time (PT) symmetric systems.

The jumps can be ignored for large quantum numbers, but by neglecting them, the noise limits of PT symmetric systems will be smaller than they would be in a full theory. For a solution $u_r(t)$ of the Floquet eigenvalue equation with quasi-energy ε_r, then for any integer n also

$$u_{r,n}(t) \equiv u_r(t)\exp\{i\omega_L nt\}$$

is a solution with quasi-energy

$$\varepsilon_{r,\,n} \equiv \varepsilon_r + \hbar\omega_L n.$$

For all integers n, the states lead to one and the same solution of the Schrödinger equation since they are copies of the wave function shifted by an integer number of periods.

4.5 Perturbation theory and rate equations

We now return to perturbation theory for an examination of what happens when we drive the atom with many colors of light at once, instead of monochromatic excitation [3]. Recall that for one color of light we had

$$P_e(t) = \frac{\Omega^2 t^2}{4}\frac{\sin^2(\Delta t/2)}{(\Delta t/2)^2}. \tag{4.82}$$

If we now shine light of many colors on the atom, and this light is completely incoherent, we may sum probabilities instead of summing amplitudes and then squaring to obtain a probability. We also use $E_0^2 = 2U(\Omega)/\epsilon_0$, where $U(\omega) = |E_0|^2/\epsilon_0$ is the energy density of the field at frequency ω. For the incoherent driving field, we then have

$$P_e(t) = \frac{|\mu_{eg}|^2 t^2}{4\epsilon_0\hbar^2}\frac{2}{t}\int_{-\infty}^{\infty} U(\Delta)\frac{\sin^2(\Delta t/2)}{(\Delta t/2)^2} d\left(\frac{\Delta t}{2}\right). \tag{4.83}$$

The definite integral has no time dependence, so the probability of a transition goes linearly with time. We have integrated all detunings and extended the lower limit of integration to $-\infty$. This is a superb approximation as $U(\Delta)$ multiplies a function that dies off to zero at large detunings, a $\sin c(x) = \sin(x)/x$ function. We next assume that the energy density of the driving field is slowly varying over the sinc function in the integral, and approximate $U(\Delta) \approx U(0)$ and pull it out of the integral. The definite integral can then be evaluated, leading to a probability of being in the excited state proportional to t instead of t^2. The final result is

$$P_e(t) = \frac{\pi|\mu_{eg}|^2}{2\epsilon_0\hbar^2}U(\omega_0)t \equiv BU(\omega_0)t, \tag{4.84}$$

where we defined the Einstein B coefficient $B = \pi|\mu_{eg}|^2/2\epsilon_0\hbar^2$. If we average over the orientation of the atom's dipole moment, we gain an additional factor of 1/3 due to a spatial averaging over the $\cos(\theta_{E,\mu})$ resulting from the dot product of the applied field and the dipole moment, resulting in $B = \pi|\mu_{eg}|^2/6\epsilon_0\hbar^2$ This result is a special case of Fermi's golden rule, where an incoherent excitation leads to a transition probability that is linear in time and proportional to the square of a matrix element of the interaction Hamiltonian. We can define the transition rate $R_{abs} = dP_e/dt = BU(\omega_0)$, which is a constant. We also have (assuming no degeneracy of ground and excited states as we have) $R_{abs} = R_{stimem}$. What if the light source is monochromatic, but the atoms in our sample have different resonant frequencies due to Doppler or pressure broadening? Then, our result still holds. We only need to have a uniform range of detunings to integrate over; hence, we can use rate equations for treatments of transitions that are subject to broadening mechanisms when pumped with monochromatic light.

Let us now consider many atoms being excited incoherently. The ground and excited-state populations change via three processes: absorption, stimulated emission, and spontaneous emission. The latter process occurs when there is no applied field. In all of our previous work, we have found transitions, but *only* when we have an applied field. We will follow Einstein's thinking and look at the rate equations for the populations in the two states:

$$\dot{N}_e = BU(\omega_0)(N_g - N_e) - AN_e \tag{4.85}$$

$$\dot{N}_g = BU(\omega_0)(N_e - N_g) + AN_e. \tag{4.86}$$

Note that the total number of atoms does not change, $dN/dt = dN_e/dt + dN_g/dt = 0$. We will follow Einstein's thinking and consider the case of thermal equilibrium. In that case, we know that the energy density of the field is given by the Planck distribution:

$$U(\omega_0) = \frac{\hbar\omega^3}{c^3\pi^2}\frac{1}{e^{\hbar\omega/kT} - 1}. \tag{4.87}$$

Also, the time derivatives are zero as we are in equilibrium. Furthermore, we know that the relative number of atoms in the ground and excited states is given by the Boltzmann distribution (assuming that the mean free path is large compared to the de Broglie wavelength of the atoms)

$$N_e/N_g = e^{-\hbar\omega_0/kT}. \tag{4.88}$$

All that is left is to find A! Rearranging, we find

$$A = \frac{\hbar\omega^3 B}{\pi^2 c^3} = \frac{\hbar\omega^3\mu_{eg}^2}{6\pi\epsilon_0\hbar^2 c^3}. \tag{4.89}$$

It is interesting to see how excluding the stimulated emission term affects the results. At the turn of the century, only absorption and spontaneous emission were observed in the laboratory. We would then have from our rate equation analysis

$$BU(\omega_0)N_g - AN_e = 0. \tag{4.90}$$

This leads to

$$U(\omega_0) = (A/B)e^{-\hbar\omega_0/kT}. \tag{4.91}$$

This is certainly not the Planck distribution at all, but the distribution derived by Wien based on Maxwell's equations. Einstein's purpose was to learn more about the Planck distribution, and he found that by adding the stimulated emission term to the two known processes, he could derive the Planck distribution from very basic principles. This was in 1905, and no one had calculated B yet! It was 20 years before the Schrödinger equation was discovered. Einstein showed that to explain the data in 1905, the process of stimulated emission must exist; he was the father of the laser. We can derive A today from a treatment that treats the field quantum mechanically, done first by Dirac,b ut where in Einstein's treatment is the field treated quantum mechanically? Answer: Using the Planck distribution, which was where $\hbar$ first appeared.

What if we ignore the spontaneous emission term? After all, we have only put it in by hand. Then we would have

$$BU(\omega_0)(N_g - N_e) = 0, \tag{4.92}$$

which leads to $N_g = N_e$, and this means that we would have got the atomic statistics wrong. One needs all three processes to get everything right.

4.6 Pauli operators and the Bloch sphere representation

It is often useful to consider a real vector space instead of a complex one. In the case of our two-level atom, we have a two-dimensional Hilbert space and a wave function

$$|\psi\rangle = C_e|e\rangle + C_g|g\rangle, \tag{4.93}$$

where C_e and C_g are complex numbers specified by four real numbers (either real and imaginary parts, or magnitudes and phases). Thus, our two-dimensional Hilbert space can be represented as a four-dimensional real vector space. We can write our wave function as

$$\begin{aligned} |\psi\rangle &= |C_e|e^{i\phi_e}|e\rangle + |C_g|e^{i\phi_g}|g\rangle \\ &= e^{i\phi_e}(|C_e||e\rangle + |C_g|e^{i(\phi_g - \phi_e)}|g\rangle) \\ &= |C_e||e\rangle + |C_g|e^{i(\phi_g - \phi_e)}|g\rangle, \end{aligned} \tag{4.94}$$

where in the last step, we have ignored the overall phase $e^{i\phi_e}$. Therefore, given the global phase invariance used here, we now have only three real numbers to describe our quantum state. We can also use the relation $|C_e|^2 + |C_g|^2 = 1$ to reduce our description of the two-level system to two real variables:

$$\begin{aligned} |\psi\rangle &= |C_e||e\rangle + \sqrt{1 - |C_e|^2}\, e^{i(\phi_g - \phi_e)}|g\rangle \\ &= \sin(\theta/2)|e\rangle + \cos(\theta/2)e^{i\phi/2}|g\rangle, \end{aligned} \tag{4.95}$$

where we have defined $C_e = \sin(\theta/2)$ and $\phi_g - \phi_e = \phi/2$. The factors of two stem from the fact that a π rotation in the Bloch sphere space is a $\pi/2$ rotation in the two-dimensional complex space in which our states and operators live in figure 4.7.

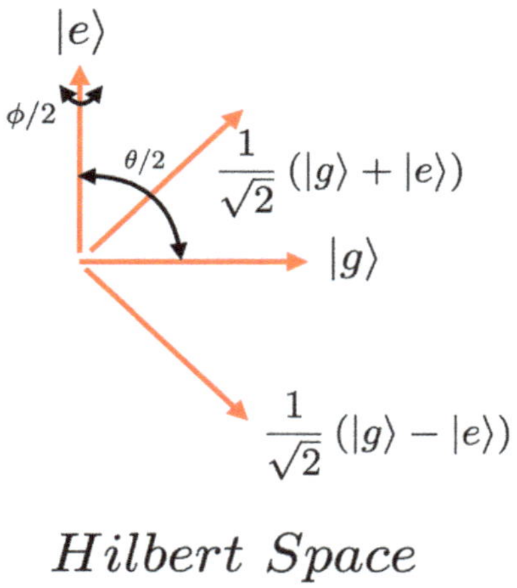

Figure 4.7. Quantum states in Hilbert space.

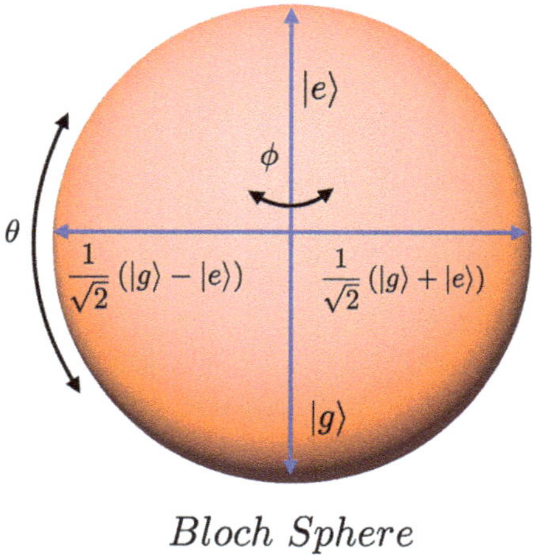

Figure 4.8. Quantum states on the Bloch sphere.

We treat θ and ϕ as a latitude and longitude, and we can represent any quantum state of a two-level system on the Bloch sphere, as shown in figure 4.8.

As our quantum state is described by a complex vector of length two, all the operators that act on the state are unitary 2×2 matrices,

$$U = \begin{pmatrix} a & b \\ c & d \end{pmatrix}, \tag{4.96}$$

with $ad - bc = \det(U) = 1$. Just as a vector can be expanded in terms of a set of linearly independent basis vectors, $\overrightarrow{V} = V_x i + V_y j + V_z k$, one can expand a matrix in terms of linearly independent basis matrices,

$$U = a\begin{pmatrix} 1 & 0 \\ 0 & 0 \end{pmatrix} + b\begin{pmatrix} 0 & 1 \\ 0 & 0 \end{pmatrix} \\ c\begin{pmatrix} 0 & 0 \\ 1 & 0 \end{pmatrix} + d\begin{pmatrix} 0 & 0 \\ 0 & 1 \end{pmatrix}. \tag{4.97}$$

OK, this is not so impressive and may seem a waste of time. However, besides the obvious choice of basis matrices above, it is useful to consider the Pauli operators,

$$\sigma_x = \begin{pmatrix} 0 & 1 \\ 1 & 0 \end{pmatrix} \tag{4.98}$$

$$\sigma_y = \begin{pmatrix} 0 & -i \\ i & 0 \end{pmatrix} \tag{4.99}$$

$$\sigma_z = \begin{pmatrix} 1 & 0 \\ 0 & -1 \end{pmatrix}. \tag{4.100}$$

These operators, plus the identity operator I, can be used as a complete set of operators for a two-level system, as any operator can be written in the form

$$O = AI + B\sigma_x + C\sigma_y + D\sigma_z. \tag{4.101}$$

The Pauli operators satisfy the following properties:

$$\mathrm{Tr}(\sigma_i) = 0 \tag{4.102}$$

$$\sigma_i\sigma_j = -\sigma_j\sigma_i \tag{4.103}$$

$$(\sigma_i)^2 = I \tag{4.104}$$

$$[\sigma_i, \sigma_j] = 2i\epsilon_{ijk}\sigma_k \tag{4.105}$$

$$(\vec{\sigma}\cdot n)^2 = 1 \tag{4.106}$$

$$[\sigma_i, \sigma_j]_+ = 2\delta_{ij}. \tag{4.107}$$

Here $\vec{\sigma}$ is a vector whose three components are the three Pauli operators, and n is any unit vector in three dimensions. In addition, ϵ_{ijk} is the Levi-Civita symbol, a fully antisymmetric tensor of rank 2. More simply, it is zero if any index repeats, $+1$ if the indices are in any cyclic order of 1, 2, 3, and -1 if 1, 2, 3 appears in a non-cyclic order, such as 3, 2, 1. These operators obey the same algebra, known as the Clifford algebra, that angular momentum operators satisfy, and in fact are the appropriate operators to deal with a spin 1/2 system. This is not surprising as a spin 1/2 system has two states: spin up and spin down. Therefore, the math for the spin 1/2 system is the same as for our two-level atom, as they are both-state systems. There is another relation that is very useful for manipulating the Pauli operators,

$$\exp(i\vec{\theta}\cdot\sigma/2) = \cos(\theta/2)I + i(\theta\cdot\vec{\sigma}/2)\sin(\theta/2), \tag{4.108}$$

which can be verified by a power series expansion of both sides. Here $\vec{\theta}$ is a vector with a magnitude θ, and some direction given by the unit vector θ. Recall that any state of or two-level system can be represented in two ways. One as a vector in a two-dimensional complex vector space or, as we have seen, in a three dimensional real space, the Bloch sphere. In either case, any operator that acts on a state to generate a new state can be viewed as a rotation of the vector, so we need to be able to manipulate rotation operators of the form equation (4.108). Let us focus on the Bloch sphere representation for now, here a position on the Bloch sphere can be determined by the latitude (θ) and longitude (ϕ). A unit vector that points at this point is

$$n = \begin{pmatrix} \sin(\theta/2)\cos(\phi/2) \\ \sin(\theta/2)\sin(\phi/2) \\ \cos(\theta/2) \end{pmatrix}. \tag{4.109}$$

We use half angles to be consistent with our earlier work. Two other operators that are variations of the Pauli operators that will be useful are

$$\sigma_{\pm} = \frac{1}{2}(\sigma_x \pm \sigma_y) \tag{4.110}$$

with matrix representations

$$\sigma_{+} = \begin{pmatrix} 0 & 1 \\ 0 & 0 \end{pmatrix} \tag{4.111}$$

$$\sigma_{-} = \begin{pmatrix} 0 & 0 \\ 1 & 0 \end{pmatrix}, \tag{4.112}$$

and with basis states

$$|e\rangle = \begin{pmatrix} \\ 0 \end{pmatrix} \tag{4.113}$$

$$|g\rangle = \begin{pmatrix} 0 \\ 1 \end{pmatrix}. \tag{4.114}$$

The action of these operators on the ground and excited states are

$$\begin{aligned} \sigma_{+}|e\rangle &= 0 \\ \sigma_{+}|g\rangle &= |e\rangle \\ \sigma_{-}|e\rangle &= |g\rangle \\ \sigma_{-}|g\rangle &= 0. \end{aligned} \tag{4.115}$$

We refer to σ_{+} as a raising operator, as it takes the atom from the ground state to the excited state. It will obviously play a role in describing the absorption of a photon. Then σ_{-} is referred to as a lowering operator and will be useful in describing the emission process.

The matrix representation of the Hamiltonian for a two-level atom in the absence of a driving field is

$$H_0 = \begin{pmatrix} E_e & 0 \\ 0 & E_g \end{pmatrix} = \frac{(E_e + E_g)}{2} I + \frac{(E_e - E_g)}{2} \sigma_z. \tag{4.116}$$

The identity term can be eliminated by redefining the zero of the energy to be midway between the excited and ground states, which leaves us with

$$H_0 = \frac{\hbar\omega_0}{2} \sigma_z, \tag{4.117}$$

where again $E_e - E_g = \hbar\omega_0$. Starting with the initial state

$$|\psi(0)\rangle = \begin{pmatrix} C_e(0) \\ C_g(0) \end{pmatrix}, \tag{4.118}$$

we can solve for the time dependence via the Schrödinger equation

$$|\psi(t)\rangle = e^{-iH_0 t/\hbar} |\psi(0)\rangle = e^{-i\omega_0 \sigma_z t/2} |\psi(0)\rangle . \tag{4.119}$$

The exponentiation of an operator has meaning via its Taylor series expansion, and as σ_z is a diagonal representation, we find that

$$e^{-iH_0 t/\hbar} = \begin{pmatrix} e^{-i\omega t/2} & 0 \\ 0 & e^{i\omega t/2} \end{pmatrix}, \tag{4.120}$$

which leads to

$$|\psi(t)\rangle = \begin{pmatrix} C_e(0) e^{-\imath\omega t/2} \\ C_g(0) e^{\imath\omega t/2} \end{pmatrix}, \tag{4.121}$$

which is the form we used earlier.

The electric dipole Hamiltonian, $H_I = -\vec{\mu} \cdot \vec{E}$ has the matrix representation

$$H_I = -\mu_{eg}^z E_z \begin{pmatrix} 0 & 1 \\ 1 & 0 \end{pmatrix} = -\mu_{eg}^z E_z \sigma_x = -\mu_{eg}^z E_z (\sigma_+ + \sigma_-), \tag{4.122}$$

where we have assumed the driving electric field to be polarized along the z direction. We can see that the raising and lowering operators are involved, dealing with the absorption and emission terms, as we alluded to earlier. We recall that in the Heisenberg picture, the time evolution is carried by the operators,

$$\begin{aligned} \frac{dO_H}{dt} &= \frac{1}{\hbar} H e^{\imath Ht/\hbar} O_H e^{-\imath Ht/\hbar} \\ &+ e^{\imath Ht/\hbar} \frac{\partial O_H}{\partial t} e^{-\imath Ht/\hbar} \\ &- \frac{\imath}{\hbar} e^{\imath Ht/\hbar} O_H e^{-\imath Ht/\hbar}. \end{aligned} \tag{4.123}$$

In our case, we are concerned with the three Pauli operators, for example

$$\begin{aligned}\frac{d\sigma_x}{dt} &= [H_0 + H_I, \sigma_+] \\ &= \frac{\iota\omega_0}{2}[\sigma_z, \sigma_x] + \frac{\iota\Omega_0}{2}e^{i\omega t}[\sigma_x, \sigma_x] \\ &= -\omega_0\sigma_y.\end{aligned} \tag{4.124}$$

Similarly, we find that

$$\frac{d\sigma_y}{dt} = \omega_0\sigma_x + 2\Omega_0 \cos\omega t\sigma_z \tag{4.125}$$

$$\frac{d\sigma_z}{dt} = -2\Omega_0 \cos\omega t\sigma_y. \tag{4.126}$$

In terms of the raising and lowering operators, we find

$$\frac{d\sigma_+}{dt} = i\omega_0\sigma_+ + i\Omega_0\cos(\omega t)\sigma_z \tag{4.127}$$

$$\frac{d\sigma_-}{dt} = -i\omega_0\sigma_- + i\Omega_0\cos(\omega t)\sigma_z \tag{4.128}$$

$$\frac{d\sigma_z}{dt} = 2i\Omega_0\cos\omega t(\sigma_+ - \sigma_-). \tag{4.129}$$

It is often convenient to go to a rotating frame via

$$\sigma_+ = \Sigma_+ e^{i\omega t} \tag{4.130}$$

$$\sigma_- = \Sigma_- e^{-i\omega t} \tag{4.131}$$

$$\sigma_z = \Sigma_z. \tag{4.132}$$

This leads to equations for the slowly varying Σ operators

$$\frac{d\Sigma_+}{dt} = i\Delta\Sigma_+ + i\frac{\Omega_0}{2}(1 + e^{-2i\omega t})\Sigma_z \tag{4.133}$$

$$\frac{d\Sigma_-}{dt} = -i\Delta\Sigma_+ - i\frac{\Omega_0}{2}(1 + e^{-\omega t})\Sigma_z \tag{4.134}$$

$$\frac{d\Sigma_z}{dt} = +i\Omega_0(\Sigma_+ - \Sigma_-), \tag{4.135}$$

where we have $\Delta = \omega_0 - \omega$ as before. If we drop rapidly oscillating terms (2ω), which is the rotating wave approximation, we find

$$\frac{d\Sigma_+}{dt} = i\Delta\Sigma_+ + i\frac{\Omega_0}{2}\Sigma_z \tag{4.136}$$

$$\frac{d\Sigma_-}{dt} = -i\Delta\Sigma_+ - i\frac{\Omega_0}{2}\Sigma_z \tag{4.137}$$

$$\frac{d\Sigma_z}{dt} = +i\Omega_0(\Sigma_+ - \Sigma_-). \tag{4.138}$$

We could have invoked the rotating wave approximation earlier by using the interaction Hamiltonian

$$H_I = -\frac{\hbar\Omega_0}{2}(\sigma_+ e^{-i\omega t} + \sigma_- e^{i\omega t}). \tag{4.139}$$

It is worth noting that an equivalent Hamiltonian would be

$$H_I = i\frac{\hbar\Omega_0}{2}(\sigma_+ e^{-i\omega t} - \sigma_- e^{i\omega t}). \tag{4.140}$$

This basically involves a phase shift of the driving field, or induced dipole moment, of $\pi/2$. This leads to the following equations:

$$\frac{d\Sigma_+}{dt} = i\Delta\Sigma_+ + \frac{\Omega_0}{2}\Sigma_z \tag{4.141}$$

$$\frac{d\Sigma_-}{dt} = -i\Delta\Sigma_+ - +\frac{\Omega_0}{2}\Sigma_z \tag{4.142}$$

$$\frac{d\Sigma_z}{dt} = -\Omega_0(\Sigma_+ - \Sigma_-). \tag{4.143}$$

The i in the new Hamiltonian is canceled by the one in the Heisenberg equation of motion, so on resonance we are dealing only with real quantities. We will use the earlier version with the i for now.

Is there a relation between the Heisenberg equations above and the $\dot{C}$ equations of the Schrödinger picture? What is the physical interpretation of these equations? To make these connections, we examine the expectation values of the Pauli operators. For example

$$\begin{aligned}\langle\sigma_z\rangle &= \langle\psi|\sigma_z|\psi\rangle \\ &= (C_e^* C_g^*)\begin{pmatrix} & 0 \\ 0 & -1\end{pmatrix}\begin{pmatrix} C_e \\ C_g\end{pmatrix} \\ &= |C_e|^2 - |C_g|^2.\end{aligned} \tag{4.144}$$

We see that the expectation value of Σ_z is the population inversion of the atom. What about the raising and lowering operators? Consider the induced dipole operator

$$\begin{aligned}\mu_{eg}\begin{pmatrix}0 & 1\\ & 0\end{pmatrix} &= \mu_{eg}\sigma_x\\ &= \mu_{eg}(\sigma_+ + \sigma_-)\\ &= \mu_{eg}(\Sigma_+ e^{i\omega t} + \Sigma_- e^{-i\omega t})\\ &= \mu_{eg}[(\Sigma_+ + \Sigma_-)\cos\omega t + i(\Sigma_+ - \Sigma_-)\sin\omega t].\end{aligned} \tag{4.145}$$

The expectation value of σ_+ is given by

$$\begin{aligned}\langle\sigma_z\rangle &= \langle\psi|\sigma_z|\psi\rangle\\ &= (C_e^* C_g^*)\begin{pmatrix}0 & 1\\ 0 & 0\end{pmatrix}\begin{pmatrix}C_e\\ C_g\end{pmatrix}\\ &= C_e^* C_g.\end{aligned} \tag{4.146}$$

This shows us that the expectation values of $\sigma_\pm$ are the average induced dipole moment of the atom. Let us refer back to the Bloch sphere picture for a bit. If the atom is in the excited state, $C_e = 1$ and $C_g = 0$, the population inversion is 1, and the dipole moment is zero. If the atom is in the ground state, the population inversion is -1 and the dipole is again zero. The dipole is maximized when the atom is halfway between the excited and ground states, maximizing the product $C_e^* C_g$, on the equator of the Bloch sphere.

The equations for the expectation values of the Pauli operators could have been obtained from the $\dot{C}$ equations via

$$\langle\dot{\sigma}_z\rangle = \dot{C}_e C_e^* + \dot{C}_e^* C_e + \dot{C}_g C_g^* + \dot{C}_g^* C_g \tag{4.147}$$

$$\langle\dot{\sigma}_+\rangle = \dot{C}_e^* C_g + C_e^* \dot{C}_g \tag{4.148}$$

$$\langle\dot{\sigma}_-\rangle = \dot{C}_g^* C_e + C_g^* \dot{C}_e. \tag{4.149}$$

We now introduce damping phenomenologically, with a decay rate γ for the population inversion, and Γ for the dipole,

$$\frac{d\Sigma_+}{dt} = (i\Delta - \Gamma)\Sigma_+ - \frac{i\Omega_0}{2}\Sigma_z \tag{4.150}$$

$$\frac{d\Sigma_-}{dt} = -(i\Delta + \Gamma)\Sigma_+ - \frac{i\Omega_0}{2}\Sigma_z \tag{4.151}$$

$$\frac{d\Sigma_z}{dt} = -\gamma(\Sigma_z + 1) + i\Omega_0(\Sigma_+ - \Sigma_-). \tag{4.152}$$

The reason for the $\Sigma_z + 1$ is that in the absence of a driving field, we require that the population inversion decay to -1, that is, the atom in the ground state. We also have $\Gamma = \gamma/2 + \gamma_{\text{phase}}$, where γ_{phase} is a rate of decay of the dipole that comes from a process that randomly disturbs the phase of the dipole but does not affect the population inversion, for example, elastic collisions between atoms in a vapor cell. The dipole decays at half the rate of the population inversion if only radiative decay (spontaneous emission) is involved. The intensity of the spontaneously emitted light is proportional to the population inversion and decays as $\exp(-\gamma t)$. The radiated field is proportional to the dipole moment and should decay at half the rate, as the field is the square root of the intensity. These equations are known as the optical Bloch equations.

These equations have the following steady-state solutions for the expectation values:

$$\langle \Sigma_+ \rangle^{ss} = -\frac{\Omega}{2\Omega_{\text{sat}}}\sqrt{\frac{\gamma}{\Gamma}}\frac{(1 - i\widetilde{\Delta})}{\left(1 + \widetilde{\Delta}^2 + \Omega^2/\Omega_{\text{sat}}^2\right)} \tag{4.153}$$

$$\langle \Sigma_- \rangle^{ss} = -\frac{\Omega}{2\Omega_{\text{sat}}}\sqrt{\frac{\gamma}{\Gamma}}\frac{(1 + i\widetilde{\Delta})}{\left(1 + \widetilde{\Delta}^2 + \Omega^2/\Omega_{\text{sat}}^2\right)} \tag{4.154}$$

$$\langle \Sigma_z \rangle^{ss} = \frac{-\left(1 + \widetilde{\Delta}^2\right)}{\left(1 + \widetilde{\Delta}^2 + \Omega^2/\Omega_{\text{sat}}^2\right)}, \tag{4.155}$$

where we have defined $\tilde{\Delta} = \Delta/\Gamma$ and $\Omega_{\text{sat}}^2 = \gamma\Gamma$. These are a dimensionless detuning and a saturation Rabi frequency. The latter is usually expressed as a saturation field, $E_{\text{sat}} = \sqrt{\gamma\Gamma}\,\hbar/\mu_{eg}$, or a saturation intensity $I_{\text{sat}} = \epsilon_0 c E_{\text{sat}}^2$. Plots of the atomic dipole and population inversion as a function of the driving field are shown in figures 4.9 and 4.10.

At the saturation intensity, the population inversion is halfway between the values at the zero driving intensity and a state with a full population inversion.

We see from these plots that the two-level atom is not a linear system, as the polarization is not linear with the driving field, but saturates. Generally, in a linear, isotropic, homogeneous system, one has $P = \epsilon_0 \chi E = N\mu_{eg}E$, where P is the polarization and χ is the permittivity of the material. In an anisotropic material, $\chi \to \overrightarrow{\chi}$, and the polarization does not lie in the same direction as the applied field, an inhomogeneous material would have a positional dependence for $\chi = \chi(\overrightarrow{r})$. For a nonlinear material, χ is not a constant but a function of the driving field. For our two-level atom, we have (on resonance)

$$\chi = \frac{N\mu_{eg}/\hbar\epsilon_0\Omega_{\text{sat}}}{1 + \Omega^2/{\Omega_{\text{sat}}}^2}. \tag{4.156}$$

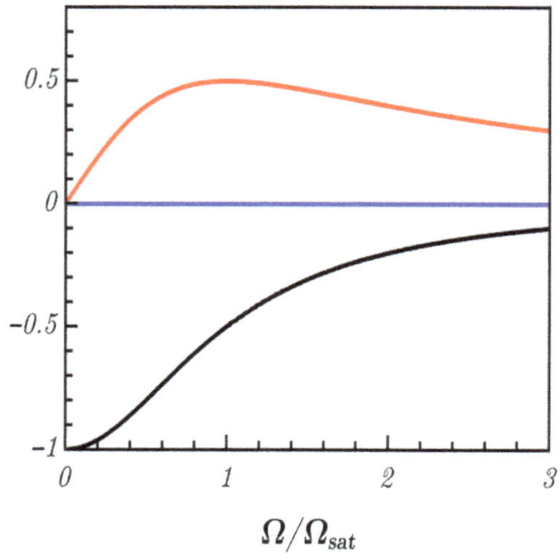

Figure 4.9. The population inversion (black), in-phase polarization (blue), and out-of-phase polarization (red), as functions of driving intensity, for zero detuning.

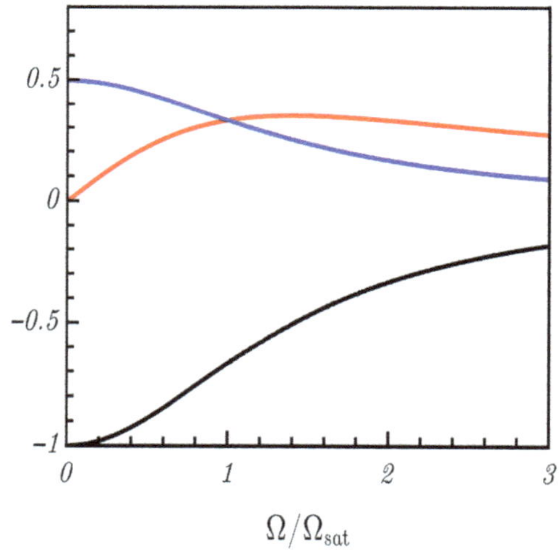

Figure 4.10. The population inversion (black), in-phase polarization (blue), and out-of-phase polarization (red), as functions of driving intensity, for nonzero detuning. Note that the in-phase polarization is nonzero.

As we can see, this is a nonlinear susceptibility that can be expanded as

$$\begin{aligned}\chi &= N\mu_{eg}/\hbar\epsilon_0\Omega_{\text{sat}} - N\mu_{eg}^3/\hbar^3\epsilon_0\Omega_{\text{sat}}^3 + \dots \\ &= \chi^{(1)} + \chi^{(3)} + \dots .\end{aligned} \tag{4.157}$$

The nonlinearity is cubic to the first order, as it should be for a medium with inversion symmetry, as a collection of atoms typically is (figure 4.11).

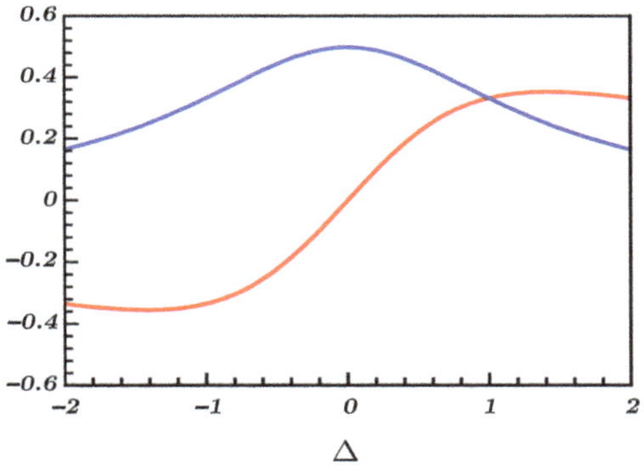

Figure 4.11. The absorption spectra (blue), and index of refraction (red), as functions of detuning.

Another form of the Bloch equations involves the so-called in- and out-of-phase quadratures $\sigma_x = U\cos(\omega t) - V\sin(\omega t)$. These are related to the polarization, of course, via

$$P = N\mu_{eg}\sigma_x. \tag{4.158}$$

These quadratures are defined via

$$U = \langle\Sigma_+\rangle + \langle\Sigma_-\rangle \tag{4.159}$$

$$V = -i(\langle\Sigma_+\rangle - \langle\Sigma_-\rangle) \tag{4.160}$$

$$W = \langle\Sigma_z\rangle. \tag{4.161}$$

These satisfy the following dynamical equations:

$$\dot{U} = \Delta V - \Gamma U \tag{4.162}$$

$$\dot{V} = -\Delta U + \Omega W - \Gamma V \tag{4.163}$$

$$\dot{W} = -\gamma(W + 1) - \Omega V. \tag{4.164}$$

The physical interpretation of these is the population inversion (W), the dipole moment $\pi/2$ out of phase with the driving field, scaled by μ_{eg} (V), and the dipole moment in phase with the driving field (U). On resonance, the steady-state dipole is $\pi/2$ out of phase with the driving field, as a damped driven simple harmonic oscillator is. In terms of latitude and longitude on the Bloch sphere, we have

$$W = \cos\theta \tag{4.165}$$

$$U = \sin\theta\sin\phi \tag{4.166}$$

$$V = \sin\theta\cos\phi \tag{4.167}$$

and without decay processes, we have $U^2 + V^2 + W^2 = 1$, the equation for the Bloch sphere itself.

4.7 Relation to the classical Lorentz model

For weak excitation, we can approximate $W = -1$ and find the equations for U and V as

$$\dot{U} = \Delta V - \Gamma U \tag{4.168}$$

$$\dot{V} = -\Delta U - \Omega - \Gamma V. \tag{4.169}$$

These are the same equations that result from the Lorentz model, where we assume the electron is bound to the nucleus by a spring of $k = m\omega_0^2$, as we saw in chapter 2, driven by an external field $E_0 \cos \omega t$, with a damping rate Γ. The Rabi frequency $\mu_{eg} E/\hbar$ is analogous to what we referred to as a classical Rabi frequency in chapter 2, $\Omega_{\text{cl}} = eE/2m\omega X_0$.

We can combine the equations for U, V, and W with the field equations derived from Maxwell's equations to write down the Maxwell–Bloch equations by defining the intercavity field amplitude

$$a = (\gamma/\sqrt{2}g)x \tag{4.170}$$

and polarization

$$\langle \sigma \rangle = -p/\sqrt{2} \tag{4.171}$$

with

$$\langle \sigma_z \rangle = -D \tag{4.172}$$

$$\Theta = \Delta_c/\kappa \tag{4.173}$$

$$\Delta = \Delta_a/\gamma \tag{4.174}$$

$$C = g^2/2\kappa\gamma \tag{4.175}$$

$$y = \sqrt{2}gJ/\kappa\gamma, \tag{4.176}$$

where y is the amplitude of the driving field multiplied by the transmission of the mirror on which the driving field is incident, and we assume no phase shift for transmission through that mirror. The Maxwell–Bloch equations can be written in their final form

$$\dot{x} = \kappa(-2Cp + y - (i\Theta + 1)x) \tag{4.177}$$

$$\dot{p} = \Gamma(-(1 + i\Delta)p + xD) \tag{4.178}$$

$$\dot{D} = \gamma(2(1 - D) - (x^*p + xp^*)), \tag{4.179}$$

where $\Gamma = \gamma/2 + \gamma_{\text{ph}}$ to include the dephasing processes.

4.8 An interlude on the form of the atom–field interaction

The standard way to introduce atom–field interactions into the quantum field theory is the minimal coupling replacement

$$\vec{p} \to \vec{\pi} = \vec{p} - \frac{e}{c}\vec{A}, \tag{4.180}$$

where $\vec{A}$ is the usual vector potential. The quantity $\vec{\pi}$ is the canonical momentum that has the operator representation $\vec{\pi} = -i\hbar\vec{\nabla}$. The usual kinetic energy term in the Hamiltonian then becomes

$$\begin{aligned} \frac{\pi^2}{2m} &\to \frac{1}{2m}\left(\pi^2 + \frac{e^2}{c^2}|A|^2 - \frac{e}{c}\vec{A}\cdot\vec{\pi} - \frac{e}{c}\vec{\pi}\cdot\vec{A}\right) \\ &= \frac{1}{2m}\left(\pi^2 + \frac{e^2}{c^2}|A|^2 - 2\frac{e}{c}\vec{A}\cdot\vec{\pi}\right), \end{aligned} \tag{4.181}$$

where we have used the Coulomb gauge condition $\vec{\nabla}\cdot\vec{A} = 0$ to show that in that gauge the vector potential and the momentum commute,

$$\begin{aligned} \vec{\pi}\cdot\vec{A}\,|\psi\rangle &= -i\hbar\,\vec{\nabla}\cdot\vec{\pi}|\psi\rangle \\ &= -i\hbar(\vec{\nabla}\cdot\vec{A})|\psi\rangle - i\hbar\vec{A}\cdot\vec{\nabla}\,|\psi\rangle \\ &= -i\hbar\vec{A}\cdot\vec{\nabla}\,|\psi\rangle \\ &= \vec{A}\cdot\vec{\pi}|\psi\rangle, \end{aligned} \tag{4.182}$$

which leads us to an interaction Hamiltonian of the form

$$H_{\text{int}} = -\frac{e}{m}\vec{A}\cdot\vec{\pi} + e^2|A|^2. \tag{4.183}$$

The spatial dependence of the components of $\vec{A}$ for a plane electromagnetic wave of wavevector $\vec{k}$ is

$$A_i = A_i(t)e^{i\vec{k}\cdot\vec{r}}. \tag{4.184}$$

In the dipole approximation, in which the field does not vary much over a wavelength of light, we can evaluate the exponential factor at the location of the atom, which we take to be $\vec{r} = 0$, then the components of $\vec{A}$ have a time dependence but no spatial dependence.

We now consider a gauge transformation [4],

$$\vec{A}' = \vec{A} - \vec{\nabla}\lambda \tag{4.185}$$

$$\phi' = \phi + \frac{\partial\lambda}{\partial t}, \tag{4.186}$$

with the choice

$$\lambda = \vec{A}\cdot\vec{r}. \tag{4.187}$$

In quantum mechanics, when one performs a gauge transformation on the vector and scalar potentials, there is an accompanying transformation of the wavefunction,

$$|\psi'\rangle = e^{ie\lambda/\hbar c}|\psi\rangle. \tag{4.188}$$

The Schrödinger equation for the new wave function $|\psi'\rangle$ is

$$\begin{aligned} i\hbar|\dot{\psi}\rangle &= i\hbar\left(\frac{ie\dot{\lambda}}{\hbar c}|\psi'\rangle + |\dot{\psi}'\rangle\right)e^{ie\vec{A}\cdot\vec{r}/\hbar c} \\ &= i\hbar\left(\frac{ie\dot{\vec{A}}\cdot\vec{r}}{\hbar c}|\psi'\rangle + \frac{ie\vec{A}\cdot\dot{\vec{r}}}{\hbar c}|\psi'\rangle + |\dot{\psi}'\rangle\right)e^{ie\vec{A}\cdot\vec{r}/\hbar c} \\ &= e^{ie\vec{A}\cdot\vec{r}/\hbar c}\left(\frac{\pi^2}{2m} + V(r)\right)|\psi'\rangle. \end{aligned} \tag{4.189}$$

Using $\vec{E} = -\dot{\vec{A}}$, and the Lorentz gauge $\vec{\nabla}\cdot\vec{A} = 0$, we find

$$i\hbar|\dot{\psi}'\rangle = \left(\frac{\pi^2}{2m} + V(r) - e\vec{r}\cdot\vec{E}\right)|\psi'\rangle. \tag{4.190}$$

Therefore, we can obtain the electric dipole form from the canonical form. This has generated a periodic debate in the literature, with claims that the two forms give different answers. The typical resolution is that someone has forgotten to use $\psi'\rangle$ with the $H_I = -\vec{\mu}\cdot\vec{E}$, instead using $|\psi\rangle$ or the other way around. The discrepancy between the two 'different' answers often is of the form of some factor of ω/ω_0 which is often small for optical problems; nevertheless, care should be taken! The advantage of the electric dipole form is that the field is gauge invariant and this Hamiltonian includes the A^2 term. Conversely, going beyond the dipole approximation is easy in the $\vec{A}\cdot\vec{\pi}$ form by using the spatial dependence of the vector potential; in the other approach, one must use a multipolar expansion. Both forms yield equivalent results if one is careful, but when mixed with other approximations like the rotating wave approximation and treating 'real' systems as two-level atoms, differences may arise.

4.9 Spontaneous emission from two coupled atoms

How can atoms interact? They must do so via an electromagnetic field. The physical process may be one atom spontaneously emitting, and a second atom absorbing that photon, resulting in an interaction Hamiltonian of the form [5, 6]

$$H_{\text{int}} \propto \left(\sigma_+^1\sigma_-^2 + \sigma_+^2\sigma_-^1\right). \tag{4.191}$$

Essentially, the field seen by one atom is the dipole field of the other, plus the vacuum. Let us assume this form for now. We will see that is valid when the two atoms are predominantly coupled to the same vacuum, such as atoms within a wavelength of one another, or in a cavity/waveguide. Let us start with the structure of our Hilbert space:

$$|gg\rangle = |g\rangle \otimes |g\rangle \tag{4.192}$$

$$|ge\rangle = |g\rangle \otimes |e\rangle \tag{4.193}$$

$$|eg\rangle = |e\rangle \otimes |g\rangle \tag{4.194}$$

$$|ee\rangle = |e\rangle \otimes |e\rangle . \tag{4.195}$$

The two atoms are a pair of pseudospins, and we use the notation of angular momentum states and define

$$|1, 1\rangle = |ee\rangle \tag{4.196}$$

$$|1, 0\rangle = \frac{1}{\sqrt{2}}(|ge\rangle + |eg\rangle) \tag{4.197}$$

$$|1, -1\rangle = |gg\rangle \tag{4.198}$$

$$|0, 0\rangle = \frac{1}{\sqrt{2}}(|ge\rangle - |eg\rangle), \tag{4.199}$$

where the $|1*\rangle$ states correspond to the spin triplet, and $|00\rangle$ corresponds to the singlet state, and our Hilbert space takes the form $|1*\rangle \otimes |00\rangle$. The labels J and M denote the eigenvalues of the z component of the angular momentum, and the square of the total angular momentum via

$$J^2|JM\rangle = J(J+1)|JM\rangle \tag{4.200}$$

$$J_z|JM\rangle = M|JM\rangle . \tag{4.201}$$

We will elaborate more in the next section on the N atom situation. At this point, let us consider the dipole matrix elements between these eigenstates:

$$\langle ee|\mu E_{\text{vac}}| + \rangle = \frac{1}{\sqrt{2}}\langle ee|((\mu_1 + \mu_2)E_{\text{vac}}(|eg\rangle + |ge\rangle)) \tag{4.202}$$

$$= \frac{1}{\sqrt{2}}[\langle e|e\rangle\langle e|\mu_1|g\rangle + \langle e|g\rangle\langle e|\mu_1|e\rangle \tag{4.203}$$

$$+ \langle |e\rangle\langle e|\mu_2|g\rangle + \langle e|g\rangle\langle e|\mu_2|e\rangle +] \tag{4.204}$$

$$= \sqrt{2}\,\langle \mu E_{\text{vac}}\rangle \tag{4.205}$$

$$= \sqrt{2\gamma}, \tag{4.206}$$

where we have taken $\mu_1 = \mu_2$ for identical atoms, a $\Delta m = 0$ transition (real μ), and $\gamma = \mu E_{\text{vac}}$ Then, the probability for a transition is

$$|\langle ee|\mu E_{\text{vac}}|+\rangle|^2 = 2\gamma. \tag{4.207}$$

Repeating this for $|\langle ee|\mu E_{\text{vac}}|-\rangle| = 0$, there are no transitions from $|ee\rangle$ to $|-\rangle$.

Then, looking at the transitions for $|+\rangle$ and $|-\rangle$ to $|gg\rangle$, we again find 2γ and 0. This result is consistent with the initial *and* final states of a transition being symmetric under atom exchange. The state $|-\rangle$ is antisymmetric under atom exchange, and thus is not connected to the triplet states:

$$\text{Hilbert space} = |1^*\rangle \otimes |0\rangle. \tag{4.208}$$

For N identical two-level atoms with ground state $|g\rangle$ and excited state $|e\rangle$, one often switches to the Dicke (collective) picture (figure 4.12) by defining

$$S^{\pm} = \sum_{j=1}^{N} \sigma_j^{\pm}, \tag{4.209}$$

$$S^z = \frac{1}{2}\sum\nolimits_{j=1}^{N} \sigma_j^z, \tag{4.210}$$

where $\sigma_j^+ = |e\rangle_j\langle g|$, $\sigma_j^- = |g\rangle_j\langle e|$ and $\sigma_j^z = |e\rangle_j\langle e| - |g\rangle_j\langle g|$. These operators obey the angular momentum $SU(2)$ algebra in the symmetric (Dicke) subspace. The Lindblad equation describing the dissipative evolution due to collective spontaneous emission is

$$\frac{d\rho}{dt} = -\frac{i}{\hbar}[H, \rho] + \Gamma\mathcal{D}[S^-]\rho, \tag{4.211}$$

where the dissipator $\mathcal{D}[\mathcal{O}]\rho$ for an operator $\mathcal{O}$ is given by

$$\mathcal{D}[\mathcal{O}]\rho = \mathcal{O}\rho\mathcal{O}^\dagger - \frac{1}{2}\{\mathcal{O}^\dagger\mathcal{O}, \rho\}. \tag{4.212}$$

We will derive this form in a later chapter. Hence, in full detail

$$\frac{d\rho}{dt} = -\frac{i}{\hbar}[H, \rho] + \Gamma\left(S^-\rho S^+ - \frac{1}{2}\{S^+S^-, \rho\}\right). \tag{4.213}$$

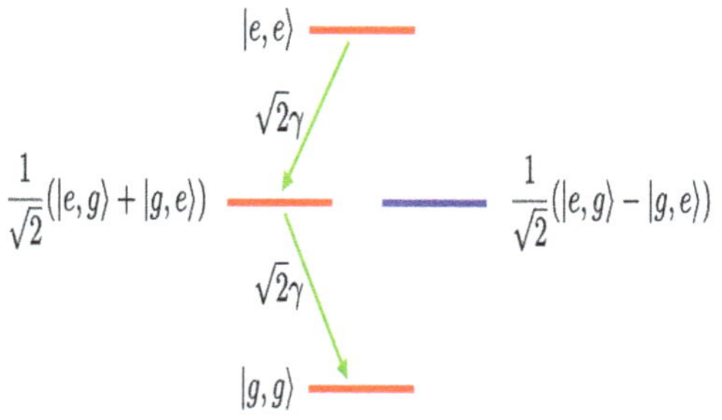

Figure 4.12. Two-atom Dicke manifold.

To see how the collective decay modifies the usual single-atom Bloch equations, one can either (i) derive the Heisenberg equations for the operators $S^{\pm}$, S^{z}, or (ii) write the equations of motion for their expectation values $\langle S^{\pm}\rangle$, $\langle S^{z}\rangle$. In the Heisenberg picture with the Hamiltonian

$$H = \hbar\omega_0 S^{z} \tag{4.214}$$

and with the Lindblad dissipator $\Gamma\mathcal{D}[S^{-}]\rho$, the operator equations become

$$\frac{d}{dt}S^{-} = -\left(i\omega_0 + \frac{\Gamma}{2}\right)S^{-} + \Gamma S^{z}S^{-} \tag{4.215}$$

$$\frac{d}{dt}S^{+} = +\left(i\omega_0 - \frac{\Gamma}{2}\right)S^{+} + \Gamma S^{+}S^{z} \tag{4.216}$$

$$\frac{d}{dt}S^{z} = -\frac{\Gamma}{2}(S^{+}S^{-} + S^{-}S^{+}). \tag{4.217}$$

These can be viewed as 'collective Bloch equations'. Note that for *single* two-level atom (i.e. $N = 1$), you would replace $S^{\pm} \mapsto \sigma^{\pm}$ and $S^{z} \mapsto \frac{1}{2}\sigma^{z}$. One recovers the familiar single-atom decay $\dot{\sigma}^{-} = -(i\omega_0 + \frac{\Gamma}{2})\sigma^{-}$. However, once $S^{\pm}$ are sums over all atoms, the extra terms (such as $\Gamma S^{z}S^{-}$) generate correlations and lead to superradiant (superfluorescent) bursts when the ensemble is initially inverted in a symmetric Dicke state. For N atoms, the Lindblad dissipator is $\Gamma\mathcal{D}[S^{-}]$ with the collective operator $S^{-} = \sum_j \sigma_j^{-}$. In the Heisenberg picture, this leads to

$$\frac{d}{dt}S^{-}(t) = -\left(i\omega_0 + \frac{\Gamma}{2}\right)S^{-}(t) + \Gamma S^{z}(t)S^{-}(t),$$

where $S^{z} = \sum_j \frac{1}{2}\sigma_j^{z}$. Here you see the product $S^{z}S^{-}$. Physically, this emerges because different atoms can stimulate each other's emissions in the same mode (i.e. 'cooperative' effects). If you expand $S^{z} = \sum_j \frac{1}{2}\sigma_j^{z}$ and $S^{-} = \sum_k \sigma_k^{-}$, you obtain cross terms such as $\sigma_j^{z}\sigma_k^{-}$. These vanish if $j = k$ in the single-atom case but now appear for $j \neq k$ (plus reordering from commutators). Hence, $\sigma_z\sigma^{-}$ product terms do not appear for a single atom's Bloch equations simply because there is no sum over multiple atoms that leads to collective or cross terms In the multi-atom Dicke model, these sums (and commutators) introduce the operator products. In the single-atom case, the interaction $-\mu E$ yields $\sigma_1 E$. In superfluorescence, the field the Nth atom sees is due to the dipoles of the other fields, and so we obtain the product terms. The equations for the expectation values are

$$\frac{d}{dt}\langle S^{-}\rangle = -\left(i\omega_0 + \frac{\Gamma}{2}\right)\langle S^{-}\rangle + \Gamma\langle S^{z}S^{-}\rangle \tag{4.218}$$

$$\frac{d}{dt}\langle S^{z}\rangle = -\frac{\Gamma}{2}\langle S^{+}S^{-} + S^{-}S^{+}\rangle \tag{4.219}$$

and similarly for $\langle S^+ \rangle$. In a full quantum treatment, one must deal with operator products such as $\langle S^z S^- \rangle$. In large-N or semiclassical approximations, one might factorize these, $\langle S^z S^- \rangle \approx \langle S^z \rangle \langle S^- \rangle$, which then closes the set into 'mean-field Bloch equations'. This approximation can qualitatively capture the onset of a superradiant pulse but misses certain quantum fluctuation effects. Even if initially $\langle S^- \rangle = 0$, the quantum fluctuations in S^- do not vanish (the system is fully inverted but has an undefined phase). The vacuum acts like a 'seed' that breaks this symmetry and initiates the emission. One can see the buildup of a macroscopic dipole $\langle S^- \rangle$ on a short timescale, culminating in the characteristic superradiant (superfluorescent (SF)) burst, with intensity $\propto \langle S^+ S^- \rangle$.

Hence, to capture the onset of SF from an initially inverted state $\rho(0) \approx |ee \dots e\rangle\langle ee \dots e|$, you *must* treat quantum noise at some level. We will look at the quantum Langevin methods. Suppose the Hamiltonian is just

$$H = \hbar\omega_0 S^z \tag{4.220}$$

and the dissipator is $\Gamma\mathcal{D}[S^-]$. One then obtains the operator equations of motion:

$$\frac{d}{dt}S^- = -\left(i\omega_0 + \frac{\Gamma}{2}\right)S^- + \underbrace{F^-(t)}_{\text{vacuum noise}} \tag{4.221}$$

$$\frac{d}{dt}S^+ = +\left(i\omega_0 - \frac{\Gamma}{2}\right)S^+ + \underbrace{F^+(t)}_{\text{vacuum noise}} \tag{4.222}$$

$$\frac{d}{dt}S^z = -\frac{\Gamma}{2}(S^+S^- + S^-S^+) + \underbrace{F^z(t)}_{\text{noise}}, \tag{4.223}$$

where $F^{\pm}(t)$, $F^z(t)$ are noise operators with specific commutation relations that ensure the correct (i) damping, (ii) commutation relations among $S^{\pm}$, S^z, and (iii) seeding from the vacuum. The only nonzero two-time correlation is

$$\langle F^-(t)F^+(t')\rangle = \Gamma\delta(t - t'), \tag{4.224}$$

and all other normally-ordered correlations vanish, e.g.

$$\langle F^+(t)F^-(t')\rangle = 0, \tag{4.225}$$

$$\langle F^{\pm}(t)F^z(t')\rangle = 0. \tag{4.226}$$

The mean values of the noise operators themselves are zero, $\langle F^{\pm}(t)\rangle = 0$.

The essential physics here is the same as that for a bosonic annihilation operator $a_{\text{in}}(t)$ in the vacuum state. A convenient picture is to represent the collective atomic state by a Bloch (or pseudospin) vector of length $S = N/2$ (figure 4.13). At $t = 0$, if all atoms are excited, we imagine the Bloch vector pointing straight 'up' along the $+z$ axis. Classically, if it were exactly along $+z$, the transverse dipole (the x–y components) would be strictly zero. However, quantum mechanics tells us to not simultaneously have $S^x = 0$ and $S^y = 0$ with no uncertainty. For a spin-$\frac{N}{2}$ pointing

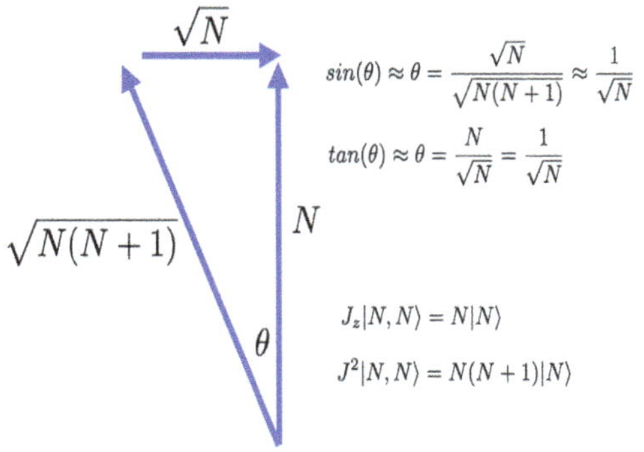

Figure 4.13. Maximum angular momentum state.

along $+z$, the transverse components S^x, S^y have uncertainty ('zero-point fluctuations') on the order of $\sqrt{N}$. More precisely,

$$\Delta S^x \approx \Delta S^y \approx \sqrt{\frac{N}{4}}, \tag{4.227}$$

so in vector form, the Bloch vector is not a perfect arrow but is distributed, with a typical spread $\sqrt{N}$ in the transverse direction. If the length of the Bloch vector is $S = N/2$, then a transverse fluctuation of order $\sqrt{N}$ corresponds to a small angular deviation from the $+z$ axis of

$$\theta_{\text{tip}} \sim \frac{\sqrt{N}}{N/2} = \frac{1}{\sqrt{N/2}} \approx \frac{1}{\sqrt{N}}. \tag{4.228}$$

Thus, we say that the Bloch vector is 'tipped' from the $+z$ axis by a small random angle $\theta_{\text{tip}} \sim 1/\sqrt{N}$. This small tilt acts as an initial seed for the transverse polarization $\langle S^- \rangle$. Once the superradiant dynamics kick in, the seed grows (non-linearly) and produces a burst of radiation.

Another closely related feature is the random phase of the dipole moment (whether it appears in the S^x or S^y direction). Because the quantum noise is isotropic in the transverse plane (no preferred direction in x–y), each realization will select a random phase for the eventual dipole. Experimentally, one sees 'shot-to-shot' fluctuations in the phase of the emitted light from the superfluorescent pulses.

In the semiclassical approach, one can define

$$\theta(t) = \cos^{-1}\left(\frac{\langle S^z(t)\rangle}{S}\right) \tag{4.229}$$

$$\phi(t) = \tan^{-1}\left(\frac{\langle S^y(t)\rangle}{\langle S^x(t)\rangle}\right) \tag{4.230}$$

and then write the approximate Bloch equations for $\langle S^x \rangle$, $\langle S^y \rangle$, $\langle S^z \rangle$. From these, you can obtain $\dot{\theta}(t)$ and $\dot{\phi}(t)$. For a large spin (large N) in a coherent or near-coherent state, these angles behave much like classical variables, albeit with small quantum noise terms that cause shot-to-shot fluctuations in θ and ϕ:

$$\frac{d}{dt}\langle S_x \rangle = \omega_0 \langle S_y \rangle + \Gamma \langle S_z \rangle \langle S_x \rangle, \tag{4.231}$$

$$\frac{d}{dt}\langle S_y \rangle = -\omega_0 \langle S_x \rangle + \Gamma \langle S_z \rangle \langle S_y \rangle, \tag{4.232}$$

$$\frac{d}{dt}\langle S_z \rangle = -\Gamma(\langle S_x \rangle^2 + \langle S_y \rangle^2)/S. \tag{4.233}$$

Now, we represent the collective spin $\langle \mathbf{S} \rangle$ in spherical coordinates on a Bloch sphere of radius S. Define

$$\langle S_x \rangle = S \sin[\theta(t)]\cos[\phi(t)] \tag{4.234}$$

$$\langle S_y \rangle = S \sin[\theta(t)]\sin[\phi(t)] \tag{4.235}$$

$$\langle S_z \rangle = S \cos[\theta(t)]. \tag{4.236}$$

For purely cooperative decay, we find

$$\dot{\theta}(t) = \Gamma S \sin[\theta(t)], \tag{4.237}$$

$$\dot{\phi}(t) = 0 \quad \text{(in the simplest symmetric solution)}. \tag{4.238}$$

Thus, if $\theta(0) \approx 0$ but with a small 'seed' tilt ($\theta(0) \neq 0$)—often from quantum noise —$\theta(t)$ grows quickly, flipping the spin from 'up' to 'down' in a characteristic superradiant burst.

What is the overall decay rate? When you expand

$$S^+S^- = \left(\sum_{j=1}^{N} \sigma_j^+\right)\left(\sum_{k=1}^{N} \sigma_k^-\right) \tag{4.239}$$

you obtain

$$S^+S^- = \sum_j \sigma_j^+ \sigma_j^- + \sum_{j \neq k} \sigma_j^+ \sigma_k^-. \tag{4.240}$$

The diagonal terms $\sigma_j^+ \sigma_j^-$ represent each atom decaying on its own (rate γ, say). The off-diagonal (cross) terms $\sigma_j^+ \sigma_k^-$ for $j \neq k$ can enhance or reduce the total decay, depending on the state (i.e. relative phases). In the fully symmetric Dicke states, these cross terms add up constructively. Hence, if all N atoms are in a symmetric configuration (e.g. all in the same mode, with equal phases), the total instantaneous emission rate can be $\sim N\gamma$. An intuitive argument is that each atom's dipole is $\sim \sqrt{\gamma}$.

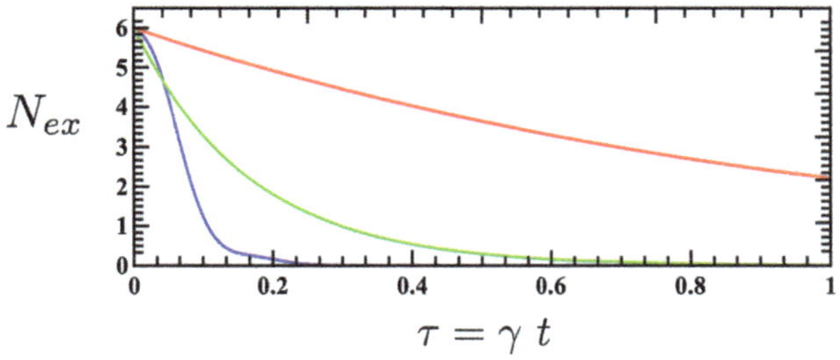

Figure 4.14. Decay of a Dicke state in blue. The red curve is the single-atom decay at rate γ. The green curve is the exponential decay at $g\gamma$ which is a good fit for early times.

If all N dipoles line up in phase (the symmetric case), the collective dipole amplitude is $\propto N\sqrt{\gamma}$. The emission power is then proportional to the square of the dipole, so that the power is $\propto N^2\gamma$. The effective rate per excited atom Γ_{eff} is then $\sim(N^2\gamma)/N = N\gamma$, as shown in figure 4.14.

References

[1] Allen L and Eberly J H 1987 *Optical Resonance and Two-Level Atoms* (New York: Dover)

[2] Loudon R 2000 *The Quantum Theory of Light* (Oxford: Oxford University Press)

[3] Sargent M, Scully M and Lamb W 1975 *Laser Physics* (Reading, MA: Addison-Wesley)

[4] Milonni P 2019 *An Introduction to Quantum Optics and Quantum Fluctuations* (Oxford: Oxford University Press)

[5] Gross M and Haroche S 1982 Superradiance: an essay on the theory of collective spontaneous emission *Phys. Rep.* **93** 301–96

[6] Dicke R H 1954 Coherence in spontaneous radiation processes *Phys. Rev.* **93** 99–110

Chapter 5

Multilevel systems

5.1 Quantum beats

Consider a three-level atom in the V or Λ configuration, as shown in figure 5.1. What is the total emitted dipole? The state of the system is given generally by [1]

$$|\psi(t)\rangle = c_a \exp(-i\omega_a t)|a\rangle + c_b \exp(-i\omega_b t)|b\rangle + c_c \exp(-i\omega_c t)|c\rangle, \tag{5.1}$$

which is true for both systems.

Now consider the polarizations in each case. For a V atom, we have

$$P_{ac} = e\langle a|r|c\rangle \tag{5.2}$$

$$P_{bc} = e\langle b|r|c\rangle. \tag{5.3}$$

For the Λ system we have

$$P_{ac} = e\langle a|r|c\rangle \tag{5.4}$$

$$P_{ab} = e\langle a|r|b\rangle. \tag{5.5}$$

We define the transition frequencies $\omega_1 = \omega_a - \omega_c$, $v_2 = \omega_b - \omega_c$ for V type atoms and $\omega_1 = \omega_a - \omega_b$, $\omega_2 = \omega_a - \omega_c$ for Λ type atoms. From a semiclassical perspective, the field radiated will then be the sum of two terms

$$E^{(+)} = E_1 \exp(-i\omega_1 t) + E_2 \exp(-i\omega_2 t). \tag{5.6}$$

The detected intensity will be

$$|E^{(+)}|^2 = |E_1|^2 + |E_2|^2 + \left\{E_1^* E_2 \exp[i(\omega_1 - \omega_2)t] + \text{c.c.}\right\}, \tag{5.7}$$

which will clearly show a beat note in both cases when we treat the field classically.

doi:10.1088/978-0-7503-6014-2ch5

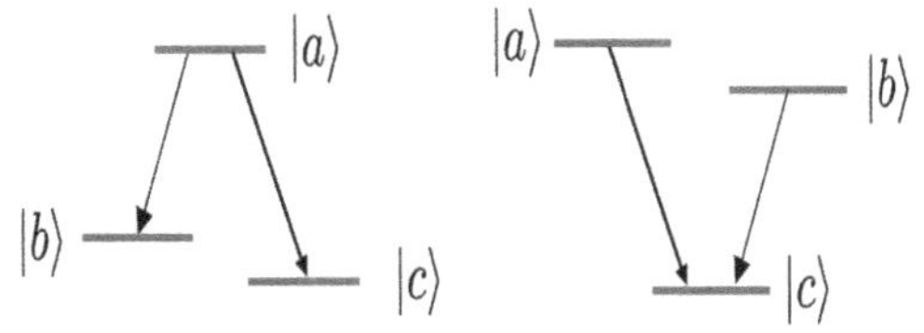

Figure 5.1. On the left is a three-level atom in the Λ configuration, on the right is the V configuration.

To quantum mechanically treat the electromagnetic field quantum in the calculations for both the Λ and V configurations, we need to describe the field using the quantized electric field operator and account for the interaction between the atom and the field in terms of the quantum states. Including one-photon states for the fields on the two transition,

$$|\psi_V(t)\rangle = c_1|c,\, 1_{\nu_1}\rangle + c_2|c,\, 1_{\nu_2}\rangle + \sum_{i=a,b,c} c_i|i,\, 0\rangle\,. \tag{5.8}$$

We find the beat term to be

$$\langle\psi_V(t)|E_1^{(-)}(t)E_2^{(+)}(t)|\psi_V(t)\rangle, \tag{5.9}$$

where $E_1^{(-)}(t)$ and $E_2^{(+)}(t)$ are proportional to the creation and annihilation operator expressions $a_1^\dagger\exp(i\omega_1 t)$ and $a_2\exp(-i\omega_2 t)$, respectively. Using $|\psi\rangle_V$, we find

$$\begin{aligned} B_V &= \langle 1_{\omega_1} 0_{v\omega_2}|a_1^\dagger a_2|0_{\omega_1} 1_{\omega_2}\rangle \exp[i(\omega_1 - \omega_2)t]\,\langle c|c\rangle\,, \\ &= \langle 1_{v_1} 0_{v_2}|a_1^\dagger a_2|0_{\omega_1} 1_{\omega_2}\rangle \exp[i(\omega_1 - \omega_2)t] \end{aligned} \tag{5.10}$$

as $\langle c|c\rangle = 1$. For Λ type atoms we have

$$\langle\psi_\Lambda(t)|E_1^{(-)}(t)E_2^{(+)}(t)|\psi_\Lambda(t)\rangle, \tag{5.11}$$

which yields

$$\begin{aligned} B_\Lambda &= \langle 1_{\omega_1} 0_{\omega_2}|a_1^\dagger a_2|0_{\omega_1} 1_{\omega_2}\rangle \exp[i(\omega_1 - \omega_2)t]\langle c|b\rangle \\ &= 0 \end{aligned} \tag{5.12}$$

as $\langle c|b\rangle = 0$. Thus, we see that while both configurations yield quantum beats in the semiclassical theory, only the V system shows beats when we quantize the field. In the quantum case for the V system, the emitted photons carry 'which-path' information (corresponding to whether the atom transitioned to $|g_1\rangle$ or $|g_2\rangle$), destroying the coherence required for the beats. We do not know the decay path to the ground state, analogous the lack of knowledge of the slit the photon went through in a double-slit experiment

5.2 Driven three-level system

The Hamiltonian of a three-level atom that is interacting with two lasers such as the one in figure 5.2 is

$$H = H_0 + H_1, \tag{5.13}$$

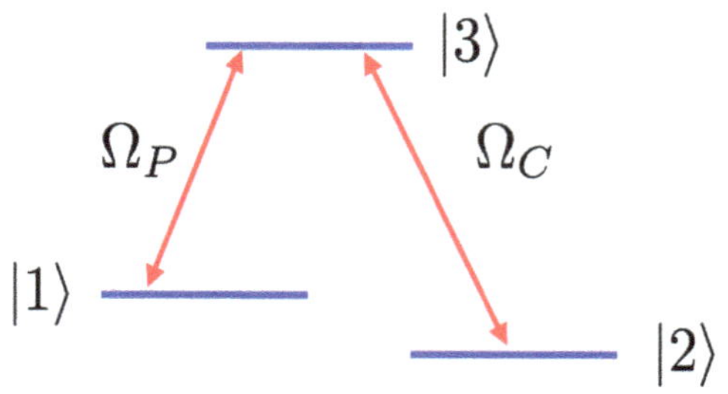

Figure 5.2. A three-level atom in a Λ configuration.

where

$$H_0 = \hbar\omega_1|1\rangle\langle 1| + \hbar\omega_2|2\rangle\langle 2| + \hbar\omega_3|e\rangle\langle e| \tag{5.14}$$

$$H_1 = -\frac{\hbar}{2}(\Omega_p e^{-i\phi_p} e^{-i\omega_p t}|3\rangle\langle 1| + \Omega_c e^{-i\phi_c} e^{-i\omega_c t}|3\rangle\langle 2|) + \text{h.c.} \tag{5.15}$$

Here $\Omega_p e^{-i\phi_p}$ and $\Omega_c e^{-i\phi_c}$ are the Rabi frequencies associated with the field modes coupled to the atomic transitions. The state vector of a three-level atom is given by

$$|\psi\rangle = C_1|1\rangle + C_2|2\rangle + C_3|3\rangle \tag{5.16}$$

and by taking the time derivative, we obtain

$$|\dot{\psi}\rangle = \dot{C}_1|1\rangle + \dot{C}_2|2\rangle + \dot{C}_3|3\rangle. \tag{5.17}$$

Our goal is to derive expressions for how the probability amplitudes change with time. To solve for these, we will need to use the Schrödinger equation

$$|\dot{\psi}\rangle = -\frac{i}{\hbar}H|\psi\rangle. \tag{5.18}$$

This results in the equations for the three probability amplitudes

$$\dot{C}_1 = -i\omega_1 C_1 + \frac{i}{2}\Omega_p e^{i\phi_p} e^{i\omega_p t} C_3. \tag{5.19}$$

Similarly, we can obtain

$$\dot{C}_2 = ega_i C_i + \frac{i}{2}\Omega_c e^{i\phi_c} e^{i\omega_c t} C_e \tag{5.20}$$

$$\dot{C}_3 = -i\omega_e C_3 + \frac{i}{2}\Omega_c e^{-i\phi_c} e^{-i\omega_c t} C_2 + \frac{i}{2}\Omega_p e^{-i\phi_p} e^{-i\omega_p t} C_1. \tag{5.21}$$

To simplify things, we move to a rotating frame by defining

$$C_1 = D_1 e^{-i\omega_1 t} \tag{5.22}$$

$$C_2 = D_2 e^{-i\omega_2 t} \tag{5.23}$$

$$C_3 = D_e e^{-i\omega_3 t}. \tag{5.24}$$

We find

$$\dot{C}_1 = \dot{D}_1 e^{-i\omega_1 t} - i\omega_1 D_1 e^{-i\omega_1 t} \tag{5.25}$$

and we have

$$(\dot{D}_1 - i\omega_1 D_1)e^{-i\omega_1 t} = -i\omega_1 D_1 e^{-i\omega_1 t} + \frac{i}{2}\Omega_p e^{i\phi_p} e^{i\omega_p t} D_e e^{-i\omega_e t}. \tag{5.26}$$

Canceling common terms, which is why we chose this frame,

$$\dot{D}_1 = \frac{i}{2}\Omega_p D_3. \tag{5.27}$$

We also find

$$\dot{D}_3 = \frac{i}{2}(\Omega_p D_1 + \Omega_c D_2) \tag{5.28}$$

$$\dot{D}_2 = \frac{i}{2}\Omega_c D_3. \tag{5.29}$$

For simplicity, we have set $\Delta = \phi = 0$, and we now write these equations in the matrix form

$$\begin{pmatrix} \dot{D}_1 \\ \dot{D}_3 \\ \dot{D}_2 \end{pmatrix} = \frac{i}{2}\begin{pmatrix} 0 & \Omega_p & 0 \\ \Omega_p & 0 & \Omega_c \\ 0 & \Omega_c & 0 \end{pmatrix}\begin{pmatrix} D_1 \\ D_3 \\ D_2 \end{pmatrix}. \tag{5.30}$$

The 3×3 matrix on the right-hand side has three eigenvalues, and there is one in particular in which we are interested. For $\lambda = 0$, the corresponding eigenstate is

$$D_{\text{dark}} = \begin{pmatrix} -\Omega_c \\ 0 \\ \Omega_p \end{pmatrix}. \tag{5.31}$$

This represents the dark state, and by looking at it, we see that as the state evolves, the probability of being in the excited state is zero. Here the population is always in a superposition of the ground states, which means that there is no absorption, even in the presence of a laser.

5.3 Electromagnetically induced transparency

Here we consider how one can make a transtition non-absorbing, across a window of frequencies, known as electromagnetically induced transparency (EIT) [2]. There are three types of three-level systems, depending on what transitions are driven. These are the Λ, V, and Ladder systems, as shown in figure 5.3. We will concentrate on the Λ system here, although the results are generalized to the others.

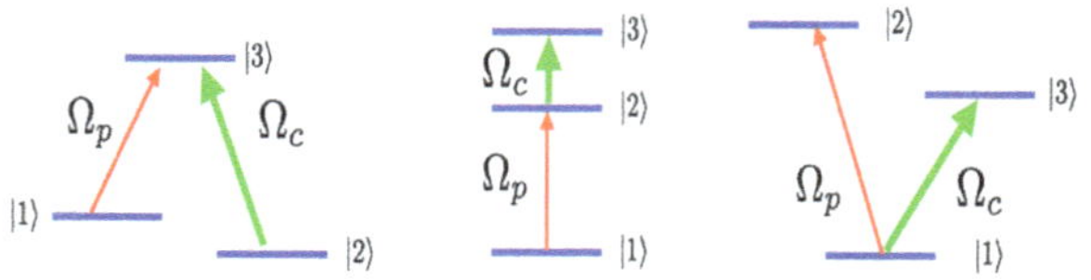

Figure 5.3. On the left is a three-level atom in the Λ configuration, in the middle is the ladder configuration, and on the right is the V configuration.

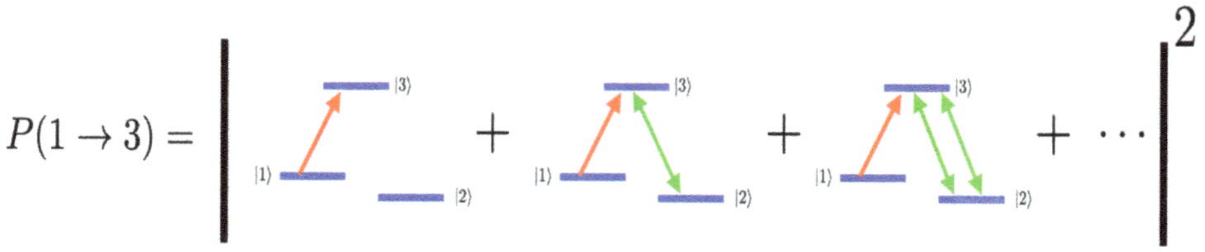

Figure 5.4. Schematic of the terms in the perturbative expansion of the absorption amplitude.

We consider the case where the atom begins in state $|1\rangle$ and wish to calculate the probability for the atom to be found some time later in the state $|3\rangle$. We simply have to add up the amplitudes for the different ways the process could occur. These are the amplitude for the atom to go from state $|1\rangle$ to state $|3\rangle$ (denoted $1 \rightarrow 3$) and stay there, the amplitude for the atom to go from $|1\rangle$ to $|3\rangle$ to $|2\rangle$ to $|3\rangle$ ($(1 \rightarrow 3)(3 \rightarrow 2)(2 \rightarrow 3)$), etc. That is, we wish to calculate

$$\langle 3|U(t, 0|1\rangle) = 1 \rightarrow 3 + (1 \rightarrow 3)(3 \rightarrow 2)(2 \rightarrow 3) + (1 \rightarrow 3)(3 \rightarrow 1)(1 \rightarrow 3) + \ldots. \tag{5.32}$$

This is represented in figure 5.4.

We want to know the form of the amplitudes $1 \rightarrow 3$. They can be calculated from Fermi's golden rule or by expanding the time evolution operator in $\langle 3|U|1\rangle$,

$$\rightarrow 3 = 3 \rightarrow 1 = -i\frac{t}{2}\Omega_p, \tag{5.33}$$

where t is the interaction time and Ω_p is the probe Rabi frequency, so we will not do that here. Similarly,

$$\rightarrow 3 = 3 \rightarrow 2 = -i\frac{t}{2}\Omega_c, \tag{5.34}$$

where Ω_c is the control Rabi frequency. The amplitudes may be combined according to

$$(1 \rightarrow 3)(3 \rightarrow 2)(2 \rightarrow 3) = \frac{1}{3!}\left(-i\frac{t}{2}\right)^3 \Omega_p^1 \Omega_c^2, \tag{5.35}$$

where the initial factorial is necessary as we must divide by the number of permutations. The total amplitude is then

$$\langle 3|U(t, 0)|1\rangle = -i\frac{t}{2}\Omega_p + \frac{1}{3!}\left(-i\frac{t}{2}\right)^3\left(\Omega_p^3 + \Omega_p^2\Omega_c\right) + \ldots. \tag{5.36}$$

If the control beam is much stronger than the probe beam,

$$\Omega_c >> \Omega_p, \tag{5.37}$$

then we can keep only terms that are first order in Ω_p. This simplifies the amplitude to

$$\begin{aligned}\langle 3|U(t,0)|1\rangle &= -i\frac{t}{2}\Omega_p + \frac{1}{3!}\left(-i\frac{t}{2}\right)^3\Omega_p\Omega_c^2 + \\ &= -i\frac{t}{2}\Omega_p\left(1 + \frac{1}{3!}\left(-i\frac{t}{2}\Omega_c\right)^2 + \ldots\right).\end{aligned} \tag{5.38}$$

Collecting the terms in this series gives

$$\langle 3|U(t,0)|1\rangle = -i\frac{t}{2}\Omega_p\sum_{n=0}^{\infty}\frac{(-i\Omega_c t/2)^{2n}}{(2n+1)!} = -\frac{\Omega_p}{\Omega_c}\sin(\Omega_c t/2), \tag{5.39}$$

which becomes arbitrarily small as $\frac{\Omega_p}{\Omega_c}$ becomes arbitrarily small. This is what is meant when it is said that EIT is the result of coherent interference between the amplitudes for different processes.

We now add damping, as in figure 5.5. The macroscopic polarization P that the atoms produce in response to the probe field is

$$P = N\langle d\rangle \approx N d_{13}\rho_{31}e^{-i\omega t} + \text{c.c.}, \tag{5.40}$$

where N is the atomic density (number of atoms per unit volume), $d_{13} = \langle 1|d|3\rangle$ is the electric-dipole matrix element for the $|1\rangle \leftrightarrow |3\rangle$ transition, ρ_{13} is the corresponding off-diagonal density-matrix element, and ω is the probe-field frequency. The polarization P is related to the applied electric field E by

$$P = \epsilon_0\chi(\omega)E, \tag{5.41}$$

where $\chi(\omega)$ is the (complex) electric susceptibility of the medium at frequency ω. Equating these,

$$\epsilon_0\chi(\omega)E = N d_{13}\rho_{31}. \tag{5.42}$$

Hence, the linear susceptibility (in the regime where ρ_{31} is linear in E) can be written as

$$\chi(\omega) = \frac{N d_{13}}{\epsilon_0 E}\rho_{31}(\omega). \tag{5.43}$$

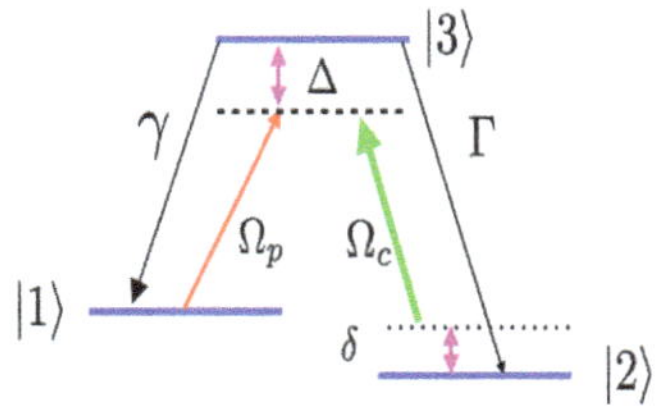

Figure 5.5. The Λ structure we consider, with drives, decays, and detunings noted.

Often, one re-expresses E in terms of the Rabi frequency $\Omega_{13} = \frac{d_{13}E}{\hbar}$. Then

$$\chi(\omega) = \frac{N\,|d_{13}|^2}{\hbar\epsilon_0}\frac{\rho_{13}(\omega)}{\Omega_{13}}. \tag{5.44}$$

These forms explicitly show how the coherence ρ_{13} of the atomic transition $|1\rangle \leftrightarrow |3\rangle$ determines the macroscopic susceptibility $\chi(\omega)$ of a gas of N atoms. In linear (weak-field) optics, the complex susceptibility $\chi(\omega)$ directly gives both the index of refraction and the absorption coefficient. Here is the usual summary, assuming $\chi(\omega)$ is not too large. When an electromagnetic wave traverses a medium with susceptibility $\chi(\omega)$, one defines the complex refractive index

$$n(\omega) = n_r(\omega) + in_i(\omega), \tag{5.45}$$

where $n_r(\omega)$ is the usual (real) index of refraction, and $n_i(\omega)$ is the extinction coefficient describing the absorption. The effects of a complex index of refraction are dispersion and gain/loss:

$$\exp[in(k_0x)] = \exp i([(n_r)k_0 + i(in_ik_0)])x = \exp(in_rk_0x)\exp(-n_ik_0x). \tag{5.46}$$

The relation between n and χ is

$$n(\omega) = \sqrt{1 + \chi(\omega)}. \tag{5.47}$$

If $\chi(\omega) = \chi'(\omega) + i\chi''(\omega)$ is small compared to 1, one can expand in a Taylor series and obtain the approximate relations

$$n(\omega) \approx 1 + \frac{1}{2}\chi'(\omega), \quad k(\omega) \approx \frac{1}{2}\chi''(\omega). \tag{5.48}$$

The group velocity of light in a material is related to the real part of the susceptibility

$$v_g = \frac{c}{n + \omega\dfrac{dn}{d\omega}} \approx \frac{c}{1 + \omega 2\dfrac{d}{d\omega}\mathrm{Re}[\chi(\omega)]}. \tag{5.49}$$

The absorption coefficient is determined by the imaginary part of the susceptibility

$$\alpha(\omega) = \frac{2\omega}{c}k(\omega) \approx \frac{\omega}{c}\chi''(\omega) \tag{5.50}$$

in the same small-χ approximation where $n(\omega) \approx 1$. An active medium, such as a laser, would have a negative imaginary part of the susceptibility/index.

In a gas of N atoms, if the weak probe field couples levels $|1\rangle$ and $|3\rangle$, then the macroscopic polarization P at frequency ω is

$$P(\omega) = N\langle d\rangle \approx Nd_{13}\rho_{13}(\omega)e^{-i\omega t} + \text{c.c.}, \tag{5.51}$$

where d_{13} is the dipole matrix element and ρ_{13} is the corresponding density-matrix coherence. By the definition of susceptibility,

$$P(\omega) = \epsilon_0\chi(\omega)E(\omega). \tag{5.52}$$

Hence, in the linear response limit ($\rho_{13} \propto E$),

$$\chi(\omega) = \frac{Nd_{13}}{\epsilon_0 E(\omega)}\rho_{31}(\omega) \approx \frac{N\,|d_{13}|^2}{\hbar\epsilon_0}\frac{\rho_{31}(\omega)}{\Omega_{13}(\omega)}. \tag{5.53}$$

Therefore, we must use quantum mechanics to find the density-matrix elements we want to find, specifically ρ_{13}.

In the interaction picture, we have

$$H_{\text{int}} = -\hbar\left[\frac{\Omega_p}{2}|3\rangle\langle 1|e^{-i\omega_p t} + \frac{\Omega_c}{2}|3\rangle\langle 2|e^{-i\omega_c t} + \text{h.c.}\right] + (\text{detuning terms}). \tag{5.54}$$

This results in the following equations for the matrix elements (where the Rabi frequencies are considered real)

$$\dot{\rho}_{11} = \gamma\rho_{33} + i\frac{\Omega_p}{2}(\rho_{31} - \rho_{13}) \tag{5.55}$$

$$\dot{\rho}_{22} = \Gamma\rho_{33} + i\frac{\Omega_c}{2}(\rho_{32} - \rho_{23}) \tag{5.56}$$

$$\dot{\rho}_{33} = -(\gamma + \Gamma)\rho_{33} + i\frac{\Omega_p}{2}(\rho_{13} - \rho_{31}) + i\frac{\Omega_c}{2}(\rho_{23} - \rho_{32}) \tag{5.57}$$

$$\dot{\rho}_{13} = -\left(i\Delta + \frac{\gamma + \Gamma}{2}\right)\rho_{13} + i\frac{\Omega_p}{2}(\rho_{11} - \rho_{33}) + i\frac{\Omega_c}{2}\rho_{12} \tag{5.58}$$

$$\dot{\rho}_{23} = -\left(i(\Delta + \delta) + \frac{\gamma + \Gamma}{2}\right)\rho_{23} + i\frac{\Omega_c}{2}(\rho_{22} - \rho_{33}) + i\frac{\Omega_p}{2}\rho_{21} \tag{5.59}$$

$$\dot{\rho}_{12} = -(\gamma + i\delta)\rho_{12} + i\frac{\Omega_p}{2}\rho_{32} - i\frac{\Omega_c}{2}\rho_{13}. \tag{5.60}$$

In the EIT regime, almost all the atoms remain in the ground state(s) because the probe Ω_p is very weak ($\Omega_p \ll \gamma$), so it does not saturate the excited state. The coupling Ω_c is large but will not saturate the excited state (since γ is the main relaxation). Instead, Ω_c mostly drives the coherent oscillations between $|2\rangle$ and $|e\rangle$ and creates a stable dark-state superposition with $|1\rangle$. Hence, to obtain an approximation, we set

$$\rho_{11} \approx 1, \quad \rho_{22} \approx 0, \quad \rho_{33} \approx 0. \tag{5.61}$$

We also did note that $\rho_{23} \sim \Omega_p$, and we set this to zero as well,

$$\rho_{23} \approx 0. \tag{5.62}$$

We are left with equations for the two nonzero coherences, which are correct to first order in Ω_p,

$$\dot{\rho}_{13} = -\Gamma_{12}\rho_{21} + i\Omega_c\rho_{31} \tag{5.63}$$

$$\dot{\rho}_{12} = -\Gamma_{13}\rho_{31} + i\Omega_{\mathrm{c}}\rho_{21} + i\Omega_{\mathrm{p}}, \tag{5.64}$$

with

$$\Gamma_{12} = \gamma - i\Delta \tag{5.65}$$

$$\Gamma_{13} = \Gamma - i(\Delta + \delta). \tag{5.66}$$

In the steady state, we have

$$\rho_{12} = \frac{\Delta\Omega_p}{\Omega_c^2 - \Delta^2 - i\gamma} \tag{5.67}$$

$$\rho_{13} = \frac{i\Omega_p\Omega_c}{\Omega_c^2 - \Delta^2 - i\gamma\Delta}. \tag{5.68}$$

We now assume two-photon resonance ($\delta = 0$) and that there is no decay on the coupling transition ($\Gamma = 0$). The latter will maximize the effect, as decay on that transition spoils the Rabi flopping. With this we have

$$\rho_{13} = \frac{i\Omega_p\Omega_c}{\Omega_c^2 - \Delta^2 - i\gamma\Delta} \tag{5.69}$$

$$\rho_{12} = \frac{\Delta\Omega_p}{\Omega_c^2 - \Delta^2 - i\gamma\Delta}. \tag{5.70}$$

Finding real and imaginary part results in

$$\rho_{12} = \frac{\Delta\Omega_p\left(\Omega_c^2 - \Delta^2\right)}{\left(\Omega_c^2 - \Delta^2\right)^2 + (\gamma\Delta)^2} + i\frac{\gamma\Delta^2\Omega_p}{\left(\Omega_c^2 - \Delta^2\right)^2 + (\gamma\Delta)^2} \tag{5.71}$$

and

$$\rho_{13} = \frac{-\gamma\Delta\Omega_p\Omega_c}{\left(\Omega_c^2 - \Delta^2\right)^2 + (\gamma\Delta)^2} + i\frac{\left(\Omega_c^2 - \Delta^2\right)\Omega_p\Omega_c}{\left(\Omega_c^2 - \Delta^2\right)^2 + (\gamma\Delta)^2}. \tag{5.72}$$

The absorption seen by the probe is proportional to

$$S_{\mathrm{abs}} = \mathrm{Im}\left[\rho_{13}(\Delta)\right] \tag{5.73}$$

and the index of refraction (dispersion) is

$$n(\Delta) = \mathrm{Re}\left[\rho_{13}(\Delta)\right]. \tag{5.74}$$

From our previous discussion, we expect a dip in the probe-field absorption at $\Delta = \delta = 0$ when

$$\gamma > \Omega_c > \Omega_p. \tag{5.75}$$

The atomic response to the weak probe exhibits a transparency window of width

$$\Gamma_{\mathrm{EIT}} \approx \frac{2\Omega_c^2}{\gamma}. \tag{5.76}$$

Inside this narrow window, the medium becomes transparent and shows a steep dispersion. This window must be smaller than γ as the EIT regime requires $\Omega_c/\gamma\langle 1$. This dip in absorption at the line center is the defining characteristic of EIT. This indicates that the medium becomes transparent at a frequency to which it was previously opaque to. The real part of the polarizability is related to the dispersive response of the medium, which determines the speed of propagation. If one increases the control field to γ or larger, one enters the Autler–Townes regime, where we have a flat band at line zero. This is an AC Stark shift, similar to the Mollow sidebands in resonance fluorescence (in chapter 10). Essentially, we have two different oscillators whose energy has been split beyond the linewidth by the strong coupling $\Omega_c > \gamma$. In many experimental situations, both effects can occur, with one dominating or both of equal importance. The key is to look at the strength of the coupling field with respect to the decay rates.

5.4 Slow light

Referring back to the plots above, we see that in the EIT regime, the slope of the imaginary part of the susceptibility can be quite large, resulting in a much lower group velocity [3]. An approximate result for the group velocity is

$$\frac{v_g}{c} \approx \frac{\Omega_c^2}{\left(\frac{N\mid\mu_{13}\mid^2}{2\epsilon_0\hbar}\omega_p\right)} \ll 1. \tag{5.77}$$

Let us now examine some plots (figures 5.6–5.11). In the plots, we used the full master equation, with no approximations.

In figure 5.6, we see an absorption dip at the line center.

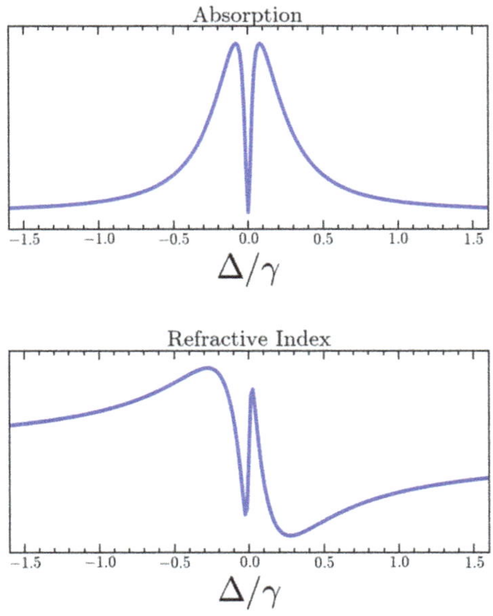

Figure 5.6. $\Omega_p = 0.\ 1$.

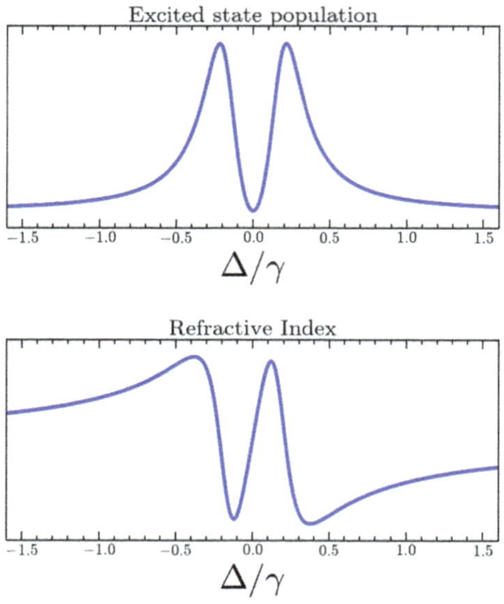

Figure 5.7. $\Omega_p = 0.\ 3$.

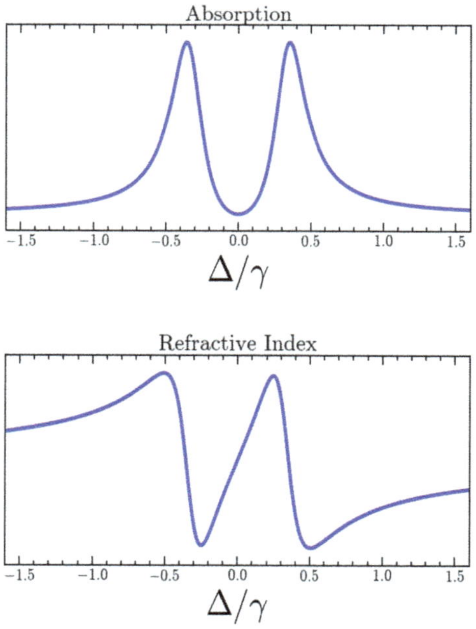

Figure 5.8. $\Omega_p = 0.\ 5$.

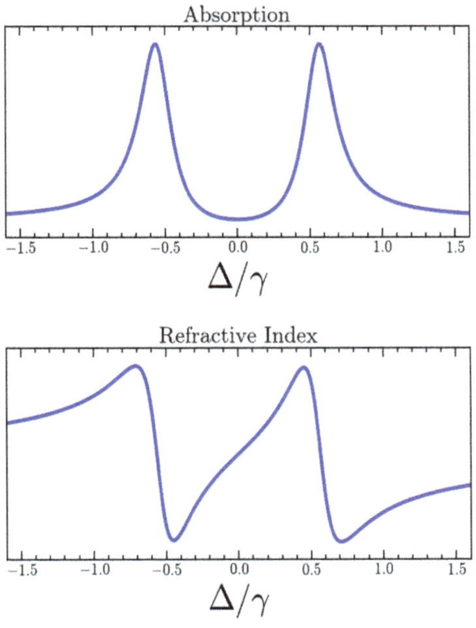

Figure 5.9. $\Omega_p = 0.8$.

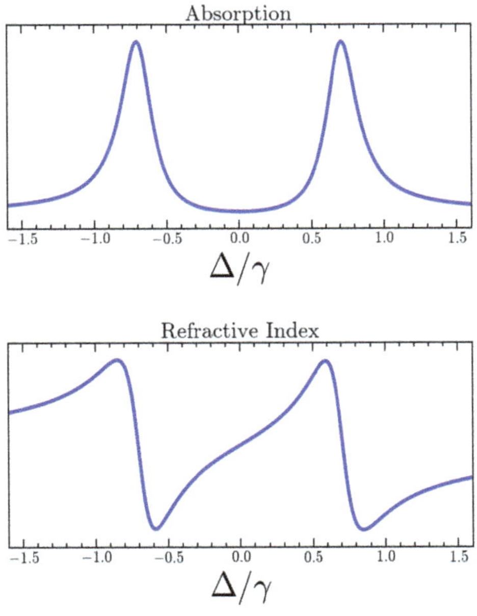

Figure 5.10. $\Omega_p = 1$.

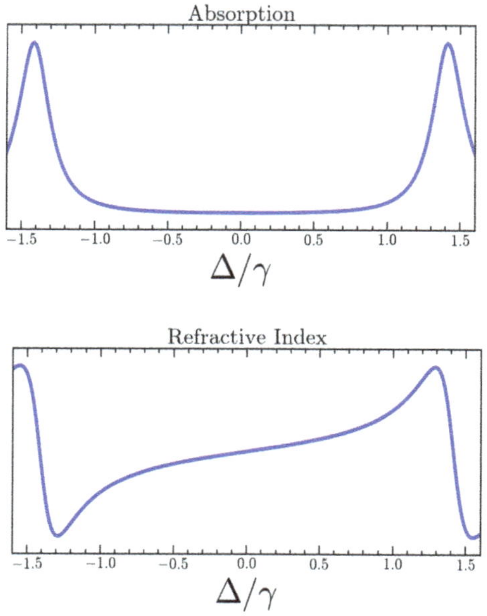

Figure 5.11. $\Omega_p = 2$.

5.5 Quantum memory

In 2000, M Fleischhauer and M D Lukin proposed a mechanism to store and retrieve single photons in an atomic ensemble by exploiting EIT. Their idea relies on the ability of EIT to slow and 'stop' light in a medium and transfer its quantum state onto collective atomic excitations (often referred to as spin waves). Upon demand, the stored excitation can be converted back into a photon with high fidelity. This process, therefore, forms the basis for an all-optical quantum memory or photon memory scheme. Consider an ensemble of N identical atoms, each with three relevant states forming a Λ configuration, we relabel $|g\rangle$ is the ground state, $|s\rangle$ is a metastable state, and $|e\rangle$ is the excited state. We assume that each atom is initially in $|g\rangle$. Two fields couple these levels. A quantum (signal) field $\mathcal{E}(z, t)$, with frequency ω, coupling $|g\rangle \leftrightarrow |e\rangle$. $\mathcal{E}(z, t)$ is an operator in the Heisenberg picture (or equivalently the slowly varying envelope of the quantized signal mode). The coupling constant (single-photon Rabi frequency) is often denoted by g. This could easily be a classical Rabi frequency; we will not use explicitly the properties of a quantum field operator, nor will we need to. There is also a classical control field with Rabi frequency $\Omega(t)$, coupling $|e\rangle \leftrightarrow |s\rangle$. $\Omega(t)$ is typically strong and can be switched on/off or varied in time.

The goal is to use EIT to coherently map the photon state $\mathcal{E}$ onto a collective spin-wave excitation σ_{gs} (an atomic coherence between $|g\rangle$ and $|s\rangle$) and later retrieve it. In the interaction picture (and under the usual rotating-wave approximation), the relevant part of the atom–field interaction Hamiltonian can be written as

$$H_{\mathrm{I}} = -\hbar\sum_{j=1}^{N}\left[g\mathcal{E}(z_j)\sigma_{eg}^{(j)} + \Omega(t)\sigma_{es}^{(j)} + \text{h.c.}\right],$$

where $\sigma_{\alpha\beta}^{(j)} = |\alpha\rangle_{jj}\langle\beta|$ acts on the jth atom, z_j is the position of the jth atom along the propagation axis z, g is the single-atom coupling rate between $\mathcal{E}$ and the atomic transition $|g\rangle \leftrightarrow |e\rangle$, and $\Omega(t)$ is the (time-dependent) Rabi frequency of the classical control field. We will be interested in the wave equation for the probe field. Under the slowly varying envelope approximation (SVEA), we have the one-dimensional wave equation

$$\left(\frac{\partial}{\partial z} + \frac{1}{c}\frac{\partial}{\partial t}\right)\Omega_p(z, t) = i\eta P(z, t),$$

where $P(z, t)$ is proportional to the atomic polarization on the $|g\rangle \leftrightarrow |e\rangle$ transition, and η is a coupling constant that depends on the atomic density, dipole matrix element, etc, ($\eta = \frac{g^2 N}{\Gamma c}$).

We typically assume two-photon resonance, i.e. the signal and control are tuned such that the transition $|g\rangle \leftrightarrow |s\rangle$ is on resonance. Next we define the slowly varying atomic coherences (in one dimension, treating the atomic density as continuous):

$$\sigma_{ge}(z, t) \approx \frac{1}{N_z}\sum_j e^{-ikz_j}\sigma_{ge}^{(j)}, \tag{5.78}$$

$$\sigma_{gs}(z, t) \approx \frac{1}{N_z}\sum_j e^{-ikz_j}\sigma_{gs}^{(j)}, \tag{5.79}$$

where N_z is the number of atoms per unit length at position z, and k is the wavenumber of the signal field. One can similarly define σ_{es}:

$$\dot{\sigma}_{ge} = i\Omega_p\,(\sigma_{gg} - \sigma_{ee}) + i\,\Omega_c\,\sigma_{gs} - \Gamma_{ge}\,\sigma_{ge}, \tag{5.80}$$

$$\dot{\sigma}_{se} = i\Omega_c\,(\sigma_{ss} - \sigma_{ee}) + i\,\Omega_p\,\sigma_{sg} - \Gamma_{se}\,\sigma_{se}, \tag{5.81}$$

$$\dot{\sigma}_{gs} = i\Omega_p\,\sigma_{es} - i\Omega_c\,\sigma_{ge}. \tag{5.82}$$

Under the usual EIT assumptions (low excitation, large atomic ensemble, weak quantum field), the Bloch equations for these coherences simplify. For instance, we take the approximation that the ground state population is unity, and the other two levels have no population (to first order in Ω_p)

$$\frac{\partial}{\partial t}\sigma_{ge}(z, t) \approx ig\mathcal{E}(z, t) - i\Omega(t)\sigma_{gs}(z, t) - \Gamma_{ge}\sigma_{ge}(z, t).$$

If the control Rabi frequency Ω_c is large (compared to the probe Rabi frequency Ω_p) we can set

$$\dot{\sigma}_{ge} \approx 0, \quad \dot{\sigma}_{se} \approx 0$$

and solve for σ_{ge} and σ_{se}. One finds that

$$\sigma_{ge} \approx -\frac{\Omega_p}{\Omega_c}\sigma_{gs}, \quad \sigma_{se} \approx 0 \quad \text{(under ideal EIT)}$$

or more precisely,

$$\sigma_{se} \approx -\frac{\Omega_p^2}{\Omega_c^2}\sigma_{gs}$$

is small if $\Omega_p \ll \Omega_c$. Inserting σ_{ge} into the Maxwell equation yields a coupling between Ω_p and the spin coherence σ_{gs}:

$$\frac{\partial}{\partial t}\sigma_{gs}(z, t) \approx i\Omega^*(t)\sigma_{ge}(z, t) - \Gamma_{gs}\sigma_{gs}(z, t).$$

Here, Γ_{ge} and Γ_{gs} can represent (small) decoherence rates for σ_{ge} and σ_{gs}, respectively. In the ideal EIT scenario, $\Gamma_{gs} \ll \Gamma_{ge}$ and is often negligible on short time scales. The quantum field $\mathcal{E}(z, t)$ obeys a 1D wave equation, we have taken the slowly varying envelope approximation

$$\frac{1}{c}\frac{\partial}{\partial t}\mathcal{E}(z, t) + \frac{\partial}{\partial z}\mathcal{E}(z, t) = i\frac{gN}{c}\sigma_{ge}(z, t),$$

where N is the (effective) density of atoms along the propagation axis,

$$\frac{\partial}{\partial z}\mathcal{E}(z, t) = i\frac{gN}{c}\sigma_{ge}(z, t).$$

When $\Omega(t)$ is large, $\theta \approx 0$ and the polariton is mostly photonic; when $\Omega(t) \to 0$, $\theta \to \frac{\pi}{2}$ and the polariton is mostly a spin-wave excitation. The dark-state polariton operator is

$$\Psi(z, t) = \cos\theta(t)\mathcal{E}(z, t) - \sin\theta(t)\sqrt{N}\sigma_{gs}(z, t).$$

By combining the Maxwell and Bloch equations and adiabatically eliminating the fast variables $\Psi(z, t)$ obeys a simple propagation-like equation, of the form

$$\left(\frac{\partial}{\partial t} + v_g(t)\frac{\partial}{\partial z}\right)\Psi(z, t) = 0$$

in the ideal lossless limit. Here,

$$v_g(t) = c\cos^2\theta(t)$$

is the group velocity of the polariton. Thus, if $\Omega(t) \neq 0$ is constant and large, $\cos^2\theta \approx 1$, so $v_g \approx c$. If we adiabatically lower $\Omega(t)$ to zero, $\cos\theta \to 0$, hence $v_g \to 0$. The polariton becomes purely atomic (no photon traveling in the medium), effectively stopping the light. The protocol is as follows. With $\sigma_{ge} \approx -(\Omega_p/\Omega_c)\sigma_{gs}$, the wave equation

$$\left(\frac{\partial}{\partial z} + \frac{1}{c}\frac{\partial}{\partial t}\right)\Omega_p(z, t) = i\eta\sigma_{ge}(z, t).$$

The hallmark of the Fleischhauer–Lukin method is the introduction of the dark-state polariton, a collective normal mode that is a superposition of the light field $\mathcal{E}$ and the atomic spin-wave σ_{gs}. One defines a mixing angle θ via

$$\tan\theta(t) = \frac{g\sqrt{N}}{\Omega(t)}.$$

Here is the protocol for storage and retrieval:

1. A single-photon wave packet enters the atomic ensemble. The control field $\Omega(t)$ is on and large at first, creating a wide EIT transparency window and a slow group velocity inside the medium.
2. While the photon is inside the medium, $\Omega(t)$ is gradually reduced to zero.
3. As $\Omega \to 0$, $\theta \to \frac{\pi}{2}$, the polariton Ψ becomes dominated by σ_{gs}.
4. The photon's quantum state is thus mapped onto the collective atomic spin-wave $\sigma_{gs}(z)$.

Once the photon is stored, the system is primarily in the atomic coherence σ_{gs}. If the metastable state $|s\rangle$ has a long coherence time, the photon can be 'held' for microseconds, milliseconds, or even seconds. The retrieval protocol is:

1. To retrieve the photon, $\Omega(t)$ is adiabatically turned on again.
2. The spin-wave σ_{gs} is converted back into $\mathcal{E}$. The polariton returns to being photonic and propagates out of the medium.
3. The resulting photon will ideally have the same quantum state (amplitude and phase) as the initial photon, enabling faithful quantum storage and retrieval.

A physical analogy for this is two coupled pendulums. One pendulum is in motion, and we couple it to the other. After some period, depending on the strength of the coupling all of the energy has been transferred to the originally motionless one. If at that point we remove the coupling, we have 'stored' energy in the second pendulum. Put the coupling back on until the energy is transferred back and then remove the coupling. Of course as the pendulums are damped, there is a loss that increases with time.

5.6 Classical analog of EIT

We can view the two driven transitions as a pair of coupled oscillators (figure 5.12). We can model this with the following electrical circuit:

$$\begin{aligned}\ddot{x}_1(t) + \gamma_1\dot{x}_1(t) + \omega^2 x_1(t) - \Omega_r^2 x_2(t) &= \frac{F}{m}e^{-i\omega_s t}\\ \ddot{x}_2(t) + \gamma_2\dot{x}_2(t) + \omega^2 x_2(t) - \Omega_r^2 x_1(t) &= 0,\end{aligned} \tag{5.83}$$

where we set $\phi_s = 0$ for the probe force without loss of generality. We also defined $\Omega_r^2 = K/m$, the frequency associated with the coherent coupling between the pumping oscillator and the oscillator modeling the atom, γ_1 the friction constant

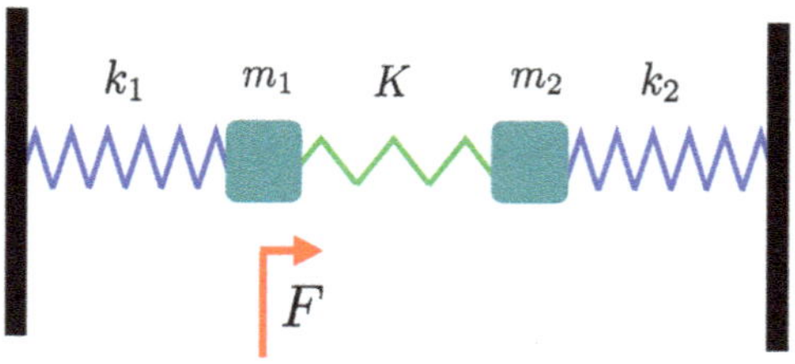

Figure 5.12. Mechanical oscillator analog of EIT.

associated with the energy dissipation rate acting on particle 1 (which simulates the spontaneous emission from the atomic excited state), and γ_2, the energy dissipation rate of the pumping transition. Because we are interested in the power absorbed by particle 1 from the probe force, we will seek a solution for $x_1(t)$. Let us suppose that $x_1(t)$ is of the form

$$x_1(t) = A(t)e^{-i\omega_s t}, \tag{5.84}$$

where $A(t)$ is a time-varying amplitude, albeit varying slowly compared to ω_s. After taking a similar expression for $x_2(t)$ and substituting it in equation (5.83), we find

$$x_1(t) = \frac{\left(\omega^2 - \omega_s^2 - i\gamma_2\omega_s\right)Fe^{-i\omega_s t}}{m\left[\left(\omega^2 - \omega_s^2 - i\gamma_1\omega_s\right)\left(\omega^2 - \omega_s^2 - i\gamma_2\omega_s\right) - \Omega_r^4\right]}. \tag{5.85}$$

Now, by computing the mechanical power $P(t)$ absorbed by particle 1 from the probe force $\mathcal{F}_s$,

$$P(t) = Fe^{-i\omega_s t}\dot{x}_1(t) \tag{5.86}$$

we find for the power absorbed during one period of oscillation of the probe force

$$P_s(\omega_s) = -\frac{2\pi iF^2\omega_s\left(\omega^2 - \omega_s^2 - i\gamma_2\omega_s\right)}{m\left[\left(\omega^2 - \omega_s^2 - i\gamma_1\omega_s\right)\left(\omega^2 - \omega_s^2 - i\gamma_2\omega_s\right) - \Omega_r^4\right]}. \tag{5.87}$$

To make contact with the atomic case, we use an SVEA. We use $x_1 = A_1e^{-i\omega_s t}$ and $x_2 = A_2e^{-i\omega_s t}$, and the approximation

$$\ddot{x}_1 \approx -2i\omega_s\dot{A}_1e^{-i\omega_s t}, \quad \dot{x}_1 \approx -i\omega_s A_1e^{-i\omega_s t}, \tag{5.88}$$

and we obtain

$$2i\omega_s\dot{A}_1 - i\gamma_1\omega_s A_1 + \left(\omega^2 - \omega_s^2\right)A_1 - \Omega_p^2A_2 = \frac{F}{m}. \tag{5.89}$$

Similarly, for A_2 we find

$$2i\omega_s\dot{A}_2 - i\gamma_2\omega_s A_2 + \left(\omega^2 - \omega_s^2\right)A_2 - \Omega_p^2A_1 = 0. \tag{5.90}$$

The absorbed power is the real part of P above,

$$P_{\text{real}} = \frac{2\pi F^2\omega_s\left(\omega^2 - \omega_s^2\right)}{|D|}, \tag{5.91}$$

with $|D|$ just a raging mess of a polynomial. The key thing to note is that no power is absorbed on the line center, as in EIT.

5.7 Stationary light

It is also possible to stop the light without turning it into an atomic excitation with fields propagating in both directions [4].

We give a sketch of the process in figure 5.13. The total electric field in the medium can be decomposed into a strong coupling (or 'control') field, nearly resonant with $|1\rangle \leftrightarrow |2\rangle$,

$$E_c(t) = \frac{1}{2}\left[\mathcal{E}_+^c(z,\, t)e^{-i(\omega_c t - k_c z)} + \mathcal{E}_-^c(z,\, t)e^{-i(\omega_c t - k_c z)}\right] + \text{c.c.}, \tag{5.92}$$

and a weak probe field, nearly resonant with $|1\rangle \leftrightarrow |3\rangle$. To allow for two counter-propagating probe components, we write

$$E_p(z,\, t) = \frac{1}{2}[E_+(z,\, t)e^{i(k_p z - \omega_p t)} + E_-(z,\, t)e^{-i(k_p z + \omega_p t)}] + \text{c.c.}, \tag{5.93}$$

where $E_+(z, t)$ is the slowly varying envelope of the forward-propagating probe component, and $E_-(z, t)$ is the slowly varying envelope of the backward-propagating probe component. Under the usual SVEA, the Maxwell equation for each component of the probe field is

$$\left(\frac{\partial}{\partial z} + \frac{1}{c}\frac{\partial}{\partial t}\right)E_+(z,\, t) = i\frac{\omega_p}{2\epsilon_0 c}\mathcal{P}_+(z,\, t) \tag{5.94}$$

$$\left(-\frac{\partial}{\partial z} + \frac{1}{c}\frac{\partial}{\partial t}\right)E_-(z,\, t) = i\frac{\omega_p}{2\epsilon_0 c}\mathcal{P}_-(z,\, t). \tag{5.95}$$

Here, $\mathcal{P}_+$ and $\mathcal{P}_-$ are the corresponding slowly varying envelopes of the polarization on the $|1\rangle \leftrightarrow |3\rangle$ transition.

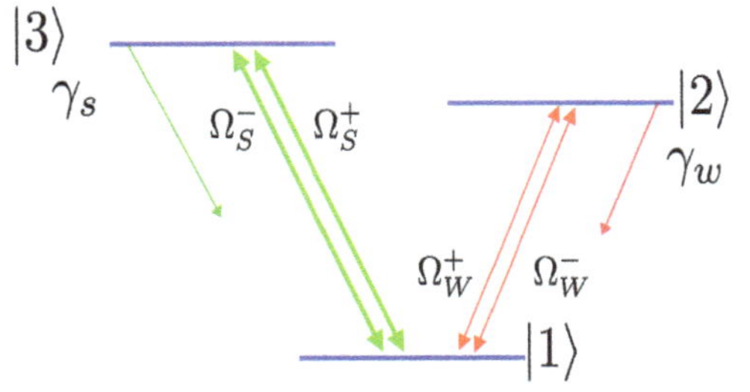

Figure 5.13. Stationary light set-up.

$$\mathcal{P}_{\pm}(z, t) = Nd_{13}\rho_{31}(z, t)\exp\left[\mp i(k_p z - \omega_p t)\right]. \tag{5.96}$$

We introduce a coupling constant $\kappa \equiv \omega_p N\,|d_{13}|^2/2\epsilon_0\hbar c$ and write the Maxwell equations in terms of the Rabi frequency $\Omega_+ \equiv d_{13}E_+/\hbar$ (and similarly for Ω_-). The two coupled Maxwell equations for the forward and backward probe envelopes are

$$\left(\frac{\partial}{\partial z} + \frac{1}{c}\frac{\partial}{\partial t}\right)\Omega_+(z, t) = i\kappa\rho_{31}(z, t), \tag{5.97}$$

$$\left(-\frac{\partial}{\partial z} + \frac{1}{c}\frac{\partial}{\partial t}\right)\Omega_-(z, t) = i\kappa\rho_{31}(z, t), \tag{5.98}$$

and we also have

$$\frac{\partial}{\partial t}\rho_{32} = -(\gamma_{32} + i(\Delta_3 - \Delta_2))\rho_{32} + i\Omega_c^*\rho_{31} - \ldots. \tag{5.99}$$

As with slow light, let us define two polaritons

$$\Psi_\pm = \cos\theta E_\pm - \sin\theta\rho_{31}, \tag{5.100}$$

where θ is a mixing angle chosen so that $\Psi_\pm$ diagonalizes the linearized equations (in the ideal EIT limit). One can then form further symmetric/antisymmetric combos of Ψ_+ and Ψ_- to isolate the 'stationary' solution,

$$\Psi_s = \frac{\Psi_+ + \Psi_-}{\sqrt{2}}, \quad \Psi_d = \frac{\Psi_+ - \Psi_-}{\sqrt{2}}. \tag{5.101}$$

Since we now have two probe components (+ and $^-$), plus the atomic polarization ρ_{31}, we can define 'symmetric' (Ψ_s) and 'antisymmetric' (Ψ_d) combinations. A common choice is as follows: if the forward and backward probe components have equal amplitudes and phases, one can form a standing wave inside the medium. In that case, the 'dark' combination can have an even smaller group velocity—ideally going to zero, producing stationary light. After some algebra, we arrive at a pair of first-order (in time) PDEs coupling the dark (Ψ_D) and bright (Ψ_B) polaritons:

$$\partial_t\Psi_B = -\Gamma_B\Psi_B - iv_B\partial_z\Psi_B \tag{5.102}$$

$$\partial_t\Psi_D = -iv_D\partial_z\Psi_D. \tag{5.103}$$

The dark polariton is rapidly damped and $v_B \approx c$. The bright polariton has no damping (approximately) and has a group velocity of

$$v_g \simeq c\cos^2\theta \ll c. \tag{5.104}$$

OK, what is happening physically? The counterpropagation beams set up an intensity grating that forms an effective photonic band gap, where light cannot propagate. So one stores an excitation as in the stored light, turns off the fields and turns on the counterpropagating control and probe fields, and the excitation is released into the gap.

One interesting thing about this scenario is that the excitation behaves like a particle in a well. This can be formalized by expanding the polariton dispersion relation beyond the linear term near the two-photon resonance

$$\omega(k) = \omega_0 + v_g(k - k_0) + \frac{1}{2}\frac{d^2\omega}{dk^2}\bigg|_{k_0} (k - k_0)^2 + \dots. \tag{5.105}$$

The quadratic piece $\frac{1}{2}\left(\frac{d^2\omega}{dk^2}\right)(k - k_0)^2$ plays the role of an effective mass for the polariton. Writing $k - k_0 = -i\partial_z$, etc, one sees that to second order in small deviations from resonance, the wave equation acquires a term such as

$$i\frac{\partial}{\partial t}\Psi \approx -\frac{1}{2m_{\text{eff}}}\frac{\partial^2}{\partial z^2}\Psi + \dots, \tag{5.106}$$

which is formally identical to a 1D Schrödinger equation with $\Psi(z, t)$ as the 'wavefunction'. Here,

$$m_{\text{eff}} \sim \left[\frac{d^2\omega}{dk^2}\right]^{-1}. \tag{5.107}$$

5.8 Cavity-induced transparency

This section has been reproduced with permission from [5].

The system we consider is shown in figure 5.14. Here there is no coupling laser between the two upper levels, but instead the atom–field coupling g [5]. I guess if one wanted to be cute we could call this 'vacuum induced transparency', but we won't.

It consists of a single two-level atom inside an optical cavity, with dissipation present in the form of spontaneous emission to non-cavity modes at rate $\gamma/2$ and a cavity field decay rate of κ. The atom–field coupling strength is $g = \mu\sqrt{\hbar\omega/2\epsilon_0 V}$. We assume that the atom is at an antinode of the cavity mode function; μ is the induced dipole strength, and V is the volume of the cavity mode. Typically, to obtain a large g a small mode volume is necessary, requiring a small physical volume or a highly focused degenerate mode cavity.

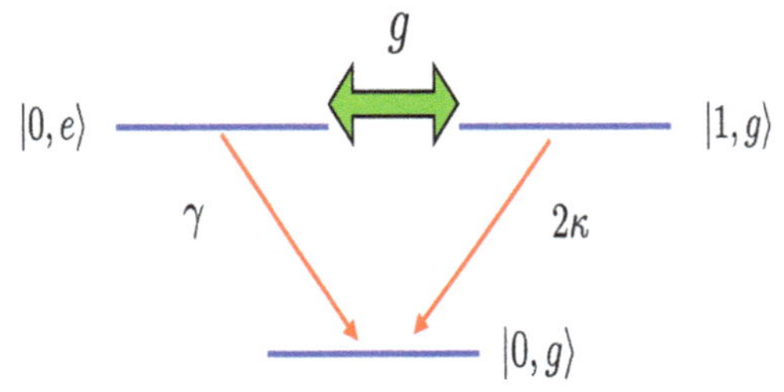

Figure 5.14. Lower energy levels of an atom–cavity system.

The Hamiltonian for this system is

$$H = \frac{1}{2}\hbar\omega_a\sigma_z + \hbar\omega_c a^\dagger a + i\hbar g(a^\dagger\sigma_- - a\sigma_+) \\ + i\hbar\mathcal{E}(\sigma_- e^{-i\omega_p t} - \sigma_+ e^{i\omega_p t}). \tag{5.108}$$

Here we have the Jaynes–Cummings Hamiltonian with the addition of the last term, which is a probe field (taken to be weak throughout the paper) coupled directly to the atoms. The cavity is assumed to be tuned to the atomic resonance, $\omega_a = \omega_c = \omega_0$. The corresponding master equation in the Born–Markoff approximation and in a frame rotating at the frequency ω_0 is

$$\dot{\rho} = -(i/\hbar)[H, \rho] + \kappa(2a^\dagger\rho a - a^\dagger a\rho - \rho a^\dagger a) \tag{5.109}$$

$$+(\gamma/2)(2\sigma_+\rho\sigma_- - \sigma_+\sigma_-\rho - \rho\sigma_+\sigma_-), \tag{5.110}$$

where $\Delta = \omega_p - \omega_0$ is the detuning of the weak probe field from the simultaneous resonant frequency of the atom and cavity. Note that the only external field in the model is the weak probe field, which we will use to study the probe absorption of the atom–cavity 'molecule'; the resonator is initially empty.

We are mainly interested in the limiting case of a weak probe, taken to be so weak that we need only consider the zero- and one-photon states of the system. These states are

$$|0\,-\rangle,\ |0\,+\rangle,\ |1\,-\rangle\,. \tag{5.111}$$

The numerical index labeling the state refers to the number of photons in the cavity field, and the $+(-)$ sign refers to the atom in the excited (ground) state, respectively. We are interested in a specific parameter regime where $\gamma > g > \kappa$. This corresponds roughly to the 'good cavity limit' where the atomic dynamics can be adiabatically eliminated. The latter requires the more stringent condition that $\gamma/2 \gg \kappa(1 + 2C)$. The effect we consider here persists in the good cavity limit, but we stress that adiabatic elimination of the atomic dynamics is not needed for an absorption dip at the line center. Again, g is the intermediate rate, and hence we are not working in a regime where vacuum-Rabi splitting occurs; the eigenvalues of the systems dynamics are real. We can write down the density-matrix element equations for our system in the three-state basis

$$\dot{\rho}_{0-:0-} = \gamma\rho_{0+:0+} + 2\kappa\rho_{1-:1-} + \mathcal{E}(\rho_{0+;0-} + \rho_{0-;0+}), \tag{5.112}$$

$$\dot{\rho}_{0-;0+} = -(\gamma/2 - i\Delta)\rho_{0-;0+} - g\rho_{0-;1-} \\ + \mathcal{E}(\rho_{0+;0+} - \rho_{0-;0-}) \tag{5.113}$$

$$\dot{\rho}_{0-;1-} = -(\kappa - i\Delta)\rho_{0-:1-} + g\rho_{0-;0+} + \mathcal{E}\rho_{0+;1-} \tag{5.114}$$

$$\dot{\rho}_{0+;0+} = -\gamma\rho_{0+;0+} - \mathcal{E}(\rho_{0-;0+} + \rho_{0+;0-}) \\ - g(\rho_{1-;0+} + \rho_{0+;1-}) \tag{5.115}$$

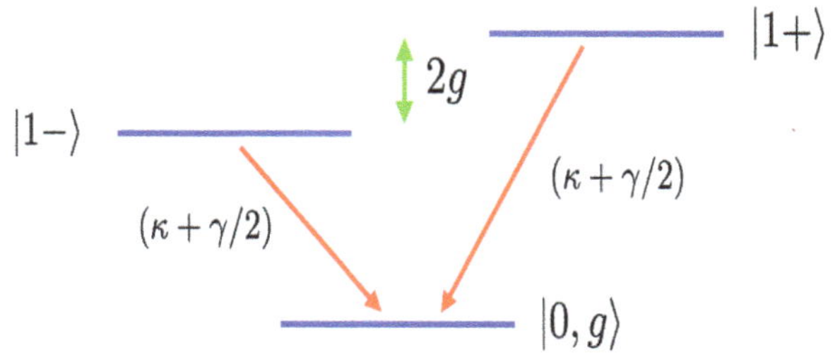

Figure 5.15. Dressed states of an atom–cavity system.

$$\dot{\rho}_{0+;1-} = -(\gamma/2+\kappa)\rho_{0+;1-} \\ -g(\rho_{1-;1-}-\rho_{0+;0+})-\mathcal{E}\rho_{0-;1-} \tag{5.116}$$

$$\dot{\rho}_{1-;1-} = g(\rho_{1-;0+}+\rho_{0+;1-})-2\kappa\rho_{1-;1-}. \tag{5.117}$$

In the weak-field limit, we set $\rho_{0-;0-}=1$; one finds that the other populations are proportional to $\mathcal{E}^2$, and that the off-diagonal matrix elements are proportional to $\mathcal{E}$. Keeping terms that are of the lowest order in $\mathcal{E}$, we find

$$\dot{\rho}_{0-;0+} = -g\rho_{0-;1-}-(\gamma/2)(1-2i\Delta/\gamma)\rho_{0-;0+}-\mathcal{E} \tag{5.118}$$

$$\dot{\rho}_{0-;1-} = \rho_{0-;0+}-\kappa(1-i\Delta/\kappa)\rho_{0-;1-}. \tag{5.119}$$

Consider the system shown in figure 5.15, it is equivalent to that of a three-level atom with a strong pump coupling the two upper levels, 2 and 3. Under the conditions that $\gamma_2 > E > \gamma_3$, it has been shown that this system exhibits a dip in absorption at the line center, a phenomenon known as EIT. It is important that γ_3 be small for the effect to occur, and further that $E\langle\gamma_2$ so that the dressed state splitting is small compared to the linewidth of those states. The latter is to rule out Stark shifts as the cause of the absorption dip. The atom in a cavity in the weak-field limit can be thought of as a three-level system. In the EIT case, the coupling field E_c has to be large; in quantum terms $E_c = g\sqrt{n}$, and for an atom in free space EIT is accomplished by a large photon number n. In the cavity enhanced transparency case, it is g that is large so that the interference effect persists in the single-photon limit. Again, one must have κ small, but $\gamma > g$ so that one is not merely observing large vacuum-Rabi splittings that take the atom–cavity molecule out of resonance with the probe field.

In our case, the two-level atom in a cavity is a three-level system, and no external fields whatsoever are required to induce the transparency. For a stronger probe field, one cannot truncate the basis states, and the system is not a three-level system.

5.9 Two-photon absorption

Consider the three-level atom in figure 5.16, a ladder configuration.

We would like to examine the two-photon absorption to the upper level. Ignoring the damping to start, we have

$$|\psi(t)\rangle = C_g(t)|g\rangle + C_i(t)|i\rangle + C_e(t)|e\rangle \tag{5.120}$$

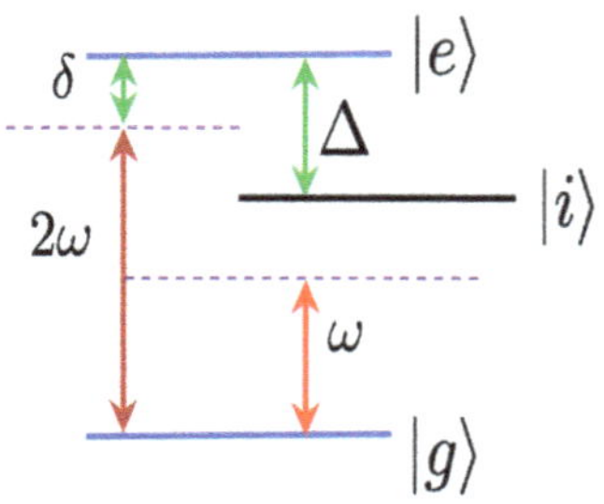

Figure 5.16. Two-photon absorption.

and the Hamiltonian

$$H = \begin{bmatrix} 0 & \frac{\hbar\Omega_1}{2} & 0 \\ \frac{\hbar\Omega_1}{2} & \hbar\Delta & \frac{\hbar\Omega_2}{2} \\ 0 & \frac{\hbar\Omega_2}{2} & \hbar(\delta + 2\omega) \end{bmatrix}. \tag{5.121}$$

For a three-level system, we cannot easily use Pauli operators, but we write this in terms of the Gell-Mann operators introduced for $SU(3)$ in the context of strong force physics:

$$H = \frac{\hbar\Delta}{2}\left(\mathbb{I} - \lambda_3 - \frac{\lambda_8}{\sqrt{3}}\right) + \frac{\hbar(\delta + 2\omega)}{3}(\mathbb{I} + \sqrt{3}\,\lambda_8) + \frac{\hbar\Omega_1}{2}\lambda_1 + \frac{\hbar\Omega_2}{2}\lambda_6 \tag{5.122}$$

We will not use this, but it is good you know it exists! The resulting equations for the probability amplitudes are

$$i\hbar\frac{dC_g}{dt} = \frac{\hbar\Omega_1}{2}C_i \tag{5.123}$$

$$i\hbar\frac{dC_i}{dt} = \frac{\hbar\Omega_1}{2}C_g + \hbar\Delta C_i + \frac{\hbar\Omega_2}{2}C_e \tag{5.124}$$

$$i\hbar\frac{dC_e}{dt} = \frac{\hbar\Omega_2}{2}C_i + \hbar(\delta + 2\omega)C_e. \tag{5.125}$$

Let us now assume that the intermediate level is off resonance (large Δ) and adiabatically eliminate that level

$$i\hbar\frac{dC_i}{dt} = \frac{\hbar\Omega_1}{2}C_g + \hbar\Delta C_i + \frac{\hbar\Omega_2}{2}C_e. \tag{5.126}$$

Since Δ is large, C_i changes rapidly, and we can approximate $dC_i/dt \approx 0$, giving

$$0 = \frac{\Omega_1}{2}C_g + \Delta C_i + \frac{\Omega_2}{2}C_e \tag{5.127}$$

leading to

$$C_i \approx -\frac{\Omega_1}{2\Delta}C_g - \frac{\Omega_2}{2\Delta}C_e. \tag{5.128}$$

This leaves us with two remaining equations

$$i\hbar\frac{dC_g}{dt} = -\frac{\hbar\Omega_1^2}{4\Delta}C_g - \frac{\hbar\Omega_1\Omega_2}{4\Delta}C_e \tag{5.129}$$

$$i\hbar\frac{dC_e}{dt} = -\frac{\hbar\Omega_1\Omega_2}{4\Delta}C_g - \frac{\hbar\Omega_2^2}{4\Delta}C_e + \hbar(\delta + 2\omega)C_e. \tag{5.130}$$

This is a two-level system, which would be described by the Hamiltonian

$$H_{\text{eff}} = H_{\text{AC}} + H_{\text{eff-coupling}} + H_{\text{shift}} \tag{5.131}$$

with an AC Stark shift term

$$H_{\text{AC}} = -\frac{\hbar}{8\Delta}\left(\Omega_1^2 + \Omega_2^2\right)I - \frac{\hbar}{8\Delta}\left(\Omega_2^2 - \Omega_1^2\right)\sigma_z, \tag{5.132}$$

a coupling term

$$H_{\text{eff-coupling}} = -\frac{\hbar\Omega_1\Omega_2}{4\Delta}\sigma_x, \tag{5.133}$$

and a detuning term

$$H_{\text{shift}} = \frac{\hbar(\delta + 2\omega)}{2}(I + \sigma_z). \tag{5.134}$$

The identity terms can be used to shift the zero of energy, previously the ground state energy

$$E_0 = \frac{\hbar(\delta + 2\omega)}{2} - \frac{\hbar}{8\Delta}\left(\Omega_1^2 + \Omega_2^2\right) \tag{5.135}$$

with a resulting Hamiltonian of

$$H_{\text{eff}}' = \frac{\hbar}{2}\left[(\delta + 2\omega) - \frac{\Omega_2^2 - \Omega_1^2}{4\Delta}\right]\sigma_z - \frac{\hbar\Omega_1\Omega_2}{4\Delta}\sigma_x. \tag{5.136}$$

Now we wish to deal with the dissipative terms, we use the effective Hamiltonian with a dissipator for the upper level

$$\frac{d\rho}{dt} = -\frac{i}{\hbar}\left[\frac{\hbar}{2}\tilde{\Delta}\sigma_z - \frac{\hbar\tilde{\Omega}}{2}\sigma_x, \rho\right] + \gamma\left(\sigma_-\rho\sigma_+ - \frac{1}{2}\{\sigma_+\sigma_-, \rho\}\right). \tag{5.137}$$

OK, to check the validity of this, let us go back to the original three level Hamiltonian. The total Hamiltonian is

$$H = \hbar\Delta|i\rangle\langle i| + \hbar(\delta + 2\omega)|e\rangle\langle e| + \frac{\hbar\Omega_1}{2}(|i\rangle\langle g| + |g\rangle\langle i|) + \frac{\hbar\Omega_2}{2}(|e\rangle\langle i| + |i\rangle\langle e|). \tag{5.138}$$

We use this to find a master equation, adding dissipators for the upper and intermediate levels

$$\sum_k \mathcal{D}[L_k]\rho = \mathcal{D}[L_e]\rho + \mathcal{D}[L_i]\rho. \tag{5.139}$$

The resulting density-matrix equations are

$$\frac{d\rho_{gg}}{dt} = i\frac{\Omega_1}{2}(\rho_{gi} - \rho_{ig}) + \gamma_i \rho_{ii} \tag{5.140}$$

$$\frac{d\rho_{ii}}{dt} = i\frac{\Omega_1}{2}(\rho_{ig} - \rho_{gi}) + i\frac{\Omega_2}{2}(\rho_{ei} - \rho_{ie}) - \gamma_i \rho_{ii} + \gamma_e \rho_{ee} \tag{5.141}$$

$$\frac{d\rho_{ee}}{dt} = i\frac{\Omega_2}{2}(\rho_{ie} - \rho_{ei}) - \gamma_e \rho_{ee} \tag{5.142}$$

$$\frac{d\rho_{gi}}{dt} = -\left(i\Delta + \frac{\gamma_i}{2}\right)\rho_{gi} + i\frac{\Omega_1}{2}(\rho_{gg} - \rho_{ii}) + i\frac{\Omega_2}{2}\rho_{ge} \tag{5.143}$$

$$\frac{d\rho_{ge}}{dt} = -\left(i(\delta + 2\omega) + \frac{\gamma_e}{2}\right)\rho_{ge} + i\frac{\Omega_1}{2}\rho_{ie} - i\frac{\Omega_2}{2}\rho_{gi} \tag{5.144}$$

$$\frac{d\rho_{ie}}{dt} = -\left(i(\delta + 2\omega - \Delta) + \frac{\gamma_e + \gamma_i}{2}\right)\rho_{ie} + i\frac{\Omega_2}{2}(\rho_{ii} - \rho_{ee}) + i\frac{\Omega_1}{2}\rho_{ge}. \tag{5.145}$$

We approximate $d\rho_{i\alpha}/dt \approx 0$, meaning that the intermediate state's population and coherence reach a steady state rapidly. Using the equation for ρ_{gi}

$$\frac{d\rho_{gi}}{dt} = -\left(i\Delta + \frac{\gamma_i}{2}\right)\rho_{gi} + i\frac{\Omega_1}{2}(\rho_{gg} - \rho_{ii}) + i\frac{\Omega_2}{2}\rho_{ge} = 0. \tag{5.146}$$

Solving for ρ_{gi} we find

$$\rho_{gi} \approx \frac{i}{2}\frac{\Omega_1}{\Delta}(\rho_{gg} - \rho_{ii}) + \frac{i}{2}\frac{\Omega_2}{\Delta}\rho_{ge}. \tag{5.147}$$

Similarly, solving for ρ_{ie},

$$\frac{d\rho_{ie}}{dt} = -\left(i(\delta + 2\omega - \Delta) + \frac{\gamma_e + \gamma_i}{2}\right)\rho_{ie} + i\frac{\Omega_2}{2}(\rho_{ii} - \rho_{ee}) + i\frac{\Omega_1}{2}\rho_{ge} = 0. \tag{5.148}$$

Approximating $\delta + 2\omega - \Delta \approx -\Delta$, we solve for ρ_{ie}

$$\rho_{ie} \approx \frac{i}{2}\frac{\Omega_2}{\Delta}(\rho_{ii} - \rho_{ee}) + \frac{i}{2}\frac{\Omega_1}{\Delta}\rho_{ge}. \tag{5.149}$$

Neglecting small terms, this simplifies to

$$\rho_{ii} \approx \frac{\Omega_1^2}{\Delta^2}\rho_{gg} + \left(\frac{\Omega_2^2}{\Delta^2} + \frac{\gamma_e}{\gamma_i}\right)\rho_{ee}. \tag{5.150}$$

Putting this all together, we find

$$\frac{d\rho_{ee}}{dt} = -\gamma_e \rho_{ee} + i\frac{\Omega_2}{2}(\rho_{ie} - \rho_{ei}) \tag{5.151}$$

$$\frac{d\rho_{gg}}{dt} = \Gamma_g \rho_{gg} + \Gamma_e \rho_{ee} \tag{5.152}$$

$$\frac{d\rho_{ge}}{dt} = -\left(i(\delta + 2\omega) + \frac{\gamma_e}{2} + \frac{\Omega_1\Omega_2}{4\Delta}\right)\rho_{ge} - i\frac{\Omega_1\Omega_2}{4\Delta}(\rho_{gg} - \rho_{ee}), \tag{5.153}$$

with effective decay rates

$$\Gamma_g = \gamma_i \frac{\Omega_1^2}{\Delta^2}, \quad \Gamma_e = \gamma_i\left(\frac{\Omega_2^2}{\Delta^2} + \frac{\gamma_e}{\gamma_i}\right). \tag{5.154}$$

We again retrieve an effective two-level system. Before we stated that Δ was 'large', now we have seen that $\Delta > > \gamma_i$ is the condition for adiabatic elimination. After adiabatic elimination, the effective two-level system between $|g\rangle$ and $|e\rangle$ is described by

$$H_{\text{eff}} = \frac{\hbar\tilde{\delta}}{2}\sigma_z - \frac{\hbar\tilde{\Omega}}{2}\sigma_x, \tag{5.155}$$

where

$$\tilde{\delta} = (\delta + 2\omega) - \frac{\Omega_1^2 - \Omega_2^2}{4\Delta} \tag{5.156}$$

$$\tilde{\Omega} = \frac{\Omega_1\Omega_2}{2\Delta}. \tag{5.157}$$

This describes a two-level system with a modified detuning and effective coupling. The transition rate for two-photon absorption goes as I^2 as the effective Rabi frequency $\tilde{\Omega}_{2p}$ goes as the product of two fields, or the square if they are the same. This is dependent on the photon statistics of the source though. One defines an effective two-photon Rabi frequency. Taking the expectation value of the field operators, one defines

$$\Omega_{2p} = g\langle a^2\rangle. \tag{5.158}$$

For a given light state, the value of $\langle a^2\rangle$ determines the effective strength of the two-photon coupling. For a coherent state $|\alpha\rangle$ (with $\alpha = \sqrt{N}$ and N proportional to the intensity I), we have

$$\langle a^2\rangle = \alpha^2. \tag{5.159}$$

Thus,

$$\Omega_{2p}^{(\text{coh})} = g\alpha^2 \propto I. \tag{5.160}$$

Since the transition rate is proportional to $|\Omega_{2p}|^2$, the TPA rate becomes

$$R_{\text{TPA}}^{(\text{coh})} \propto \left| \Omega_{2p}^{(\text{coh})} \right|^2 \propto I^2. \tag{5.161}$$

For a squeezed vacuum state (which has $\langle a \rangle = 0$ but nonzero $\langle a^2 \rangle$ as we shall soon see), we obtain

$$\langle a^2 \rangle = \langle a^{\dagger 2} a^2 \rangle \approx 2\, N^2 + N = -\sinh r \cosh r e^{i\theta}, \tag{5.162}$$

where r is the squeezing parameter and $N = \sinh^2 r$ is the mean photon number (intensity proportional to N). In the low-intensity (small r) regime,

$$\sinh r \approx r, \quad \cosh r \approx 1, \tag{5.163}$$

so that

$$\langle a^2 \rangle \approx -r \quad \text{and} \quad |\langle a^2 \rangle|^2 \propto r^2 \propto N \propto I. \tag{5.164}$$

Hence, the effective two-photon Rabi frequency is

$$\Omega_{2p}^{(\text{sqz})} = g \langle a^2 \rangle, \tag{5.165}$$

and the corresponding transition rate becomes

$$R_{\text{TPA}}^{(\text{sqz})} \propto \left| \Omega_{2p}^{(\text{sqz})} \right|^2 \propto I. \tag{5.166}$$

For a thermal state, the two-photon absorption rate is enhanced by a factor related to the photon bunching. In fact, one typically finds

$$g^{(2)}(0) = \langle a^{\dagger 2} a^2 \rangle = \alpha \langle n \rangle^2, \tag{5.167}$$

so that if the TPA rate scales as

$$R_{\text{TPA}} \propto \alpha^2, \tag{5.168}$$

which is enhanced by a factor α but still quadratic in intensity and not linear as with squeezed light.

References

[1] Scully M O and Zubairy M S 1997 *Quantum Optics* (Cambridge: Cambridge University Press)

[2] Finkelstein R, Bali S, Firstenberg O and Novikova I 2023 A practical guide to electromagnetically induced transparency in atomic vapor *New J. Phys.* **25** 035001

[3] Milonni P W 2004 *Fast Light, Slow Light, and Left-Handed Light* (London: Routledge)

[4] Bajcsy M, Zibrov A S and Lukin M D 2003 Stationary pulses of light in an atomic medium *Nature* **426** 638–41

[5] Rice P R and Brecha R J 1996 Cavity induced transparency *Opt. Commun.* **126** 230–5

IOP Publishing

An Introduction to Quantum Optics (Second Edition)
An open systems approach
Perry Rice

Chapter 6

Quantum fields I

6.1 Maxwell equations again

Recall that these are

$$\vec{\nabla}\cdot\vec{E} = \frac{\rho}{\epsilon_0} \tag{6.1}$$

$$\vec{\nabla}\cdot\vec{B} = 0 \tag{6.2}$$

$$\vec{\nabla}\times\vec{E} = \frac{-\partial\vec{B}}{\partial t} \tag{6.3}$$

$$\vec{\nabla}\times\vec{B} = \mu_0\vec{J} + \frac{1}{c^2}\frac{\partial\vec{E}}{\partial t}. \tag{6.4}$$

In addition, we introduced the vector potential $\vec{A}$ via $\vec{B} = \vec{\nabla}\times\vec{A}$, and the scalar potential ϕ via $\vec{E} = -\vec{\nabla}\phi - \partial\vec{A}/\partial t$.

6.2 Quantization of the electromagnetic field

Let us now explicitly prove how the electromagnetic field becomes quantized [1, 2]. We start with the equation for the vector potential developed in chapter 3:

$$-\nabla^2\vec{A} + \frac{1}{c^2}\frac{\partial^2\vec{A}}{\partial t} = \mu_0\vec{J}_T. \tag{6.5}$$

doi:10.1088/978-0-7503-6014-2ch6

We consider charge/current free space where $\overrightarrow{J_T} = \rho = \phi = 0$. The wave equation for $\overrightarrow{A}$ is then

$$-\nabla^2 \overrightarrow{A} + \frac{1}{c^2}\frac{\partial^2 \overrightarrow{A}}{\partial t^2} = 0. \tag{6.6}$$

The solution to the wave equation for $\overrightarrow{A}$, using a plane wave expansion is

$$\overrightarrow{A} = \sum_{k,l}\left(A_{k,l}(0)e^{i(\overrightarrow{k}\cdot\overrightarrow{r}-\omega_k t)} + A^*_{k,\,l}(0)e^{-i(\overrightarrow{k}\cdot\overrightarrow{r}-\omega_k t)}\right)\epsilon_{k,l}, \tag{6.7}$$

where $k_x = 2\pi n_x/L$ and similarly for k_y and k_z. These are appropriate choices for box quantization, where we have assumed that the field must vanish at the boundaries of a cube of sides L. For a rectangular cavity, one would simply use different lengths for each k_i and, of course, other geometries are possible. In one's first course on electromagnetism, one usually solves Poisson's equations for a variety of geometries, and is introduced to the jungle of special functions. The sum over k is a sum over all directions in space, all possible frequencies $\omega_k = ck$, and all polarizations. Now, since $\overrightarrow{\nabla}\cdot\overrightarrow{A} = 0$, this means that $\overrightarrow{k}\cdot\overrightarrow{A_k} = \overrightarrow{k}\cdot\epsilon_{k,l} = 0$. The two orientations for the polarization are two unit vectors ϵ_{k1} and ϵ_{k2}, which together with $\overrightarrow{k}$ form a right-handed set of orthogonal vectors for each $\overrightarrow{k}$. As $\overrightarrow{A}$ is driven by the transverse part of the current density $\overrightarrow{J_T}$, we only have transverse solutions. Instead of plane waves, we could use an expansion in terms of cylindrical or spherical waves; we could also use complex polarization vectors $\epsilon_{R,L} = (1/\sqrt{2})(\epsilon_{k,1} \pm \epsilon_{k,2})$ for right and left circular polarizations.

The electric field is then given by

$$\begin{aligned}\overrightarrow{E} &= -\frac{\partial \overrightarrow{A}}{\partial t} \\ &= \sum_{k,l} - i\omega_k\left(A_{k,l}(0)e^{i(\overrightarrow{k}\cdot\overrightarrow{r}-\omega_k t)}A^*_{k,\,l}(0)e^{-i(\overrightarrow{k}\cdot\overrightarrow{r}-\omega t)}\right) \\ &= \sum_{k,i} E_{k,l}\epsilon_{k,i},\end{aligned} \tag{6.8}$$

with

$$E_{k,\,l} = -i\omega_k A_{k,\,l}(0)e^{i\left(\overrightarrow{k}\cdot\overrightarrow{r}-\omega_k t\right)} + \text{c. c.}, \tag{6.9}$$

where c.c. stands for the complex conjugate. The magnetic field is then given by

$$\overrightarrow{B} = \sum_{k,i} B_{k,l}(k \times \epsilon_{k,l}) \tag{6.10}$$

with

$$B_{k,\,l} = ikA_{k,\,l}e^{i\left(\overrightarrow{k}\cdot\overrightarrow{r}-\omega_k t\right)} + \text{c. c.} \tag{6.11}$$

We see that $|B_k| = \frac{1}{c}|E_k|$, so the energy of the kth mode time averaged (denoted by the bar) is

$$\overline{\varepsilon_k} = \frac{1}{2}\int\left(\epsilon_0 \bar{E}_k^2 + \frac{1}{\mu_0}\bar{B}_k^2\right)dV$$
$$= 2\epsilon_0 V\omega_k^2 \overrightarrow{A_k}\cdot\overrightarrow{A}_k^*. \tag{6.12}$$

Now, we transform the above equations into a harmonic oscillator with associated transverse polarization vectors $\overrightarrow{\epsilon_k}$ so that the vector potential and its complex conjugate can be written as

$$\overrightarrow{A_{k,l}} = \left(4\pi\epsilon_0 V\omega_k^2\right)^{1/2}(\omega_k Q_{k,l} + iP_{k,l})\epsilon_{k,l} \tag{6.13}$$

$$\overrightarrow{A}^*_{k,\,l} = \left(4\pi\epsilon_0 V\omega_k^2\right)^{1/2}(\omega_k Q_{k,l} - iP_{k,l})\epsilon_{k,l}. \tag{6.14}$$

The time average of the energy of a given mode is then

$$\overline{\epsilon_{k,l}} = \frac{1}{2}\left(P_{k,\,l}^2 + \omega_{k},^2\, l2\right). \tag{6.15}$$

For a quantum harmonic oscillator, the Hamiltonian is

$$H = \frac{1}{2}(p^2 + \omega^2 q^2) \tag{6.16}$$

and we know that the appropriate commutation relation is

$$[q, p] = -i\hbar. \tag{6.17}$$

We define a set of operators that obey the following:

$$a = (2\hbar\omega)^{1/2}(\omega q + ip) \tag{6.18}$$

$$a^\dagger = (2\hbar\omega)^{-1/2}(\omega q ip) \tag{6.19}$$

$$q = \left(\frac{\hbar}{2\omega}\right)^{1/2}(a + a^\dagger) \tag{6.20}$$

$$p = i\left(\frac{\hbar\omega}{2}\right)^{1/2}(a - a^\dagger), \tag{6.21}$$

which satisfy the commutation relation

$$[a, a^\dagger] = 1. \tag{6.22}$$

Considering the product $a^\dagger a$, we find that

$$a^\dagger a = \frac{1}{2\hbar\omega}(p^2 + \omega^2 q^2 + i\omega qp - i\omega pq)$$
$$= \frac{1}{\hbar\omega}\left(H - \frac{1}{2}\hbar\omega\right), \tag{6.23}$$

which leaves us with

$$H = \hbar\omega\left(a^\dagger a + \frac{1}{2}\right). \tag{6.24}$$

Let us consider the energy eigenstates, or number states, also known as Fock states, $|n\rangle$. Here, the number refers to the number of photons in the mode

$$\begin{aligned} H|n\rangle &= E_n|n\rangle \\ &= \left(a^\dagger a + \frac{1}{2}\right)\hbar\omega|n\rangle. \end{aligned} \tag{6.25}$$

What is the effect of the operator $a^\dagger$ on eigenstates $|n\rangle$. We act on $a^\dagger|n\rangle$ with H,

$$\begin{aligned} \hbar\omega\left(a^\dagger a a^\dagger + \frac{1}{2}a^\dagger\right)|n\rangle &= \hbar\omega\left(a^\dagger a^\dagger a + a^\dagger + \frac{1}{2}a^\dagger\right)|n\rangle \\ &= \hbar\omega a^\dagger\left(a^\dagger a + \frac{3}{2}\right)r|n\rangle \\ &= a^\dagger(H + \hbar\omega)|n\rangle \\ &= (E_n + \hbar\omega)a^\dagger|n\rangle. \end{aligned} \tag{6.26}$$

We see that $a^\dagger|n\rangle$ is also an energy eigenstate with $E = E_n + \hbar\omega$ This means that $a^\dagger$ raises the state to an energy level $\hbar\omega$ higher, and we refer to $a^\dagger$ as a raising operator. Likewise, it can be easily shown a is a lowering operator represented by

$$H(a|n\rangle) = (E_n - \hbar\omega)(a|n\rangle). \tag{6.27}$$

As the lowest state is defined to be $|0\rangle$, we know that $a|0\rangle = 0$, so that we can determine the ground state energy:

$$\begin{aligned} H|0\rangle &= \hbar\omega\left(a^\dagger a + \frac{1}{2}\right)|0\rangle \\ &= \frac{1}{2}\hbar\omega|0\rangle \\ &= E_0|0\rangle. \end{aligned} \tag{6.28}$$

This yields the ground state energy $E_0 = \hbar\omega/2$. Another state, $|1\rangle$, with energy $E_1 = 3\hbar\omega/2$ is obtained by raising the ground state. Are there any states between $|0\rangle$ and $|1\rangle$? Our notation would suggest not, and indeed there is none. If there was a state with energy $\hbar\omega/2 < E < 3\hbar\omega/2$, we could lower it and obtain a state with energy below that of the ground state. Since by construction the ground state is the state of lowest energy, we would have a contradiction. Therefore, we know that our energy eigenstates are equally spaced, with energies $E_n = \hbar\omega(n + 1/2)$. We note that the ground state energy is not zero; the energy $E_0 = \hbar\omega/2$ is referred to as the zero-point energy.

Let us now look at exactly how the raising and lowering operators act on our number states:

$$\begin{aligned} a|n\rangle &= C_n|n-1\rangle \\ \langle n|a^\dagger &= \langle n-1|C_n^* \\ \langle n|a^\dagger a|n\rangle &= |C_n|^2\langle n-1|n-1\rangle \\ &= |C_n|^2 \\ &= n \\ C_n &= \sqrt{n}. \end{aligned} \tag{6.29}$$

Similarly, we find that

$$a^\dagger|n\rangle = \sqrt{n+1}\,|n+1\rangle. \tag{6.30}$$

So, how do we quantize the field? We simply let the momentum P_k, which was originally a scalar, goes to p_k which is an operator, as well as replacing Q_k by q_k. Then, $A_k \rightarrow a_{k,l}$ and $A^*_{k,\,l} \rightarrow a^\dagger_{k,\,l}$.

This gives us the quantized electric and magnetic fields that are still given by the sums of equations (6.8) and (6.10) with mode coefficients that are operators,

$$E_{k,\,l} = -i\left(\frac{\hbar\omega_k}{2\epsilon_0 V}\right)^{1/2}\left[a_{k,\,l}e^{i\left(\vec{k}\cdot\vec{r}-\omega t\right)} - a^\dagger_{k,\,l}e^{-i\left(\vec{k}\cdot\vec{r}-\omega t\right)}\right] \tag{6.31}$$

$$B_{k,\,l} = \frac{i}{c}\left(\frac{\hbar\omega_k}{2\epsilon_0 V}\right)^{1/2}\left[a_{k,\,l}e^{i\left(\vec{k}\cdot\vec{r}-\omega t\right)} - a^\dagger_{k,\,l}e^{-i\left(\vec{k}\cdot\vec{r}-\omega t\right)}\right]. \tag{6.32}$$

6.3 Single-mode quantized fields

Let us focus on just one mode of the electric field for now. How can I create a situation in the laboratory where the use of a single mode is appropriate? In general, one needs to consider waves propagating in all directions in space, at all wavelengths/frequencies, and two polarizations. The simplest method to reduce the number of modes is to use a pair of mirrors to make an optical cavity. For mirrors made from a perfect conductor, the boundary conditions are such that the field must vanish at the mirrors. As we stated before, this gives rise to a quantization of the wavenumbers for the modes in the cavity, $k_n = 2\pi nc/L$, that denote what we refer to as *longitudinal modes*. These have a spatial dependence of $U_n = \cos(kz)$ or $\sin(kz)$, where we have taken the z axis perpendicular to the cavity mirrors. The transverse $x-y$ dependence is determined by a solution to Maxwell's equations, and various transverse modes emerge, typically the transverse-electromagnetic (TEM) modes labeled TEM_{ij}. The structure of these modes for a rectangular cavity is proportional to the Hermite polynomials. The losses for the higher-order modes are typically larger than those of the TEM_{00}, or that can be arranged with an intracavity element such as a pinhole. In addition, a polarizing element may be placed in the cavity to allow only one polarization to propagate. The mirrors are typically curved for no other reason than to facilitate alignment and to keep the beam from walking out of the cavity. The $1/e$ points of the transverse beam profile define the waist of the beam. If we place atoms in our cavity that have a linewidth that is small compared to the spacing between the longitudinal modes, we can restrict ourselves to the consideration of a single mode. Of course, there are modes that propagate parallel to the mirrors, which an atom might spontaneously emit into. This can be suppressed if we have a cavity structure in those directions that is shorter than $\lambda/2$. There are various other geometries one can consider such as a microdisk. Here the thickness is less than $\lambda/2$ which will suppress coupling to modes perpendicular to the disk, and the total internal reflection confines the modes that propagate around the disk, the so-called whispering gallery modes. The grazing incidence of these waves and the index mismatch between the semiconductor and air can lead to relatively large reflectivities, and hence to the cavity quality factors. However, for any cavity, some light is transmitted. Thus, our field couples to other modes, and we will consider this in more detail when we consider open quantum systems in chapter 10.

For our mathematical representation of a single-mode field, we take

$$E_{k,l} \to E = \mathcal{E}_0(ae^{i\theta} + a^\dagger e^{-i\theta}), \tag{6.33}$$

where we have absorbed a factor of $i = e^{i\pi/2}$ into θ for convenience, and $\mathcal{E}_0 = \sqrt{\hbar\omega/2\epsilon_0 V}$, where V is the volume of the cavity mode. Absorbing the factor of i is not a problem as the absolute phase is not measurable, only the relative phases. The cavity mode volume is formally defined by setting the energy stored in the field equal to $\epsilon_0|E_{\text{max}}|^2 V_{\text{eff}}$, or

$$V_{\text{eff}} = \frac{1}{|E_{\text{max}}|^2}\int |\vec{E}(\vec{r})|^2 dV. \tag{6.34}$$

For a sinusoidal mode with a Gaussian profile,

$$E = E_{\text{max}} \sin(kz) e^{\frac{-(x^2+y^2)}{w_0^2}}, \tag{6.35}$$

where w_0 is the beam waist, and the effective cavity mode volume is

$$V = \frac{\pi w_0^2}{4} L, \tag{6.36}$$

where we have assumed that the beam waist is constant (often it is not, due to the hourglass shape of some modes), and that the field vanishes at the mirror's spaced a length L apart. In a material, one must take the index of refraction into account, $V \to V/n^3$, or one must include the spatial dependence of the dielectric constant in a more complicated cavity, perhaps composed of a photonic band gap material. More problematic may be that most mirrors are not perfect, or even imperfect, conductors where the field nearly vanishes at the mirror. Most real mirrors that are used to obtain large reflectivities use interference effects to obtain large reflectivities at a particular narrow range of frequencies. These mirrors are alternating layers of a dielectric, and the field mode can penetrate into that dielectric stack. Therefore, one must be more careful not to use the above formula incautiously in that type of situation. Generally, one must model the mirror, and solve for the true classical modes, and then integrate that up to determine the effective volume. If for a cavity of dielectric mirrors one uses the formula for conducting mirrors and uses the spacing between the top layers of the stack, one will be using a volume that is too small, which results in overestimating the electric field per photon.

Often, it is useful to consider the field written as a sum of quadratures,

$$E = \mathcal{E}_0(X\cos(\theta) + Y\sin(\theta)), \tag{6.37}$$

where the quadrature operators are defined by

$$X = \frac{1}{2}(a^\dagger + a) \tag{6.38}$$

$$Y = \frac{i}{2}(a^\dagger - a). \tag{6.39}$$

The energy per volume in this mode is given by

$$\bar{\varepsilon} = \frac{\epsilon \mathcal{E}_0^2}{2}(X^2 + Y^2) \tag{6.40}$$

and so it is appropriate to refer to $\mathcal{E}_0$ as the electric field per photon. In some small cavities of a micron in length, $\mathcal{E}_0$ can be 1000–10 000 V m^{-1}, which is similar to the electric field across an AC power outlet, 100 V dropped over roughly a centimeter. Thus, the presence or absence of a single photon can change the electric field in a cavity by a significant amount, which is definitely macroscopic and measurable.

Recall that the Hamiltonian for an electromagnetic field mode resembles that of a harmonic oscillator,

$$\frac{p^2}{2m} + \frac{kx^2}{2} \leftrightarrow \frac{\epsilon E^2}{2} + \frac{B^2}{2\mu_0}. \tag{6.41}$$

For the harmonic oscillator, we have an uncertainty principle

$$\begin{aligned} \Delta x \Delta p &\geqslant \frac{1}{2}|\langle [x, p] \rangle| \\ &\geqslant \frac{\hbar}{2}. \end{aligned} \tag{6.42}$$

Apparently there might be an uncertainty principle between the electric and magnetic fields. How does this square with the fact that for a plane wave mode, $|B| = |E|/c$? We must recall that when we quantize classical objects and promote a classical c-number to an operator (say q), we can identify the variable conjugate to q, which we denote by p. Then, we have the relation $[q, p] = i\hbar$. In the case of the electromagnetic fields, E and B are not conjugate variables, but E and A are, recall that $E \propto \dot{A}$ as $p \propto \dot{x}$. These are both functions of x, and as such their commutation relation is of the form

$$[E(x), A(x')] \propto \delta_T(x - x'), \tag{6.43}$$

where $\delta_T(x - x')$ is a transverse delta function related to the transverse nature of the vector potential. Let us continue to work with the quadrature operators X and Y. These operators have a commutation relation given by

$$[X, Y] = \frac{-i}{2} \tag{6.44}$$

which leads us to the uncertainty relation

$$\Delta X \Delta Y \geqslant \frac{1}{4}. \tag{6.45}$$

For a classical field, X and Y could take on a well-defined value, effectively a point in a type of phase space. For example, a field given by $E = X\cos(\omega t) + Y\sin(\omega t)$ would be described as a phasor, with well-defined phase and amplitude. For a quantum field, we must replace that phasor by a phasor for the mean field plus a type of quantum 'fuzzball'. For now the 'fuzzball' is a schematic representation of the uncertainty, or (we will use the term noise) in the state. This 'fuzzball' takes on a formal meaning when we discuss the quasi-probability amplitudes in chapter 12. There we will see that we can perform quantum optics in a manner similar to classical statistical mechanics in phase space, using probability distributions. However, these distributions will have some bizarre properties, such as being allowed to be negative. The quantum weirdness must show up somehow!

6.4 Number states

In the number state or Fock state, the expectation value of the quadrature amplitudes is zero,

$$\langle n|(a + a^\dagger)|n\rangle = \sqrt{n+1}(\langle n+1|n\rangle + \text{h.c.}) = 0. \tag{6.46}$$

We can easily show the following with the properties of the raising and lowering operators:

$$\langle X\rangle = 0 \tag{6.47}$$

$$\langle X^2\rangle = (1/2)(n + 1/2) \tag{6.48}$$

$$\langle Y\rangle = 0 \tag{6.49}$$

$$\langle Y^2\rangle = (1/2)(n + 1/2), \tag{6.50}$$

which leads to

$$\Delta X \Delta Y = (1/2)(n + 1/2) \geqslant 1/4, \tag{6.51}$$

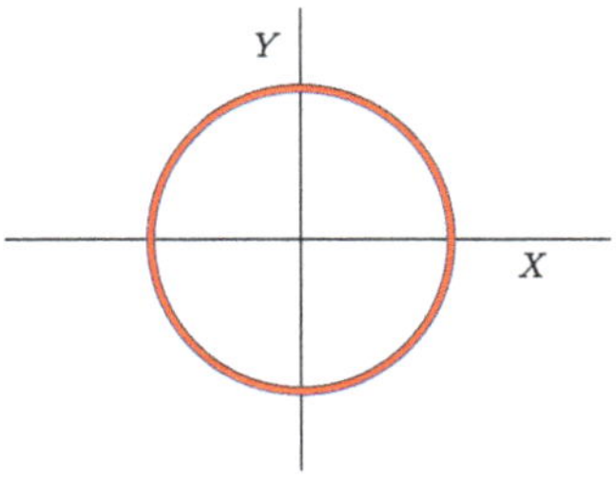

Figure 6.1. Cartoon of the quadrature distribution of a Fock state.

where the minimum uncertainty occurs for the vacuum state $|0\rangle$. The ground state of a harmonic oscillator is a Gaussian in the position representation, which is a minimum uncertainty state. Pictorially, we may illustrate the number state in figure 6.1, where it seems that the amplitude of the field is perfectly well known, and the phase is completely unknown.

Also, as we have

$$X^2 + Y^2 = a^\dagger a + 1/2 = N + 1/2, \tag{6.52}$$

where we use the notation $N \equiv a^\dagger a$ to distinguish it from its eigenvalues n, the number of photons in the Fock state. The Fock states obviously satisfy

$$N|n\rangle = n|n\rangle \tag{6.53}$$

$$N^2|n\rangle = n^2|n\rangle, \tag{6.54}$$

which then leads to $\Delta N = 0$, and of course the photon number is perfectly known.

This leads us to consider the question of amplitude and phase operators. Can we write the electric field as

$$E = A\exp(i\phi) \tag{6.55}$$

with A and ϕ Hermitian operators that can then be measured? The answer is no. Dirac initially considered operators of this form where $A = \sqrt{N}$, and the commutation relations for these operators are

$$[e^{i\phi}, N] = e^{i\phi}. \tag{6.56}$$

This would be satisfied if we have

$$[N, \phi] = \frac{i}{2} \tag{6.57}$$

which would yield an uncertainty relation

$$\Delta N \Delta\phi \geqslant \frac{1}{2}. \tag{6.58}$$

So far this seems OK, but what if we consider

$$\langle n|[N, \phi]|n'\rangle = i\delta_{n,n'}, \tag{6.59}$$

but if we let the number operator N act on the states, we find

$$(n - n')\langle n|\phi|n'\rangle = i\delta_{n,n'}, \tag{6.60}$$

which leads to a contradiction for $n = n'$, unless the matrix elements of the phase operator in the number basis do not exist, leading to the conclusion that ϕ can not be a well-defined Hermitian operator. One can also show that the following relations hold:

$$(e^{i\phi})(e^{\text{ß}\phi})^\dagger = 1 \tag{6.61}$$

$$(e^{\text{ß}\phi})^\dagger(e^{i\phi}) \neq 1. \tag{6.62}$$

If ϕ were a Hermitian operator, $e^{i\phi}$ should be unitary, but it is not.

The failure to define a Hermitian phase operator stems from the difficulty of defining a phase for the ground state $|0\rangle$, which (at our level of rigor so far) is described by a circular 'fuzzball' around the origin [3]. The notion of a 'fuzzball' will be formalized in chapter 12, where we discuss the quasi-probability distributions. Also, as for an oscillating field we have $\phi = \omega t$, which would mean that if we *could* define a phase operator, we would also have a time operator. In nonrelativistic quantum mechanics, time is merely a parameter. More sophisticated ways to deal with time in quantum mechanics must and do show up in a relativistic formulation. There are various versions of amplitude and phase operators that work on some subset of states, even all states except the vacuum state, or operators that can be used to construct a probability distribution for a phase that has the correct properties, but there is no rigorous amplitude/phase uncertainty relation. Susskind and Glogower [4] considered operators of the form

$$e = \widehat{e^{i\phi}} = (N + 1)^{-1/2}a \tag{6.63}$$

$$e^\dagger = \widehat{e^{-i\phi}} = a^\dagger(N + 1)^{-1/2}. \tag{6.64}$$

These new operators are not unitary either as

$$e^\dagger e = 1 - |0\rangle\langle 0| \tag{6.65}$$

$$ee^\dagger = 1, \tag{6.66}$$

but there are useful operators

$$C = \frac{1}{2}(e + e^\dagger) \tag{6.67}$$

$$S = \frac{1}{2i}(e - e^\dagger) \tag{6.68}$$

with commutators

$$[C, N] = iS \tag{6.69}$$

$$[S, N] = -iC, \tag{6.70}$$

which leads to uncertainties

$$\Delta N \Delta C \geqslant \frac{1}{2} |\langle S \rangle| \tag{6.71}$$

$$\Delta N \Delta S \geqslant \frac{1}{2} |\langle C \rangle|. \tag{6.72}$$

More recent work by Pegg and Barnett [5], Mandle [6], and others [7], have helped describe the problem of the phase operator and phase distributions, with a focus on measurable quantities. A cartoon of the quadrature distribution of a Fock state is shown in figure 6.1.

The photon number distribution for a Fock state is rather boring, as we have a unitary probability for $n = N_0$, and zero otherwise, as shown in figure 6.2.

The time evolution of the photon number is shown in figure 6.3. The blue curve denotes the expectation value, while the red shaded region denotes the variance.

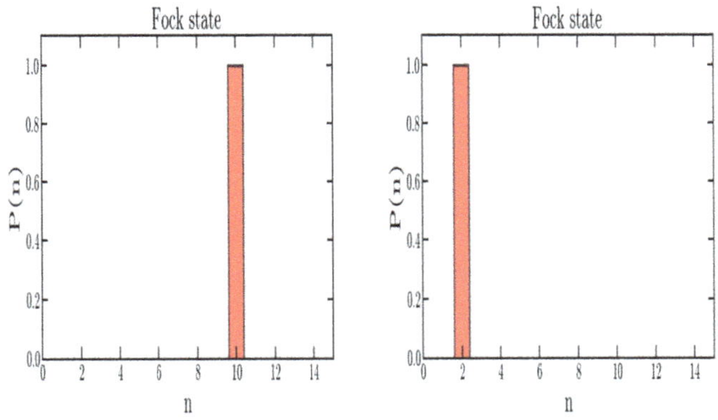

Figure 6.2. Photon number probability distribution for Fock states $|n = 2\rangle$ and $|n = 10\rangle$.

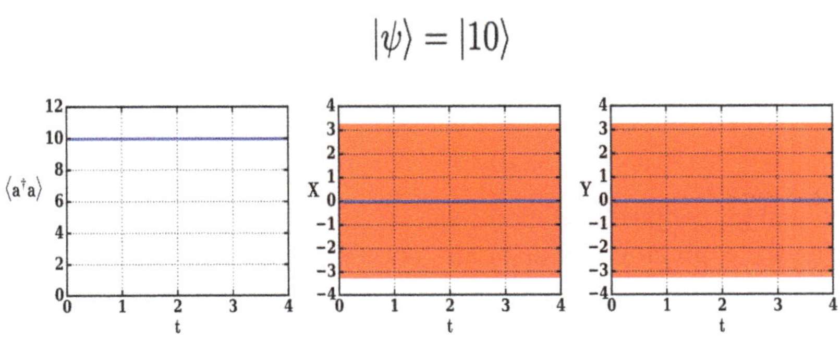

Figure 6.3. Time evolution of the average (blue curve) and variance (red shaded region) of a number state $|n = 10\rangle$.

There is no time dependence other than the factor $e^{-in\omega_0 t}$ which has a magnitude of unity.

Finally, we consider generating Fock states from the ground state. Using the properties of the raising operator, we have

$$\begin{aligned} |1\rangle &= a^\dagger|0\rangle \\ |2\rangle &= \frac{a^\dagger}{\sqrt{2}}|1\rangle \\ |3\rangle &= \frac{a^\dagger}{\sqrt{3}}|2\rangle . \end{aligned} \tag{6.73}$$

We can see then how we can form

$$|n\rangle = \frac{a^{\dagger n}}{\sqrt{n!}}|0\rangle . \tag{6.74}$$

Because any state can be decomposed into a superposition of Fock states, we can find an operator that generates any state from the ground state,

$$\begin{aligned} |\psi\rangle &= \sum_n C_n|n\rangle \\ &= \sum_n \frac{C_n a^{\dagger n}}{\sqrt{n!}}|0\rangle \\ &= S(C_n)|0\rangle , \end{aligned} \tag{6.75}$$

where we define the state-generating operator

$$S(C_n) = \sum_n \frac{C_n a^{\dagger n}}{\sqrt{n!}} . \tag{6.76}$$

6.5 Coherent states

We begin by finding states that have a well-defined expectation value of the electric field that might be related to a classical field. Consider $\langle E\rangle = \langle (a^\dagger + a)\rangle$. If we could find a state of the field that was an eigenstate of the annihilation operator, $a|\alpha\rangle = \alpha|\alpha\rangle$, we would have

$$\langle 2X\rangle = \langle (a^\dagger + a)\rangle = (\alpha + \alpha^*) = 2\,\mathrm{Re}(\alpha) \tag{6.77}$$

$$\langle 2Y\rangle = i\langle (a^\dagger - a)\rangle = i(\alpha^* - \alpha) = 2\,\mathrm{Im}(\alpha), \tag{6.78}$$

where we used the conjugate relation $\langle\alpha|a^\dagger = \langle\alpha|\alpha^*$. This state is even more special, considering the expectation value of any normally ordered product of a and $a^\dagger$,

$$\langle\alpha|a^{\dagger n}a^m|\alpha\rangle = \alpha^{*n}\alpha^m . \tag{6.79}$$

Normal ordering refers to having all annihilation operators on the right and all creation operators on the left. apparently for these types of products, one can replace the operator a with α, and $a^\dagger$ with α^*, replacing an operator with a c number. Thus, it seems that there may be some relation with a classical field. Any polynomial function of a and $a^\dagger$ can be put in the form of a sum of normally ordered products using the commutation relation $[a, a^\dagger] = 1$; therefore, we can always use these states, called coherent states, to calculate all quantities of interest. For example, the quantity $aa^\dagger + a^\dagger a = 2aa^\dagger a + 1$, and so for a coherent state we would have

$$\langle aa^\dagger + a^\dagger a\rangle = \langle 2a^\dagger a + 1\rangle = 2|\alpha|^2 + 1. \tag{6.80}$$

Of course, for a classical field, we would not need the various extra factors due to the non-commutative nature of a and $a^\dagger$. Therefore, while one can always use the coherent states for calculation, it is not always the case that operators are replaced by c numbers; rather, one must use the commutation relations first. Therefore, the quantum nature of the field can still 'peek' through.

How do we find these coherent states? Consider the expansion

$$|\alpha\rangle = \sum_n C_n|n\rangle. \tag{6.81}$$

Using the definition of the coherent state, we have

$$\begin{aligned} a|\alpha\rangle &= \sum_n C_n a|n\rangle \\ &= \sum_n C_n\sqrt{n}\,|n-1\rangle \\ &= \sum_n \alpha C_n|n\rangle. \end{aligned} \tag{6.82}$$

Performing an index shift, we find

$$C_{n+1}\sqrt{n+1} = \alpha C_n, \tag{6.83}$$

which leads to

$$|\alpha\rangle = C_0\sum_n \frac{\alpha^n}{\sqrt{n!}}|n\rangle. \tag{6.84}$$

The overall factor C_0 is determined by requiring that the coherent states be normalized

$$\begin{aligned} \langle\alpha|\alpha\rangle &= |C_0|^2\sum_{n,m}\frac{\alpha^n\alpha^{*m}}{\sqrt{n!m!}}\langle n|m\rangle \\ &= |C_0|^2\sum_n\frac{|\alpha|^{2n}}{n!} \\ &= |C_0|^2\exp(|\alpha|^2) \\ &= 1, \end{aligned} \tag{6.85}$$

which gives us $C_0 = \exp(-|\alpha|^2/2)$ where we have absorbed an arbitrary phase and chosen C_0 to be real. The annihilation operator a is not Hermitian, so it is not guaranteed to have real eigenvalues or orthogonal eigenstates. The coherent states are nonorthogonal states characterized by a complex number α, and form an overcomplete set.

What about the fluctuations in the coherent state? If we define $\alpha = |\alpha|e^{i\phi}$, we have

$$\langle X\rangle = |\alpha|\cos(\theta) \tag{6.86}$$

$$\langle Y\rangle = |\alpha|\sin(\theta) \tag{6.87}$$

$$\langle X^2\rangle = \frac{1}{4} + |\alpha|^2\cos^2(\theta) \tag{6.88}$$

$$\langle Y^2\rangle = \frac{1}{4} + |\alpha|^2\sin^2(\theta) \tag{6.89}$$

$$\langle N\rangle = |\alpha|^2 \tag{6.90}$$

$$\langle N^2\rangle = |\alpha|^4 + |\alpha|^2 \tag{6.91}$$

$$\Delta N = |\alpha| \tag{6.92}$$

$$\Delta X = \frac{1}{2} \tag{6.93}$$

$$\Delta Y = \frac{1}{2}. \tag{6.94}$$

There are various interesting results here. The first is that the coherent state is a minimum uncertainty state satisfying $\Delta X \Delta Y = 1/4$. The second is that the fractional uncertainty in the photon number is

$$\frac{\Delta N}{\langle N\rangle} = \frac{1}{\sqrt{\alpha}}. \tag{6.95}$$

Thus, by increasing the value of $|\alpha|$ we can reduce the fractional uncertainty in amplitude to an arbitrarily small number. Also, just via the relation for arc length, we can write down $\Delta\phi \approx 1/(2|\alpha|) = 1/2\bar{n}$ which is only valid for $|\alpha| \gg 1$. As in our general discussion of a phase operator, we know that problems with the definitions of a phase operator and the accompanying uncertainty relations arise when the average field amplitude is small.

If a measurement of the photon number is made on a coherent state, the probability of obtaining the result n is just

$$P_n = |C_n|^2 = e^{-|\alpha|^2/2}\frac{\alpha^n}{\sqrt{n!}}, \tag{6.96}$$

which is a Poisson distribution. We could have surmised this, as for a coherent state the variance $(\Delta N)^2 = \langle N \rangle$, that is, the variance is equal to the mean, which is the hallmark of the Poisson distribution. However, we need to look at the entire distribution as we have done to confirm a suspicion that the photon number distribution is Poissonian in nature. A fuzzball cartoon of the quadrature distribution of a coherent state is shown in figure 6.4. We see that the amplitude fluctuation is on the order of $\sqrt{\langle n \rangle} = \alpha$. There is a notion of phase, with a variance of $\Delta\phi = 1/|\alpha|$, coherent states are useful for interferometry, with a phase uncertainty that decreases the larger the amplitude $|\alpha|$, making interferometry possible with strong classical lasers. The quantum noise limit for a coherent state is a limiting factor in metrology, particularly in gravitational wave detection.

In figure 6.5 we reproduce the cartoon from figure 6.4, but note the quadrature uncertainties. Unlike the phase uncertainty, which is heuristic, these are well-defined quantum variances.

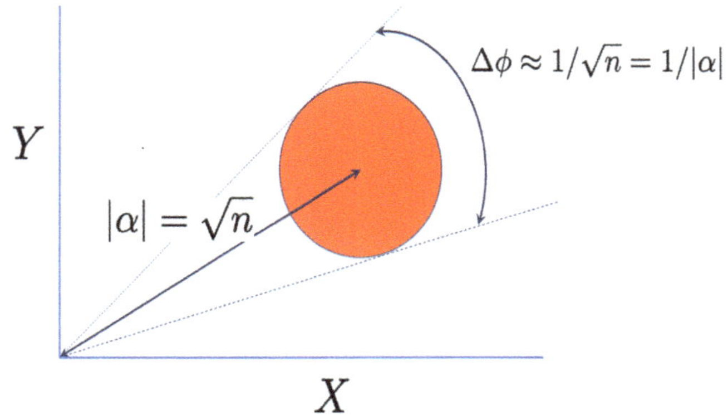

Figure 6.4. Cartoon of the quadrature distribution of a coherent state, noting phase and amplitude uncertainties.

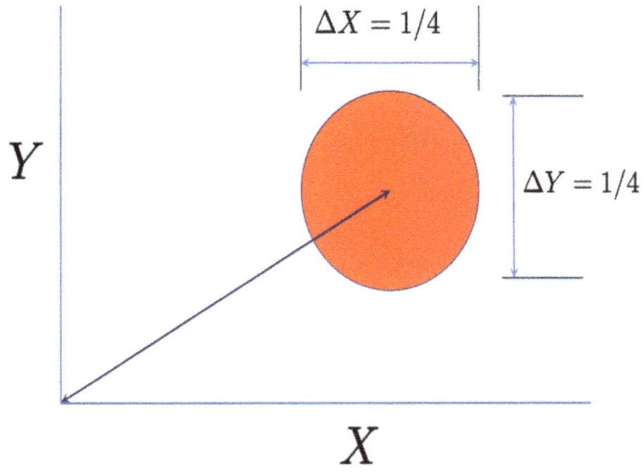

Figure 6.5. Cartoon of the quadrature distribution of a coherent state, noting the X and Y quadrature uncertainties.

We have seen that, at least formally, we can generate any field state by applying an operator onto the ground state. What is that operator for the coherent state? We have

$$\begin{aligned} |\alpha\rangle &= \sum_n e^{-|\alpha|^n} \frac{\alpha^2 a^{\dagger n}}{n!} |0\rangle \\ &= e^{-|\alpha|^2} e^{\alpha a^\dagger} |0\rangle . \end{aligned} \tag{6.97}$$

As the action of the lowering operator on the ground state yields zero, so does any function of the lowering operator. We can use $e^{-\alpha * a}|0\rangle = |0\rangle$, which leads to

$$\begin{aligned} |\alpha\rangle &= e^{-|\alpha|^2} e^{\alpha a^\dagger} e^{-\alpha * a} |0\rangle \\ &= e^{\alpha a^\dagger - \alpha * a} |0\rangle \\ &= D(\alpha)|0\rangle , \end{aligned} \tag{6.98}$$

where we define the displacement operator $D(\alpha)$. We have had to be careful with commutation relations here, recall that for operators

$$\begin{aligned} e^A e^B &= (1 + A + \frac{1}{2}A^2 + \dots)(1 + B + \frac{1}{2}B^2 + \dots) \\ &= 1 + (A + B) + \frac{1}{2}(A^2 + B^2 + 2AB) + \dots \end{aligned} \tag{6.99}$$

while we also have

$$e^{A+B} = 1 + (A + B) + \frac{1}{2}(A^2 + B^2 + AB + BA) + \dots , \tag{6.100}$$

which shows that for operators $\exp(A)\exp(B) \neq \exp(A + B)$. If the commutator of A and B commutes with both operators, that is if $[A, [A, B]] = [B, [A, B]] = 0$, then a version of the Baker–Campbell–Hausdorff theorem yields

$$e^A e^B = e^{A+B+\frac{1}{2}[A, B]}. \tag{6.101}$$

We used this to write the displacement operator in the form

$$D(\alpha) = e^{-|\alpha|^2} e^{\alpha a^\dagger} e^{-\alpha * a} = e^{\alpha a^\dagger - \alpha * a}. \tag{6.102}$$

As $D(\alpha)$ is an exponentiation of i times a Hermitian operator, it is then a unitary operator, $D^\dagger(\alpha)D(\alpha) = D(\alpha)D^\dagger(\alpha) = 1$. Let us write $D(\alpha)$ not in terms of creation and annihilation operators, but in terms of position and momentum operators, for real values of α,

$$\begin{aligned} D(\alpha) &= \exp\left(i\alpha \left(\frac{2}{\hbar\omega} \right)^{1/2} p \right) \\ &= \exp\left(-\alpha \left(\frac{2\hbar}{\omega} \right) \frac{d}{dx} \right) \\ &= \exp\left(x_\alpha \frac{d}{dx} \right), \end{aligned} \tag{6.103}$$

where we have gone to the position representation and defined a displacement distance x_0. Now, consider the form of a wave function $\psi(x)$ by some amount x_α, using a Taylor series expansion,

$$\begin{aligned}\psi &= \psi(x) + x_\alpha \left.\frac{d\psi}{dx}\right|_x + \frac{1}{2}x_\alpha^2 \left.\frac{d^2\psi}{dx^2}\right|_x + \ldots \\ &= \exp\left(x_\alpha \frac{d}{dx}\right)\psi(x),\end{aligned} \tag{6.104}$$

where we have summed the series. Therefore, it seems that the displacement operator exactly does that it picks up the wave function and moves it over to a new position. The state that is displaced in this case to form a coherent state is the ground or vacuum state. Thus, we can see that the fact that the 'noise' in the coherent state is the same as that in the vacuum state can be more easily explained.

We can also see how the displacement operator acts on the operators,

$$D^\dagger(\alpha) a D(\alpha) = a + \alpha \tag{6.105}$$

$$D^\dagger(\alpha) a^\dagger D(\alpha) = a^\dagger + \alpha^*. \tag{6.106}$$

Similarly, we have

$$D^\dagger(\alpha) X D(\alpha) = X + \mathrm{Re}(\alpha) \tag{6.107}$$

$$D^\dagger(\alpha) Y D(\alpha) = Y + \mathrm{Im}(\alpha). \tag{6.108}$$

These relations, when used when acting on the vacuum state, yield all the formulas for the averages and fluctuations we have derived earlier. For example,

$$\begin{aligned}\langle \alpha | N | \alpha \rangle &= \langle 0 | D^\dagger(\alpha) N D(\alpha) | 0 \rangle \\ &= \langle 0 | D^\dagger(\alpha) a^\dagger D(\alpha) D^\dagger(\alpha) a D(\alpha) | 0 \rangle \\ &= \langle 0 | |\alpha|^2 + \alpha a^\dagger + \alpha^* a + a^\dagger a | 0 \rangle \\ &= |\alpha|^2.\end{aligned} \tag{6.109}$$

What about the time dependence of the coherent state? The time dependence of a number state is just $|n(t)\rangle = \exp(-iHt/\hbar)|n(0)\rangle = \exp(-i(n + 1/2)\omega t)|n(0)\rangle$ The time dependence of the coherent state is then

$$\begin{aligned}|\alpha(t)\rangle &= e^{-i\omega t/2} \sum_n \frac{\alpha^n e^{-in\omega t}}{\sqrt{n!}} |n(0)\rangle \\ &= \sum_n \frac{(\alpha e^{-i\omega t})^n}{\sqrt{n!}} |0\rangle \\ &= \sum_n \frac{\alpha(t)}{\sqrt{n!}} |0\rangle,\end{aligned} \tag{6.110}$$

where we have suppressed a time-dependent but overall phase. Thus, we see that the time dependence of the coherent state is merely the replacement of $\alpha(t) = \alpha(0)e^{-i\omega t}$, a coherent state is ground state quantum noise superimposed over a classical-like phasor. In terms of a mechanical oscillator in a quadratic potential well, the coherent state is a Gaussian of ground state width that merely oscillates back and forth in the well, without spreading. In many ways, the coherent state is the most classical-like state of the oscillator, where the classical mass oscillating as $x = x_0 \cos(\omega t + \phi)$ is replaced by our quantum 'fuzzball' (or Gaussian wave function).

The time dependence of a coherent state in terms of a wave function oscillating in a well is shown in figure 6.6. We have a Gaussian wave function that is displaced from the origin and which maintains its shape as it propagates, much like a classical mass on a spring. How does this arise? We know that states of different energies should have different velocities; indeed, this leads to dispersion of a pulse during propagation in free space. In the case of the quadratic potential, the higher energy parts of the wave travel faster but have a larger amplitude of oscillation. The two factors balance one another, and a Gaussian wave function will maintain its shape as it oscillates. In terms of the quadrature amplitudes, we have a Gaussian fuzzball that rotates in the $X - Y$ plane at frequency ω.

The photon number distribution for a coherent state is a Poissonian distribution, with the mean equal to the variance, as shown in figure 6.7.

The noise, or variance, in a coherent state remains constant as it evolves over time; recall that it remains a coherent state with a different complex phase. In figure 6.8 we show the time evolution of the photon number (constant) and the quadrature amplitudes X and Y.

It is not terribly convenient to use the coherent states as a basis because they are not orthonormal,

$$\langle \alpha | \alpha' \rangle = e^{\frac{-1}{2}|\alpha - \alpha'|^2}, \tag{6.111}$$

and indeed form an overcomplete set. They can resolve the identity,

$$I = \frac{1}{\pi} \int_{-\infty}^{\infty} |\alpha\rangle \langle \alpha | d^2\alpha. \tag{6.112}$$

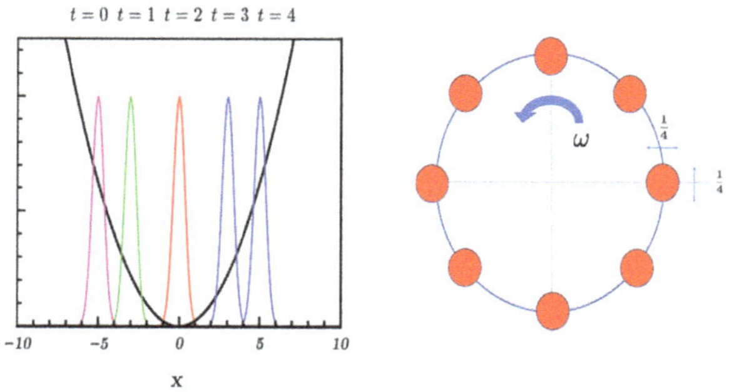

Figure 6.6. Time evolution of a coherent state wave function and quadrature distribution cartoon.

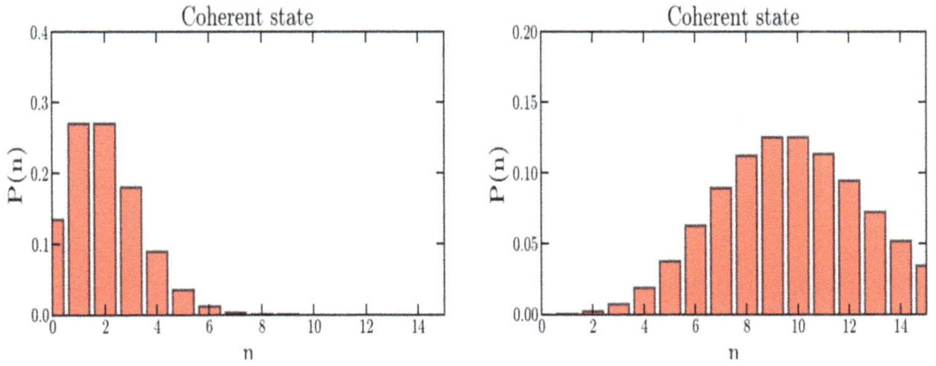

Figure 6.7. Photon probability distribution $P(n)$ for coherent states with average photon numbers 2 and 10.

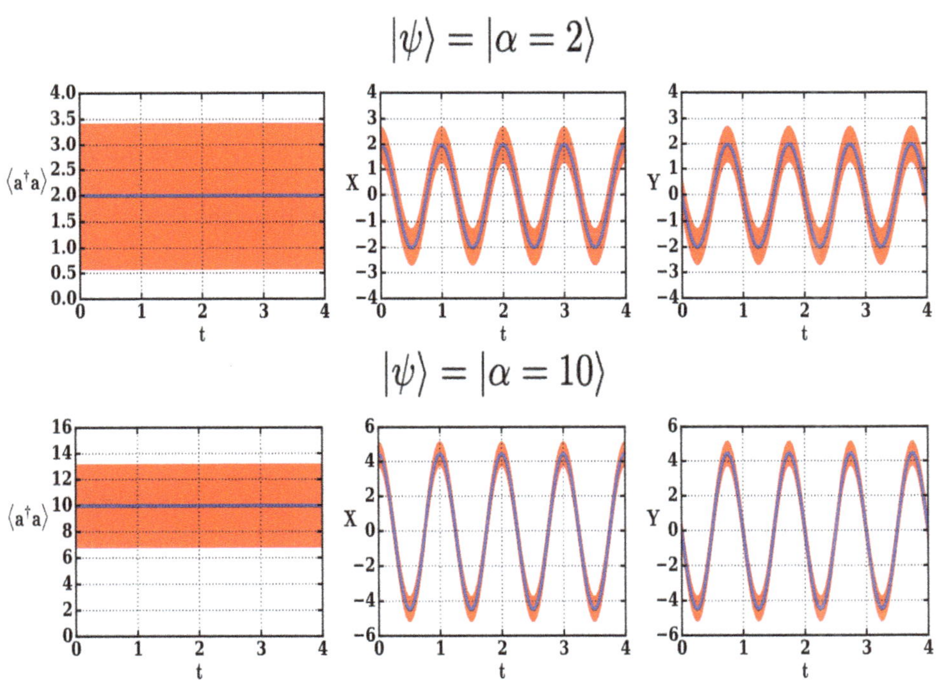

Figure 6.8. Time evolution of the average (blue curve) and variance (red shaded region) of coherent states with average photon numbers of 2 and 10, respectively.

How does one generate a coherent state in the laboratory? If we have a Hamiltonian of the form

$$H = i\hbar(\beta a^\dagger - \beta^* a) \tag{6.113}$$

then the time evolution of the vacuum for some time T would yield

$$\psi(t) = D(\beta T)\psi(0), \tag{6.114}$$

which would be a coherent state with $\alpha = \beta T$. This Hamiltonian can come from the linear coupling of two field modes, such as one field incident on an imperfect end mirror of a cavity. The interaction time would be of the order of $1/\kappa$ and the field inside the cavity would reach a steady state described by a coherent state. Of course, the real Hamiltonian for a linear coupling between the field modes would be

$$H = i\hbar(ba^\dagger - b^\dagger a), \tag{6.115}$$

where b and $b^\dagger$ are annihilation and creation operators for the driving mode. This kind of interaction kills a photon in the driving mode and creates one in the cavity being driven. The other term represents the reverse process, which would be light leaking out of the cavity from the same mirror on which the driving field was incident. We regain the Hamiltonian needed to generate a coherent state if we can replace $b \to \beta$ and $b^\dagger \to \beta^*$. This would be appropriate for a strong coherent state, but we are trying to *make* a coherent state, so this becomes circular! A strong steady oscillating current will generate a coherent state to a good approximation, but a laser well above the threshold produces a coherent state to a very good approximation, with the addition of phase diffusion noise due to the spontaneous emission noise in the laser. So in the end, we get a laser, run it well above the threshold, and conduct our experiments on a time scale small compared to the phase diffusion rate, and then we will have a coherent state.

Another way to generate a coherent state can be seen from the canonical form of the atom–field interaction,

$$H_{\text{int}} = -\frac{e}{m}\vec{A}\cdot\vec{p}. \tag{6.116}$$

The current density for charges moving at velocity v is $\vec{J} = e\vec{v} = (e/m)\vec{p}$, which leads to an interaction Hamiltonian

$$H_{\text{int}} = -\vec{J}\cdot\vec{A}. \tag{6.117}$$

Recall the form of the quantized vector potential

$$A = A_0(ae^{-i\omega t} + a^\dagger e^{i\omega t}). \tag{6.118}$$

If we have an oscillatory current density in one dimension $J = J_0\cos(\omega t)$, we are led to a time evolution operator

$$U(t) = e^{-iA_0J_0(a+a^\dagger)t}, \tag{6.119}$$

which is in the form of a displacement operator that will generate a coherent state.

6.6 Squeezed states

The coherent states are minimum uncertainty states that have symmetric noise, in the sense that any quadrature $a_\theta = (1/2)(ae^{i\theta} + a^\dagger e^{-i\theta})$ satisfies $\Delta a_\theta = 1/2$. Here we wish to consider a class of states that are minimum uncertainty states but do not have symmetric noise. We have number squeezed states and phase squeezed states, which could be more useful for measurements of amplitude and phase than coherent states, as we shall see. States of this type are not generated in a linear interaction, as we saw that would produce a coherent state. Let us consider an analogy of the displacement operator, but one that involves a^2 and $a^{\dagger 2}$, the squeezing operator

$$S(r) = \exp\left(\frac{1}{2}(ra^2 - r^* a^{\dagger 2})\right). \tag{6.120}$$

Let us examine the action of this operator on a vacuum state, which generates what is known as a squeezed vacuum state. One can use a Taylor series approximation of the squeezing operator, and take r to be real for now, to order α^2,

$$S(r) = 1 + \frac{r}{2}(a^2 - a^{\dagger 2}) + \frac{r^2}{4}(a^4 + a^{\dagger 4} - a^2 a^{\dagger 2} - a^{\dagger 2} a^2) + \ldots \tag{6.121}$$

and generate

$$S(r)|0\rangle = |0\rangle - \frac{1}{\sqrt{2}} r|2\rangle + \frac{r^2}{2}|0\rangle + 6r^2|2\rangle + \ldots. \tag{6.122}$$

Here we see that due to the $a^{\dagger 2}$ terms, we find photons in pairs in the sense that only even numbered number states show up in the squeezed vacuum state. This points to the fragility of the squeezed vacuum to loss, a coherent state is of the form (to order α^2)

$$|\alpha\rangle \approx (1 - |\alpha|^2)|0\rangle + \alpha|1\rangle + \frac{\alpha^2}{\sqrt{2}}|2\rangle \tag{6.123}$$

and so losing photons can turn a squeezed vacuum into something more like a coherent state, and the noise will be symmetrized. Using the Baker–Hausdorff theorem, we can show

$$b = S^\dagger(r) a^\dagger S(r) = a^\dagger \cosh(|r|) - ae^{-i\phi} \sinh(|r|) \tag{6.124}$$

$$b^\dagger = S^\dagger(r) a S(r) = a \cosh(|r|) - a^\dagger e^{-i\phi} \sinh(|r|), \tag{6.125}$$

where we have written the squeezing parameter $r = |r| \exp^{(i\phi)}$. This is known as a Bogoliubov transformation and can be written in matrix form as

$$\begin{pmatrix} b \\ b^\dagger \end{pmatrix} = \begin{pmatrix} \cosh(r) & -\sinh(r) \\ -\sinh(r) & \cosh(r) \end{pmatrix} \begin{pmatrix} a \\ a^\dagger \end{pmatrix}. \tag{6.126}$$

The quadratures also transform under the same transition

$$\begin{pmatrix} X' \\ Y' \end{pmatrix} = \begin{pmatrix} \cosh(r) & -\sinh(r) \\ -\sinh(r) & \cosh(r) \end{pmatrix} \begin{pmatrix} X \\ Y \end{pmatrix}. \tag{6.127}$$

You have perhaps seen a transformation similar to this before in special relativity,

$$\begin{pmatrix} x' \\ xt' \end{pmatrix} = \begin{pmatrix} \gamma & -\gamma v/c \\ -\gamma v/c & \gamma \end{pmatrix} \begin{pmatrix} x \\ ct \end{pmatrix}, \tag{6.128}$$

the Lorentz transformation with $\theta = v/c$ defined as the rapidity. The Boglioubov transformation is a Lorentz transformation with $\tanh(r) = v/c$ playing the role of the rapidity. The creation and annihilation operators become mixed, and hence so do the absorption and emission.

In general, we can expand the squeezed vacuum in a superposition of the number states,

$$|r\rangle = S(r)|0\rangle = \sum_n D_n |n\rangle . \tag{6.129}$$

It will be useful to consider

$$\begin{aligned} 0 &= a|0\rangle \\ &= S(r)a|0\rangle \\ &= S(r)aS^\dagger(r)S(r)|0\rangle \\ &= S(r)aS^\dagger(r)|r\rangle \\ &= (a\cosh(|r|)) - a^\dagger e^{-i\phi}\sinh(|r|)|r\rangle . \end{aligned} \tag{6.130}$$

Then, we can write

$$\begin{aligned} 0 &= a\cosh(|r|) + a^\dagger e^{-i\phi}\sinh(|r|)\sum_n D_n|n\rangle \\ &= \sum_n \cosh(|r|)D_n\sqrt{n}\,|n-1\rangle + \sum_n \sinh(|r|)e^{-i\phi}\sqrt{n+1}\,|n+1\rangle \\ &= \sum_n (\cosh(|r|)D_{n+1} + \sinh(|r|)e^{-i\phi}D_{n-1}), \end{aligned} \tag{6.131}$$

which leads to a recursion relation of the form

$$D_{n+1} = -e^{i\phi}\tanh(|r|)\sqrt{\frac{n}{n+1}}\,D_{n-1}, \tag{6.132}$$

with D_0 to be determined from normalization and $D_1 = 0$. By requiring $\langle r|r\rangle = 1$ and the above recursion relation, we find

$$|r\rangle = \sqrt{|r|}\sum_n \frac{\sqrt{(2n)!}}{n!}\left(\frac{-\tanh(|r|)e^{i\phi}}{2}\right)^n |2n\rangle . \tag{6.133}$$

The squeezed vacuum state indeed contains only even numbered Fock states, as we anticipated. These states are also called two-photon coherent states; a squeezed vacuum is a two-photon coherent state, albeit with a zero coherent amplitude. We will consider squeezed states *with* a coherent amplitude shortly. For now, we will consider r real. For the squeezed vacuum, we have

$$\langle X\rangle = 0 \tag{6.134}$$

$$\langle Y\rangle = 0 \tag{6.135}$$

$$\langle N \rangle = \sinh^2(r) \tag{6.136}$$

$$\langle N^2 \rangle = 3\sinh^4(r) + 2\sinh^2(r) \tag{6.137}$$

$$\Delta N = \sqrt{2\langle N \rangle(\langle N \rangle + 1)}. \tag{6.138}$$

Squeezed vacuum can be a misnomer; it is generated by the squeezing operator on the vacuum, but we see that there is indeed a nonzero photon number. The quadrature variances are given by

$$(\Delta X)^2 = \frac{1}{4}e^{-2r} \tag{6.139}$$

$$(\Delta Y)^2 = \frac{1}{4}e^{2r} \tag{6.140}$$

with

$$\Delta X \Delta Y = \frac{1}{4}, \tag{6.141}$$

which shows that the squeezed vacuum is a minimum uncertainty state with asymmetric noise. In general, if the squeezing parameter is complex (nonzero ϕ, meaning the squeeze does not occur along the X axis), the variances are mixed, and the 'quiet' quadrature is not X, but the quadrature defined by a_ϕ. The variances for X and Y are

$$(\Delta X)^2 = \frac{1}{4}(e^{-2|r|}\cos^2(\phi/2) + e^{2|r|}\sin^2(\phi/2)) \tag{6.142}$$

$$(\Delta Y)^2 = \frac{1}{4}(e^{2|r|}\cos^2(\phi/2) + e^{-2|r|}\sin^2(\phi/2)). \tag{6.143}$$

This is shown in figure 6.9.

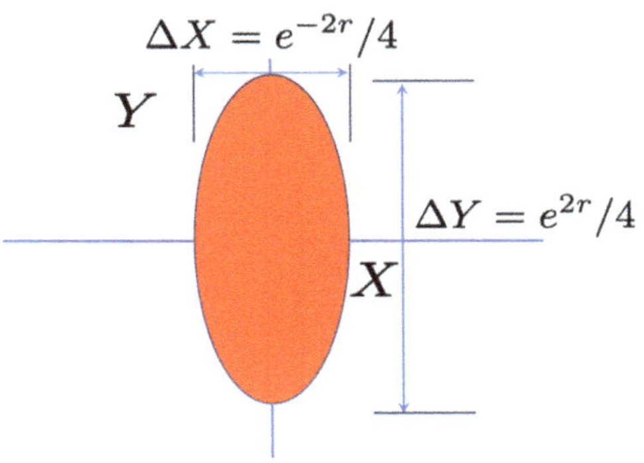

Figure 6.9. Cartoon of the quadrature distribution of a squeezed state, noting the X and Y quadrature uncertainties.

The uncertainty principle $\Delta X \Delta Y \geqslant 1/4$ limits the available physical states. This is shown in figure 6.10. Here, all the coherent states lie at the point $\Delta X = \Delta Y = 1/4$, For the squeezed vacuum, one can have less noise in one quadrature, below the coherent state limit of 1/4 with a corresponding increase of noise in the other quadrature. States that live between the red and blue curves are all referred to as squeezed states, as one quadrature has noise $\Delta X(Y) \leqslant 1/4$. The squeezed vacuum states lie on the blue curve. These types of states were first observed by Slusher *et al* [8].

In figure 6.11 we show a sketch of the quadrature distribution for a squeezed state created via the displacement of a squeezed vacuum state.

The actions of squeezing and displacement do not commute, but they can be related,

$$D(\alpha)S(r) = S(r)D(\beta), \tag{6.144}$$

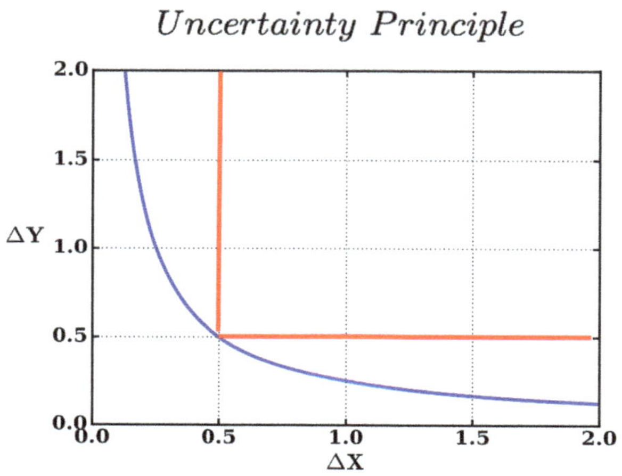

Figure 6.10. The uncertainty principle allows physical states above the blue curve. All states between the red and blue curves are referred to as squeezed states. The squeezed vacuum states lie on the blue curve.

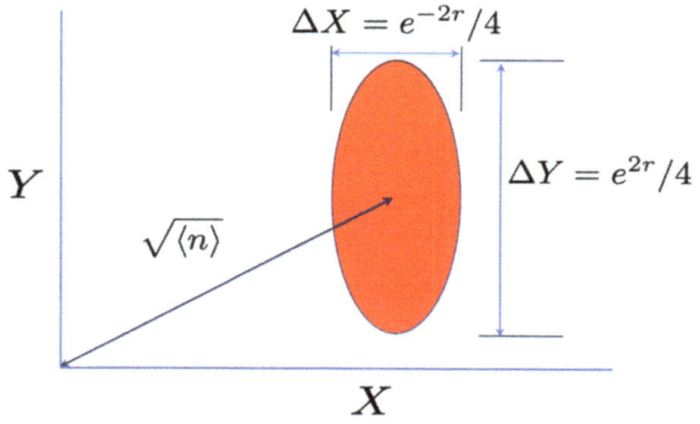

Figure 6.11. Cartoon of the quadrature distribution of a squeezed state, noting the X and Y quadrature uncertainties.

with the displacements α and β related,

$$\beta = \alpha \cosh(r) + \alpha^* \sinh(r), \tag{6.145}$$

essentially a Bogoloubov transformation of the displacements.

We have cheated a bit so far, we have characterized the various quadratures, and have primarily talked about the phases $\theta = 0$ and $\pi/2$. As the phase of the field tends to have a contribution ωt, the various properties of the squeezed vacuum are time dependent. At $t = \pi/2\omega$, the quiet phase that has variance $e^{-2r}/4$ at $t = 0$ will be the noisy phase. So to always 'see' the quiet phase, we must just look at every half cycle; there are measurement schemes that do that precisely. This is shown in figure 6.12. In terms of a wave packet in a quadratic potential well, at $t = 0$ if the wave packet is at the origin and is narrower than the coherent state width, at a quarter of a cycle, the wave packet will be larger than the coherent state width. A half cycle later, it is back at the origin as it was initially, then it repeats this motion to the other side of the well as shown in figure 6.12.

The time evolution of this fuzzball cartoon is shown in figure 6.13.

The squeezed vacuum only has a nonzero occupation probability for even photon numbers. As the squeezing operator depends on a^2, applying that to a vacuum state to generate the squeezed vacuum only even photon numbers are obtained. Two examples are shown in figure 6.14.

How does one generate a squeezed vacuum? The initial state is trivial, being the vacuum state. The squeezing operator can be generated by a Hamiltonian

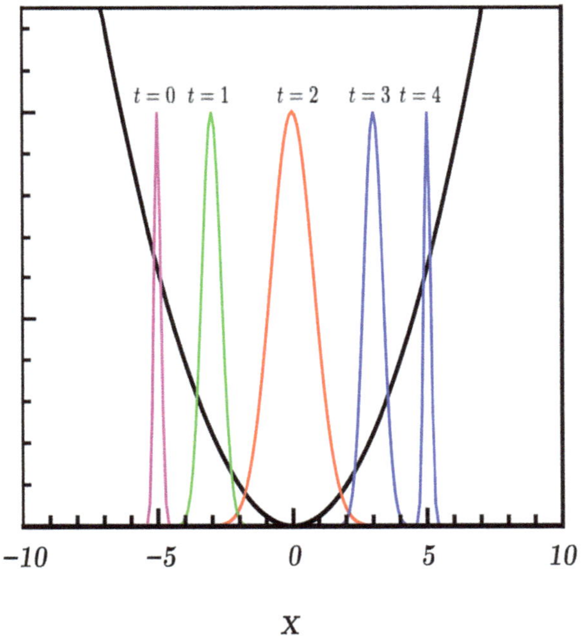

Figure 6.12. Time evolution of a squeezed-state wave function.

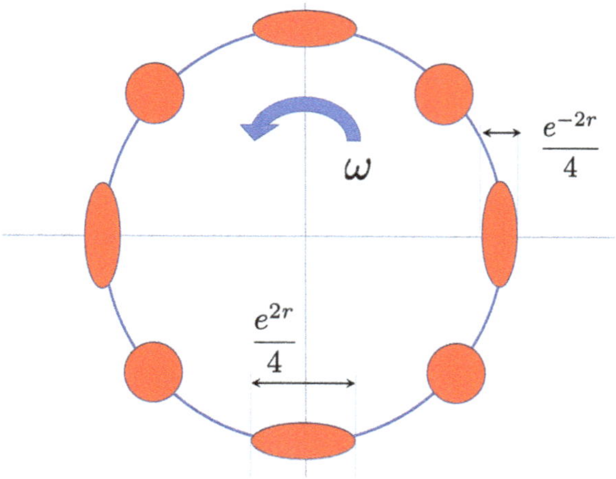

Figure 6.13. Time evolution of the quadrature distribution of a squeezed state, essentially a quadrupole oscillation.

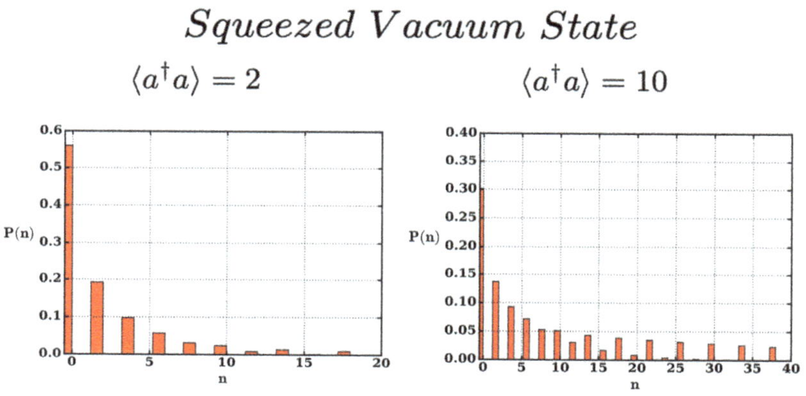

Figure 6.14. Photon probability distribution $P(n)$ for squeezed vacuum states with average photon numbers 2 and 10.

$$H = \frac{i\hbar}{2}(sa^2 - s^*a^{\dagger 2}), \tag{6.146}$$

which would lead to the state

$$\psi(t) = e^{\frac{1}{2}(sa^2 - s*a^{\dagger 2})T}\psi(0) \tag{6.147}$$

after time evolution for $t = T$, which is a squeezed state with $r = sT$. What physical process would correspond to such a Hamiltonian? Well one that creates or destroys two photons at a time (figure 6.15). What if we have a material that has a nonlinear polarization $P = \epsilon_0 \chi^{(2)} E^2$, while not pretending to do a nonlinear version of a quantized field; this would look something like

$$P = \epsilon_0 \chi^{(2)} \mathcal{E}_0^2 (a^2 + a^{\dagger 2} + a^\dagger a + aa^\dagger). \tag{6.148}$$

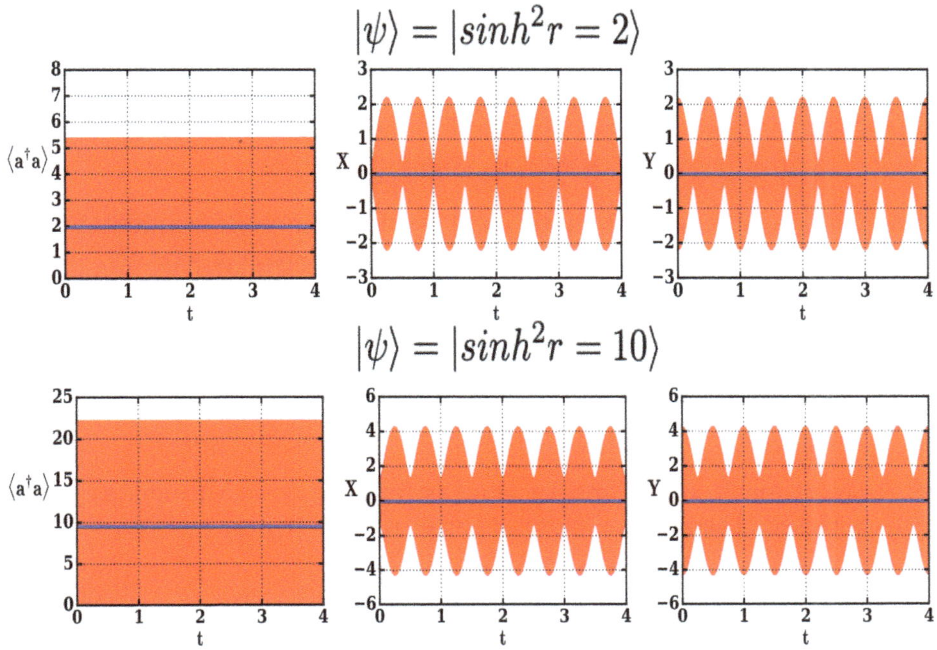

Figure 6.15. Time evolution of the average (blue curve) and variance (red shaded region) of squeezed states with average photon numbers of 2 and 10, respectively.

This looks promising, and indeed a material that has a second-order nonlinearity can be used to generate a squeezed vacuum. These types of materials will take a laser of one frequency ω, absorbing two photons, and then output one photon at twice the frequency 2ω, a process known as second harmonic generation. This process results from the $a^{\dagger 2}$ term in the Hamiltonian, where s is proportional to the incident field amplitude at the fundamental frequency. The opposite process is the parametric down conversion, where a material with a second-order nonlinearity will absorb a photon at ω and emit two photons at frequencies ω_1 and ω_2, which due to energy conservation need satisfy $\omega = \omega_1 + \omega_2$. If this is done with a nonlinear crystal inside a cavity, we can have a system that reaches a steady state, where the gain of the parametric process equals the loss through the end mirrors, just as in a laser. This is generally known as parametric oscillation. If $\omega_1 = \omega_2 = \omega/2$, this is known as degenerate parametric oscillation. To generate a squeezed vacuum, we pump a nonlinear system in a cavity with light at 2ω, and light at ω will spontaneously arise at frequency ω (assuming the cavity has a mode at ω), which is in a squeezed vacuum state. Of course, there are technical problems that arise that we are completely ignoring here. We will discuss how to detect the squeezed vacuum and how to formalize quantum 'fuzzballs' in chapter 12.

If we take our optical parametric oscillator and input a coherent state of light at frequency ω, we would generate a state

$$|SS\rangle = S(r)D(\alpha)|0\rangle . \tag{6.149}$$

These states will be minimum uncertainty states with asymmetric noise and also have a coherent amplitude. Mathematically, we obtain states that are easier to deal with if we reverse the order of the operators to generate the ideal squeezed coherent states,

$$|r,\, \alpha\rangle = D(\alpha)S(r)|0\rangle\,, \tag{6.150}$$

they are a squeezed vacuum that is then displaced. They are different from the states generated by squeezing a coherent state, as the squeezing operator and the displacement operator do not commute. Again, we examine the effect of transforming operators instead of states:

$$D^{\dagger}(\alpha)S^{\dagger}(r)a^{\dagger}S(r)D(\alpha) = a^{\dagger}\cosh(|r|) - ae^{-i\phi}\sinh(|r|) + \alpha \tag{6.151}$$

$$D^{\dagger}(\alpha)S^{\dagger}(r)aS(r)D(\alpha) = a\cosh(|r|) - a^{\dagger}e^{-i\phi}\sinh(|r|) + \alpha^{*}. \tag{6.152}$$

The variances for X and Y are the same as for a squeezed vacuum, as they are merely displaced. The average photon number is

$$\langle N\rangle = \sinh^{2}(|r|) + |\alpha|^{2} \tag{6.153}$$

with variance

$$\begin{aligned}(\Delta N)^{2} = {} & |\alpha|^{2}(e^{2|r|}\sin^{2}(\theta - \phi/2) + e^{-2|r|}\cos^{2}(\theta - \phi/2)) \\ & + 2\sinh^{2}(|r|)(\sinh^{2}(|r|) + 1).\end{aligned} \tag{6.154}$$

Our original proposition for generating squeezed light with a coherent amplitude generates states of the sort we have considered, if we consider the relations

$$D(\alpha)S(r) = S(r)D(\alpha') \tag{6.155}$$

with

$$\alpha' = \alpha r + r\alpha. \tag{6.156}$$

When we squeeze and then displace, we just get the squeezed vacuum plus a coherent amplitude, as we have seen. Using the above relations, we can show that if we displace and then squeeze, the squeezing ellipse is rotated, as well as displaced along an arc too. However, they are still squeezed coherent states of the form we considered in detail, albeit with different parameters,

$$S(r)D(\alpha)|0\rangle = D(\alpha')S(r)|0\rangle = |r,\, \alpha'\rangle. \tag{6.157}$$

More schematically, we can talk about phase and amplitude squeezed states, in which the squeezing is parallel or perpendicular to the coherent amplitude; $\theta = \phi/2$ for phase squeezing and $\theta - \phi/2 = \pi/2$ for amplitude squeezed light (figure 6.16). This is shown in figure 6.17 for a displaced squeezed vacuum state with an average photon number of 2, $\sinh^{2}(r) = 1$ and $|\alpha|^{2} = 1$. In figure 6.18 we have an average photon number of 10, $\sinh^{2}(r) = 5$ and $|\alpha|^{2} = 5$.

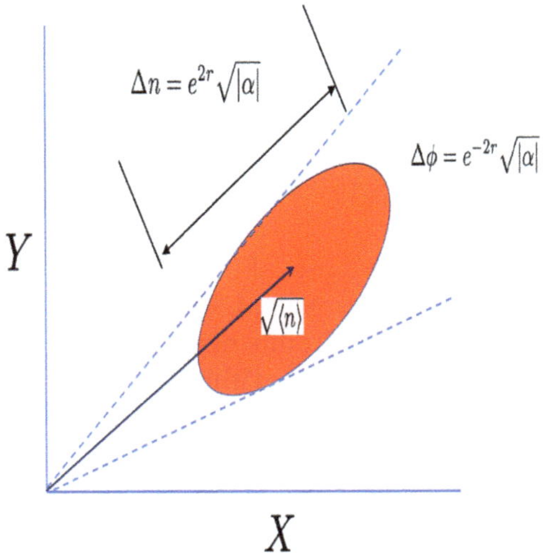

Figure 6.16. Cartoon of the quadrature distribution for a phase squeezed state.

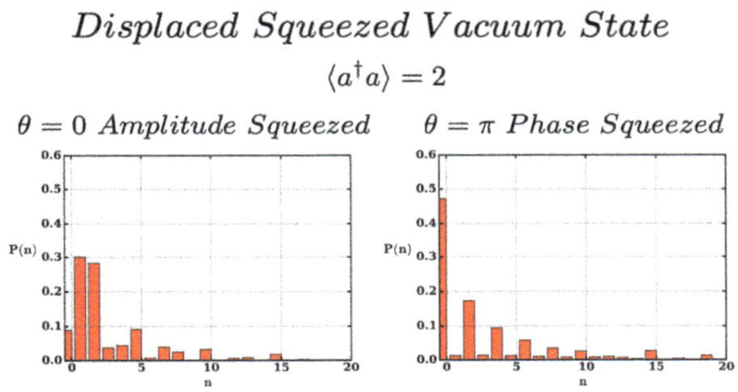

Figure 6.17. Photon number distribution for an average photon number of 2, $\sinh^2(r) = 1$, and $|\alpha|^2 = 1$. Shown are $\theta = 0$ and $\theta = \pi$.

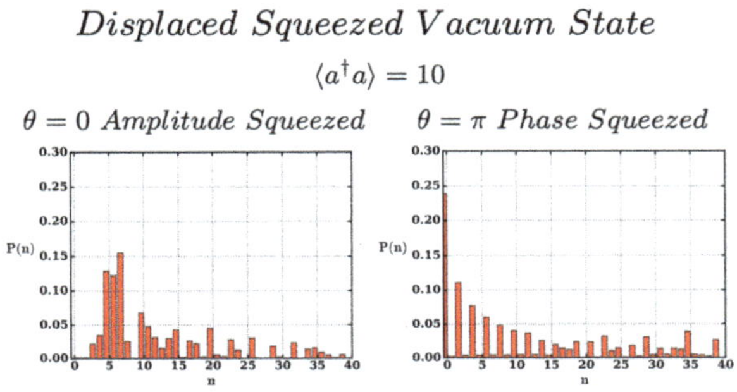

Figure 6.18. Photon number distribution for an average photon number of 10, $\sinh^2(r) = 5$, and $|\alpha|^2 = 5$. Shown are $\theta = 0$ and $\theta = \pi$.

6.7 Cat states

Coherent states are essentially classical states with added vacuum noise. Recall the Schrödinger cat thought experiment where

$$|\Psi\rangle = \frac{1}{\sqrt{2}}(|\text{live}\rangle + |\text{dead}\rangle). \tag{6.158}$$

Representing a cat that is, in some sense, alive and dead at the same time. The state of the cat is not known until it is measured. We remark that it is very difficult to isolate quantum states from their environments. For a real box, if death is triggered by a radioisotope decay inside the box, within a lifetime or two the cat is dead. Furthermore, for a real box, one would eventually notice the cat is dead due to a strong odor.

We can superimpose two classical-ish states, a coherent state $|\alpha\rangle$, and one of similar amplitude but π out of phase (displaced in a negative direction) $|\alpha\rangle$. In this case, we know that the photon number distribution is Poissonian for each, but what is the result when they are superposed? In figure 6.19 we show the resulting photon distribution for the cat state

$$|\text{Cat}\rangle = \frac{1}{\sqrt{2}}(|\alpha\rangle + |-\alpha\rangle). \tag{6.159}$$

In figure 6.20 we show the time evolution of the expectation values and the variance of n, X, and Y for the cat states with average photon numbers of 2 and 10.

We see behavior similar to that of a squeezed vacuum state, where only even photon numbers have nonzero probability. This makes us suspect that this state is nonclassical. It is, and when we discuss the quasi-probability distributions in chapter 10, we can make that quantitative. Do we see something different with the state?

$$|\text{Cat}\rangle = \frac{1}{\sqrt{2}}(|l\alpha\rangle - |-\alpha\rangle). \tag{6.160}$$

Not by simple quadrature and photon number measurements. We would need some type of interferometric measurement.

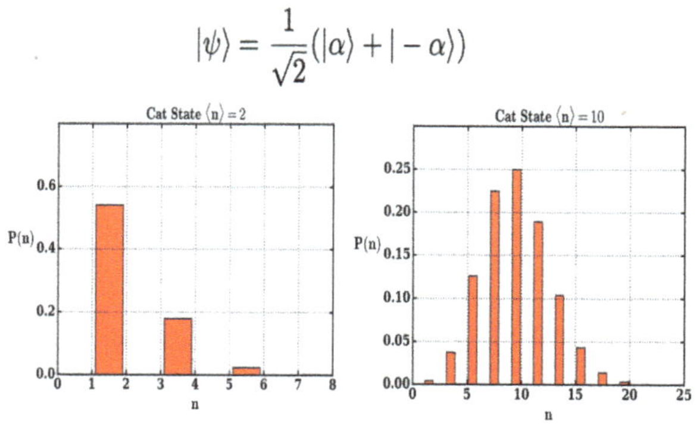

Figure 6.19. Photon number distributions for the cat states with average photon numbers of 2 and 10.

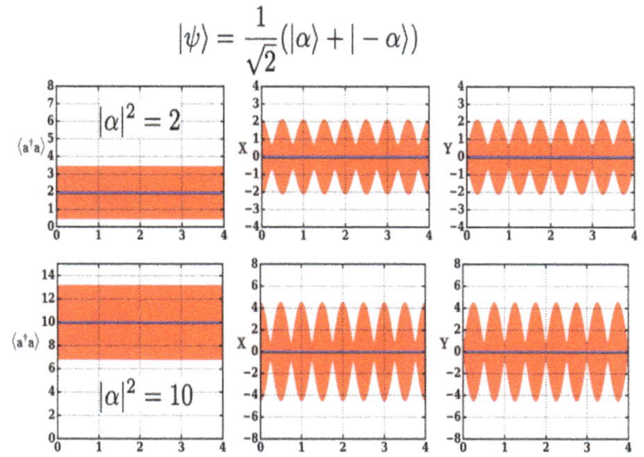

Figure 6.20. Time evolution of expectation values and variance of n, X, and Y for cat states with average photon numbers of 2 and 10.

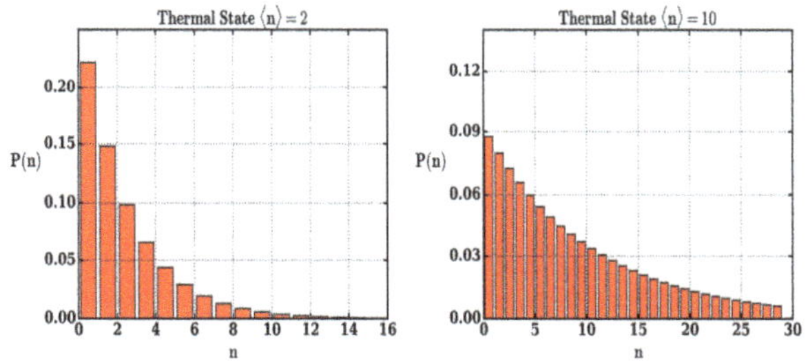

Figure 6.21. Photon number distribution for the thermal states of the average photon number of 2 and 10.

6.8 Thermal states

Finally, we examine our old friend the thermal state. A thermal state has no wave function but is described by a density matrix. The diagonal elements of ρ_{thermal} are given by the Planck distribution, and the off diagonal elements are all zero. There is no coherence in the thermal state. The population distribution of the thermal state indeed decays exponentially, as displayed in figure 6.21.

The complete lack of any phase coherence in the thermal state can be seen in the plot of the expectation values and variances of the photon number and the X and Y quadratures. Nothing 'wavelike' is manifest. Examples are presented in figure 6.22.

6.9 Twin beam states

There is another crucial class of squeezed states, the so-called twin-beam states. These are generated by the operator

$$T = e^{\eta(ab + a^\dagger b^\dagger)}. \tag{6.161}$$

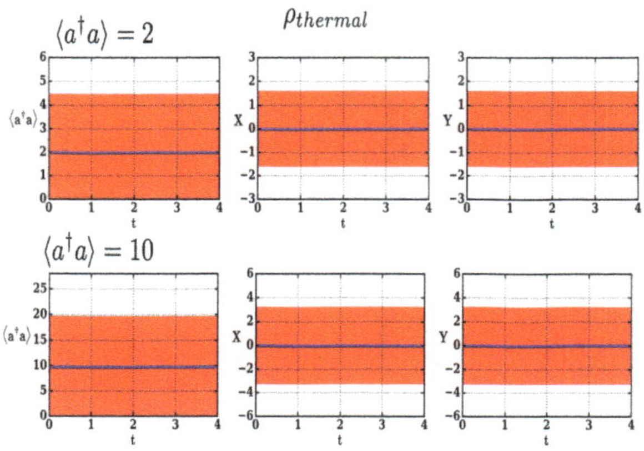

Figure 6.22. Time evolution of expectation values and variance of n, X, and Y for thermal states with average photon numbers of 2 and 10.

This generates states where the two modes are correlated. They may individually be as noisy as the dickens, but they will have the *same* noise. These are quite useful for quantum key distribution and other applications.

References

[1] Cohen-Tannoudji C, Grynberg G and Dupont-Roc J 1997 *Photons and Atoms: Introduction to Quantum Electrodynamics* (New York: Wiley)

[2] Cohen-Tannoudji C, Grynberg G and Dupont-Roc J 1992 *Atoms-Photon Interactions: Basic Processes and Applications* (New York: Wiley)

[3] Dirac P A M 1982 *The Principles of Quantum Mechanics* (Oxford: Clarendon)

[4] Susskind L and Glogower J 1964 Quantum mechanical phase and time operator *Phys. Physique Fizika* **1** 49

[5] Pegg D T and Barnett S M 1989 Phase properties of squeezed states of light *Phys. Rev.* A **39** 1665

[6] Noh J W, Fougeres A and Mandel L 1991 Measurement of the quantum phase by photon counting *Phys. Rev. Lett.* **67** 1426

[7] Barnett S and Vaccaro J 2007 *The Quantum Phase Operator: A Review* (London: Taylor and Francis)

[8] Slusher R E, Hollberg L W, Yurke B, Mertz J C and Valley J F 1985 Observation of squeezed states generated by four-wave mixing in an optical cavity *Phys. Rev. Lett.* **55** 2409

IOP Publishing

An Introduction to Quantum Optics (Second Edition)
An open systems approach
Perry Rice

Chapter 7

Quantum fields II

7.1 Vacuum fluctuations, beam splitters, and interferometers

Consider the beam splitter shown in figure 7.1, with fields E_1 and E_2 incident and fields E_3 and E_4 transmitted. These satisfy the relations

$$E_3 = rE_1 + tE_2 \tag{7.1}$$

$$E_4 = t'E_1 + r'E_2. \tag{7.2}$$

Because the intensities are proportional to the square of the field, $I = \epsilon_0 c|E|^2$, these are

$$I_3 = |r|^2 I_1 + |t|^2 I_2 + 2rt\epsilon_o cE_1E_2 \tag{7.3}$$

$$I_4 = |r'|^2 I_2 + |t'|^2 I_1 + 2r't'\epsilon_0 cE_1E_2. \tag{7.4}$$

For energy to be conserved, we must have the relations

$$|r|^2 + |t|^2 = 1 \tag{7.5}$$

$$|r'|^2 + |t'|^2 = 1 \tag{7.6}$$

$$rt + r't' = 0. \tag{7.7}$$

These can be satisfied by taking $r = r'$ and $t' = -t$, and we see that there must be a relative π phase shift between the two transmitted parts. Now consider the quantum-mechanical version; instead of fields, we will use operators. We have modes described by a and b incident on the beam splitter, with modes c and d in the outputs. We then have

$$c = ra + tb \tag{7.8}$$

$$d = -ta + rb. \tag{7.9}$$

doi:10.1088/978-0-7503-6014-2ch7

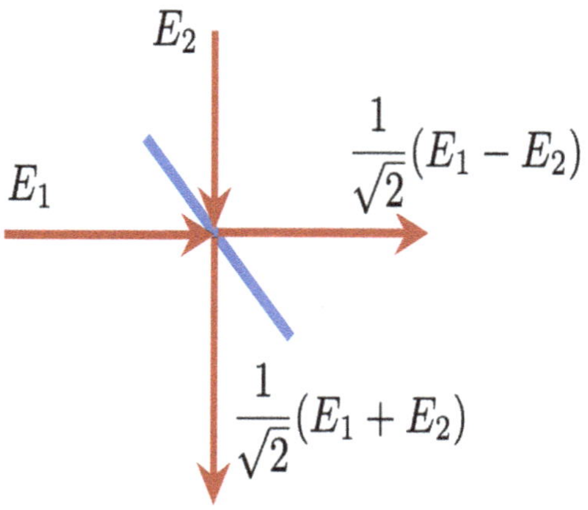

Figure 7.1. Inputs and outputs of the beam splitter.

We can write this in the matrix form as

$$\begin{pmatrix} c \\ d \end{pmatrix} = \begin{pmatrix} r & t \\ t & r \end{pmatrix} \begin{pmatrix} a \\ b \end{pmatrix}. \tag{7.10}$$

Let us consider the commutation relations for c and $c^\dagger$,

$$\begin{aligned} [c, c^\dagger] &= |r|^2[a, a^\dagger] + |t|^2[b, b^\dagger] + 2rt^*[a, b^\dagger] + 2r^*t[a^\dagger, b] \\ &= |r|^2 + |t|^2 \\ &= 1. \end{aligned} \tag{7.11}$$

Similarly $[d, d^\dagger] = 1$. What if we have no field input on the beam splitter in the b mode? Can we just set $b = 0$? If we do that, we would have the commutation relations

$$[c, c^\dagger] = |r|^2 \neq 1 \tag{7.12}$$

$$[d, d^\dagger] = |t|^2 \neq 1, \tag{7.13}$$

which clearly would be a problem. We are reminded once again that there can be no absence of the electric field quantum mechanically! The average value of the field $\langle a \rangle$ can be 0, but of course there are vacuum fluctuations, from $\langle a^2 \rangle \neq 0$. When one considers quantum noise limits in interferometry, an open port (a term meaning one with just vacuum 'noise' incident) is a source of noise, as the zero point fluctuations of the vacuum field leak through the beam splitter.

How do states transform via a beam splitter? We can form states by letting a function of a and $a^\dagger$ acting on the vacuum. Let us consider a particular input state to the beam splitter,

$$|\psi\rangle_{\text{in}} = |0_1 1_2\rangle = a_2^\dagger |0_1 0_2\rangle, \tag{7.14}$$

which refers to a vacuum state for mode 1 and a single photon state for 2. We take the beamsplitter to have $r = t = 1/\sqrt{2}$. Now our output state can be determined by considering

$$a^{\dagger} = \frac{1}{\sqrt{2}}(c^{\dagger} - d^{\dagger}) \tag{7.15}$$

so

$$\begin{aligned} |\psi\rangle_{\text{out}} &= \frac{1}{\sqrt{2}}(c^{\dagger} - d^{\dagger})|0_3 0_4\rangle \\ &= \frac{1}{\sqrt{2}}(|1_3 0_4\rangle - |0_3 1_4\rangle). \end{aligned} \tag{7.16}$$

This is a very interesting state, where we definitely have one photon in one arm and none in the other. This is an example of what we will call an entangled state in later chapters, meaning that the state cannot be written as a tensor product of states for the individual output modes. For these types of states, a measurement of system one yields some information about the other system. This state is what we call maximally entangled in that when we find a photon in mode 3 we *know* that there will be no photon in mode 4 and vice versa; we obtain total information about system 2 if we measure system 1.

What if we try two photons in one arm and none in the other? Let us see:

$$|\psi\rangle_{\text{in}} = |0_1 2_2\rangle = \frac{1}{\sqrt{2}} a_2^{\dagger\,2} |0_1 0_2\rangle . \tag{7.17}$$

This produces an output of the form

$$\begin{aligned} |\psi\rangle_{\text{out}} &= \frac{1}{2}(c^{\dagger 2} + d^{\dagger 2} - 2c^{\dagger}d^{\dagger})|0_3 0_4\rangle \\ &= \frac{1}{2}(\sqrt{2}\,|2_3 0_4\rangle - |1_3 1_4\rangle + \sqrt{2}\,|0_3 2_4\rangle). \end{aligned} \tag{7.18}$$

So sometimes both photons go into mode 3, sometimes, both go into mode 4, and with half the probability, one goes into each mode. Here the state generated is entangled, but is not maximally entangled. There is something special about the one-photon state that is useful for something called entanglement, which we discuss in chapter 16.

What if we put one photon in each input arm?

$$|\psi\rangle_{\text{in}} = |1_1 1_2\rangle = a^{\dagger} b^{\dagger} |0_1 0_2\rangle . \tag{7.19}$$

The output state is then

$$\begin{aligned} |\psi_{\text{out}}\rangle &= \frac{1}{2}(c^{\dagger 2} + d^{\dagger 2} + c^{\dagger}d^{\dagger} - c^{\dagger}d^{\dagger})|0_3 0_4\rangle \\ &= \frac{1}{\sqrt{2}}(|2_3 0_4\rangle + |0_3 2_4\rangle). \end{aligned} \tag{7.20}$$

Here we again find a maximally entangled state, in that we know there are two photons in one output mode and none in the other. However, one can easily show that an input state of $|n_1 n_2\rangle$ does not give a simple state of the form $|(2n)_3 0_4\rangle + |0_3, (2n)_4\rangle$ but a distribution of all possible states of the form $|2n - m, m\rangle$. Therefore, there is indeed something special about single photon states. There are two paths to generate one photon in either arm, which are indistinguishable. The amplitudes have the opposite sign and destructively interfere. This is shown schematically in figure 7.2.

If we take the (spatially separate) outputs of a down converter and direct them to the two inputs of a beam splitter, we never see a coincidence count if the two path lengths are equal. One can unbalance the arms, and plotting the coincident counts versus the path difference one sees a dip when the arms are balanced, that is, no phase difference, as shown in figure 7.3. This is known as the Hong–Ou–Mandel effect [13], or the HOM dip. If we do not have perfect single photon states at each input, we may still see a dip. A classical stochastic field can produce a dip of 50%.

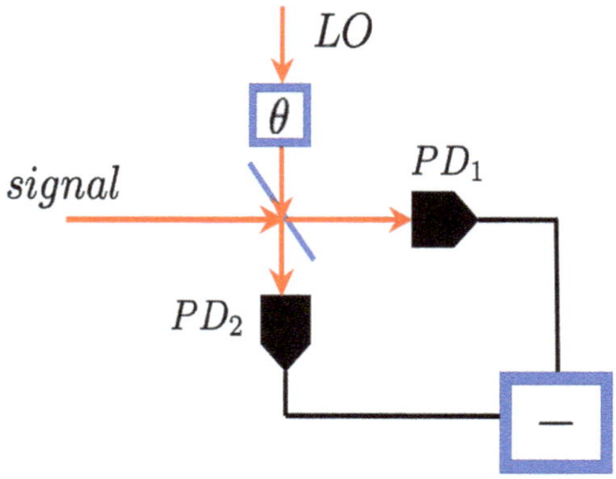

Figure 7.2. The two paths that lead to one photon in each output of the beam splitter, for a single photon incident in each input.

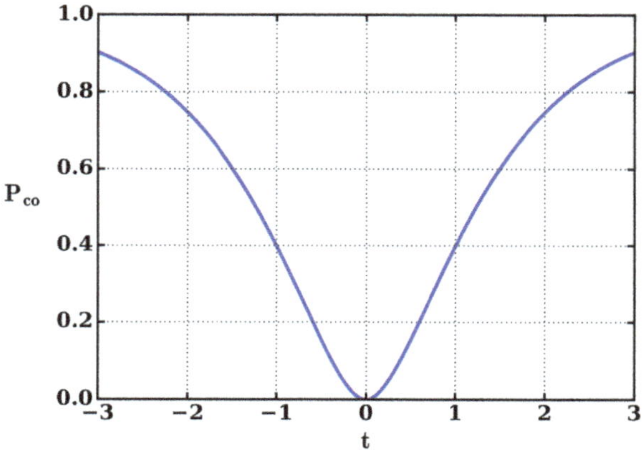

Figure 7.3. Coincidence counts in the Hong–Ou–Mandel interferometer versus detuning.

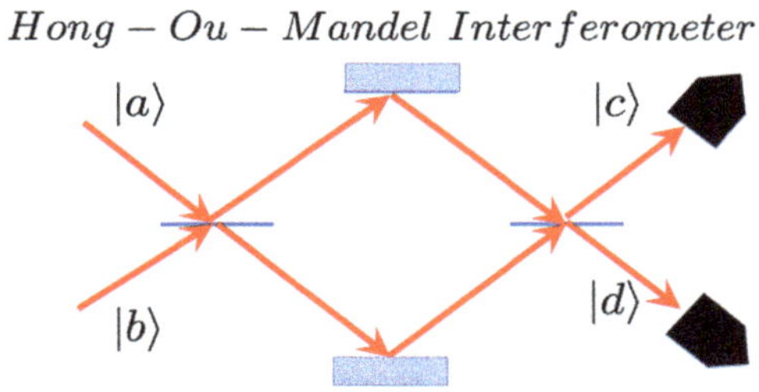

Figure 7.4. Hong–Ou–Mandel interferometer.

A HOM dip greater than 50% is a nonclassical effect. One can construct an interferometer as shown in figure 7.4. When the two interferometer paths are balanced, there is a rise in coincident counts.

What if we have a vacuum in mode 1 and a coherent state in mode 2? This would mean an input state of

$$|\psi\rangle_{\text{in}} = |0_1, \alpha_2\rangle = D_2(\alpha)|0_1, 0_2\rangle \tag{7.21}$$

and it can be shown that

$$|\psi\rangle_{\text{out}} = |\alpha/\sqrt{2}, -\alpha/\sqrt{2}\rangle, \tag{7.22}$$

which means that the amplitude is equally split between the two output modes, just as if we had a classical field in one input mode and nothing in the other. What if our inputs are a squeezed vacuum and a normal vacuum? As squeezing results from the correlations of pairs of photons, we would expect random parsing at the beam splitter to degrade the squeezing, as looking at one output port is the same as looking at the squeezed input port with random deletion of photons.

This would mean an input state of

$$|\psi\rangle_{\text{in}} = |0_1, r_2\rangle = S_2(r)|0_1, 0_2\rangle. \tag{7.23}$$

If we look at the output mode d,

$$d = t\,a - r\,b, \tag{7.24}$$

we find that

$$\begin{aligned}\Delta(c + c^{\dagger})^2 &= \langle((c + c^{\dagger})^2)\rangle + \langle(c + c^{\dagger})\rangle \\ &= t^2 e^{-r} + r^2\end{aligned} \tag{7.25}$$

and we see that our output has only partially squeezed fluctuations, as there is an admixture of vacuum noise from the input port with no import. As is often the case in quantum mechanics, a vacuum is not 'nothing' and cannot just be ignored.

There are two other interferometers of interest. First is the Franson interferometer [9] (see figure 7.5). Here a signal and idler are generated in a nonlinear crystal, and each is sent to an interferometer. Consider the initial entangled state of two photons, say in a maximally entangled state (also called a Bell state):

$$|\Psi\rangle = \frac{1}{\sqrt{2}}(|H\rangle_1|V\rangle_2 + |V\rangle_1|H\rangle_2), \tag{7.26}$$

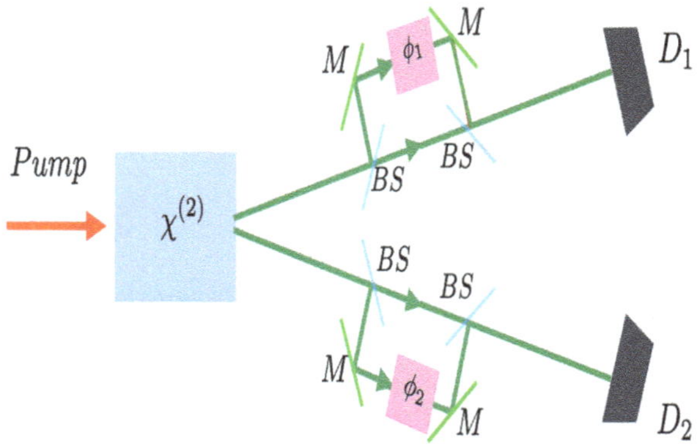

Figure 7.5. Franson interferometer.

where $|H\rangle$ and $|V\rangle$. We find that we obtain detection coincidences when the signal and idler both take the long or short legs of their interferometers. If one blocks the long leg of one and the short length of the other, there are no coincidence counts. The addition of the path information wipes out the fringe. After their generation, the two photons do not interact in any way.

Each photon passes through a beam splitter (BS) and is sent down one of the two possible paths. For simplicity, let us denote the paths of photon 1 and photon 2 as A_1, B_1, and A_2, B_2, respectively. The interference effect arises when the photons recombine at the second beam splitter after traveling through different optical paths. The phase shifts in these paths, denoted as ϕ_1 for photon 1 and ϕ_2 for photon 2, determine the interference. For each photon, the path length can introduce a phase shift, path A_1 and B_1 for photon 1, with a phase shift ϕ_1; path A_2 and B_2 for photon 2, with a phase shift ϕ_2. Thus, the total phase shift experienced by each photon depends on the relative distance traveled in the interferometer arms In general, the phase shift between the two arms for photon 1 is ϕ_1, and for photon 2 is ϕ_2. When the photons recombine at the final beam splitter, their interference pattern depends on the relative phase difference between the two arms for each photon. The output state at the detectors is a superposition of the possible polarization states. Using the fact that the probability of detecting coincidence events at both detectors D_1 and D_2 is related to the interference pattern, we can write the probability $P(\phi_1, \phi_2)$ of detecting coincidences as a function of the phase shifts ϕ_1 and ϕ_2. This probability can be expressed as

$$P(\phi_1, \phi_2) = \frac{1}{2}[1 + \cos(\phi_1 - \phi_2)]. \tag{7.27}$$

The key thing is that the fringe pattern for the joint system can be affected by a phase adjustment on one of the arms, no matter how long they are. We will revisit this in chapter 16 when we talk about Bell inequality violations.

In the Wang–Zou–Mandel interferometer [18] (see figure 7.6) we obtain coincident counts at D_1 and D_2, but if we close the shutter, the fringes disappear. Closing

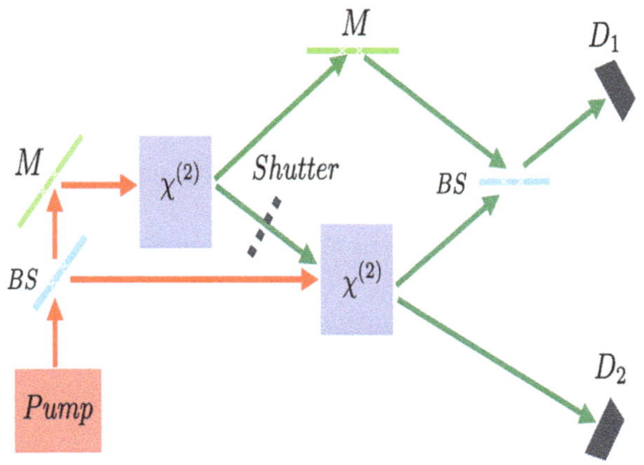

Figure 7.6. Zou–Wang–Mandel interferometer.

the shutter means we know that the counts at D_2 must have come from the crystal behind the shutter. This which-path information wipes out the interference. These are analogous to blocking a slit in the two-slit interferometer wiping out an interference pattern in single counts. The Zou–Wang–Mandel is a striking demonstration that quantum interference depends on the indistinguishability of all potential paths, even if those paths are never directly observed or populated. This principle is central to a range of 'which-path' and 'quantum eraser' phenomena.

Here we try to quantify 'which-path' information [8]. Consider a single quantum particle (e.g. a single photon or atom) that can travel along one of two paths, labeled $|1\rangle$ and $|2\rangle$. For simplicity, assume we have a pure state such that, before detection, the system can be written as

$$|\psi\rangle = \alpha|1\rangle + \beta|2\rangle, \quad \text{with} \quad |\alpha|^2 + |\beta|^2 = 1, \tag{7.28}$$

where α and β are complex amplitude, and $|1\rangle$ means 'the particle took path 1', and $|2\rangle$ means 'the particle took path 2'. When the paths are recombined on a beamsplitter or some final mixing optics, one measures the interference pattern in the detection probability

$$P(\phi) = |\alpha + \beta e^{i\phi}|^2 = |\alpha|^2 + |\beta|^2 + 2\,\mathrm{Re}\,[\alpha^* \beta e^{i\phi}]. \tag{7.29}$$

With ϕ the controllable phase difference between the two paths (e.g. by changing a mirror position). The interference term is proportional to $\alpha^*\beta$, whose magnitude determines the magnitude of the fringe oscillations can be. The maximum and minimum probabilities are, taking α and β to be real,

$$P_{\max} = |\alpha + \beta|^2 \tag{7.30}$$

$$P_{\max} = |\alpha - \beta|^2. \tag{7.31}$$

Hence, the classical definition of fringe visibility becomes

$$\mathcal{V} = \frac{P_{\max} - P_{\min}}{P_{\max} + P_{\min}} \tag{7.32}$$

or

$$\mathcal{V} = 2|\alpha||\beta|. \tag{7.33}$$

Distinguishability $\mathcal{D}$ (or 'which-path predictability') is a measure of how well one can guess which path the particle traveled if one were given the best possible measurement designed for path identification. For a pure two-path superposition $\alpha|1\rangle + \beta|2\rangle$, the optimal strategy is to guess 'path 1' if $|\alpha|^2 \geqslant |\beta|^2$, and guess 'path 2' otherwise. The probability of guessing correctly is then $\max\{|\alpha|^2, |\beta|^2\}$. The commonly used definition of distinguishability is

$$\mathcal{D} = ||\alpha|^2 - |\beta|^2|, \tag{7.34}$$

where $\mathcal{D} = 0$ means you have no idea which path (they have equal amplitudes), and $\mathcal{D} = 1$ means you are certain which path (one amplitude is 0). From the above definitions,

$$\mathcal{D} = ||\alpha|^2 - |\beta|^2|, \quad \mathcal{V} = 2|\alpha||\beta|. \tag{7.35}$$

If we square each and add

$$\mathcal{D}^2 = (|\alpha|^2 - |\beta|^2)^2 = |\alpha|^4 - 2|\alpha|^2|\beta|^2 + |\beta|^4 \tag{7.36}$$

$$\mathcal{V}^2 = 4|\alpha|^2|\beta|^2, \tag{7.37}$$

we find

$$\mathcal{D}^2 + \mathcal{V}^2 = |\alpha|^4 + 2|\alpha|^2|\beta|^2 + |\beta|^4 = (|\alpha|^2 + |\beta|^2)^2 = 1. \tag{7.38}$$

Thus, for a pure superposition with no external which-path measurement, we obtain

$$\mathcal{D}^2 + \mathcal{V}^2 = 1. \tag{7.39}$$

The higher the fringe visibility, the lower the distinguishability, and vice versa. In more realistic situations, where the particle may become entangled with an environment or a partial which-path detector, the state is no longer a perfect superposition $\alpha|1\rangle + \beta|2\rangle$. Instead, one typically obtains

$$\mathcal{D}^2 + \mathcal{V}^2 \leqslant 1.$$

7.2 Gravitational wave detection

In recent years, gravitational waves have been detected at the Laser Interferometric Gravitational Observatory (LIGO) [1]. This promises to open up a new view of the cosmos. Previously, all of our information was spectroscopic in nature; now we can hope to investigate the motion of large-scale objects that are charge neutral and do not emit light. The two facilities, in Livingston, LA, and Hanford, WA, are interferometers 4 km on a side, in a hard vacuum, with quantum noise limited lasers (at least over some bandwidth). This work resulted in the award of the 2017 Nobel Prize in Physics to Rainier Weiss, Kip Thorne, and Barry Barish. Recent

Advanced LIGO investigations have included the injection of squeezed light, so it is appropriate that we look into this. Other interferometers are coming online, and hopefully we will have one in space, the Laser Interferometer Space Antennae (LISA).

We start by looking at the noise sources of the detection system based on a Michaelson interferometer. In practice, the arms of the Michaelson are Fabry–Perot cavities to increase the measurement/interaction time. The laser light is injected at a beam splitter, entering both arms. The system is set up so that there are no detections at the output port (a null fringe). When a gravitational wave hits the interferometer, the length of the arms will change, and we will obtain some signal. The light that enters the interferometer comes out of the input port. As maximizing the laser power is helpful, Ronald Drever suggested putting a recycling mirror to send that light right back into the interferometer. There are many technical problems involving temperature and seismic fluctuations and other classical noise sources; either to mitigate or eliminate them. There are two interferometers oriented at 45 degrees to one another, so both could get a signal. If a graduate student bumps into a mirror, that only affects one interferometer. But what about Schrödinger's grad student, in two places at once? We will save that for the next book if there is one.

One crucial advance in understanding the noise properties of these devices was the insight of Carl Caves that the source of the relative radiation-pressure fluctuations in the two arms is due to the vacuum fluctuations entering the undriven input port. The abstract of the paper that brought this to light is my favorite of all time. The ending is quoted below:

> The interferometers now being developed to detect gravitational waves work by measuring small changes in the positions of free masses. There has been a controversy regarding whether quantum-mechanical radiation-pressure fluctuations disturb this measurement. This Letter resolves the controversy: They do. [4].

He further suggested the injection of squeezed light into that port to increase sensitivity; this is currently a part of the Advanced LIGO project [5].

Notice that we are careful to use the phrase 'gravitational wave' as it is analogous to an electromagnetic wave (light), moving at the speed of light. Instead of photons, we have gravitons (we think), the gravitational force carrier. Because there is one type of mass and gravity is always attractive, the gravitons should have a spin of 2 instead of the spin 1 for a photon. As for the photon, the spin is either 1 or -1, the graviton has a spin of 2 or -2. For a massless particle, the intermediate J_z states are not present. A gravity wave would be a literal wave on the ocean, say.

Before diving in, let us take a very quick look at the physics of the gravitational waves themselves. The metric corresponding to the presence of a gravitational wave is

$$g_{\mu\nu} = \eta_{\mu\nu} + h_{\mu\nu}, \tag{7.40}$$

with the flat space Minkowski metric

$$\eta_{\mu\nu} = \begin{pmatrix} 1 & 0 & 0 & o \\ 0 & 1 & 0 & 0 \\ 0 & 0 & 1 & 0 \\ 0 & 0 & 0 & 1 \end{pmatrix} \tag{7.41}$$

and the metric induced by the gravitational wave

$$\eta_{\mu\nu} = \begin{pmatrix} 0 & 0 & 0 & o \\ 0 & a & b & 0 \\ 0 & b & -a & 0 \\ 0 & 0 & 0 & 0 \end{pmatrix}. \tag{7.42}$$

The two parameters in the above represent the amplitudes of two polarizations, the + and × modes (see figure 7.7). These are quadrapole modes, reflecting the graviton spin of 2. The motion of the test masses arranged in an interferometer set-up are shown below. The interferometers are arranged to maximize the detection of a particular polarization; to fully determine the wave, we need two interferometers with a 45 relative rotation.

If we assume one polarization, the metric can be written as

$$ds^2 = -dt^2 + (1 + h\,dx^2) = 0. \tag{7.43}$$

The zero is a result of the waves propagating at the speed of light. This means there is a time delay between the propagation between the two mirrors of

$$\Delta t = hL/2c = h\tau_{\text{stor}}. \tag{7.44}$$

We have a time delay for the interferometer output of twice this in each leg of the interferometer, and both arms move, resulting in a total delay of

$$\Delta\tau = 2hL/c = 2h\tau_{\text{stor}}. \tag{7.45}$$

If we use Fabry–Perot cavities as arms, this will result in multiple trips between the mirrors, which we can model as $L = bL$, where b is the number of round trips. This time delay results in a phase difference in the output fringe of the interferometer:

$$\Delta\phi = h\tau_{\text{stor}}\frac{2\pi c}{\lambda}. \tag{7.46}$$

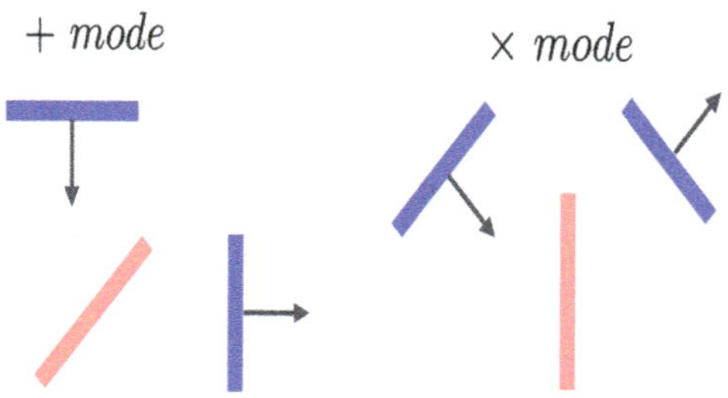

Figure 7.7. Gravitational wave modes.

This is the same result as if the arm length is changed by $\Delta L/L = h/2$. OK, let us start by considering a Michaelson interferometer, where the mirrors are attached to large suspended masses.

We are effectively making a position measurement on the mirror position(s) by bouncing photons off the mirror. We wish to locate the position of the mirror at two different times, before the arrival of the wave and immediately afterwards:

$$x(T) = x + pT/m. \tag{7.47}$$

This means that the two position measurements do not commute

$$[x(0), x(T)] = i\hbar T/m. \tag{7.48}$$

Assuming the two noises are uncorrelated, they combine as

$$\begin{aligned}\Delta x(T) &= \sqrt{(\Delta x(0))^2 + \frac{T^2(\Delta p(0))^2}{2m}} \\ &= \sqrt{(\Delta x(0))^2 + \frac{T^2\hbar^2}{2m(\Delta x(0))^2}}.\end{aligned} \tag{7.49}$$

This is minimized for

$$\Delta x_{\mathrm{SQL}} = \sqrt{\hbar T/m}. \tag{7.50}$$

Now let us separately consider the fundamental noise sources, shot noise and radiation pressure. We are effectively making a position measurement on the mirror position(s) by bouncing photons off the mirror. There will be noise in that measurement if there is an uncertainty in the photon number. As

$$\Delta x = \frac{c}{2\omega}\Delta\phi \tag{7.51}$$

with $\Delta\phi$ being the phase difference in the interferometer before and after the wave interacts, and for large N ΔN $\Delta\phi \approx 1$, we have

$$\Delta x_{\mathrm{SN}} = \frac{c}{2\omega}\frac{1}{\sqrt{N}}. \tag{7.52}$$

There will also be noise due to the photons reflecting off the mirror

$$\Delta p = \frac{2(\hbar\omega)}{2m} \tag{7.53}$$

that leads to a position noise from the photon recoil of $2p = 2\hbar\omega/c$,

$$\Delta x_{\mathrm{RP}} = \frac{\hbar\omega}{mc}\sqrt{N}. \tag{7.54}$$

Assuming that these noises are independent, we find the total noise

$$(\Delta x)^2 = (\Delta x_{\rm SN})^2 + (\Delta x_{\rm RP})^2 \tag{7.55}$$

$$= \frac{c^2}{4\omega^2}\frac{1}{N} + \frac{\hbar^2\omega^2 T^2}{m^2c^2}N. \tag{7.56}$$

Minimizing this with respect to the average photon number N yields the optimum operating point

$$N_{\rm opt} = \frac{mc^2}{4\hbar\omega^2 T}, \tag{7.57}$$

which recovers the standard quantum limit

$$\Delta x_{\rm SQL} = \sqrt{\hbar T/m}. \tag{7.58}$$

In terms of photon number, this gives us

$$\Delta x_{\rm SQL} = \frac{c}{\omega}\frac{1}{\sqrt{N}}. \tag{7.59}$$

The first argument based solely on the uncertainty principle is quite solid. The second approach was controversial for a bit, do not both mirrors see the same radiation pressure? It turns out they do, but the noise limit remains. This is due to the presence of vacuum fluctuations entering the unused port of the interferometer.

To see how this goes, we perform a full quantum treatment of the Michelson interferometer [11]. The SQL derived above assumed that the radiation pressure and shot noises are uncorrelated. This suggests that we put a field into the previously unused input port that has correlated noise between two quadratures: squeezed light. We assume no noise sources other than quantum ones, no losses, perfect mirrors, beam splitters, and detectors. As the signal from the exit port jitters, the amplitude and phase noise couple, and we would need to jitter the error ellipse of the squeezed light. More on that later, but for now we ignore that coupling, as for large N it is quite small.

We sketch the interferometer in figure 7.8.

Typically, a will be a strong input field in a coherent state, and b a vacuum or squeezed-vacuum field. In practice, the beams are misaligned to spatially separate them, and Faraday rotators (optical diodes) eliminate back reflections.

We are interested in the relative phase between the output fields c' and d' in the figure, which can be accessed by looking at the intensity of b'

Both c and d have been split from the same incident beam at the beam splitter, which means that, classically, all their fluctuations would be perfectly correlated, going back to fluctuations in a. As it turns out, this is not the case for a quantum field.

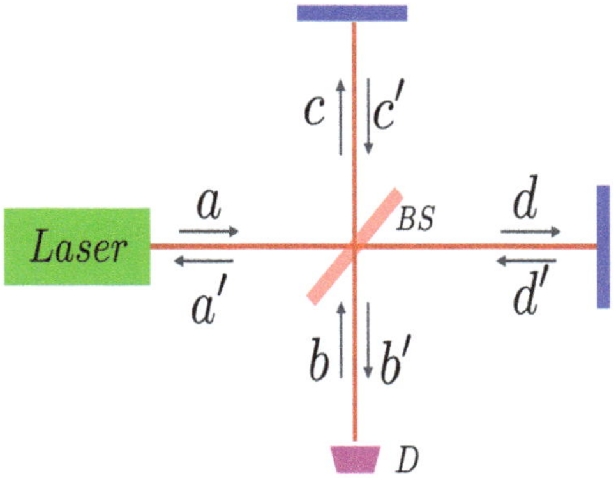

Figure 7.8. Michaelson interferometer for gravitational wave detection.

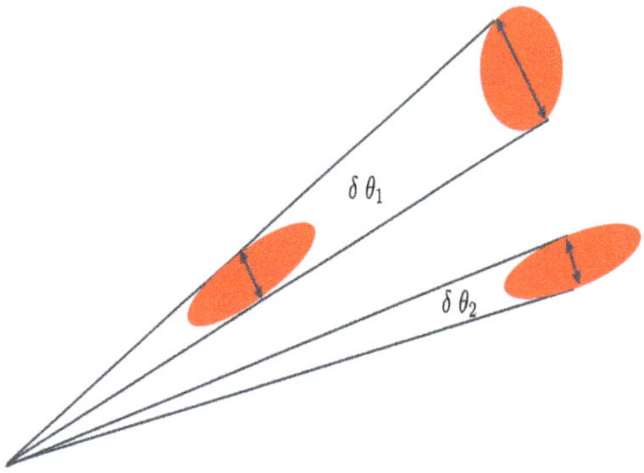

Figure 7.9. With squeezed input at b, the same resolution can be achieved with a less powerful laser, or at the same power we find enhanced resolution.

The origin of these uncorrelated fluctuations may be understood by considering the light marked as b in figure 7.9: the light entering the interferometer from the other port of the beam splitter. Classically, there is no amplitude in b, but there are vacuum fluctuations, 'nothing' is not nothing. We let the beam splitter combine the mode operators,

$$c = (a + b)/\sqrt{2} \tag{7.60}$$

$$d = (a - b)/\sqrt{2}. \tag{7.61}$$

While the fluctuations in the field a are indeed correlated in c and d as expected, so that they cancel when the two fields are subtracted and we operate at a null, the fluctuations in b are anticorrelated in c and d, so that they add when the two fields are subtracted. No combination of c and d can be made free from fluctuations, since, of course, the fluctuations of b are totally independent of those of a.

We may then think of the fluctuations in the difference between the fields c and d as being entirely due to the fluctuations in b, the vacuum mode of the field that enters the interferometer through the unused port of the beam splitter. Perhaps if we use 'quieter light' in some sense than vacuum noise, i.e. squeezed light, we might obtain a lower sensitivity to gravitational waves. This is essentially the proposal of Caves. Just adding amplitude *or* phase squeezed light will not work; we need to add light with some squeezing in both quadratures to correlate the radiation-pressure noise and the shot noise.

We formalize this (now) familiar language of squeezed states using 'amplitude' and 'phase' quadratures for an electric field whose phasor (in the complex-amplitude picture) lies along the common direction of $\langle a\rangle$, $\langle d\rangle$ and (with our choice of phase shifts at the beam splitter) $\langle c\rangle$. Let that direction be δ, where

$$\langle a\rangle = A\exp(i\delta); \tag{7.62}$$

the 'amplitude' and 'phase' quadratures a_1 and a_2 are, respectively,

$$a_{\rm amp} = \frac{1}{2}[a + a^\dagger] \tag{7.63}$$

and

$$a_{\rm ph} = \frac{1}{2i}[a - a^\dagger] \tag{7.64}$$

so that

$$a = (a_1 + ia_2) \tag{7.65}$$

$$\begin{aligned} b' &= \frac{1}{\sqrt{2}}\left(\frac{1}{\sqrt{2}}(a + b)e^{i\phi_c} - \frac{1}{\sqrt{2}}(a - b)e^{i\phi_d}\right) \\ &= \frac{1}{2}(2i\sin(\varphi/2)a + 2\cos(\varphi/2)b)e^{i(\phi_c + \phi_d)}, \end{aligned} \tag{7.66}$$

with ϕ_c and ϕ_d the phase changes for the round trip beams c and d, and $\varphi = \phi_c - \phi_d$. What is measured in an experiment with a photodetector is the photon number, $\langle b'^\dagger b'\rangle$, and the noise can be interpreted as shot noise in b, but it is driven by noise in a and b. The source of that noise is both the amplitude and phase noise in c and d. The signal that the photodetector detects is

$$b'^\dagger b' = \sin^2(\varphi/2)a^\dagger a + \cos^2(\varphi/2)b^\dagger b - i\sin(\varphi/2)\cos(\varphi/2)(a^\dagger b - b^\dagger a). \tag{7.67}$$

We consider a to be in a coherent state. Then, the expectation value of the square of the last term in equation (7.67) is

$$\begin{aligned} \langle(a^\dagger b - b^\dagger a)^2\rangle &= \langle a^\dagger a\rangle\langle((b) - b^\dagger)^2\rangle - \langle b^\dagger b\rangle \\ &= 4\langle n_a\rangle\left\langle\Delta b_2^2\right\rangle - \langle n_b\rangle, \end{aligned} \tag{7.68}$$

where n_a and n_b are the photon-number operators in modes a and b, respectively. The dominant part of the fluctuations is indeed proportional to the fluctuations in the phase quadrature, b_2, of mode b. If b is in a vacuum state, the first term has the value $4\langle n_a\rangle(1/4) = \langle n_a\rangle$, which leads to the standard shot noise limit. It is this term that we want to 'squeeze', by putting b in a state for which $\langle \Delta b_2^2\rangle \ll 1/4$,

$$\begin{aligned}\langle(\Delta b'^{\dagger}b')^2\rangle = \sin^4(\varphi/2)\left\langle \Delta n_a^2\right\rangle + \cos^4(\varphi/2)\left\langle \Delta n_b^2\right\rangle \\ + \sin^2(\varphi/2)\cos^2(\varphi/2)\langle(a^{\dagger}b - b^{\dagger}a)^2\rangle.\end{aligned} \tag{7.69}$$

When a is in a coherent state, one naturally has $\langle \Delta n_a^2\rangle = \langle n_a\rangle$. The other averages follow from the properties of squeezed-vacuum states

$$\left\langle \Delta n_b^2\right\rangle = 2\cosh^2 r \sinh^2 r \tag{7.70}$$

$$\left\langle \Delta b_1^2\right\rangle = \frac{1}{4}\exp(2r), \tag{7.71}$$

$$\left\langle \Delta b_2^2\right\rangle = \frac{1}{4}\exp(-2r), \tag{7.72}$$

$$\langle n_b\rangle = \sinh^2 r. \tag{7.73}$$

The fluctuations in the detected number of photons (the 'shot noise') are the noise terms shown in equation (7.68), which, by making use of equations (7.70)–(7.72), and keeping only the dominant contribution to each term (in an expansion in powers of φ_0), may be written as

$$N \simeq \left[\frac{1}{16}\varphi_0^4\langle n_a\rangle + 2\cosh^2 r \sinh^2 r + \frac{1}{4}\varphi_0^2\langle n_a\rangle \exp(-2r) + \frac{1}{4}\varphi_0^2 \sinh^2 r\right]^{1/2}. \tag{7.74}$$

When there is no squeezing ($r = 0$), we recover the SQL. The third term will be the dominant term:

$$\delta\varphi > \frac{\exp(-r)}{\sqrt{<n_a>}} = \frac{1}{\sqrt{\langle n_a\rangle n_{\rm sq}}}. \tag{7.75}$$

We would need to relax some of our assumptions, but if $\langle n_a\rangle \approx n_{\rm sq} = \sinh^2 r = N$ we approach the Heisenberg limit

$$\delta\varphi \approx 1/N. \tag{7.76}$$

It is difficult to operate at a null, so often an offset bias is imposed. However, the current method is to modulate the signal and look at the sidebands. For the squeezed enhancement, one must use two-mode squeezing, where the noise in the sidebands is correlated. Shot noise is dominant for higher frequencies, and radiation pressure for lower ones. To maximize sensitivity, one should adjust the angle of the squeezing ellipse so that one is injecting amplitude squeezed light for higher frequencies, and

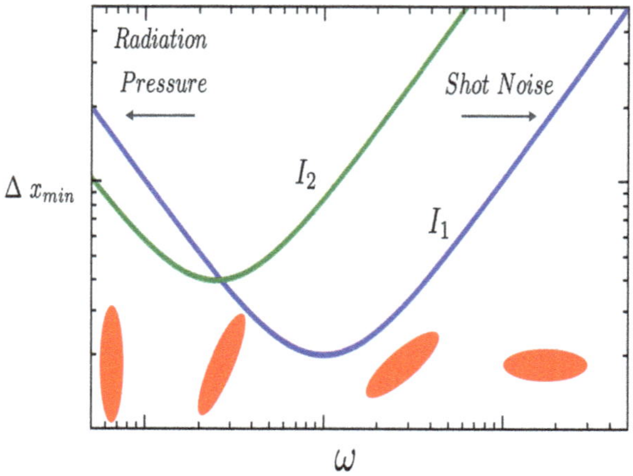

Figure 7.10. Noise curves, motivating the use of frequency-dependent squeezing.

phase squeezed light for lower frequencies. This can be accomplished using frequency-dependent squeezing (see figure 7.10).

7.3 Amplifier noise

Can one have a noiseless amplifier? If we have a gain G, this would require [6]

$$b = \sqrt{G}a, \tag{7.77}$$

where G is the intensity gain, b is the output mode, and a is the input mode. This is not possible as it would require

$$[b, b^\dagger] = G[a, a^\dagger] = G. \tag{7.78}$$

Let us consider a narrow-band amplifier (see figure 7.11), where

$$b = \sqrt{G} + F, \tag{7.79}$$

where F is a noise term that is needed to preserve the commutation relations. The requirement that $[b, b^\dagger] = 1$ results in

$$F = \sqrt{G-1}\,c^\dagger, \tag{7.80}$$

where $c^\dagger$ is an operator that obeys bosonic commutation relations. The operator $c^\dagger$ appears instead of c as we take $[c, c^\dagger] = 1$. Thus, we see that the amplifier must satisfy

$$b = \sqrt{(G)}a + \sqrt{(G-1)}c^\dagger. \tag{7.81}$$

We have input noise

$$(\Delta x)^2 = \frac{1}{2}\langle aa^\dagger + a^\dagger a\rangle \ = n_{\text{in}} + \frac{1}{2} \tag{7.82}$$

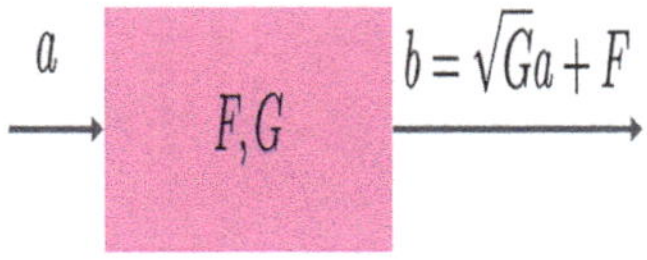

Figure 7.11. Linear amplifier.

for $G \gg 1$, and $\langle c^{\dagger}c\rangle = 0$ we have output noise

$$(\Delta x_{\text{out}})^2 = \frac{1}{2}\langle bb^{\dagger} + b^{\dagger}b\rangle = \frac{G}{2}\langle\{a + c^{\dagger}, a^{\dagger} + c\}\rangle = G\left(n_{\text{in}} + \frac{1}{2} + \frac{1}{2}\right) \quad (7.83)$$

where $G(n_{\text{in}} + 1/2)$ is the amplification of the input field plus vacuum fluctuations, and there is an additional noise introduced of $G/2$. Thus, there can be no noise-free amplifiers. To amplify light via interaction with atoms, we must have an inverted atomic system, and as such there will be spontaneous emission. This is the physical origin of the added noise. This is also related to the no-cloning theorem of quantum mechanics, which we discuss in chapter 16.

There is another effect of spontaneous emission in a laser. The intensity of the laser field is given by

$$dn = (A - C)n - Bn^2, \quad (7.84)$$

with A the linear gain, C the loss at the output mirror, and B the saturation parameter. This leads to a steady-state photon number

$$n_{\text{ss}} = \frac{A - C}{B}. \quad (7.85)$$

This is borne out from a full quantum theory. The laser turns on when the gain equals the loss, and increasing the linear gain (via increasing pumping) The spontaneous emission kicks cause the laser phase to diffuse in the plane (see figure 7.12). The deviation in the phase is given by

$$(\delta\phi)^2 = \frac{\langle\sin^2\theta\rangle}{n}. \quad (7.86)$$

The rate of these kicks is the gain of the laser A, which is equal to the loss C leading to

$$\Delta\omega = \frac{C}{2n_{\text{ss}}}, \quad (7.87)$$

where we have used $\langle\sin^2\theta\rangle = 1/2$. This is known as the Schawlow–Townes limit [15]. Thus, the laser has a finite linewidth due to spontaneous emission. There are many more noise sources for a real laser, and there are active locking schemes that can reduce the linewidth below the Schawlow–Townes limit.

When gain = loss, $A = C$, there is a second-order phase transition according to semiclassical theory [19]. In the fully quantum theory, the phase transition blurs due to the quantum noise. For small microlasers, the phase transition is quite blurry, and

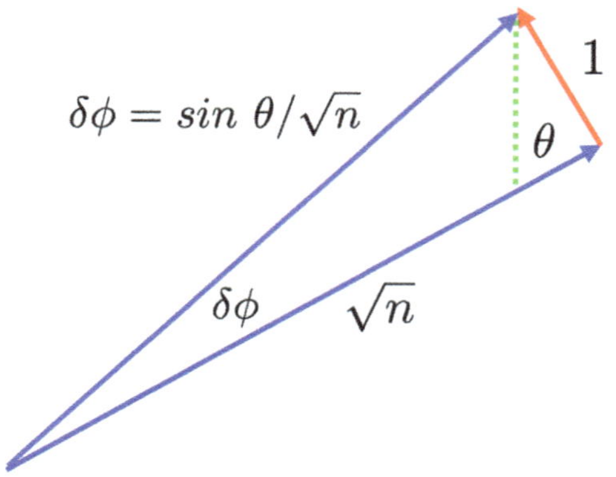

Figure 7.12. Laser phase diffusion due to spontaneous emission.

for a laser with an average intracavity photon number on the order of one, the phase transition is gone. There is no threshold [20].

7.4 More vaccuum effects

One interesting effect is the Casimir effect [3]. Two conducting plates, effectively forming an optical cavity, will attract one another! The physical reason is that inside the cavity, only certain distinct modes exist, each with their corresponding vacuum fluctuations. Outside the cavity, there are fluctuations of all frequencies, as shown in figures 7.13 and 7.14. Each frequency/mode exerts a pressure due to its non-zero energy content, and the pressure is greater on the outsides of the cavity than on the inside, hence there is an attraction. We are essentially comparing the difference between two infinite numbers, which is quite fast and loose of course. We refer the interested reader to the nice treatment of Milonni *et al* [21]. The result of careful computation results in

$$\frac{F_{\text{cas}}}{A} = -\frac{\hbar c \pi^2}{240 a^4}, \tag{7.88}$$

where a is the plate spacing. Experiments have confirmed this effect's existence, albeit with a lot of theoretical work to go with heroic experiments. It seems that real metal plates are not perfect conductors and have all kinds of imperfections, but we seem to have a good understanding of the basic effect as well as detailed calculations for particular experiments.

If we have two perfectly conducting plates separated by a distance a, the boundary conditions allow only the standing wave modes

$$k_n = \frac{n\pi}{a}, \quad n = 1, 2, 3, \ldots. \tag{7.89}$$

The corresponding frequencies are

$$\omega_n = \frac{c\pi n}{a}. \tag{7.90}$$

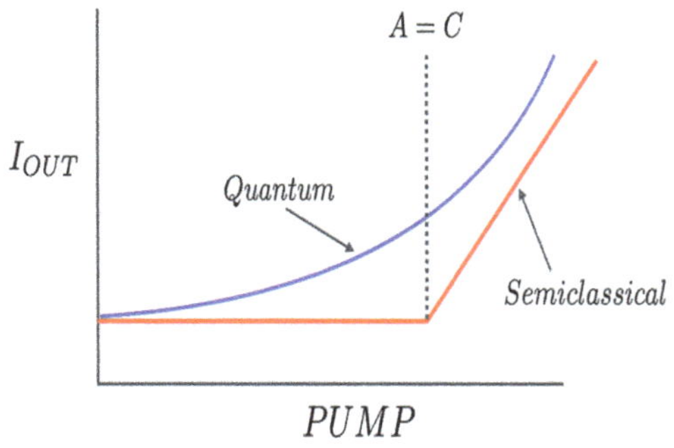

Figure 7.13. Laser output versus pump: semiclassical versus full quantum model.

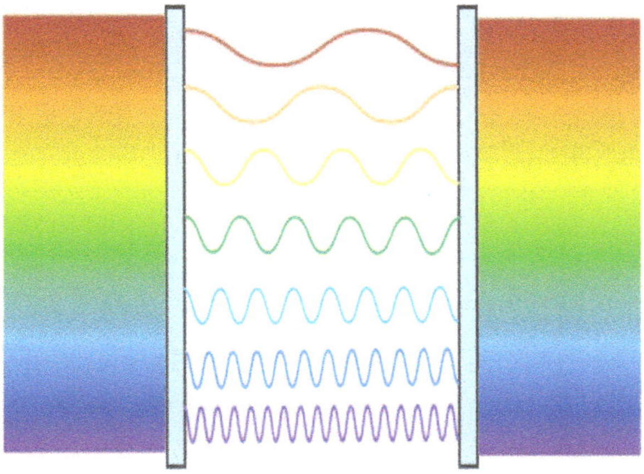

Figure 7.14. Unbalanced vacuum fluctuations in the Casimir effect.

The total vacuum energy per unit area between the plates is given by summing all the allowed modes

$$E_{\text{inside}} = \frac{1}{2}\sum_{n=1}^{\infty}\hbar\frac{c\pi n}{a}. \tag{7.91}$$

Similarly, the vacuum energy without the plates (free space) is given by an integral over all possible wave vectors

$$E_{\text{outside}} = \frac{1}{2}\int_{0}^{\infty} dn\hbar c\pi n. \tag{7.92}$$

The difference between the energies inside and outside gives the net effect. This sum is divergent, but using zeta function regularization

$$\sum_{n=1}^{\infty} n = -\frac{1}{12}. \tag{7.93}$$

The regularized energy per unit area is

$$E = -\frac{\hbar c\pi}{24a^3}. \tag{7.94}$$

The attractive force per unit area (pressure) is obtained by differentiating the energy with respect to the plate separation a:

$$F = -\frac{\partial E}{\partial a} = -\frac{\hbar c\pi}{8a^4}. \tag{7.95}$$

Expressing in terms of the standard form

$$F = -\frac{\hbar c}{240\pi a^4}. \tag{7.96}$$

Uh, what are these zeta function renormalizations? That is an infinite sum in the intro calculus! While $\sum_{n=1}^{\infty} n = -1/12$ is not valid in elementary mathematics, it makes sense in the context of more sophisticated mathematics. The idea of zeta function regularization is to assign finite values to divergent series using the Riemann zeta function and its analytic continuation. The Riemann zeta function is defined for s with a real part greater than 1 as

$$\zeta(s) = \sum_{n=1}^{\infty}\frac{1}{n^s}. \tag{7.97}$$

For example $\zeta(2) = \pi^2/6, \zeta(3) \approx 1.202$, -$\zeta(4) = \frac{\pi^4}{90}$. This series diverges for $s = -1$, but we can extend $\zeta(s)$ to other values using analytic continuation. To find $\zeta(-1)$, we use the following functional equation:

$$\zeta(s) = 2^s\pi^{s-1}\sin\left(\frac{\pi s}{2}\right)\Gamma(1 - s)\zeta(1 - s). \tag{7.98}$$

Setting $s = -1$, we obtain

$$\zeta(-1) = -\frac{1}{12}. \tag{7.99}$$

This is not the value of the original sum, but an assigned regularized value. Remember that the integral for the outside also diverges. But the difference in those two infinity can be measured, and they have been. The force can be repulsive in certain geometries, particularly with dielectrics. This includes designer metamaterials. As with many (most? all?) quantum effects, causation can be a tricky thing! The Casimir effect can alternatively be seen as a macroscopic extension of van der Waals forces, which are interactions between neutral atoms or molecules due to correlated dipole fluctuations. The Lifshitz theory [14] generalizes this to dielectric materials, showing that the force arises from the interaction of the fluctuating dipoles in the materials rather than from the vacuum energy. In this view, the Casimir forces are the result of the retarded dispersion forces in the continuous media.

There is also a dynamic Casimir effect [7, 17] If we take our two conducting plates and set one into oscillation, the energy that drives the oscillation can promote vacuum fluctuations into real particles (photons in this case). This requires that the velocity of the oscillating plate be a significant fraction of the speed of light. Therefore, this has yet to be observed. However, an electrical analog using a Josephson junction has verified the effect, it is far easier to modulate an electrical signal than to vibrate a plate!

Another interesting quantum effect is Unruh radiation [10, 16]. An accelerated photodetector exhibits excitations as if it were in a thermal field with temperature

$$T = \frac{\hbar a}{2\pi c k_B}, \tag{7.100}$$

where a is the acceleration of the detector and k_B is Boltzmann's constant. There is some controversy, but the detector absorbing a photon is well accepted. The concept of Unruh radiation, whereby an unaccelerated detector also sees photons, is not. However, if the acceleration of the detector promotes vacuum fluctuations to a pair of particles, the second detection seems reasonable. This is similar to Hawking radiation [12], whereby black holes emit radiation and hence decay as their mass decreases. One way to envision this process is that an observer far from the black hole horizon will see photons (and other particles), which result from pair fluctuations near the horizon. One of the pair falls into the black hole, while the other escapes, taking energy away from the black hole and reducing its mass. The particles cannot be localized on the horizon. Bekenstein observed that the area of a black hole could be considered a form of entropy [2]. He was led to this conclusion by the fact that the area of a black hole merger A is greater than the sum of the individual areas of the two original holes. He related the area to the entropy as

$$S_{\rm BH} = \frac{k_B A}{l_P^2}, \tag{7.101}$$

where l_p is the Planck length $l_p = \sqrt{G\hbar/c^3}$ Hawking questioned this, realizing that this would require [12]

$$dE = T\ dS \tag{7.102}$$

and the black hole would have a temperature

$$T_H = \frac{\hbar c^3}{8\pi G M k_B}. \tag{7.103}$$

One can think of this as the Unruh effect merged with the equivalence principle, in which the acceleration and gravitational fields are indistinguishable. Hot objects radiate, so a black hole should radiate a blackbody spectrum. Hawking proceeded to calculate the probability of this occurring using quantum field theory on a static Schwarzschild metric and found that the black hole will indeed radiate. The calculation uses quantum field theory and classical general relativity, and as such does not involve elements of quantum gravity, which would require a quantum description of the metric. This leads to the information paradox. Objects thrown into a black hole contain information about their composition. As the hole decays, the only information that is available via the radiation is the temperature of the hole, so information must be missing, it would seem. The resolution of this paradox is pending or trivial, depending on whom one talks to.

References

[1] Barish B C and Weiss R 1999 Ligo and the detection of gravitational waves *Phys. Today* **52** 44–50

[2] Bekenstein J D 2019 Black holes and the second law *Lett. Nuovo Cimento* **4** 737–40

[3] Casimir H B G 1948 On the attraction between two perfectly conducting plates *Indag. Math.* **10** 261–3

[4] Caves C M 1980 Quantum-mechanical radiation-pressure fluctuations in an interferometer *Phys. Rev. Lett.* **45** 75–9

[5] Caves C M 1981 Quantum-mechanical noise in an interferometer *Phys. Rev.* D **23** 1693–708

[6] Caves C M 1982 Quantum limits on noise in linear amplifiers *Phys. Rev.* D **26** 1817–39

[7] Dodonov V 2020 Fifty years of the dynamical Casimir effect *Physics* **2** 67–104

[8] Englert B-G 1996 Fringe visibility and which-way information: an inequality *Phys. Rev. Lett.* **77** 2154–7

[9] Franson J D 1989 Bell inequality for position and time *Phys. Rev. Lett.* **62** 2205–8

[10] Fulling S A 1973 Nonuniqueness of canonical field quantization in Riemannian space-time *Phys. Rev.* D **7** 2850–62

[11] Gea-Banacloche J and Leuchs G 1987 Squeezed states for interferometric gravitational-wave detectors *J. Mod. Opt.* **34** 793–811

[12] Hawking S W 1974 Black hole explosions? *Nature* **248** 30–1

[13] Hong C K, Ou Z Y and Mandel L 1987 Measurement of subpicosecond time intervals between two photons by interference *Phys. Rev. Lett.* **59** 2044–6

[14] Lifshitz E M 1956 The theory of molecular attractive forces between solids *Sov. Phys.—JETP* **2** 73–83

[15] Schawlow A L and Townes C H 1958 Infrared and optical masers *Phys. Rev.* **112** 1940–9

[16] Unruh W G 1976 Notes on black-hole evaporation *Phys. Rev.* D **14** 870–92

[17] Wilson C M, Johansson G, Pourkabirian A, Simoen M, Johansson J R, Duty T, Nori F and Delsing P 2011 Observation of the dynamical Casimir effect in a superconducting circuit *Nature* **479** 376–9

[18] Zou X Y, Wang L J and Mandel L 1991 Induced coherence and indistinguishability in optical interference *Phys. Rev. Lett.* **67** 318–21

[19] DeGiorgio V and Scully M O 1970 Analogy between the laser threshold region and a second-order phase transition *Phys. Rev.* A **2** 1170–7

[20] Rice P R and Carmichael H J 1994 Photon statistics of a cavity-QED laser: A comment on the laser–phase-transition analogy *Phys. Rev.* A **50** 4318–29

[21] Milonni P W, Cook R J and Goggin M E 1988 Radiation pressure from the vacuum: Physical interpretation of the Casimir force *Phys. Rev.* A **38** 1621–23

IOP Publishing

An Introduction to Quantum Optics (Second Edition)

An open systems approach

Perry Rice

Chapter 8

Two-level atom coupled to a quantized field

8.1 Atom–field interaction in quantum optics

The atom–field interaction we consider is an electric dipole interaction with the same form we considered in the semiclassical theory of chapter 4, with an interaction energy given by

$$H_{\text{int}} = -\vec{d}\cdot\vec{E}, \tag{8.1}$$

where $\vec{d}$ is the dipole moment of the atom and $\vec{E}$ is the electric field. Recall that the details of the dipole moment of the atom depend on the atomic wavefunctions,

$$\vec{d} = \begin{pmatrix} \langle e|e\vec{r}|e\rangle & \langle e|e\vec{r}|g\rangle \\ \langle g|e\vec{r}|e\rangle & \langle g|e\vec{r}|g\rangle \end{pmatrix}, \tag{8.2}$$

and the electric field is an operator

$$\vec{E} = \sqrt{\frac{\hbar\omega}{\epsilon_0 V}}(a + a^\dagger)\sin kz = \mathcal{E}_0(a + a^\dagger), \tag{8.3}$$

where V is the volume of the mode and k is the wavenumber of the mode. Assuming that the atom has no permanent dipole moment, the diagonal elements of the matrix dipole moment operator vanish, and we can write the operator in terms of the Pauli matrices

$$\begin{pmatrix} 0 & \mu_{eg} \\ \mu_{eg} & 0 \end{pmatrix} = \mu_{eg}(\sigma_+ + \sigma_-), \tag{8.4}$$

where μ_{eg} is the (real) dipole matrix element between the ground and excited states of the atom and σ_+ and σ_-, the same as in chapter 3. The transition matrix element is real for transitions that satisfy $\Delta m = 0$.

doi:10.1088/978-0-7503-6014-2ch8

To obtain the operator form of the atom–field interaction, we use the dipole moment operator and electric field operator above. This gives

$$\begin{aligned} H_{\text{int}} &= \hbar g_0 \sin kz(a + a^\dagger)(\sigma_+ + \sigma_-) \\ &= \hbar g_0 \sin kz(a\sigma_+ + a^\dagger\sigma_- + a\sigma_- + a^\dagger\sigma_+), \end{aligned} \tag{8.5}$$

with the atom–field coupling $g_0 \equiv \sqrt{(\mu_{eg}^2 \omega / 2\epsilon_0 V)}$. Considering each term of the interaction, we see that $a\sigma_+$ destroys a photon and excites the atom, $a^\dagger\sigma_-$ creates a photon and lowers the atom, $a\sigma_-$ destroys a photon and lowers the atom, and $a^\dagger\sigma_+$ creates a photon and excites the atom. The first two terms correspond to absorption and emission and conserve energy, whereas the last two terms do not conserve energy and describe virtual processes. Although these virtual processes produce small energy level shifts, they are not important in this work and thus their contributions will be neglected. They correspond to the rapidly oscillating terms we neglected in the rotating wave approximation in our work that treated the field classically in chapter 4. This model was first considered by Jaynes and Cummings and is therefore known as the Jaynes–Cummings model. A most excellent reference for the Jaynes–Cummings model and variants is [1]. The final form of the interaction Hamiltonian is then

$$H_{\text{int}} = \hbar g_0 \sin kz(a\sigma_+ + a^\dagger\sigma_-) \tag{8.6}$$

or, on changing the phase and mode function,

$$H_{\text{int}} = \imath\hbar g_0 \cos kz(a\sigma_+ - a^\dagger\sigma_-). \tag{8.7}$$

There is another way to see how the energy nonconserving terms can be neglected; let us consider the interaction picture. Here, we decompose the Hamiltonian into a 'bare' part that is independent of time plus a time-dependent interaction term,

$$H = H_0 + H_I(t). \tag{8.8}$$

Recall from chapter 3 that the wave function evolves in time via the interaction Hamiltonian,

$$\dot{\psi} = -\frac{i}{\hbar} H_I(t)\psi, \tag{8.9}$$

while operators also carry a time dependence

$$O(t) = e^{iH_0 t/\hbar} O(0) e^{-iH_0 t/\hbar}. \tag{8.10}$$

For the annihilation and creation operators for the field, we have

$$a(t) = e^{i\omega a^\dagger a t} a(0) e^{-i\omega a^\dagger a t} = a(0)e^{-i\omega t} \tag{8.11}$$

$$a^\dagger(t) = e^{i\omega a^\dagger a t} a^\dagger(0) e^{-i\omega a^\dagger a t} = a^\dagger(0)e^{i\omega t}. \tag{8.12}$$

To derive these results, we use the Baker–Campbell–Hausdorff theorem. This theorem is actually considerably broader than what we will use it for. Consider $e^{aA}Be^{-aA}$ with A and B operators and a as a scalar, and Taylor expand:

$$\begin{aligned} e^{aA}Be^{-aA} &= (1 + aA + a^2A^2/2 + \dots)B(1 - a^A + a^2A^2/2 + \dots) \\ &= B + a(AB - BA) + (a^2/2)(A^2B - BA^2) + \dots \\ &= B + a[A, B] + (a^2/2)(A^2B - ABA + ABA - BA^2) + \dots \\ &= B + a[A, B] + (a^2/2)[A, [A, B]] + \dots. \end{aligned} \tag{8.13}$$

It is also sometimes helpful to consider the commutator of a function $f(a^\dagger)$ with a. To begin, consider

$$\begin{aligned} [a, a^{\dagger 2}] &\equiv aa^{\dagger 2} - a^{\dagger 2}a \\ &= aa^{\dagger 2} - a^\dagger aa^\dagger + a^\dagger aa^\dagger - a^{\dagger 2}a \\ &= [a, a^\dagger]a^\dagger + a^\dagger[a, a^\dagger] \\ &= 2a^\dagger. \end{aligned} \tag{8.14}$$

Continuing in this manner, one can show

$$[a, a^{\dagger 3}] = 3a^{\dagger 2} \tag{8.15}$$

$$[a, a^{\dagger n}] = na^{\dagger n-1} \tag{8.16}$$

$$[a, f(a, a^\dagger)] = \frac{df}{da^\dagger} \tag{8.17}$$

and one can work out similar relations for commutators of the form $[a^\dagger, f(a)]$. Similarly, for the Pauli operators, we can show that

$$\sigma_+(t) = e^{i\omega\sigma_z t/2}\sigma_+(0)e^{-i\omega\sigma_z t/2} = \sigma_+(0)e^{i\omega t} \tag{8.18}$$

$$\sigma_-(t) = e^{i\omega\sigma_z t/2}\sigma_-(0)e^{-i\omega\sigma_z t/2} = \sigma_-(0)e^{-i\omega t}. \tag{8.19}$$

Using these relations, the interaction Hamiltonian in the interaction picture is

$$\begin{aligned} H_I(t) &= e^{i\omega(a^\dagger a+\sigma_z/2)t}H_I^{-i\omega(a^\dagger a+\sigma_z/2)t} \\ &= \hbar g_0 cos(kz)(a\sigma_+ e^{-i\Delta t} + a\sigma_- e^{i(\omega+\omega_0)t} \\ &\quad + a^\dagger\sigma_+ e^{i(\omega+\omega_0)t} + a^\dagger\sigma_- e^{-i\Delta t}), \end{aligned} \tag{8.20}$$

where as earlier we have $\Delta = \omega - \omega_0$. We see that dropping the rapidly oscillating terms is the same as dropping the energy nonconserving terms. We shall see in what follows that these terms do not yield any contribution to transition probabilities.

What about the form of the wave function? We have seen that the quantized electromagnetic field is essentially a harmonic oscillator whose state can be expanded as a superposition of Fock states, and the atom has a two-state system with excited and ground states, so we can write

$$|\Psi\rangle = \sum_n (C_{g,n}(t)e^{-iE_{g,n}t/\hbar}|g, n\rangle + C_{e,n}(t)e^{-iE_{e,n}t/\hbar}|e, n\rangle). \tag{8.21}$$

As we have taken the zero of the atoms energy to lie between the two states, we have $E_{g,\,n} = \hbar\omega_0(n - 1/2)$ and $E_{e,\,n} = \hbar\omega_0(n + 1/2)$ where for now we have assumed that the atom and field are both resonant at ω_0. We have separated out the time dependence that occurs with no interaction (the exponential dependence) and that due to the interaction (the time dependence of the probability amplitudes). In this case, the Schrödinger equation

$$|\dot{\psi}\rangle = -\frac{i}{\hbar}H|\psi\rangle \tag{8.22}$$

gives us, for an atom at $z = 0$,

$$\dot{C}_{e,n} = -g\sqrt{n+1}\,C_{g,n+1} \tag{8.23}$$

$$\dot{C}_{g,n+1} = g\sqrt{n+1}\,C_{e,n}. \tag{8.24}$$

Taking the time derivative of equation (8.23) and substituting it into equation (8.24), we find

$$\ddot{C}_{e,n} = -(n+1)g^2 C_{e,n} \tag{8.25}$$

with solutions

$$C_{e,\,n}(t) = C_{e,\,n}(0)\cos(g\sqrt{n+1}\,t) - C_{g,\,n+1}(0)\sin(g\sqrt{n+1}\,t) \tag{8.26}$$

$$C_{g,\,n+1}(t) = C_{g,\,n+1}(0)\cos(g\sqrt{n+1}\,t) + C_{e,\,n}(0)\sin(g\sqrt{n+1}\,t). \tag{8.27}$$

One obtains the same equations if one uses the wave function

$$|\psi\rangle_I = \sum_n (C_{g,\,n}|g,\,n\rangle + C_{e,\,n}|e,\,n\rangle) \tag{8.28}$$

and the interaction Hamiltonian equation (8.20).

For an atom initially in the ground state, $C_{g,n}(0) = 1$, we have

$$C_{e,n-1} = -\sin(g\sqrt{n}\,t) \tag{8.29}$$

and we see that at a later time, the probability of the atom being in the excited state while the field loses a photon is not zero; we have absorption. Note that if $n = 0$, there is no absorption. If the atom is initially in the excited state, $C_{e,n}(0) = 1$, we have

$$C_{g,n+1} = \sin(g\sqrt{n+1}\,t). \tag{8.30}$$

A seemingly extraordinary thing occurs; if the atom is in the excited state at $t = 0$ with no photons around ($n = 0$), there is still a probability for the atom to be in the ground state at a later time. By treating the field as a quantum operator, we find that spontaneous emission is contained automatically in the model.

Recall our semiclassical treatment of chapter 4,

$$\dot{C}_g = \frac{-i\Omega}{2}C_e \tag{8.31}$$

$$\dot{C}_e = \frac{-i\Omega}{2} C_g, \tag{8.32}$$

with $\Omega = \mu_{eg} E_0/\hbar$. One can make a connection via $\Omega/2 \to \Omega_{n+1}/2 \equiv g\sqrt{(n+1)} \approx \Omega_n/2$, which is consistent with $\hbar g/\mu_{eg}$ having units of electric field. We have ignored the difference between n and $n+1$, the semiclassical treatment is generally valid for large photon numbers, with the important distinction of needing to add spontaneous emission. It appears that we use $\Omega \to 2g\sqrt{n}$ for absorption, and $\Omega \to 2g\sqrt{n+1}$ for emission. The ratio of the stimulated emission to the spontaneous emission is n: 1, and in a rate equation limit, one has a total emission rate of $n+1$ and a total absorption rate of n.

By detuning the driving laser from atomic resonance, $\Delta = \omega_L - \omega_0$, we have

$$\dot{C}_{e,n} = -g\sqrt{n+1}\, C_{g,n+1} - \imath\Delta C_{e,n} \tag{8.33}$$

$$\dot{C}_{g,n+1} = g\sqrt{n+1}\, C_{e,n} - \imath\Delta C_{g,n}, \tag{8.34}$$

which leads to oscillatory solutions with frequencies $\Omega'_{n+1} = \sqrt{g^2(n+1) + \Delta^2}$, a generalized Rabi frequency.

For short times, we can approximate our results above or use first-order perturbation theory. For example, if we start in the excited state of the atom with n photons, we find

$$C_{g,n+1} = g\sqrt{n+1}\, e^{-i\Delta t/2} \frac{\sin(\Delta t/2)}{\Delta t/2} t, \tag{8.35}$$

which on resonance gives us

$$C_{g,n+1} = g\sqrt{n+1}\, t, \tag{8.36}$$

where again we see that spontaneous emission occurs with no photons in the field. If instead we start in the ground state of the atom with n photons in the field, we find

$$C_{e,n-1} = -g\sqrt{n}\, e^{-i\Delta t/2} \frac{\sin(\Delta t/2)}{\Delta t/2} t, \tag{8.37}$$

where again we find no spontaneous absorption

8.2 Wigner–Weisskopf approximation

Let us now consider the interaction of a two-level atom with many quantized field modes, as would be the case for an atom interacting with the vacuum modes around it [2]. In the interaction picture, we have the Hamiltonian

$$H_{\text{int}} = i\hbar \sum_k \left(g_k \sigma_+ a_k e^{i\Delta_k t} - g_k^* a^\dagger \sigma_- e^{-i\Delta_k t} \right), \tag{8.38}$$

where k denotes all fields modes to which the atom is coupled to. With the atom in the excited state initially and all the field modes unoccupied, the only other states accessible via the above Hamiltonian are the atom in the ground state, and *one* of the field modes occupied with one photon. We then take the state vector

$$|\psi(t)\rangle = C_{e,\,0}(t)|e,\,0\rangle + \sum_k C_{g,\,1k}|g,\,1k\rangle, \tag{8.39}$$

with initial conditions $C_{e,0} = 1$ and $C_{g,1k} = 0$. The Schrödinger equation (in the interaction picture) yields

$$\dot{C}_{e,0} = -\sum_k g_k^* e^{-i\Delta_k t} C_{g,1k} \tag{8.40}$$

$$\dot{C}_{g,1k} = g_k e^{i\Delta_k t} C_{e,0}. \tag{8.41}$$

We can formally solve the second of these equations,

$$C_{g,1k} = g_k \int_0^t dt' e^{i\Delta_k t'} C_{e,0}(t'). \tag{8.42}$$

We then substitute this into the equation for $C_{e,1}$, and find

$$\dot{C}_{e,0} = -\sum_k |g_k|^2 \int_0^t e^{-i\Delta_k(t-t')} C_{e,0}(t'). \tag{8.43}$$

So far, we have not made any approximations beyond our usual ones (dipole, rotating wave). We now must address the mode distribution of the field. As we are interested in the case of free space, we realize that the spacing between modes $c/2L$ goes to zero and we have a continuum. The discrete sum over modes can be turned into an integral and we switch to polar coordinates so that $d^3\overrightarrow{k} = 2dk_x dk_y dk_z = k^2 \sin(\theta) d\theta d\phi dk$ where the factor of 2 is due to the two possible polarizations. We have $V = L^3$ as the quantization volume. This leaves us with

$$\dot{C}_{e,0} = -\frac{2V|g_k|^2\omega}{(2\pi)^3} \int_0^{2\pi} d\phi \int_0^{\pi} d\theta \sin(\theta)\cos^2(\theta) \int_0^{\infty} dk k^2 |g_k|^2 \int_0^t e^{i\Delta(t-t')} C_{g,1}(t'). \tag{8.44}$$

The factor of $(L/2\pi)^3 = V/(2\pi)^3$ stems from the volume of the unit cell in phase space. The ϕ integral merely yields a factor of 2π, the θ integral is an average of $\cos^2(\theta)$ over three dimensions, which yields 1/3. This leaves us with

$$\dot{C}_{e,\,0} = -\frac{4|\mu_{eg}|^2}{6(2\pi)^2\epsilon_0\hbar c^3} \int_0^{\infty} d\omega\omega^3 \int_0^t e^{i\Delta(t-t')} C_{e,\,0}(t'), \tag{8.45}$$

where we have used the relations $\omega = ck$ and $g^2 = \omega|\mu_{eg}|^2/2\hbar\epsilon_0 V$. Now we make an approximation, taking the function ω^3 to be smoothly varying over the range of frequencies in the integral (basically the frequencies involved in the time development of $C_{e,0}(t)$ and the complex exponential. We replace it with ω_0^3 which then leaves us with

$$\dot{C}_{e,0} = -\frac{4|\mu_{eg}|^2\omega_0^3}{6(2\pi)^2\epsilon_0\hbar c^3} \int_0^{\infty} d\omega \int_0^t e^{-i\Delta(t-t')} C_{e,0}(t'). \tag{8.46}$$

The factor of $\cos^2(\theta)$ stems from the dot product between the field mode and the induced dipole moment. At this point, we assume that $C_e(t)$ is slowly varying compared to the oscillating exponential and pull it outside the integral. The frequency integral is then

$$\int_{-\infty}^{\infty} d\omega e^{i\Delta(t-t')} = 2\pi\delta(t - t'), \tag{8.47}$$

where we have extended the lower limit of the integral to $-\infty$. The δ function kills the time integral, leaving us with

$$\dot{C}_{e,0} = -\frac{\gamma}{2}C_{e,0}, \tag{8.48}$$

with

$$\gamma = \frac{\omega_0^3|\mu_{eg}|^2}{6\pi\epsilon_0\hbar c^3} \tag{8.49}$$

which leads to exponential decay.

$$C_{e,0}(t) = e^{-\gamma t/2} \tag{8.50}$$

for the probability amplitude, or for the probability we have

$$P_{e,0} = e^{-\gamma t}. \tag{8.51}$$

This result is the same that we obtained when we considered spontaneous emission via the rate equations in chapter 3, where we found the Einstein A coefficient,

$$A = \gamma = \frac{\omega_0^3|\mu_{eg}|^2}{6\pi\epsilon_0\hbar c^3}. \tag{8.52}$$

This treatment has a theme that we will see again when we consider a more general method for treating spontaneous emission and other dissipative processes in chapter 7. The density of states is proportional to an integral of ω^2 over ω, and as such it tends to blow up, i.e. diverge. This simply stems from the fact that there are an infinite number of modes. The coupling of the atom to each mode is small, however, as $g^2 \propto 1/V \to 0$, and many of the modes have some amount of detuning. Therefore, we have a large number (number of modes) multiplied by a small number (the coupling to each individual mode) that results in a finite result. In our formal work in chapter 9, we will see that this is a type of Born approximation. Also, we are making assumptions about the fact that the atom changes slowly over the time scale that the environment (the vacuum modes) can respond, that the environment reacts very fast, and this is known as the Markov approximation. This is a superb approximation for systems that oscillate at optical frequencies, and the typical theory of open (i.e. dissipative) systems uses the combined Born–Markov approximations.

8.3 Cavity-modified spontaneous emission

If we modify the density of states from that of the free-space distribution, we can alter the spontaneous emission rate from its free-space value. In a laser system, we might wish to enhance spontaneous emission into the cavity mode (which generates stimulated emission) and get rid of spontaneous emission out the sides of the cavity (which is an energy loss mechanism). The simplest change in the density of states is to make it zero! This could result from a cavity where all linear dimensions are less the $\lambda_0/2$, and so there are no modes for the atom to radiate into. This would result in a complete suppression of the spontaneous emission.

8.4 Dressed states reprise

One choice of basis for dealing with atom–field systems is the tensor product of the Fock states with the two atomic states. These are known as the bare states and are shown in figure 8.1.

Recall that for an atom interacting with a classical field, we considered the Hamiltonian

$$H = \frac{\hbar\Omega}{2}\sigma_z + i\hbar\Omega(\sigma_+ - \sigma_-), \tag{8.53}$$

with $\Omega = |\mu|_{eg}\, E_0/\hbar$ and found

$$|+\rangle = \cos(\Theta/2)||e\rangle + \sin(\Theta/2)|g\rangle \tag{8.54}$$

$$|-\rangle = -\sin(\Theta/2)|e\rangle + \cos(\Theta/2)|g\rangle, \tag{8.55}$$

with

$$\tan(\Theta) = \frac{\Omega}{\Delta} \tag{8.56}$$

with energies

$$E_\pm = \frac{\hbar}{2}(-\Delta \pm \Omega'). \tag{8.57}$$

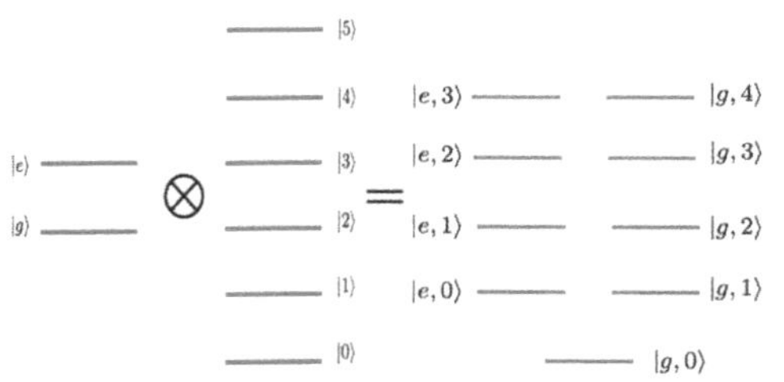

Figure 8.1. Bare and dressed states of a field–atom system.

Can we also consider the eigenstates of the full atom–field Hamiltonian with a quantized field? Sure! The Hamiltonian is, ignoring the constant zero point energy of the field,

$$H = \hbar\omega a^\dagger a + \frac{\hbar\omega_0}{2}\sigma_z + i\hbar g(a\sigma_+ - a^\dagger\sigma_-). \tag{8.58}$$

The matrix representation in the basis defined by $|g,\, n\rangle$ and $|e,\, n\rangle$ is block diagonal. The ground state energy is still $\hbar\omega/2$ even with the interaction. and so that is the upper left entry. The rest of the Hamiltonian is 2×2 blocks, as the interaction Hamiltonian only couples states $|g,\, n+1\rangle$ and $|e,\, n\rangle$ for $n \geqslant 1$. These blocks are

$$H = \hbar\begin{pmatrix} n\omega + \omega_0/2 & g\sqrt{n+1} \\ g\sqrt{n+1} & (n+1)\omega - \omega_0/2 \end{pmatrix}. \tag{8.59}$$

The eigenvalues (allowed energies) are

$$E_\pm = (n + 1/2)\hbar\omega \pm \frac{1}{2}\hbar\sqrt{\Delta^2 + 4g^2(n+1)}. \tag{8.60}$$

This is equivalent to the semiclassical theory if we make the replacement $2g\sqrt{n+1} \to \Omega$ and ignore the energy of the field mode $n\hbar\omega$ as before. The eigenstates are

$$|n,\, +\rangle = \cos(\Theta_n/2)|e,\, n\rangle + \sin(\Theta_n/2)|g,\, n+1\rangle \tag{8.61}$$

$$|n,\, -\rangle = -\sin(\Theta_n/2)|e,\, n\rangle + \cos(\Theta_n/2)|g,\, n+1\rangle, \tag{8.62}$$

with

$$\tan(\Theta_n) = \frac{\Omega_n}{\Delta} = \frac{2g\sqrt{n+1}}{\Delta}, \tag{8.63}$$

which once again can be turned into the semiclassical results with $\Omega_n \approx \Omega_{n+1} \to \Omega$. Again, on resonance, we have

$$|n,\, +\rangle = \sqrt{\frac{1}{2}}(|e,\, n\rangle + |g,\, n+1\rangle) \tag{8.64}$$

$$|n,\, -\rangle = \sqrt{\frac{1}{2}}(|g,\, n+1\rangle - |e,\, n\rangle). \tag{8.65}$$

The energy levels, known as the Jaynes–Cummings ladder, are shown in figure 8.2. Of particular note is that the two one-excitation states have a splitting of $\Delta E = 2\hbar g$. Recall that an atom in the excited state with no photons in the field would be represented by the state

$$|e,\, 1\rangle = \sqrt{\frac{1}{2}}(|1,\, +\rangle - |1,\, -\rangle) \tag{8.66}$$

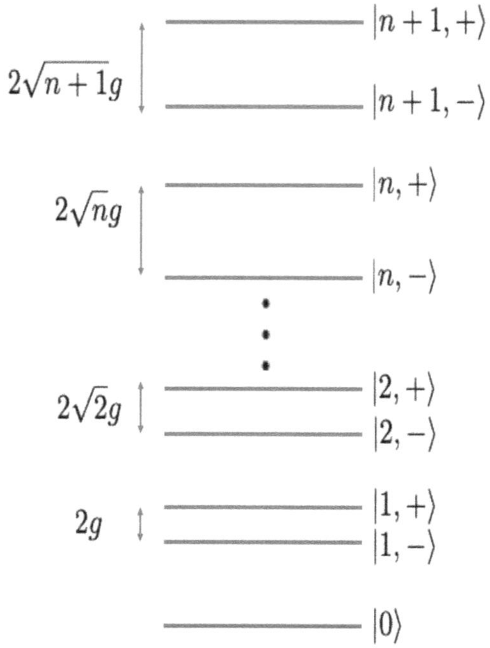

Figure 8.2. Dressed state ladder of the Jaynes–Cummings states.

and so it is in a superposition of the two one-excitation states. Spontaneous emission from this state would result in *two* spectral lines, which are known as the vacuum-Rabi doublet, or vacuum-Rabi splitting, $\Delta\omega = 2g$. The 'vacuum' part of this moniker is because the initial state of the field has no photons in it, it is in the vacuum state. However, note that if the atom were in the ground state initially with one photon in the field, the splitting would still occur, and so there is no magical mystical property of the vacuum involved. In fact, as we will show later, the phenomena of vacuum-Rabi splitting can be obtained from coupled oscillators. For these reasons, we prefer the term 'normal-mode splitting'. Now if one observes splittings higher up the Jaynes–Cummings ladder, say $\Delta\omega = g(2\sqrt{2} - 2)$ in addition to the $\Delta\omega = 2g$ splitting, then one has observed the discrete nature of the coupled atom–field system when it is quantized. This has been observed in the time domain by the group of Haroche. As the normal-mode splitting is essentially a classical phenomenon, it can be observed in solids at room-temperature.

8.5 Heisenberg equations of motion

Let us consider the quantized atom–field Hamiltonian

$$H = \hbar\omega a^\dagger a + \frac{\hbar\omega_0}{2}\sigma_z + i\hbar g(a^\dagger\sigma_- - a\sigma_+). \tag{8.67}$$

Let us now calculate the Heisenberg equations for the operators of interest, using

$$\dot{O} = \frac{i}{\hbar}[H, O]. \tag{8.68}$$

Let us start with the annihilation operator for the field,

$$\begin{aligned} \dot{a} &= [i\omega[a^{\dagger}a, a] - g[a^{\dagger}, a]\sigma_{-}] \\ &= -i\omega a + g\sigma_{-}, \end{aligned} \tag{8.69}$$

similarly we find

$$\dot{a}^{\dagger} = i\omega a^{\dagger} + g\sigma_{+} \tag{8.70}$$

$$\dot{\sigma}_{-} = i\omega_0\sigma_{-} + ga^{\dagger}\sigma_{+} \tag{8.71}$$

$$\dot{\sigma}_z = -2g(a^{\dagger}\sigma_{-} - a\sigma_{+}). \tag{8.72}$$

Evidently, we must continue to consider quantities such as $d(a\sigma_+)/dt$. This then couples to other operator products, for example

$$\frac{d(a\sigma_+)}{dt} = -i(\omega + \omega_0)a\sigma_+ + 2ga\sigma_z, \tag{8.73}$$

and then on and on we go. The system of equations for the operators does not close, so to solve things we would need some type of approximation or truncation scheme. Of interest is the equation of motion for the photon number,

$$\frac{d(a^{\dagger}a)}{dt} = g(a^{\dagger}\sigma_{-} - a\sigma_{+}), \tag{8.74}$$

where we note that we then have

$$\frac{d(a^{\dagger}a + \sigma_z/2)}{dt} = \frac{dC_1}{dt} = 0, \tag{8.75}$$

which is a statement of energy conservation. There is another conserved operator,

$$C_2 = \frac{\Delta\sigma_z}{2} + ig(a^{\dagger}\sigma_{-} - a\sigma_{+}). \tag{8.76}$$

Now we can relate these back to the Maxwell–Bloch equations of chapter 3, by considering the expectation values of the operator equations above, along with a factorization of the operator products

$$\langle\dot{a}^{\dagger}\rangle = i\omega\langle a^{\dagger}\rangle + g\langle\sigma_{+}\rangle - \kappa\langle a\rangle \tag{8.77}$$

$$\langle\dot{\sigma}_{-}\rangle = i\omega_0\langle\sigma_{-}\rangle + g\langle a^{\dagger}\rangle\langle\sigma_{+}\rangle - \frac{\Gamma}{2}\langle\sigma_{-}\rangle \tag{8.78}$$

$$\langle\dot{\sigma}_z\rangle = -2g(\langle a^{\dagger}\rangle\langle\sigma_{-}\rangle - \langle a\rangle\langle\sigma_{+}\rangle) - \gamma(\langle\sigma_z\rangle + 1), \tag{8.79}$$

where we have approximated $\langle a^\dagger \sigma_- \rangle \approx \langle a^\dagger \rangle \langle \sigma_- \rangle$ and added the phenomenological decay terms that we will derive later. If we make the correspondence $g\langle a \rangle = g\langle a^\dagger \rangle \approx \Omega/2$, we have the Maxwell–Bloch equations we discussed earlier. Of course, this factorization results in the semiclassical model, but is indeed incorrect as one must use the fully quantum mechanical model.

8.6 Collapse and revival of population inversion

Let us consider the population inversion of a two-level atom interacting with a quantized field [3]. This is given by

$$w(t) = \sum_n (|C_{e,n}(t)|^2 - |C_{g,n}(t)|^2), \tag{8.80}$$

where the summation is over the number of states of the field. If we have a definite number of photons in the field, $n = m$, and the initial state of the atom is the excited state, then this simplifies to

$$\begin{aligned} w(t) &= \cos^2(g\sqrt{n+1}\,t) - \sin^2(g\sqrt{n+1}\,t) \\ &= \cos(2g\sqrt{n+1}\,t), \end{aligned} \tag{8.81}$$

where we see Rabi oscillations due to the exchange of energy between the field and the atom. If instead the state of the field is not a Fock state, but initially has some distribution over Fock states, independent of the atoms state. In that case, the initial state vector is

$$|\psi(0)\rangle = \sum_n (C_n|n\rangle)(C_e|e\rangle + C_g|g\rangle), \tag{8.82}$$

we then have

$$w(t) = \sum_n P_n \cos(2g\sqrt{n+1}\,t), \tag{8.83}$$

where we have taken $P_n = |C_n|^2$. A plot of the inversion versus time is shown in figure 8.3, for an initial photon distribution for a coherent state. The early time evolution showing the collapse is presented in figure 8.4.

We notice that the inversion oscillates at a mean Rabi frequency $\Omega = g\sqrt{\bar{n}} = g|\alpha|$, but eventually decays, or 'collapses' to zero. At a later time these oscillations recur, or 'revive'. This is due entirely to the quantized nature of the field. The sum in equation (8.83) is the sum of cosines whose frequencies are not integral multiples of one another. Another way to say this is that equation (8.83) is *not* a Fourier series, and the population is not a periodic function. The different terms oscillate at different frequencies and, after some time, they are out of phase and destructive interference occurs. This would occur whether we used a sum over discrete photon numbers (as quantum mechanics tells us we must) or an integral to sum over a continuum of photon numbers (as some type of classical theory with no quantization of field amplitudes would allow). However, at a later time, constructive

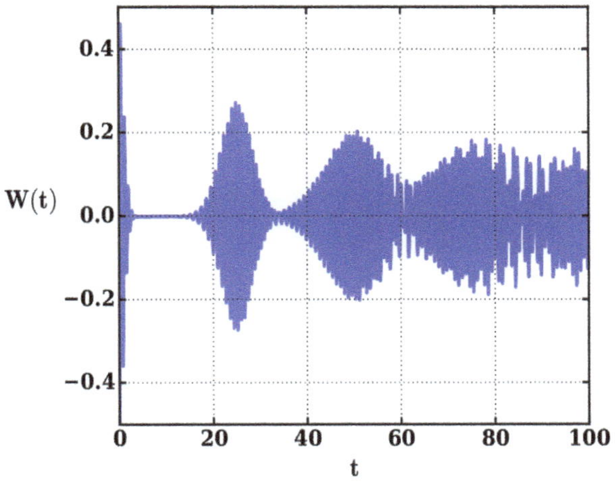

Figure 8.3. Collapse and revival of atomic inversion for initial coherent field state $\alpha = 4$.

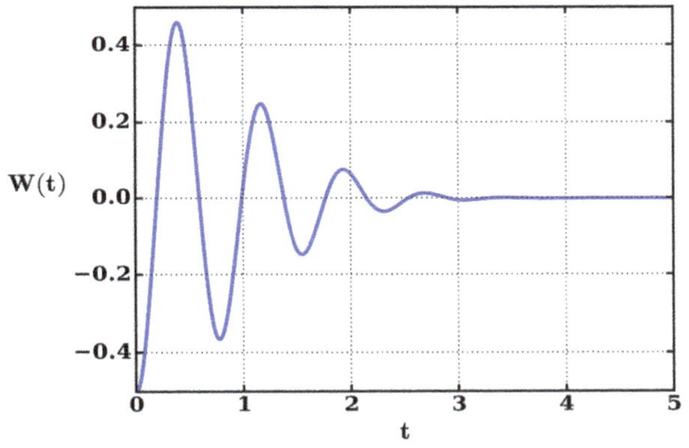

Figure 8.4. Collapse and revival of atomic inversion for initial coherent field state $\alpha = 4$. Only early times show collapse.

interference occurs, and this revival is a signature of a quantized field, i.e. that only discrete values of the photon number are allowed in the sum. To see this, let us examine the time scale of the collapses and revivals.

We can obtain a rough approximation for the collapse time by considering two terms of the series, one that corresponds to $n = \bar{n} + \Delta n$ and the other to $n = \bar{n} - \Delta n$. These give us two Rabi frequencies that bracket the range of frequencies in the sum of equation (8.83). When the phase difference between the two contributions is π, then we will start to see cancellations in the sum, and a 'collapse' occurs at time t_c, given by

$$2gt_c(\sqrt{\bar{n} + \Delta n + 1} - \sqrt{\bar{n} - \Delta n + 1}) = \pi. \tag{8.84}$$

Now we assume that the average photon number is large, and expand the square roots to obtain

$$\begin{aligned}\pi &= 2gt_c\sqrt{\bar{n}}\left(1+\frac{\Delta n}{\bar{n}}+\frac{1}{\bar{n}}-1+\frac{\Delta n}{\bar{n}}-\frac{1}{\bar{n}}\right)\\ &= 4gt_c\frac{\Delta n}{\sqrt{\bar{n}}}.\end{aligned} \tag{8.85}$$

For a coherent state, we have $\Delta n = |\alpha| = \sqrt{\bar{n}}$ and we have

$$t_c = \frac{\pi}{4g}, \tag{8.86}$$

which is independent of $\bar{n} = |\alpha|^2$. The revivals occur when the frequencies involved are all commensurate again, meaning that the phase difference between adjoining terms is an integer times 2π, at the mth revival time t_r^m:

$$\begin{aligned}2\pi m &= 2gt_r^m(\sqrt{\bar{n}+1}-\sqrt{\bar{n}})\\ &= 2gt_r^m\sqrt{\bar{n}}(\sqrt{1+1/\bar{n}}-1)\\ &\approx 2gt_r^m\sqrt{\bar{n}}\left(1+\frac{1}{2\bar{n}}-1\right)\\ &= gt_r^m/\sqrt{\bar{n}},\end{aligned} \tag{8.87}$$

which leads to a revival time of the order

$$t_r^m = 2\pi m\sqrt{\bar{n}}/g, \tag{8.88}$$

which agrees with the results of the full summation in figure 8.3. We have expanded the square root, which is appropriate for large $\bar{n}$. We can perform a more rigorous derivation, let us write $n = \bar{n} + \delta n$, where δn is the difference of n from the average number and is not to be confused with the standard deviation of the photon distribution Δn. The main part of this is the expansion of the square root,

$$\begin{aligned}\sqrt{n+1} &= \sqrt{\bar{n}+\delta n+1}\\ &= \sqrt{\bar{n}}\sqrt{1+\frac{\delta n}{\bar{n}}+\frac{1}{\bar{n}}}\\ &= \sqrt{\bar{n}}\left(1+\frac{\delta n}{2\bar{n}}\right).\end{aligned} \tag{8.89}$$

The inversion then becomes

$$\begin{aligned}w(t) &= e^{-\bar{n}}\sum_n\frac{\bar{n}^n}{n!}\cos\left(g\sqrt{\bar{n}}\left(1+\frac{\delta n}{2\bar{n}}\right)\right)\\ &= \frac{1}{2}e^{-\bar{n}}\sum_n(e^{2igt\sqrt{\bar{n}+1}}e^{igt(n-\bar{n})/\sqrt{\bar{n}}}+e^{-2igt\sqrt{\bar{n}+1}}e^{-igt(n-\bar{n})/\sqrt{\bar{n}}}).\end{aligned} \tag{8.90}$$

The sums can be done using

$$\sum_n\frac{\bar{n}^n}{n!}e^{igt(n-\bar{n})/\sqrt{\bar{n}}} = \exp(\bar{n}e^{igt/\sqrt{\bar{n}+1}}). \tag{8.91}$$

For short times, we find that

$$e^{igt/\sqrt{\bar{n}+1}} \simeq 1 + \frac{igt}{\sqrt{\bar{n}+1}} - \frac{g^2t^2}{2(\bar{n}+1)} \tag{8.92}$$

and we then have for the inversion (for short times)

$$w(t) - \cos(2gt\sqrt{\bar{n}+1})\exp\left(-\frac{g^2t^2\bar{n}}{2(\bar{n}+1)}\right), \tag{8.93}$$

which has a Gaussian dependence with decay time

$$t_c = \frac{1}{g}\sqrt{\frac{2(\bar{n}+1)}{\bar{n}}} \approx \frac{\sqrt{2}}{g}, \tag{8.94}$$

which is consistent with our previous estimate. In figure 8.3 we show an example of a collapse and revivals for $\alpha = 4$. A blow-up of early times to show the collapse is shown in figure 8.4.

In figure 8.5 we illustrate the quantum collapse and revival with a phasor diagram. All of the phasors have incommensurate frequencies, but they are distinct in the quantum case. As indicated above, a time will come when they do indeed arrive back in phase, leading to a revival.

In contrast, in the classical case, figure 8.6, all frequencies occur, and after the collapse there is never a time when all of the phasors recur, hence no revival. The revival is a direct indication of the quantum nature of the electromagnetic field.

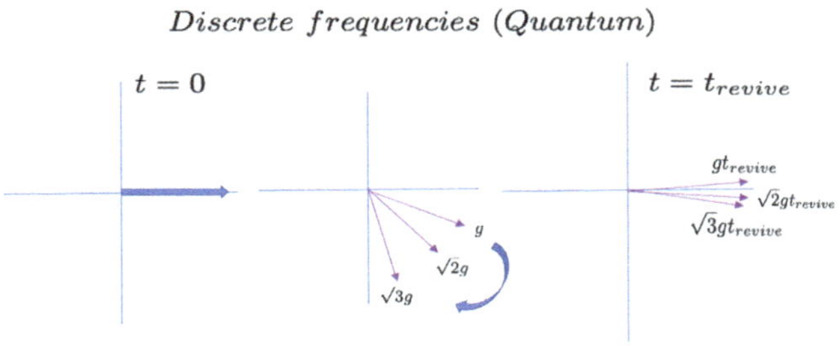

Figure 8.5. Phasor diagram of a few quantum states rotating at distinct incommensurate frequencies, leading to a collapse and revival.

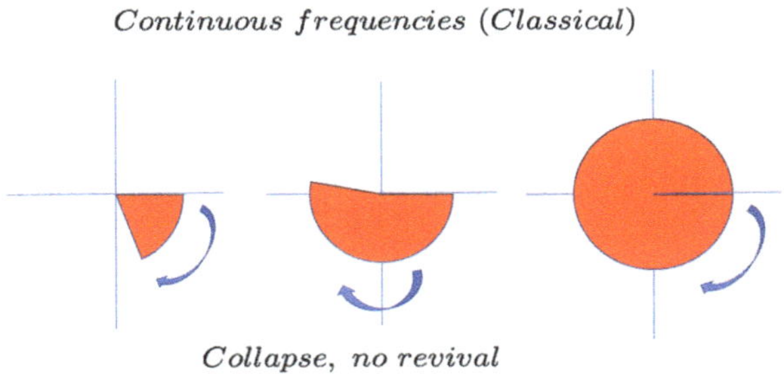

Figure 8.6. Phasor diagram of a few quantum states rotating at distinct incommensurate frequencies, leading to a collapse and revival.

8.7 Vacuum fluctuations and the radiation reaction

Are the vacuum fluctuations real? Yes indeed. An electron is surrounded by a cloud of virtual electron–positron pairs that pop into and out of existence. This is allowed by the time–energy uncertainty principle, $\Delta E \Delta t \geqslant \hbar/2$. To create such a pair, one must have an energy fluctuation of at least $2mc^2$, so the longest the fluctuation can last is $\Delta t = \hbar/2mc^2$. Pairs with kinetic energy last for a shorter time. When one combines special relativity with quantum mechanics, one is immediately led to various new effects. The first is the intrinsic spin of the electron; this fact falls directly out of the construction of the Dirac equation. Second, the Dirac equation also predicts antimatter, particles of opposite charge but otherwise identical. They are created in pairs and annihilate to yield two counterpropagating photons. Finally, the necessity of antimatter and the time–energy uncertainty principle means that we must have this virtual cloud of particle–antiparticle pairs, as shown in figure 8.7. The marriage of special relativity and quantum mechanics is known as quantum field theory, and there are no single-particle states. This virtual cloud acts as a dielectric medium in which the electron finds itself in. The effects of this cloud show up as Lamb shifts, small changes in the resonant frequency of a transition. Naive calculations show that the mass and charge of this single electron (known as a bare electron) need to be infinite. However, the charge and mass we measure in the laboratory is of course not infinite. However, in the lab we see the electron and the cloud, the real electron as it were. One can find a logically consistent procedure for performing calculations with the measured charge and mass. This means we should *start* with a theory for the electron plus cloud, and not pretend that single particles exist. For very high-energy probes, you do not measure the same values as at low energies; you probe deeper into the cloud essentially. These Lamb shifts are typically not mentioned in quantum optics. We cannot see the bare transition, so we take the experimental location of the spectral line to be the bare transition frequency plus the Lamb shift. One can predict the size of these shifts for simple atoms such as hydrogen, and obtain agreement to nine decimal places or more. So it is an ugly procedure in some ways, but it is the best theory we have! This leads to atomic clocks that may only lose a fraction of a second over the lifetime of the Universe. There are

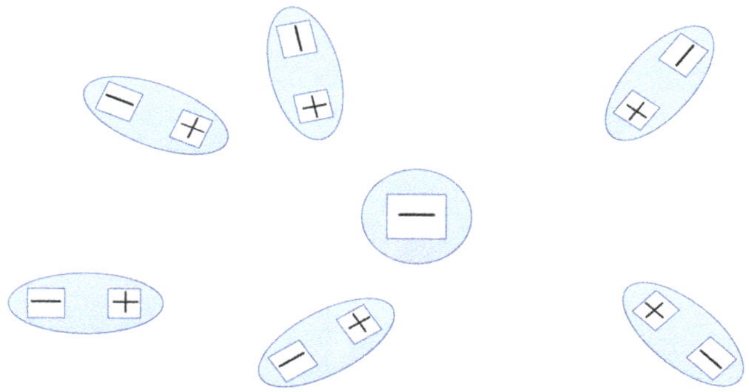

Figure 8.7. A cloud of virtual pairs around a bare particle.

also virtual pairs of W^+ and W^- and pairs of Z^0 particles, the exchange bosons for the weak force. The electromagnetic and weak interactions are unified in a gauge theory using U(1) $\otimes$ SU(2). The particle we call the photon is actually a mixture of the U(1) boson and the uncharged SU(2) boson. Similarly the Z^0 ia a mixture of the two fundamental bosons. If that were not the case, the theories would not be unified, but just stapled together. The most precise measurements of the mass of the vector bosons $W^\pm$, Z^0 come from precise measurements of the Lamb shifts.

What about the spontaneous emission? Why does an excited atom eventually decay? If it is due to the 'jostling' of vacuum fluctuations, why is there no spontaneous absorption? To give an answer to this, we consider the roles of vacuum fluctuations and radiation reaction in spontaneous processes [4, 5].

Let us consider the interaction of a two-level atom with a collection of modes, which are in their vacuum state:

$$H = \hbar\omega_0\sigma_z + \sum_k \hbar\omega_k a_k^\dagger a_k + \sum_k g_k(\sigma_+ a_k + a_k^\dagger \sigma_-), \tag{8.95}$$

where the index k runs over all wavevectors and polarizations. The resonant frequency of the atom is ω_0, and the frequency of the field modes is denoted by ω_k. We have written this in normal ordering, the annihilation operators to the right, and the creation operators to the left. Since the field and atom operators commute at equal times, there is absolutely no difference in the final result. We will preserve this order throughout the calculation; different orderings will give the same result for the Einstein A coefficient, but the (possible) interpretation will change.

The Heisenberg equations of motion are, preserving the order,

$$\dot{\sigma}_- = -i\omega_0\sigma_- + \sum_k g_k a_k^\dagger \sigma_z \tag{8.96}$$

$$\dot{\sigma}_z = -2\sum_k g_k(\sigma_+ a - a_k^\dagger \sigma_-) \tag{8.97}$$

$$\dot{a}_k = -i\omega_k + g_k(\sigma_+ + \sigma_-). \tag{8.98}$$

We formally integrate the field operator equation to find

$$a_k(t) = a_k(0)e^{-i\omega_k t} + g_k \int_0^t dt'\sigma_- e^{-i\omega_k(t-t')} \tag{8.99}$$

and the lowering operator to find

$$\sigma_-(t) = -i\omega_0\sigma_- - \sum_k g_k \int_0^t dt' a_k e^{-i\omega_-(t-t')}. \tag{8.100}$$

These solutions enter into the equation for the inversion operator multiplied by g_k. Therefore, to keep the calculation at second order altogether in g_k, for purposes of substitution into that equation we must use

$$\sigma_-(t') = \sigma_-(t)e^{-i(\omega_0(t'-t))} \tag{8.101}$$

and

$$a(t') = a(t)e^{-i(\omega_k(t'-t))} \tag{8.102}$$

to find

$$\begin{aligned} a_k(t) &= a_k(0)e^{-i\omega_k t} + g_k\sigma_-(t)\int_0^t dt' e^{-i(\omega_k - \omega_0)(t-t')} \\ &= a_k(0)e^{-i\omega_k t} + \pi g_k\sigma_-(t) \end{aligned} \tag{8.103}$$

and

$$\sigma_-(t) = \sigma_-(0)e^{-i\omega_0 t} - \sum_k g_k a_k(t)\int_0^t dt' e^{-i(\omega_k - \omega_0)(t-t')} \tag{8.104}$$

$$= -i\omega_0\sigma_- - \pi\sum_k g_k a_k(t). \tag{8.105}$$

This is the same approximation we used in deriving the Wigner–Weisskopf result. Substituting these solutions into the equation for the inversion operator, we find

$$\begin{aligned} \dot{\sigma}_z = \Bigg[& -2\sum_k(\sigma_-(0)a_k(0)e^{-i(\omega_0 - \omega_k)t}) \\ & -2\pi^2\sum_k \pi^2 g_k^2(a_k^\dagger(t)a(t)) \\ - & -2\pi^2\sum_k \pi^2 g_k^2(\sigma_+(t)\sigma_-(t))\Bigg] + \text{h.c.}. \end{aligned} \tag{8.106}$$

Taking an expectation value over the vacuum field state and using product rules for the Pauli operators, we find

$$\dot{\sigma}_z = 0 - 4\beta(\sigma_z + 1) \tag{8.107}$$

with

$$\beta = \pi^2\sum_k g_k^2. \tag{8.108}$$

Converting the sum to an integral over the field couplings in the continuum limit, we find

$$\dot{\sigma}_z = 0 - \gamma(\sigma_z + 1), \tag{8.109}$$

where γ is the usual free-space spontaneous emission rate, the Einstein A coefficient. Apparently the vacuum field makes no contribution as $\langle a_k^\dagger a_k\rangle = 0$, hence the spontaneous emission comes from the radiation reaction, $\gamma = \gamma_{\text{RR}}$.

Let us consider using

$$\dot{\sigma}_z = -2\sum_k g_k(a\sigma_+ - a_k^\dagger\sigma_-)a^\dagger. \tag{8.110}$$

As the field and atom operators commute, we will of course obtain the same result. Going through the previous procedure, we find

$$\begin{aligned}\dot{\sigma}_z = \Big[& -2\sum_k (\sigma_-(0)a_k(0)e^{-i(\omega_0 - \omega_k)t}) \\ & -2\pi^2\sum_k \pi^2 g_k^2 (a_k(t)a_k^\dagger(t)) \\ & -2\pi^2\sum_k \pi^2 g_k^2 (\sigma_-(t)\sigma_+(t)) \Big] + \text{h.c..}\end{aligned} \tag{8.111}$$

Taking the expectation values with respect to the vacuum field yields

$$\dot{\sigma}_z = 0 - \gamma(\sigma_z - 1) - 2\gamma\sigma_z, \tag{8.112}$$

where the first term comes from $\sigma_-\sigma_+ = (1/2)(\sigma_z - 1)$ and $\langle a_k^\dagger a_k \rangle = 1$.

The final result is of course

$$\dot{\sigma}_z = -\gamma(\sigma_z + 1). \tag{8.113}$$

But now we have $\gamma = 2\gamma_{\text{VF}} - \gamma_{\text{RR}}$, and so our interpretation can be different; spontaneous emission comes from an interplay of the radiation reaction and vacuum fluctuations. As Dailbard, Dupon-Roc, and Cohen-Tannoudji found [6], using a symmetrical ordering for field operators yields $\gamma = \gamma_{\text{VF}}$ and it seems that spontaneous emission is purely a vacuum fluctuation result. They argue that because symmetrically ordered operators are Hermitian, one could in principle measure them separately, and that the vacuum fluctuations are best considered as the source for spontaneous emission. The interpretation of the cause of spontaneous emission can change with different initial choices of operator ordering for commuting variables, but the result for the total rate is the same. Quantum mechanics does not specify the intermediate process as that is not monitored. What we do know for sure is that we obtain the Einstein A coefficient no matter what order we use for commuting atom and field operators!

8.8 Connections to supersymmetry

One might wonder if having a bosonic mode (a, $a^\dagger$) and a fermionic mode (c, $c^\dagger$) means the system is supersymmetric. The Jaynes–Cummings is not, but it is similar. We ignore damping to close the system. The system should have a supersymmetric Hamiltonian of the form

$$H = \{Q, Q^\dagger\}, \tag{8.114}$$

where Q and $Q^\dagger$ are supercharges that satisfy the SUSY algebra. The SUSY algebra is

$$\{Q, Q^\dagger\} = H, \quad Q^2 = Q^{\dagger 2} = 0, \tag{8.115}$$

where Q and $Q^\dagger$ are fermionic operators that transform bosonic states into fermionic ones and vice versa. The supercharges are typically constructed using the fermionic annihilation and creation operators ψ, $\psi^\dagger$, along with the usual bosonic ladder operators a, $a^\dagger$:

$$Q = a\psi^\dagger, \quad Q^\dagger = a^\dagger\psi, \tag{8.116}$$

where a, $a^\dagger$ satisfy the bosonic commutation relation, $[a, a^\dagger] = 1$, and ψ, $\psi^\dagger$ satisfy the fermionic anticommutation relation, $\{\psi, \psi^\dagger\} = 1$. From the supercharges, the Hamiltonian is

$$H = a^\dagger a + \frac{1}{2}(\psi^\dagger\psi - \psi\psi^\dagger), \tag{8.117}$$

which includes both the bosonic and fermionic contributions. A key property of the Jaynes–Cummings model is the excitation-number conservation, as the total excitation-number operator

$$N = a^\dagger a + \frac{1}{2}(\sigma_z + 1) \tag{8.118}$$

commutes with the Hamiltonian. A natural choice for a supercharge would be

$$Q = a\sigma_+, \quad Q^\dagger = a^\dagger\sigma_-. \tag{8.119}$$

These satisfy the fermionic anticommutation properties

$$Q^2 = Q^{\dagger 2} = 0. \tag{8.120}$$

If we compute the anticommutator

$$H_{\text{SUSY}} = \{Q, Q^\dagger\} \tag{8.121}$$

we obtain

$$H_{\text{SUSY}} = QQ^\dagger + Q^\dagger Q. \tag{8.122}$$

Expanding, we find

$$QQ^\dagger = a\sigma_+ a^\dagger\sigma_-, \tag{8.123}$$

$$Q^\dagger Q = a^\dagger\sigma_- a\sigma_+. \tag{8.124}$$

Using commutation relations for bosonic and fermionic operators

$$a\sigma_+ a^\dagger\sigma_- = aa^\dagger\sigma_+\sigma_-, \quad a^\dagger\sigma_- a\sigma_+ = a^\dagger a\sigma_-\sigma_+ \tag{8.125}$$

and recall

$$\sigma_+\sigma_- = \frac{1}{2}(\sigma_z + 1), \quad \sigma_-\sigma_+ = \frac{1}{2}(\sigma_z - 1), \tag{8.126}$$

we obtain

$$H_{\text{SUSY}} = a^\dagger a + \frac{1}{2}\sigma_z. \tag{8.127}$$

This closely resembles the Jaynes–Cummings Hamiltonian without the interaction term, meaning that the standard Jaynes–Cummings model is not quite supersymmetric on its own. The missing interaction term in the Jaynes–Cummings Hamiltonian, $g(a^\dagger\sigma_- + a\sigma_+)$, suggests that the supercharges should include an explicit coupling term. A familiar approach is to define, for instance,

$$Q_- = \alpha(N+1)a\sigma_+ + \beta(N+1)a^\dagger\sigma_-, \tag{8.128}$$

$$Q_+ = Q_-^\dagger, \tag{8.129}$$

where $\alpha(N+1)$ and $\beta(N+1)$ are chosen so that

$$\{Q_-, Q_+\} = H_{\rm JC} \pm C$$

with C some constant. In that more general factorization, one picks up the $(a^\dagger\sigma_- + a\sigma_+)$ term inside the anticommutator. The price is that the supercharges look 'dressed' by functions of the number operator N. In the resonant case ($\omega = \omega_0$), one simple choice is (up to overall constants)

$$\alpha(N) = \sqrt{N}, \tag{8.130}$$

$$\beta(N) = \sqrt{N}, \tag{8.131}$$

where $N = a^\dagger a$. For nonresonant cases ($\omega \neq \omega_0$), or if one wants a factorization exactly matching the usual Jaynes–Cummings form without leftover constants, the required α, β become more complicated. One choice is

$$\alpha(N+1) = F(N+1), \tag{8.132}$$

$$\beta(N+1) = G(N+1), \tag{8.133}$$

where F and G are chosen to capture the 'Rabi splitting' $\sqrt{(\omega-\omega_0)^2 + 4g^2(N+1)}$, using

$$E_n^{(\pm)} = \hbar(n+1)\omega \pm \hbar g\sqrt{n+1+\left(\frac{\omega-\omega_0}{2g}\right)^2}$$

and the functions F, G are picked so that

$$Q_-\left|E_n^{(\pm)}\right\rangle = (\text{constant})\left|E_{n-1}^{(\pm)}\right\rangle$$

with $Q_-|E_0^{(\pm)}\rangle = 0$. Solving such step-by-step 'annihilation' conditions determines (up to phases) the explicit formulas for $\alpha(N)$, $\beta(N)$.

References

[1] Larson J and Mavrogordatos T 2021 *The Jaynes-Cummings Model and Its Descendants* (Bristol: Institute of Physics Publishing)

[2] Weisskopf V and Wigner E 1930 Calculation of the natural brightness of spectral lines on the basis of Dirac's theory *Z. Phys.* **63** 54

[3] Eberly J H, Narozhny N B and Sanchez-Mondragon J J 1980 Periodic spontaneous collapse and revival in a simple quantum model *Phys. Rev. Lett.* **44** 1323
[4] Milonni P W 1976 Semiclassical and quantum-electrodynamical approaches in nonrelativistic radiation theory *Phys. Rep.* **25** 1–81
[5] Milonni P W 1994 *The Quantum Vacuum: An Introduction to Quantum Electrodynamics* (New York: Academic)
[6] Dalibard J, Dupont-Roc J and Cohen-Tannoudji C 1982 Vacuum fluctuations and radiation reaction: identification of their respective contributions *J. Phys. France* **43** 1617–38

Chapter 9

Coherence and detection

9.1 Detectors

In quantum mechanics, we know nothing unless we measure it, and the process of measurement must generally be specified if we wish to predict a result. In this chapter, we provide some details of how photodetectors work and discuss what they actually measure. The simplest model is a photodetector that absorbs a photon from the field. This could be a two-level atom initially in its ground state, and when it absorbs a photon it is in the ground state. A potential can be applied that is sufficient to ionize the atom in the excited state but not in the ground state. This then provides a very weak current pulse (due to the one ionized electron), which can then be amplified. Most photodetectors work on some version of this process to convert a photon (or bunch of photons) into a macroscopic burst of electrons that yields a measurable current pulse. Let us start with a simple version of the measurement process. Assume we have a photodetector that works by removing a photon from the field and yields a current pulse that is spatially and temporally localized. We will refer to this as the 'click' of the detector, which is reminiscent of the noise the Geiger counter [1] makes in science fiction movies, as well as real Geiger counters! The effect of the detection is then to transform the state

$$|\psi\rangle_{\mathrm{AD}} = a|\psi\rangle_{\mathrm{BD}}, \tag{9.1}$$

where AD and BD represent after and before detection. Note that this model of measurement is not a unitary process and does not preserve the length of the state vector.

It is tempting, based on the relation

$$a|n\rangle = \sqrt{n}|n-1\rangle, \tag{9.2}$$

to think that the application of a to a quantum state represents taking away a photon from the field, removing $\hbar\omega$ of energy in the process. The most general state

doi:10.1088/978-0-7503-6014-2ch9

is a superposition of the number states. Consider the action of a on a coherent state, recall that we have

$$a|\alpha\rangle = |\alpha\rangle \tag{9.3}$$

and it does not even change the state, the average photon number and variance are unchanged. As a further example, consider the state

$$|\psi\rangle = \sqrt{1-\epsilon^2}|0\rangle + \epsilon|2\rangle, \tag{9.4}$$

which has a mean photon number of $\langle n\rangle = 2\epsilon^2$. When a acts on this state, we find

$$a|\psi\rangle = \sqrt{2}\epsilon|1\rangle \to |1\rangle, \tag{9.5}$$

where we have normalized the state after the action of a. Here we see that we have a state with a definite photon number of one. The application of a actually increased the average photon number! Thus, one must take care with our interpretation of what a does to a state! We will have more to talk about this when we consider the quantum trajectory theory in chapter 11.

9.2 Noiseless classical signal

Let us assume that we have a classical electromagnetic wave of intensity $\bar{I}$ with no fluctuations and that we have a detector that generates a signal proportional to that intensity. Let us say we have a detector that has a probability of giving a 'click', or photodetection event, over some time interval dt given by $sIdt$. At time $t + dt$, if we have detected m photons, with probability $P_m(t + dt)$, this could have occurred in two ways. We detected $m - 1$ photons in the interval $[0, t]$ and one photon in the interval $[t, t + dt]$. It is also possible that one detected m photons in the interval $[0, t]$ and none in the following interval $[t, t + dt]$. This leads us to

$$P_m(t + dt) = P_{m-1}(t)sIdt + P_m(t)(1 - sIdt), \tag{9.6}$$

where the probability of *not* firing is $(1 - sIdt)$. Rearranging a bit, we have

$$\frac{P_m(t + dt) - P_m(t)}{dt} = (P_{m-1}(t) - P_m(t))sI \tag{9.7}$$

and take the limit $dt \to 0$ to find

$$\frac{dP_m}{dt} = sI\,(P_{m-1}(t) - P_m(t)). \tag{9.8}$$

The solution is

$$P_m(t) = \frac{(sIt)^m}{n!}e^{-sIt}, \tag{9.9}$$

where we have used as boundary conditions the fact that the probability of detecting a photon in a time interval of zero length is zero,

$$P_0(t + 0) = 1 \tag{9.10}$$

$$P_m(t + 0) = 0,\ m \geq 0. \tag{9.11}$$

This is a Poisson distribution with the average photon number $\bar{m} = sIt$, and a variance of $\sqrt{\bar{m}}$. This uncertainty, or noise, in the photocurrent arises from the random nature of the detector firing, and we will come to call this 'shot noise'. This noise occurs for a light source with a constant intensity and no fluctuations. Hence, this noise is due to the detection process, and later we will find a relation to the vacuum fluctuations. This is essentially a coherent state as discussed in chapter 6. If we count photons for a time T and allow the intensity to change in time, the above formulas hold with

$$\bar{I} = \frac{1}{T}\int_0^T I(t)dt. \tag{9.12}$$

To obtain the statistical photon distribution, one generally counts photons for a time T and then repeats it for many such intervals. This is then given by an ensemble average,

$$P_m(t) = \left\langle \frac{\lambda^m}{m!} e^{-\lambda} \right\rangle, \tag{9.13}$$

with $\lambda = s\bar{I}T$, which is known as the Mandel counting formula [2].

Using this distribution, we can calculate the mean number of counts in time T,

$$\begin{aligned} \langle m \rangle &= \sum_{m=0} m P_m \\ &= \sum_{m=0} \frac{\lambda^m}{(m-1)!} e^{-\lambda} \\ &= \lambda \sum_{m=1} \frac{\lambda^{m-1}}{(m-1)!} e^{-\lambda} \\ &= \lambda. \end{aligned} \tag{9.14}$$

Similarly, one can show that

$$\langle m(m-1) \rangle = \lambda(\lambda - 1). \tag{9.15}$$

We can then calculate

$$\begin{aligned} (\Delta m)^2 &= \langle m^2 \rangle - \langle m \rangle^2 \\ &= \lambda, \end{aligned} \tag{9.16}$$

which is the usual result for a Poissonian distribution, with the variance equal to the mean. Therefore, we see that for a classical field with no fluctuations at all, there is uncertainty, or noise, in the distribution of counts. For many intervals T long, we obtain a range of photocounts on the order of $\lambda - \sqrt{\lambda}$ to $\lambda + \sqrt{\lambda}$, roughly speaking. What is the source of this noise? As there are no fluctuations in the light wave, apparently the noise comes from the detection process itself, indeed from the random nature of whether the detector fires. This type of noise is a form of what is referred to as 'shot' noise. For a field with fluctuations, we have the variance

$$(\Delta m)^2 = \langle \lambda \rangle + (\langle \lambda^2(t, T) \rangle - \langle \lambda \rangle^2), \tag{9.17}$$

where again the first term is the 'shot' noise contribution.

It is often useful to consider a waiting time distribution, that is, the distribution of times until the *next* photon is detected. We break up the time interval $T = Ndt$ and consider that in all N of those time intervals there is no photodetection, but in the interval $N + 1$ there is one. Then, we have

$$\begin{aligned} w(T)dt &= (1 - s\bar{I}dt)^N s\bar{I}dt \\ &= (1 - s\bar{I}T/N)^N s\bar{I}dt. \end{aligned} \tag{9.18}$$

In the limit that $N \to \infty$, we have

$$w(T) = e^{-s\bar{I}T} s\bar{I}T. \tag{9.19}$$

Let us now consider a light field in thermal equilibrium at temperature T. The probability of the field to have n photons in it is

$$P_n^{\text{th}} = \frac{1}{Z} e^{-E_n/kT}, \tag{9.20}$$

where $Z = \sum_n P_n$. Recall that the energy of the nth energy level is $E_n = (n + 1/2)\hbar\omega$. Ignoring the zero point energy, which cancels in P_n anyway, we can sum

$$Z = \sum_n e^{-n\hbar\omega/kT} = \frac{1}{1 - e^{-\hbar\omega/kT}}. \tag{9.21}$$

The average photon number is then

$$\begin{aligned} \langle n \rangle &= \sum_n n P_n \\ &= (1 - e^{-\hbar\omega/kT}) \sum_n n e^{-E_n/kT} \\ &= (1 - e^{-\hbar\omega/kT}) \frac{\partial}{\partial \beta} \sum\nolimits_n e^{-E_n/kT} \\ &= (1 - e^{-\hbar\omega/kT}) \frac{\partial}{\partial \beta} \frac{(-e^{-\hbar\omega/kT})}{(1 - e^{-\hbar\omega/kT})^2} \\ &= \frac{1}{e^{\hbar\omega/kT} - 1}. \end{aligned} \tag{9.22}$$

This allows us to write the photon probability as

$$P_n = \frac{\langle n \rangle^n}{(1 + \langle n \rangle)^{1+n}}. \tag{9.23}$$

For large $\langle n \rangle$, this can be approximated as follows:

$$P_n = \frac{1}{\langle n \rangle} e^{-n/\langle n \rangle}, \tag{9.24}$$

where this is only really appropriate for large photon numbers and we treat n as a continuous variable. In this regime, we can also use the intensity $I \approx n$,

$$P(I) = \frac{1}{\langle I \rangle} e^{-I/\langle I \rangle}. \tag{9.25}$$

9.3 Complex analytic signal

To consider the effect of fluctuations on the detected signal, we first consider the Fourier decomposition of the field,

$$E(t) = \frac{1}{2\pi} \int_{-\infty}^{\infty} d\omega \mathcal{E}(\omega) e^{-i\omega t} d\omega. \tag{9.26}$$

Generally, in our applications, the frequency content of the field, as given by $e(\omega)$, consists of frequencies small compared to the central oscillation frequency ω_0. Since the field is a real quantity, $E(t) = E^*(t)$, we have

$$\begin{aligned} \frac{1}{2\pi} \int_{-\infty}^{\infty} d\omega \mathcal{E}(\omega) e^{-i\omega t} &= \frac{1}{2\pi} \int_{-\infty}^{\infty} d\omega \mathcal{E}^*(\omega) e^{i\omega t} \\ &= \frac{1}{2\pi} \int_{-\infty}^{\infty} d\omega \mathcal{E}^*(-\omega) e^{-i\omega t} \end{aligned} \tag{9.27}$$

or more succinctly, $](\omega) = \mathcal{E}^*(-\omega)$. Then, we can define

$$E(t) = E^{(+)}(t) + E^{(-)}(t) \tag{9.28}$$

with

$$E^{(+)}(t) = \frac{1}{2\pi} \int_{0}^{\infty} \mathcal{E}(\omega) e^{-i\omega t} d\omega \tag{9.29}$$

and

$$\begin{aligned} E^{(-)}(t) &= \frac{1}{2\pi} \int_{-\infty}^{0} \mathcal{E}(\omega) e^{-i\omega t} d\omega \\ &= \frac{1}{2\pi} \int_{0}^{\infty} \mathcal{E}(-\omega) e^{i\omega t} d\omega \\ &= \frac{1}{2\pi} \int_{0}^{\infty} \mathcal{E}(\omega) e^{i\omega t} d\omega. \end{aligned} \tag{9.30}$$

We then have

$$E(t) = 2\,\mathrm{Re}(E^{(+)}(t)) = 2\,\mathrm{Re}(E^{(-)}(t)). \tag{9.31}$$

We usually refer to $E^{(+)}(t)$ as the positive frequency part of the field and also as the complex analytic signal. This is often useful in that we can consider the field to be decomposed into a series of complex exponentials $e^{-i\omega t}$ for only positive frequencies, then we merely take the real part at the end of any sequence of linear operations. In the case of a classical wave of unvarying amplitude, $E(t) = E_0 \cos(\omega_0 t + \phi)$, then we have

$$E^{(+)}(t) = \frac{E_0}{2} e^{-i(\omega_0 t + \phi)} \tag{9.32}$$

$$E^{(-)}(t) = \frac{E_0}{2} e^{i(\omega_0 t + \phi)}. \tag{9.33}$$

For such a case, the Fourier transform of the field consists of two delta functions, one at $\omega = \omega_0 = \omega$ and another at $\omega = -\omega_0$. In many cases of interest, the bandwidth of the light $\Delta\omega$ is negligible compared to the carrier frequency ω_0. In this case, we can approximate the light field by

$$E^{(+)}(t) = \frac{E_0(t)}{2} e^{-i(\omega_0 t + \phi(t))} \tag{9.34}$$

$$E^{(-)}(t) = \frac{E_0(t)}{2} e^{i(\omega_0 t + \phi(t))}, \tag{9.35}$$

where $E_0(t)$ and $\phi(t)$ are the slowly varying amplitude and phase of the light field, in the same sense as in our earlier chapters, $\dot{E}_0 \ll \omega_0 E_0$ and $\dot{\phi} \ll \omega_0 \phi$. What about the intensity of the light field? We have

$$\begin{aligned} I &= \epsilon_0 c |E(t)|^2 \\ &= \epsilon_o c \frac{E_0^2}{4}(2 + e^{2i(\omega t + \phi)} + e^{-2i(\omega t + \phi)}) \\ &= \epsilon_0 c \frac{E_0^2}{2}(1 + \cos(2(\omega t + \phi))) \\ &= \epsilon_0 c E_0^2 \cos^2(\omega t + \phi) \\ &= \frac{\epsilon_0 c}{4}(|E^{(+)}(t)|^2 + |E^{(-)}(t)|^2 + E^{(+)}(t)E^{(-)}(t) + E^{(+)}(t)E^{(-)}(t)+). \end{aligned} \tag{9.36}$$

Taking a time average, we find

$$\begin{aligned} \bar{I} &= \epsilon_0 c \frac{E_0^2}{2} \\ &= \frac{\epsilon_0 c}{4}(E^{(+)}(t)E^{(-)}(t) + E^{(+)}(t)E^{(-)}(t)+) \\ &= \frac{\epsilon_0 c}{2} E^{(+)}(t)E^{(-)}(t), \end{aligned} \tag{9.37}$$

where we have used the fact that $E^{(+)}(t)E^{(-)}(t) = E^{(-)}(t)E^{(+)}(t)$. When we treat the field quantum mechanically, we will find (actually we already have...) that $E^{(+)}(t)$ and $E^{(-)}(t)$ are noncommuting operators.

9.4 Semiclassical photodetection theory

We must specify how the detector actually works. The model we will consider here is an atom initially in the ground state that can be excited into an excited state or a continuum of excited states. At that point, a voltage is applied that will ionize the detector atom and yield a current due to one electron. Then some type of amplification occurs that turns that one electron into a macroscopic burst of electrons. This of course requires energy, and often the detector must recharge that energy, and the detector cannot 'fire' for some time interval, often referred to as 'dead time'. In this section, we treat the atom quantum mechanically, but not the field. We will treat the field quantum mechanically in the next section.

We assume that there is an electric dipole interaction between the field to be detected and the detector atoms

$$\begin{aligned} H_{\text{int}} &= -\vec{\mu}\cdot\vec{E} \\ &= -\mu_{eg}(\sigma_- e^{-i\omega_{if}t} + \sigma_+ e^{i\omega_{if}t})E_0(t)\cos(\omega_0 t) \\ &= -\frac{\hbar\Omega(t)}{2}(\sigma_- e^{-i(\omega_{if}-\omega_0)t} + \sigma_- e^{i(\omega_{if}-\omega_0)t}), \end{aligned} \tag{9.38}$$

as in chapter 3, although at this point we are in the interaction picture, $\Omega = \mu_{eg}E_0/\hbar$ and $\omega_{if} = \omega_i - \omega_f$ is the frequency difference between the ground (initial) and excited (final) states. We also take the incident field to be of the form $E(t) = E_0(t)\cos(\omega_0 t)$. Terms oscillating at $\omega_0 + \omega_{if}$ have been neglected—the rotating wave approximation. The formal solution to the Schrödinger equation in the interaction picture is

$$|\psi(t)\rangle = -\frac{i}{\hbar}\int_0^t dt' H_I(t')|\psi(t')\rangle. \tag{9.39}$$

In the first-order perturbation theory, we have

$$|\psi(t)\rangle = -\frac{i}{\hbar}\int_0^t dt' H_I(t')|\psi(0)\rangle. \tag{9.40}$$

The probability of finding the system in a final state $|f\rangle$ given that it started in an initial state $|i\rangle$ is given by

$$P_{i\to f} = |\langle f|\psi(t)\rangle|^2, \tag{9.41}$$

which can be written

$$\begin{aligned} P_{i\to f} &= |-\frac{i}{\hbar}\int_0^t \langle i\, H_{\text{int}}(t')|f\rangle|^2 \\ &= \frac{1}{\hbar^2}\int_0^t dt' \int_0^{t'} dt'' \langle i|H_{\text{int}}(t')|f\rangle\langle f|H_{\text{int}}(t'')|i\rangle. \end{aligned} \tag{9.42}$$

At this point, we assume that there is a range of final states and calculate

$$P_{\text{click}} = \sum_f P_{i\to f}, \tag{9.43}$$

where we use the sum of probabilities. We extend the sum from one over all final states accessible from the initial state to all states of the atom. This adds nothing to the sum, as the states that are not accessible have a zero matrix element between the initial state and the interaction Hamiltonian. We can then use

$$\sum_{\text{all states}} |f\rangle\langle f| = 1 \tag{9.44}$$

to find

$$P_{\text{click}} = \frac{1}{\hbar^2}\int_0^t dt' \int_0^{t'} dt'' \langle i|H_{\text{int}}(t')H_{\text{int}}(t'')|i\rangle . \tag{9.45}$$

We must also integrate over the density of the final states,

$$P_{\text{click}} = \frac{1}{\hbar^2}\int_0^t dt' \int_0^{t'} dt'' \int_{\omega_{if}}^{\infty} d\omega\rho(\omega_{if})\langle i|H_{\text{int}}(t')H_{\text{int}}(t'')|i\rangle , \tag{9.46}$$

where the lower limit is the energy gap between the initial state and the lowest final state, and extend the upper limit to ∞. Writing $\cos(\omega_0 t) = (1/2)(\exp(i\omega_0 t) + \exp(-i\omega_0 t))$ and putting in the form of the interaction Hamiltonian, the expectation value in equation (9.45) takes the form

$$\begin{aligned}\frac{\Omega(t')\Omega(t'')}{4} \quad &\langle i|\sigma_-^2 e^{i(\omega_0 - \omega_{if})(t'+t'')} + \sigma_+^2 e^{-(\omega_0 - \omega_{if})(t'+t'')} \\ &+ \sigma_-\sigma_+ e^{i(\omega_0 - \omega_{if})(t'-t'')} + \sigma_+\sigma_- e^{-i(\omega_0 - \omega_{if})(t'-t'')}|i\rangle\end{aligned} \tag{9.47}$$

with the initial state being the ground state and, using $\sigma_-^2 = \sigma_+^2 = 0$, we find

$$P_{\text{click}} = \frac{1}{4}\int_0^t dt' \int_0^{t'} dt'' \int_{\omega_{if}}^{\infty} d\omega\Omega(t')\Omega(t'')\rho(\omega_{if})e^{i(\omega_0 - \omega_{if})(t'-t'')}. \tag{9.48}$$

The density of the final states is a slowly varying function compared to the complex exponential, and we pull it out of the frequency integral, which can then be performed via

$$\int_{-\infty}^{\infty} d\omega e^{i(\omega_0 - \omega_{if})(t'-t'')} = 2\pi\delta(t' - t''), \tag{9.49}$$

which leaves us with

$$P_{\text{click}} = \frac{\pi\rho(\omega_0)}{2}\int_0^t dt'(\Omega(t'))^2. \tag{9.50}$$

We now perform an ensemble average over and use the complex analytic signal to write

$$P_{\text{click}} = \frac{\pi\rho(\omega_0)\mu_{eg}^2}{2}\int_0^t dt' \langle E^{(-)}(t')E^{(+)}(t')\rangle . \tag{9.51}$$

We can define an instantaneous rate via

$$R \equiv \frac{dP_{\text{click}}}{dt} = \frac{\pi\rho(\omega_0)\mu_{eg}^2}{2}\langle E^{(-)}(t)E^{(+)}(t)\rangle . \tag{9.52}$$

Please note the ordering of the operators, the annihilation operator is to the right, and the creation is to the left. This is called normal ordering and represents the absorptive nature of the detector considered. Earlier in chapter 4, when we considered the interaction of a two-level atom with a classical field, we also observed a constant excitation rate; in that case, this was due to the many frequencies on the two-level atom. This resulted in a frequency integral similar to the integration over the final atomic states above. We remarked that a constant result would also result from shining monochromatic light on an ensemble of Doppler or pressure-broadened atoms. Here, we see that a constant rate results from shining monochromatic light on an atom with many possible excited states that form a continuum. This is quite similar to the photoelectric effect, and with good reason. It *is* the photoelectric effect! Recall that the photoelectric effect occurs when monochromatic light is incident on a metal; the metal has a filled valence band and a slightly occupied conduction band. In the photoelectric effect, it was found experimentally that the light could kick out electrons and that the emitted electrons had energy

$$E = \hbar\omega - \Phi, \tag{9.53}$$

where ω is the frequency of the emitted light and Φ is the work function of the metal, the energy required to get the electron to the conduction band or ionize it. If the frequency is less than $\Phi/\hbar$ no electrons are emitted. Classically, the energy of the incident light beam is proportional to the intensity. It was initially puzzling that it was the frequency of the light that determined the energy of the ejected electron. Einstein won the Nobel prize, not for relativity, but for his interpretation of the photoelectric effect [3], that the light–matter interaction in this case is such that light exhibits a particle-like behavior. In many introductory 'modern' physics courses, this is presented as evidence for the photon nature of the field. That is, light is composed of particles of energy $E = \hbar\omega$, which can cause an electron to be emitted if the photon has enough energy to promote the electron above the gap. In this section, we consider the electric field a classical number that oscillates in time. Where are the photons? We have found an excitation rate that is constant, which implies that there can be a zero delay time for the electron emission. Also, the frequency of light must be such that $\omega \geq \omega_0$. The energy of the emitted electron varies linearly with the frequency of the input light. In short, there are no photons in this, in the sense of a proof that light is composed of particles called photons with energy $E = \hbar\omega$. Does this mean there are no photons? Of course not. This means that the photoelectric effect and the photodetectors do not provide direct indisputable evidence for photons, but their explanation is certainly consistent with photons. The quantum nature shows up in the fact that the photoelectron can be ejected instantaneously; for a classical field this would not be possible until the beam had supplied the requisite energy.

9.5 Quantum detection theory

Here, we model the interaction of a quantum field with the atoms in our detector. To this end, we use the interaction Hamiltonian in the interaction picture

$$H_{\text{int}} = \hbar g(a\sigma_{+}e^{-i\Delta t} + a^{\dagger}\sigma_{-}e^{-i\Delta t}), \tag{9.54}$$

where we have taken the detector atom to be at an antinode of the field and made the rotating wave approximation. The click rate is then given by

$$P_{\text{click}} = \frac{1}{\hbar^2}\int_0^t dt' \int_0^{t'} dt'' \langle i|H_{\text{int}}(t')H_{\text{int}}(t'')|i\rangle, \tag{9.55}$$

where the $\langle\rangle$ refers to a quantum expectation value in the initial state, which now must exist in the Hilbert space of the atom and the field. The quantum treatment follows the same steps as the semiclassical approach, which led to

$$P_{\text{click}} = \frac{\pi\rho(\omega_0)\mu_{eg}^2}{2}\int_0^t dt' E^{(-)}(t')E^{(+)}(t')(\langle g|\sigma_{-}\sigma_{+}|g\rangle + \langle g|\sigma_{+}\sigma_{-}|g\rangle). \tag{9.56}$$

In this case, the initial state is not $|g\rangle$, the atom(s) in the ground state, but $|\psi_F, g\rangle$, where $|\psi_F\rangle$ is the state of the field, and again the atom is in the ground state. In addition, instead of $E^{(-)}(t')E^{(+)}(t')$, we have a factor of $\mathcal{E}_0^2$, and with each σ_{-} operator is paired an $a^{\dagger}$, and with each σ_{+} is paired an a. These operators must of course appear *inside* the expectation value, which leads to

$$\begin{aligned} P_{\text{click}} = \frac{\pi\rho(\omega_0)\mu_{eg}^2\mathcal{E}_0^2}{2}\int_0^t dt'(\langle\psi_F, g|a^{\dagger}\sigma_{-}\sigma_{+}a|\psi_F, g\rangle \\ + \langle\psi_F, g|a\sigma_{+}\sigma_{-}a^{\dagger}|\psi_F, g\rangle), \end{aligned} \tag{9.57}$$

which gives us

$$P_{\text{click}} = \frac{\pi\rho(\omega_0)\mu_{eg}^2\mathcal{E}_0^2}{2}\int_0^t dt' \langle\psi_F|a^{\dagger}a|\psi_F\rangle. \tag{9.58}$$

Now this is no shock, perhaps, that the click rate of our detector is proportional to the average photon number of the field, $\langle n\rangle = \langle a^{\dagger}a\rangle$. *But* this result depends crucially on the fact that the atoms of the detector are in the ground state initially, and that the atomic and field raising and lowering operators are paired in the way they are. If we had detectors with atoms initially excited, we would have a detector click rate proportional to $\langle aa^{\dagger}\rangle = \langle n+1\rangle$. Classically, we have the intensity proportional to the square of the field. Quantum mechanically, the field is something like $\mathcal{E}_0(ae^{-i\omega t} + a^{\dagger}e^{i\omega t})$. Squaring this field, and ignoring rapidly oscillating terms, we find that

$$I \propto E^2 = \mathcal{E}_0^2(aa^{\dagger} + a^{\dagger}a). \tag{9.59}$$

We write this as

$$I \propto \mathcal{E}_0^2(:aa^{\dagger} + a^{\dagger}a:) = 2\mathcal{E}_0^2(a^{\dagger}a), \tag{9.60}$$

where we have introduced the notation $:O:$, which we refer to as normal ordering. This means moving all a to the right and all $a^\dagger$ to the left, *without* using commutation relations. This is not to say that we are now ignoring commutation relations! It merely means that we take the classical expression involving fields and intensities and replace them with the quantum field in terms of raising and lowering the operators for the field. Then, we do not go back through perturbation theory, integrating over densities of states, etc, but merely reorder the raising and lowering operators. It is a shortcut and *not* a mere ignoring of operator ordering! And is reflective of a detector that registers absorption events.

The next example one could consider is related to the probability of detecting two photons:

$$\begin{aligned} \langle n^2 \rangle \rightarrow & \langle :n^2: \rangle \\ &= \langle :a^\dagger a a^\dagger a: \rangle \\ &= \langle a^{\dagger 2} a^2 \rangle \\ &= \langle n(n-1) \rangle . \end{aligned} \tag{9.61}$$

Note that if the field only has one photon in it, the probability of detecting two is zero. If we did *not* use normal ordering, we would have $\langle n^2 \rangle = 1$. Therefore, the normal ordering procedure seems consistent with the discrete nature of the field, that is, the photon concept.

We could have obtained this result by recalling the semiclassical result for the click rate of the detector,

$$R \equiv \frac{dP_{\text{click}}}{dt} = \frac{\pi\rho(\omega_0)\mu_{eg}^2}{2} \langle E^{(-)}(t)E^{(+)}(t) \rangle . \tag{9.62}$$

In the quantum theory of the electromagnetic field, we have

$$E^{(+)} = \mathcal{E}_0 a e^{-i\omega t} \tag{9.63}$$

$$E^{(-)} = \mathcal{E}_0 a^\dagger e^{i\omega t}. \tag{9.64}$$

In quantum theory, the order matters, and we have discovered that we use normal ordering. In the future, we may take semiclassical results, use the prescription above for negative and positive frequency operators, and then apply normal ordering. For example, the photon distribution is given by

$$P_m(t) = \frac{\lambda^m}{n!} e^{-\lambda}, \tag{9.65}$$

where now we use $\lambda = sT\mathcal{E}_0^2 a^\dagger a$ and averages are to be taken as expectation values using the state of the field. This yields

$$\langle m \rangle = sT\mathcal{E}_0^2 \langle a^\dagger a \rangle \tag{9.66}$$

$$\langle m(m-1) \rangle = s^2T^2\mathcal{E}_0^4 \langle :(a^\dagger a)^2: \rangle . \tag{9.67}$$

9.6 Optical spectra and first-order coherence

Here we consider a single-field mode, but any real mode will actually have some finite linewidth. A spectrum measurement experiment on the system would be in the form of an intensity measurement, so we calculate the intensity distribution as a function of frequency:

$$I(\omega) = E^{*}(\omega)E(\omega), \tag{9.68}$$

where the frequency distribution of the field is given by the Fourier transform of the electric field's time evolution

$$E(\omega) = \frac{1}{\sqrt{2\pi}}\int_{-\infty}^{\infty} E(t)e^{i\omega t}dt. \tag{9.69}$$

Using this, we see that equation (9.68) can be written as

$$I(\omega) = \frac{1}{2\pi}\int_{-\infty}^{\infty} dt \int_{-\infty}^{\infty} dt' E(t)E^{*}(t')e^{i\omega(t-t')} \tag{9.70}$$

$$=\frac{1}{2\pi}\int_{-\infty}^{\infty} dt \int_{-\infty}^{\infty} d\tau E(t)E^{*}(t+\tau)e^{-i\omega\tau}, \tag{9.71}$$

where $\tau = t' - t$. Assuming that the field is ergodic and that the expectation value is equivalent to a time average, we can further simplify equation (9.70) to achieve the Wiener–Khintchine theorem,

$$I(\omega) = \frac{T}{2\pi}\int_{-\infty}^{\infty} d\tau e^{-i\omega\tau}\left\langle E(t)E^{*}(t+\tau)\right\rangle, \tag{9.72}$$

where we have used

$$\left\langle E(t)E^{*}(t+\tau)\right\rangle = \frac{1}{T}\int_{-T/2}^{T/2} dt E(t)E^{*}(t+\tau). \tag{9.73}$$

T is the total time over which the field is time averaged. Ideally, this should be very long to resolve the frequency components of interest; in practice, there are limits, and the uncertainty principle tells us that our spectral resolution in the resulting spectra is $\Delta\omega \sim 1/T$. To correctly model the experiment, we must make certain that our field $E(t)$ coincides with the experimental field signal. In the physical experiment, a detector is used to collect the transmitted photons and is thus an absorptive device. This absorbtion is correctly modeled in our calculation by replacing E with its negative frequency component $E^{(-)} \quad \alpha \quad a$. So, now we can write equation (9.72) in terms of the cavity field's creation and annihilation operators:

$$I(\omega) = \frac{T}{2\pi}\int_{-\infty}^{\infty} d\tau e^{-i\omega\tau}\left\langle a^{\dagger}(t+\tau)a(t)\right\rangle. \tag{9.74}$$

We used normal ordering (creation operators to the left, annihilation operators to the right) In the above equation, coinciding with our modeling of the detector as an absorptive device.

How could one measure the two-time correlation function above? Consider a double slit with light incident on it. The resultant field at a point behind the screen is a superposition of

$$E(t) = f(E_0(t) + E_0(t + \tau)), \tag{9.75}$$

where f is a constant of proportionality (we ignore diffraction effects for the individual slits) and $\tau = \Delta L/c$, where ΔL is the path difference between the paths to the screen from the two slits. The intensity is then, ignoring constants,

$$I(t) = 2|E_0(t)|^2 + 2\,\mathrm{Re}(E_0^*(t)E_0(t + \tau)). \tag{9.76}$$

Taking an ensemble average, we find

$$\begin{aligned}\langle I \rangle &= 2\langle I_0\rangle + 2\,\mathrm{Re}\,\langle E_0^*(t)E_0(t + \tau)\rangle \\ &= 2\langle I_0\rangle + 2\,\mathrm{Re}\,\langle E_0^*(0)E_0(\tau)\rangle\,.\end{aligned} \tag{9.77}$$

We have also invoked the stationarity of the field statistics, that is,

$$\langle E(t)\rangle = \langle E(t + \tau)\rangle = \langle E(0)\rangle \tag{9.78}$$

$$\langle E^*(t)E(t + \tau)\rangle = \langle E^*(0)E(\tau). \tag{9.79}$$

The field itself is not assumed to be stationary, but the ensemble averages indicated and the higher order ones are assumed to be stationary. An interference pattern is generally observed on the screen behind the double-slits. The fringe visibility is defined as

$$V = \frac{I_{\max} - I_{\min}}{I_{\max} + I_{\min}}. \tag{9.80}$$

Typically, for oscillating fields, the correlation function will also oscillate,

$$G^{(1)}(\tau) \equiv \langle E_0^*(0)E_0(\tau)\rangle = I_0 F(\tau)\cos(\omega\tau), \tag{9.81}$$

where $F(\tau)$ has a maximal value of 1 and we have defined the first-order correlation function $G^{(1)}(\tau)$. The maximum intensity is then $I_0 F(\tau)$ and the minimum $-I_0 F(\tau)$ which leaves us with a visibility

$$V = F(\tau). \tag{9.82}$$

Consider the normalized correlation function

$$g^{(1)}(\tau) = \frac{\langle E_0^*(0)E_0(\tau)\rangle}{\langle E_0(0)\rangle\,\langle E_0^*(0)\rangle}. \tag{9.83}$$

In the situation we consider here, this yields

$$g^{(1)}(\tau) = F(\tau)e^{i\omega\tau}. \tag{9.84}$$

We can write the intensity spectrum as

$$\begin{aligned} I(\omega) &= \frac{T}{2\pi} \langle E_0(0) \rangle \langle E_0^*(0) \rangle \,\mathrm{Re} \int_{-\infty}^{\infty} d\tau e^{-i\omega\tau} g^{(1)}(\tau) \\ &= \frac{T}{2\pi} \langle I \rangle \,\mathrm{Re} \int_{-\infty}^{\infty} d\tau e^{-i\omega\tau} g^{(1)}(\tau). \end{aligned} \tag{9.85}$$

Many times there are inelastic collisions or other inelastic processes that occur in the source producing the light that leads to changes in the phase of the light but not the intensity. This will occur in a Doppler broadened medium, for example. The ensemble average for such a light source results in

$$\begin{aligned} \langle E^*(t) E^{(t+\tau)} \rangle = |E_0|^2 e^{-i\omega_0\tau} \times \\ \langle (e^{-i\phi_1(t)} + e^{-i\phi_2(t)} + e^{-i\phi_3(t)} + \dots) \\ (e^{i\phi_1(t+\tau)} + e^{i\phi_2(t+\tau)} + e^{i\phi_3(t+\tau)} + \dots) \rangle, \end{aligned} \tag{9.86}$$

where ϕ_n is the relative phase of the nth member of the ensemble. The terms dealing with $\phi_n(t) - \phi_m(t + \tau) \equiv \Delta\phi$ average to zero, leaving the terms of the form

$$\langle e^{i\Delta\phi} \rangle = \langle 1 - \frac{1}{2}\Delta\phi^2 + \frac{1}{24}\Delta\phi^2 \rangle + \dots \tag{9.87}$$

$$= e^{-\langle \Delta\phi^2 \rangle}. \tag{9.88}$$

The final result for the correlation function is

$$\langle E^*(t) E^{(t+\tau)} \rangle = E_0^2 e^{i\omega_0\tau} e^{-\gamma\tau}, \tag{9.89}$$

where $\gamma = 1/T_{\text{coll}} \equiv \langle \Delta\phi^2 \rangle$, where T_{coll} is the average time between collisions or phase-disturbing events. In general, this would be replaced by the coherence time of the light source, T_C. In terms of our double slit experiment, the larger γ is, the faster the fringe visibility will decay with increasing delay time (or path length difference). The intensity spectrum for a field with this first-order correlation function is

$$\begin{aligned} I(\omega) &= \frac{T}{2\pi} \langle I \rangle \,\mathrm{Re} \int_{-\infty}^{\infty} d\tau e^{-i(\omega-\omega_0)\tau} e^{-\gamma\tau} \\ &= \frac{T}{2\pi} \langle I \rangle \,\mathrm{Re} \frac{\gamma - i\Delta}{\gamma^2 + \Delta^2} \\ &= \frac{\gamma T}{2\pi} \langle I \rangle \,\mathrm{Re} \frac{1}{\gamma^2 + \Delta^2}, \end{aligned} \tag{9.90}$$

where we define $\Delta = \omega_0 - \omega$, and we have a Lorentzian spectrum. In the limit that $\gamma \to 0$, the spectrum becomes a delta function. Going back to the time domain via the Young's double slit experiment, a monochromatic light field has a visibility that is constant; that is, it does not change even with large delay times, and we obtain perfect fringes. Any real light field has a finite bandwidth, and then there *will* be a decay of fringe visibility for larger delay times. In the opposite limit, where $\gamma \to \infty$, the spectrum is broad and flat, and we have what is sometimes called white light; that is, equal power at all frequencies. This leads to the complete absence of fringes. This is perfectly consistent with our experience that a narrower linewidth will generally lead to better interference effects. While the correlation function may not always exhibit purely exponential decay, in any real light field there must be a finite correlation time, and for a large time delay we will see no fringes, there is no information on what the field will be at a large delay time given the initial value of the field. This is expressed mathematically by the fact that the ensemble average in the numerator factors

$$\langle E^{(-)}(0)E^{(+)}(\tau \to \infty)\rangle \to \langle E^{(-)}(0)\rangle \langle E^{(+)(0)}\rangle , \tag{9.91}$$

where we have used stationarity. This results in

$$g^{(1)}(\tau \to \infty) \to 1. \tag{9.92}$$

In terms of the positive and negative frequency parts, for a perfectly monochromatic field one has

$$g^{(1)}(\tau) = \frac{\langle E^{(-)}(0)E^{(+)}(\tau)\rangle}{|\langle E^{(-)}(0)\rangle|^2}, \tag{9.93}$$

where we have used $\langle E^{(+)}\rangle = \langle E^{(-)}\rangle^*$. Going to our quantum recipe, we have

$$g^{(1)}(\tau) = \frac{\langle a^\dagger(0)a(\tau)\rangle}{\langle a\rangle\langle a^\dagger\rangle}. \tag{9.94}$$

We note that for a coherent state, $|\alpha\rangle$, we have $g^{(1)}(\tau) = 1$.

It is often convenient to write the optical spectrum as a coherent spectrum and an incoherent spectrum.

Defining

$$\Delta a = \langle a\rangle + \Delta a \tag{9.95}$$

and similarly for $a^\dagger$ we find that the numerator of $g^{(1)}(\tau)$ becomes

$$\langle a^\dagger(0)a(\tau)\rangle = \langle a^\dagger\rangle\langle a\rangle + \langle\Delta a^\dagger(0)\Delta a(\tau)\rangle . \tag{9.96}$$

The first term is a constant and yields a δ function upon taking a Fourier transform. This is referred to as the coherent or elastic scattering. It is just Rayleigh scattering of the incident light, with no change in energy or frequency. The second term can have a non-trivial time dependence, which can give a result with a width, or even multiple peaks, upon Fourier transformation. It results from pairwise (or higher) scattering, as shown in figure 9.1.

9.7 Photon statistics and second-order coherence

You have been told since the earliest discussion of the laser you have in classes that a laser is a bright, monochromatic, directional beam of light. But what about taking a very bright searchlight. It seems fairly directional in nature, it can be quite bright,

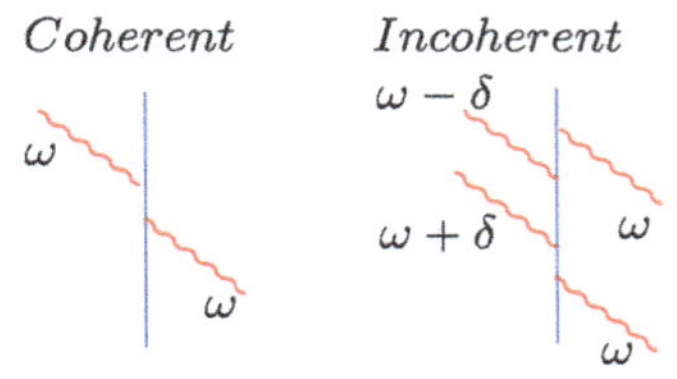

Figure 9.1. Examples of coherent and incoherent scattering processes.

and if I put a filter on the searchlight that just lets a small range of frequencies through, I have a relatively bright, monochromatic, and directional beam. As a laser is never perfectly monochromatic or directional, and as bright is a meaningless term unless we have something to compare the intensity of the light too, well then how do we distinguish laser light from thermal light from light from some other source? We simply cannot by looking just at the spectrum and mean intensity. We must look at the fluctuations. The simplest thing one can look at perhaps are the intensity fluctuations, the variance of the intensity. Consider the decomposition of the intensity into its average value plus the noise,

$$I(t) = \langle I(t)\rangle + \Delta I(t). \tag{9.97}$$

Let us now consider a quantity known as

$$g^{(2)}(0) = \frac{\langle I^2\rangle}{\langle I\rangle^2}. \tag{9.98}$$

In terms of an average and a fluctuating piece, we have

$$\begin{aligned} g^{(2)}(0) &= \frac{\langle(\langle I\rangle^2 + (\Delta I)^2 + 2\langle I\rangle\Delta I)}{\langle I\rangle^2}\rangle \\ &= 1 + \frac{(\Delta I)^2}{\langle I\rangle^2}. \end{aligned} \tag{9.99}$$

Because the variance of a classical noise process cannot be negative, we obviously have to have $g^{(2)}(0) \geq 1$. For a coherent state, we have

$$g^{(2)}(0) = \frac{|\alpha|^4}{|\alpha|^4} = 1, \tag{9.100}$$

so in some sense we see that the variance of the intensity is zero. Now this is a bit dodgy, as if we look at the 'fuzzball' pictures of a coherent state, there is noise in the electric field amplitude, and one would think that would translate into intensity noise.

Another way to look at this quantity is to write

$$\langle I\rangle(g^{(2)}(0) - 1) = \frac{(\Delta I)^2}{\langle I\rangle}, \tag{9.101}$$

that is the difference between $g^{(2)}(0) - 1$ is the difference between the variance over the mean and unity, scaled by the average intensity. Thus, we can see that for a coherent state, $g^{(2)}(0)$ must be 1, as the variance is equal to the mean. The amount of intensity noise in a coherent state is referred to as the standard quantum limit or shot noise limit. We will have more to discuss about these terms and $g^{(2)}(0)$ in chapter 10 where we talk about correlation functions and detection schemes. For now, we realize that quantum mechanically $I \propto a^\dagger a$, but for the square of the intensity we must use $I^2 \propto a^{\dagger 2}a^2 \equiv :(a^\dagger a)^2:$ where the notation $:f(a, a^\dagger):$ means to normal order the creation and annihilation operators, move all the creation operators to the left, and all the annihilation operators to the right, *without* using commutation relations. This seems like a procedure gone mad, as the hallmark of quantum mechanics is that order matters in the lab and in our mathematics! The fact that it matters in the laboratory is what forces us to use mathematics where order matters, such as differential operators, matrices, and the like. I am not saying that one just uses any ordering willy-nilly. I am saying (and we will show later) that if my detectors work by absorbing photons, as most do, then the normal ordering is the appropriate one. Using this prescription and considering the Fock states, we find

$$\begin{aligned} g^{(2)}(0) &= \frac{\langle n|(a^{\dagger 2}a^2)|n\rangle}{\langle n|a^\dagger a|n\rangle^2} \\ &= \frac{\langle n|(a^\dagger a a^\dagger a - a^\dagger a)|n\rangle}{n^2} \\ &= 1 - \frac{1}{n}. \end{aligned} \tag{9.102}$$

We see that using this (at presently not totally justified) prescription, we find that while classically the variance may never be less than the mean, for a number state it most certainly is! That is, the noise on the amplitude of the electric field corresponding to the Fock state is less than that of the coherent state. This is the beginning of how we can mathematically calculate the properties of our 'quantum fuzzballs' and make connections to the measurements. In our next chapter, we consider the interaction of a quantized electric field and a two-level atom. In looking at the photon statistics, we are interested in the intensity fluctuations of the transmitted field. The question that we ask is, if we detect a photon at some time, what is the probability of detecting a second photon after a delay time τ. τ. This conditional probability is proportional to

$$\begin{aligned} G^{(2)}(\tau) &= \langle :I(t+\tau)I(t): \rangle \\ &= \langle a^\dagger(t)a^\dagger(t+\tau)a(t+\tau)a(t)\rangle, \end{aligned} \tag{9.103}$$

where $I(t) = a^\dagger(t)a(t)$ is the intensity operator and as usual $\langle::\rangle$ denotes normal ordering. Typically, we normalize this probability to the probability for detecting two uncorrelated photons from a source of the same mean intensity as the source whose statistics we are interested in. This is also the probability of detecting two photons from our source that are separated by an extremely long delay time. This means our correlation functions are 1 in the limit $\tau \to \infty$, the numerator factorizes as the correlations have died off:

$$\begin{aligned} G^{(2)}(\infty) &= \langle a^\dagger(0)a(0)\rangle\langle a^\dagger(\tau)a(\tau)\rangle \\ &= \langle a^\dagger a\rangle^2 \end{aligned} \tag{9.104}$$

giving the normalized second-order correlation function

$$\begin{aligned} g^{(2)}(\tau) &= G^{(2)}(\tau)/G^{(2)}(\infty) \\ &= \frac{\langle a^\dagger(0)a^\dagger(\tau)a(\tau)a(0)\rangle}{\langle a^\dagger a\rangle^2}. \end{aligned} \tag{9.105}$$

A possible experimental set-up for measuring $g^{(2)}(\tau)$ is shown in figure 9.2. One waits for a detection at detector $D2$, waits a time τ and opens a shutter in front of $D1$ and records a detection or not. The second-order intensity correlation function $g^{(2)}(\tau)$ is the probability of two detections τ apart with no information on any other detections. Experimentally, it is much more convenient to let both detectors run and measure the time in between counts. No good experimentalist throws out the signal! Using a time division correlator, you can bin those times. This is called the waiting time distribution. It deals with the time until the *next* detection. For very weak fluxes, the two are equivalent for several decay lifetimes of the source. Eventually $g^{(2)}(\tau)$ goes to one after many decay times as the emissions separated by such a long delay are independent of one another, and the numerator factorizes. For long times, the waiting time distribution must go to zero, as there is no chance you have to wait until $t \to \infty$ to detect the *next* photon.

Let us begin by considering the zero time correlation,

$$g^{(2)}(0) = \frac{\langle a^{\dagger 2}a^2\rangle}{\langle a^\dagger a\rangle^2}. \tag{9.106}$$

Using commutation relations, we find that this can be written in the form

$$\begin{aligned} g^{(2)}(0) &= \frac{\langle n(n-1)\rangle}{\langle n\rangle^2} \\ &= \frac{(\Delta n)^2}{\langle n\rangle^2} + 1 - \frac{1}{\langle n\rangle} \\ &= \frac{1}{\langle n\rangle}(F-1) + 1, \end{aligned} \tag{9.107}$$

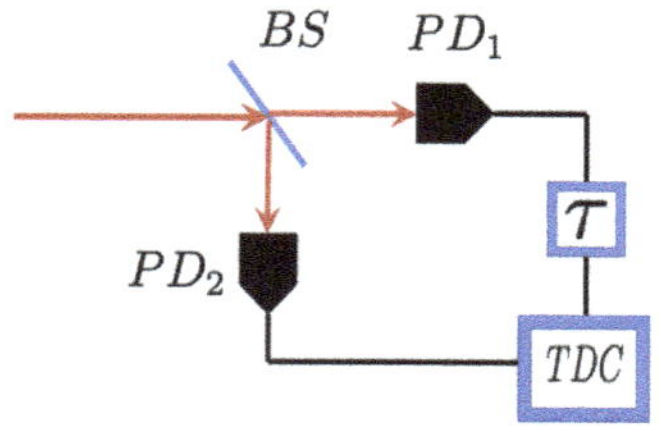

Figure 9.2. Experimental set-up for the measurement of the waiting time distribution, which is equivalent to $g^{(2)}(\tau)$ for weak fluxes.

where we defined the Fano factor [4] as the variance over the mean,

$$F = \frac{(\Delta n)^2}{\langle n \rangle}. \tag{9.108}$$

A slightly related form is

$$\langle n \rangle (g^{(2)}(0) - 1) = F - 1. \tag{9.109}$$

A related quantity is the Mandel Q parameter [5],

$$Q = F - 1. \tag{9.110}$$

Classically, we would have

$$g^{(2)}(0) = \frac{\langle I^2 \rangle}{\langle I \rangle^2}. \tag{9.111}$$

Writing $I = \langle I \rangle + \Delta I$, we find

$$\begin{aligned} g^{(2)}(0) &= \frac{\langle (\langle I \rangle + \Delta I)^2 \rangle}{\langle I \rangle^2} \\ &= \frac{\langle I \rangle^2 + \langle (\Delta I)^2 \rangle + 2\langle I \rangle \langle \Delta I \rangle}{\langle I \rangle^2} \\ &= \frac{\langle (\Delta I)^2 \rangle}{\langle I \rangle^2} + 1. \end{aligned} \tag{9.112}$$

For a classical field that has a well-defined value at any instant in time, even if it is noisy, we must have $(\Delta I)^2 \geq 0$, which leads to an inequality that must be satisfied by a classical field,

$$g^{(2)}(0) \geq 1. \tag{9.113}$$

When $g^{(2)}(0) = 1$, this denotes Poissonian statistics from a coherent state, where the variance equals the mean. When $g^{(2)}(0) > 1$, this indicates that pairs of photons are likely to be detected and is called photon bunching. In the case $g^{(2)}(0) < 1$ indicates that photons are not likely to come in pairs and that this results in photon antibunching, or amplitude squeezed light, with reduced intensity fluctuations.

How can this be? How can I have a stochastic variable, say $E(t)$ or $I(t)$, that has a negative variance? The short answer is that we *cannot*! Finding a field that has $g^{(2)}(0) < 1$ would mean finding one that has no real stochastic process associated with it. This means that we cannot consider such a field to have a well-defined field amplitude or intensity, even if it is noisy. This means that objective reality does not exist independent of our interaction with it. There is no wiggly noisy field that exists without our measurement. This is a key part of quantum weirdness, that quantum mechanics is more than just statistical in nature, like the velocities of atoms in a gas in a room. This is no weirder than realizing that the electron in a hydrogen atom has no well-defined position without our observing it at a definite position. In a later chapter, we will talk about other types of quantum weirdness to be sure. Do such nonclassical states exist? Sure! First, consider a Fock state, $|n\rangle$, which has

$$g^{(2)}(0) = \frac{n(n-1)}{n^2} = 1 - \frac{1}{n}. \tag{9.114}$$

Any Fock state other than the vacuum violates the classical inequality although the violation becomes smaller for much larger values of n. A coherent state would yield $g^{(2)}(0) = 1$, and it is second-order coherent as well as first-order coherent. In fact, we find that any normalized correlation function for a coherent state is unity. For the thermal light,

$$g^{(2)}(0) = 1 + |g^{(0)}(0)|^2 = 2, \tag{9.115}$$

and for the squeezed states, we find

$$g^{(2)}(0) = 1 + \frac{e^{-2r} - 1}{\sinh^2(r)}. \tag{9.116}$$

For resonance fluorescence, where a single two-level atom is illuminated by a monochromatic field, we find $g^{(2)}(0) = 0$ as it is impossible for a single atom to emit two fluorescent photons simultaneously, the atom must absorb in between the emission events. For N atoms, one finds that $g^{(2)}(0) = 1 - 1/N$, and the antibunching effect vanishes for a large ensemble of atoms. A violation of the inequality in equation (9.113) is referred to as photon antibunching because the photons tend not to come in pairs. Also, it means that the variance is less than the mean, which means that the intensity fluctuations are sub-Poissonian.

The time dependence of $g^{(2)}(\tau)$ can also exhibit nonclassical behavior. Using the Cauchy–Schwarz inequality, one can show that for a classical field, we have

$$g^{(2)}(\tau) \leq g^{(2)}(0). \tag{9.117}$$

This can be shown by considering a joint probability distribution function $P(I_0, 0; I_{\tau,\tau})$, which is the probability of finding an intensity $I(\tau)$ at time τ given that one had intensity $I(0)$ at $\tau = 0$. We have

$$\langle I(\tau)I(0)\rangle = \int_0^\infty dI_\tau \int_0^\infty dI_0 I_\tau I_0 P(I_0, 0; I_\tau, \tau) \tag{9.118}$$

and

$$\langle I \rangle = \int_0^\infty dIIP(I), \tag{9.119}$$

where $P(I)$ is the probability of the intensity, a single time stationary probability. Using the Cauchy–Schwarz inequality in the form

$$|\int f(x)g(x)dx|^2 \leq |\int f(x)dx||\int g(x)dx| \tag{9.120}$$

we can show that

$$\langle I(\tau)I(0)\rangle \leq \langle I^2(0)\rangle, \tag{9.121}$$

which leads to equation (9.117). A violation of this inequality was first observed in resonance fluorescence in an atomic beam by Kimble, Dagenais, and Mandel [6], and more recently for a single atom in a trap [7]. For historical reasons, a violation of this inequality is also referred to as photon antibunching. This is referred to loosely as 'a violation of the Cauchy–Schwarz inequality', but really what is going in is the conditions for the Cauchy–Schwartz inequality to be applicable are not met.

We can also show that

$$|g^{(2)}(\tau) - 1| \leq |g^{(2)}(0) - 1|. \tag{9.122}$$

This can be shown by considering $I(t) = \langle I\rangle + \Delta I(t)$. We can write

$$g^{(2)}(\tau) = 1 + \frac{\langle \Delta I(\tau)\Delta I(0)\rangle}{\langle I\rangle^2} \tag{9.123}$$

and using the Cauchy–Schwarz inequality above, but this time in terms of $\Delta I(\tau)$, resulting in

$$|\langle \Delta I(\tau)\Delta I(0)\rangle| \leq \langle(\Delta I^2)\rangle, \tag{9.124}$$

which leads to equation (9.122).

9.8 Mandel Q function

We can write $g^{(0)}(0)$ as

$$\begin{aligned} g^{(2)}(0) &= \langle a^\dagger a\rangle^{-2}\langle a^{\dagger 2}a^2\rangle \\ &= \langle a^\dagger a\rangle^{-2}[\langle(a^\dagger a)^2\rangle - \langle a^\dagger a\rangle] \\ &= 1 + \frac{(\langle n^2\rangle - \langle n\rangle^2) - \langle n\rangle}{\langle n\rangle^2}, \end{aligned} \tag{9.125}$$

where $n \equiv a^\dagger a$ is the photon number operator leading to

$$\langle n\rangle[g^{(2)}(0) - 1] = \left(\frac{(\Delta n)^2}{\langle n\rangle}\right). \tag{9.126}$$

Then

$$g^{(2)}(0) - 1 = \frac{Q}{\langle n\rangle}, \tag{9.127}$$

where the Mandel Q parameter is defined as

$$\begin{aligned} Q &\equiv \frac{(\langle n^2\rangle - \langle n\rangle^2) - \langle n\rangle}{\langle n\rangle} \\ &= \frac{\langle(\Delta n)^2\rangle - \langle n\rangle}{\langle n\rangle}. \end{aligned} \tag{9.128}$$

The classically allowed ranges of the Mandel Q parameter are

$$Q \geq 0. \tag{9.129}$$

Mandel Q parameters in the range $-1 \leq Q < 0$ would imply a nonclassical field, similarly for F, a negative value implies nonclassicality.

9.9 Balanced homodyne detection and the spectrum of squeezing

Consider the set-up in figure 9.3.

We have a signal field and a local oscillator field incident on the beamsplitter. We describe the signal field by a and the local oscillator field by b. If the beamsplitter has $r = t = 1/\sqrt{2}$, the two output modes of the beamsplitter are

$$c = \frac{1}{\sqrt{2}}(a + b) \tag{9.130}$$

$$d = \frac{1}{\sqrt{2}}(a - b). \tag{9.131}$$

The photon numbers in the two output modes are given by

$$n_c = \frac{1}{2}(a^\dagger a + b^\dagger b + ab^\dagger + ba^\dagger) \tag{9.132}$$

$$n_d = \frac{1}{2}(a^\dagger a + b^\dagger b - ab^\dagger - ba^\dagger). \tag{9.133}$$

The difference between these two photon numbers is

$$n_c - n_d = (ab^\dagger + ba^\dagger) \tag{9.134}$$

and the difference between the two detector rates is proportional to

$$\langle n_c \rangle - \langle n_d \rangle = \langle (ab^\dagger + ba^\dagger) \rangle . \tag{9.135}$$

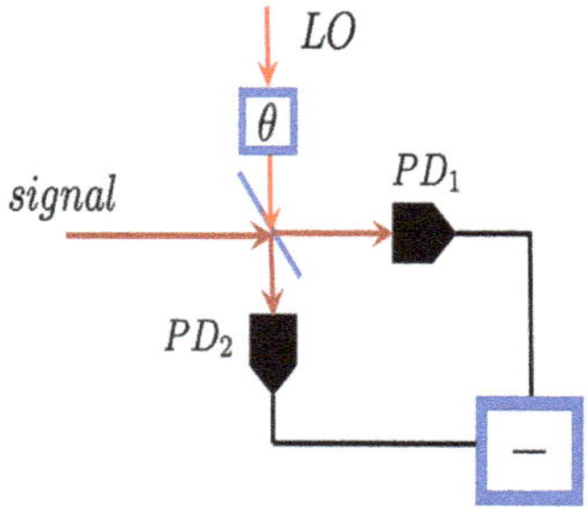

Figure 9.3. Experimental set-up for measuring the spectrum of squeezing. The difference current is Fourier transformed to obtain the result.

If the local oscillator is in a coherent state $\alpha e^{i\phi}$, we have

$$\langle n_c \rangle - \langle n_d \rangle = \alpha \left\langle (ae^{-i\phi} + e^{i\phi}a^{\dagger}) \right\rangle, \tag{9.136}$$

where we recognize this as the expectation value of the field quadrature operator,

$$\langle n_c \rangle - \langle n_d \rangle = 2\alpha \langle a_\theta \rangle. \tag{9.137}$$

The variance of the difference in the photocurrents is

$$\left\langle (\Delta n)^2 \right\rangle = 4|\alpha|^2 \left\langle (\Delta a_\theta)^2 \right\rangle. \tag{9.138}$$

Changing the phase of the local oscillator, one can measure the variance of the field quadratures, and for a squeezed state, we can find that the variance falls below the shot noise limit for the quiet quadrature, but must be above the shot noise limit for the noisier quadrature.

Any real source of light has a finite bandwidth and as such has some temporal dynamics. The above variance is a DC variance, and there are many sources of noise that can swamp the quantum noise. It is more useful to consider the correlation of the photocurrent difference $i = i_c - i_d$,

$$\left\langle i(t)i(t+\tau) \right\rangle \propto \left\langle \Delta a_\theta(0)\Delta a_\theta(\tau) \right\rangle, \tag{9.139}$$

and Fourier transform it to obtain a noise spectrum, which we refer to as the spectrum of squeezing,

$$S_\theta(\omega) = \int_{-\infty}^{\infty} d\omega e^{i\omega\tau} \left\langle \Delta a_\theta(0)\Delta a_\theta(\tau) \right\rangle. \tag{9.140}$$

There is a relation between the incoherent spectrum and the sum of the two spectra of squeezing $\pi/2$ out of phase. We will see in chapter 10 that the presence of squeezing can cause narrowing and other features in the incoherent spectra,

$$I_{\text{inc}}(\omega) = S_{\theta,\,\omega} + S_{\theta+\pi/2,\,\omega}, \tag{9.141}$$

where we have suppressed an overall scale factor. The presence of squeezing can cause narrowing and other features such as dips in the incoherent spectra.

9.10 Wave–particle duality and conditioned homodyne detection

So far, we have examined the field–field correlations, $g^{(1)}(\tau)$, and the intensity–intensity correlations, $g^{(2)}(\tau)$. In this section, we consider a normalized correlation, a field–intensity correlation introduced by Carmichael *et al* [8],

$$h(\tau) = \frac{\left\langle I(0)E(t) \right\rangle}{\left\langle I(0) \right\rangle \left\langle E(0) \right\rangle} \tag{9.142}$$

classically, and a quadrature-dependent quantum version

$$h_\theta(\tau) = \frac{\left\langle a^{\dagger}(0)a_\theta(\tau)a(0) \right\rangle}{\langle a^{\dagger}a \rangle \langle a_\theta \rangle}. \tag{9.143}$$

This is not well defined for a light source with a zero mean field for a particular quadrature, but in that event one may add a local oscillator to the source and adjust the size of that local oscillator to maximize the signal to noise ratio [9]. This correlation function can be measured by performing a balanced homodyne detection of the field and only recording data when there has been a photodetection at $\tau = 0$. As such, it is a balanced homodyne measurement *conditioned* on detection of a photon at time $\tau = 0$. This can be accomplished by the experimental set-up in figure 9.4.

With the added assumption of Gaussian statistics, which allows us to ignore third-order moments, we may then write the Fourier pair

$$S(\Omega, \theta) = 2F \int_{-\infty}^{\infty} d\tau \exp(i2\pi\Omega\tau)[\bar{h}_\theta(\tau) - 1], \tag{9.144}$$

$$\bar{h}_\theta(\tau) - 1 = \frac{1}{4\pi F} \int_{-\infty}^{\infty} d\Omega \exp(-i2\pi\Omega\tau) S(\Omega, \theta). \tag{9.145}$$

Note that the photon flux plays an inverse role in the relative sizes of the spectrum of squeezing and the intensity–field correlation function. From this, apparently for a large photon flux, nonclassical effects might be observed more readily in the measurements of the spectrum of squeezing, and for a low photon flux, in the measurements of $h_\theta(\tau)$. There is also a relationship between the time-averaged $\bar{h}_\theta(\tau)$ and the degree of squeezing at zero frequency, and between the frequency-averaged spectrum of squeezing and $\bar{h}_\theta(0)$: [10]

$$S(0, \theta) = 2F \int_{-\infty}^{\infty} d\tau [\bar{h}_\theta(\tau) - 1], \tag{9.146}$$

$$\bar{h}_\theta(0) - 1 = \frac{1}{4\pi F} \int_{-\infty}^{\infty} d\Omega\, S(\Omega, \theta). \tag{9.147}$$

In addition, in this limit of Gaussian statistics, one can show the following classical bounds that $h_\theta(\tau)$ must satisfy [9]

$$0 \leq h_\theta(0) - 1 \leq 1 \tag{9.148}$$

and

$$|h_\theta(\tau) - 1| \leq |h_\theta(0) - 1| \leq 1. \tag{9.149}$$

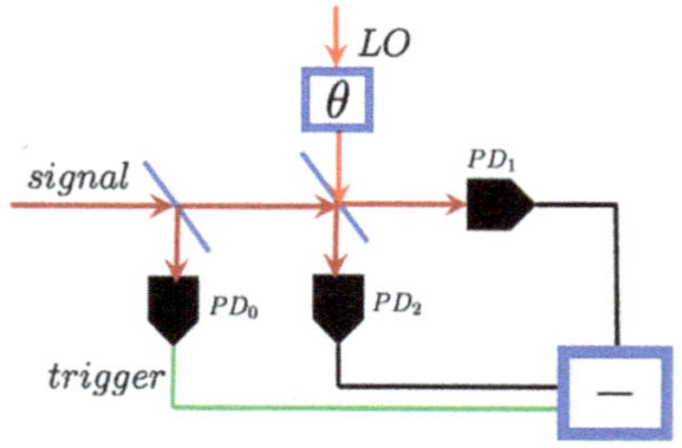

Figure 9.4. Experimental set-up for measuring the conditioned balanced homodyne signal $h_\theta(\tau)$.

The nonclassical violation of these relations was first observed by Carmichael *et al* [11, 12]. The correlation function $h_\theta(\tau)$ need not be symmetric around $\tau = 0$, in fact, an asymmetry is an indication of a breakdown of detailed balance [9].

9.11 Cross-correlation functions

It is also possible and interesting to examine the correlations between the two field modes a and b, the transmission from a cavity and the fluorescence out the side of the cavity,

$$g_{ab}^{(2)} = \frac{\langle a^\dagger(0)b^\dagger(\tau)b(\tau)a(0)\rangle}{\langle a^\dagger a\rangle_{SS}\langle b^\dagger b\rangle_{SS}}. \tag{9.150}$$

A schematic of the measurement apparatus is exhibited in figure 9.5.

The inequalities that this cross correlation satisfies for classical fields are quite different, and are

$$g_{ab}^{(2)}(0) \leq \sqrt{g_{bb}^{(2)}(0)g_{aa}^{(2)}(0)} \tag{9.151}$$

$$g_{ab}^{(2)}(\tau) - 1 \leq \sqrt{(g_{bb}^{(2)}(0) - 1)(g_{aa}^{(2)}(0) - 1)}. \tag{9.152}$$

We can also condition the homodyne detection of mode b with a detection in mode a (figure 9.6) [13, 14],

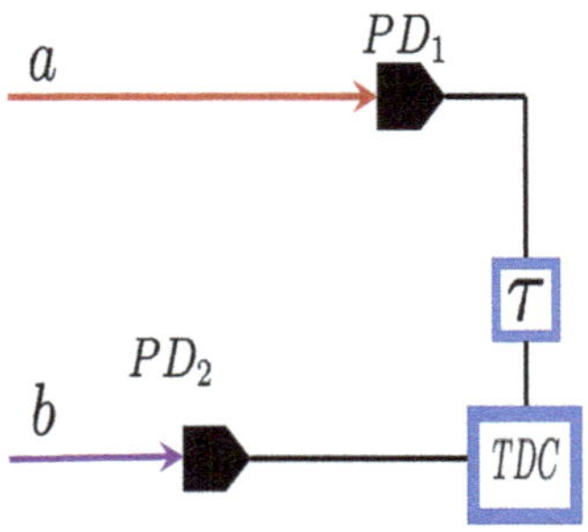

Figure 9.5. Schematic of a measurement of a cross-intensity correlation between field modes a and b.

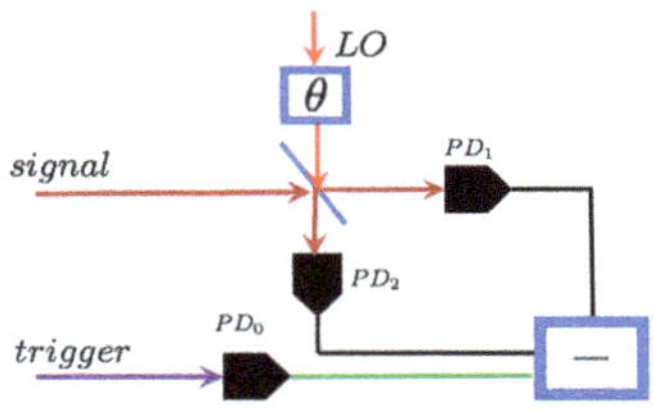

Figure 9.6. Schematic of a measurement of a homodyne detection of one field mode, conditioned by a detection in a different field mode.

$$h_{ab}^{\theta} = \frac{\langle a^{\dagger}(0)b_{\theta}(\tau)a(0)\rangle}{\langle a^{\dagger}a\rangle_{\mathrm{SS}}\langle b_{\theta}\rangle_{\mathrm{SS}}}. \tag{9.153}$$

The inequalities satisfied by this cross h_{θ} are

$$h_{ab}^{\theta}(0) \leq \sqrt{h_{bb}^{\theta}(0)h_{aa}^{\theta}(0)} \tag{9.154}$$

$$h_{ab}^{\theta}(\tau) - 1 \leq \sqrt{(h_{bb}^{\theta}(0) - 1)(h_{aa}^{\theta}(0) - 1)}\,. \tag{9.155}$$

9.12 Out of time-ordered correlations

In recent years, there has been much interest in non-time-ordered correlations to study the scrambling of quantum information and quantum chaos. The most commonly studied OTOC is a four-point correlator of the form

$$F(t) = \langle W^{\dagger}(t)V^{\dagger}(0)W(t)V(0)\rangle, \tag{9.156}$$

where $W(t)$ and $V(0)$ are the Heisenberg-picture operators. The name 'out-of-time-ordered' refers to how $W(t)$ and $V(0)$ appear in an interlaced time sequence rather than a simple chronological ordering.

In many semiclassical regimes, one finds that certain OTOCs exhibit exponential growth for an intermediate range of times, schematically:

$$F(t) \sim 1 - A\,e^{\lambda_L t}, \tag{9.157}$$

where λ_L is often called the quantum Lyapunov exponent, and A is a constant. This growth in OTOCs reflects that commutators $[W(t),\ V(0)]$ become large, meaning a small initial perturbation (encoded in V) is 'spread out' over many degrees of freedom by time t. This is akin to the classical picture of nearby trajectories diverging exponentially. An OTOC that decreases from an initial value down toward zero (or some small amplitude) indicates that the local information (the operator at time 0) has become highly 'scrambled' in the system's degrees of freedom. Black holes and chaotic spin chains are classic examples. A famous result (in the context of holography and black hole physics) is the Maldecena–Shenker–Stanford bound

$$\lambda_L \leq \frac{2\pi}{\beta}, \tag{9.158}$$

where $\beta = 1/k_B T$. This was first derived by relating quantum chaos in large-N systems to black hole dynamics (anti-de Sitter/conformal field theory), indicating that the growth rate of OTOCs is bounded from above. Black holes (and theories dual to them) saturate this bound, so they are considered 'maximally chaotic'. I The operators are arranged in a way that measures how much an initially small perturbation affects a later measurement (W) when they are out of chronological order. This cross-talk is what picks up the exponential 'butterfly effect'. Scrambling means that localized information (such as an excitation or a local operator) quickly becomes distributed across all degrees of freedom. The OTOC is sensitive to such delocalization because if $[W(t), V(0)]$ is large, it means the operator at time t no longer commutes with an operator at time 0—they significantly disturb each other. In the classical world, the Lyapunov exponent λ_{cl} quantifies how rapidly trajectories diverge in the phase space, an unambiguous hallmark of chaos. In classical Hamiltonian systems with chaotic dynamics, two initially close phase-space points $(q(0), p(0))$ and $(q(0) + \delta q(0), p(0) + \delta p(0))$ typically diverge from each other as

$$\|\delta q(t), \delta p(t)\| \approx \|\delta q(0), \delta p(0)\| e^{\lambda_{\mathrm{cl}} t}, \tag{9.159}$$

where λ_{cl} is the classical Lyapunov exponent. Physically, $\lambda_{\mathrm{cl}} > 0$ signals chaos: small perturbations are magnified exponentially in time. In a quantum system, one cannot track phase-space points in the same way, but the out-of-time-ordered correlator naturally generalizes this notion, since it probes how an operator at time 0 affects an operator at time t when inserted in a scrambled time sequence. In many treatments, one finds a relation such as

$$<[W(t), V(0)]^2> \approx 2(1 - \mathrm{Re}\,\mathrm{OTOC}(t)) \tag{9.160}$$

The gist is that if OTOC <1, then $\langle [W, V]^2 \rangle$ is greater than zero and grows as OTOC (t) decreases. If Re OTOC(t) is negative, then $1 - \mathrm{Re}\,\mathrm{OTOC}(t) > 1$, meaning that the commutator-squared is even larger. This is consistent with the strong noncommutativity and hence strong scrambling. When $F(t)$ exhibits an exponential decay (or the commutator norm grows exponentially), the system displays quantum chaos with a quantum Lyapunov exponent λ_{L}. In a classical or large- N limit, λ_{L} often matches the classical λ_{cl}. This OTOC 'scrambling' perspective is central to modern studies of quantum chaos, black hole information, and many-body localization/delocalization. In a classical limit (large quantum numbers, $\hbar \to 0$, or large N), the quantum OTOC's exponential growth rate λ_{L} often matches the system's classical Lyapunov exponent λ_{cl}. The exponential growth of the OTOC still serves as the best known proxy for chaotic dynamics and fast scrambling in quantum systems.

9.13 A note on detailed balance

In statistical mechanics and stochastic processes, detailed balance refers to a condition where every microscopic transition between states has an exact reverse process occurring at equilibrium with the same rate. Mathematically, for a system with transition rates $W(A \to B)$ between states A and B, detailed balance is given by

$$P_A W(A \to B) = P_B W(B \to A), \tag{9.161}$$

where P_A and P_B are the steady-state probabilities of states A and B. A multi-point correlation function in a steady-state system captures the statistical dependencies between variables at different times. The crucial aspect here is the transformation $\tau \to -\tau$ (i.e. time-reversal symmetry). If the system obeys a detailed balance, then the time-reversal symmetry should be preserved in the statistical properties of the system, meaning that the correlation functions should be symmetric under time reversal.

However, if the correlator is not symmetric under $\tau \rightarrow -\tau$, it suggests that there is an irreversible directional dependence in the time evolution of the system. This indicates non-equilibrium steady-state dynamics where probability currents persist even at the steady state. Such currents are a hallmark of a broken detailed balance. In non-equilibrium systems, if the system has an external drive or an internal cycle that maintains non-zero probability currents (e.g. energy dissipation or driving), the time-reversal symmetry in the correlators can break, signaling a violation of detailed balance. We consider two quantum harmonic oscillators, both damped and one driven. The Hamiltonian is

$$H = H_0 + H_{\text{int}} + H_{\text{drive}}, \tag{9.162}$$

where for the oscillators

$$H_0 = \omega_a a^\dagger a + \omega_b b^\dagger b, \tag{9.163}$$

and the interaction between the oscillators (e.g. hopping-like coupling) is

$$H_{\text{int}} = g(a^\dagger b + b^\dagger a), \tag{9.164}$$

where g is the coupling strength. There is an external drive on b

$$H_{\text{drive}} = Fe^{-i\Omega t} b^\dagger + F^* e^{i\Omega t} b, \tag{9.165}$$

where F is the drive strength and Ω is the driving frequency. Consider the four-point correlation function:

$$C(\tau) = \langle a^\dagger(0) b^\dagger(\tau) b(\tau) a(0) \rangle. \tag{9.166}$$

In equilibrium (no drive), detailed balance implies

$$C(\tau) = C(-\tau). \tag{9.167}$$

Let us explicitly calculate the correlation function

$$C(\tau) = \langle a^\dagger(0) b^\dagger(\tau) b(\tau) a(0) \rangle. \tag{9.168}$$

For the system of two coupled damped oscillators, where the oscillator b is driven, using the quantum regression theorem. The details of the Lindblad master equation and the quantum regression theorem are discussed in chapter 10, here we just assume their validity. This essentially means that in a linear system, the fluctuations satisfy the same equations as the mean values. Each oscillator is also coupled to an independent thermal bath, leading to damping and decoherence. The system evolves according to the Lindblad master equation,

$$\frac{d\rho}{dt} = -i[H, \rho] + \gamma_a \mathcal{D}[a]\rho + \gamma_b \mathcal{D}[b]\rho, \tag{9.169}$$

where the dissipators are

$$\mathcal{D}[o]\rho = o\rho o^\dagger - \frac{1}{2}\{o^\dagger o, \rho\}. \tag{9.170}$$

Here, γ_a and γ_b are the decay rates of oscillators a and b, respectively. The quantum regression theorem states that if the system's mean values obey a set of linear equations, then the two-time correlation functions obey the same equations, but with different initial conditions. Define the expectation values

$$\mathbf{v}(t) = \begin{bmatrix} \langle a(t) \rangle \\ \langle b(t) \rangle \end{bmatrix}. \tag{9.171}$$

From Heisenberg equations, the mean values satisfy

$$\frac{d}{dt}\mathbf{v}(t) = M\mathbf{v}(t) + \mathbf{f}(t), \tag{9.172}$$

where M is the drift matrix:

$$M = \begin{bmatrix} i\omega_a - \frac{\gamma_a}{2} & -ig \\ ig & -i\omega_b - \frac{\gamma_b}{2} \end{bmatrix}.. \tag{9.173}$$

The term $\mathbf{f}(t)$ represents the drive, acting only on oscillator b. Using QRT, we can solve the two-time correlation function by treating the system evolution as

$$\frac{d}{dt}C(\tau) = MC(\tau). \tag{9.174}$$

The specific correlation function we want is

$$C(\tau) = \langle a^\dagger(0)b^\dagger(\tau)b(\tau)a(0)\rangle. \tag{9.175}$$

To explicitly compute the correlation function

$$C'(\tau) = \langle a^\dagger(-\tau)b^\dagger(0)b(0)a(-\tau)\rangle. \tag{9.176}$$

The drift matrix is

$$M = \begin{bmatrix} i\omega - \frac{\gamma_a}{2} & -ig \\ ig & -i\omega - \frac{\gamma_b}{2} \end{bmatrix}. \tag{9.177}$$

The characteristic equation is

$$\det(M - \lambda I) = 0. \tag{9.178}$$

Expanding,

$$\left(-i\omega - \frac{\gamma_a}{2} - \lambda\right)\left(-i\omega - \frac{\gamma_b}{2} - \lambda\right) - g^2 = 0. \tag{9.179}$$

Expanding the terms

$$\left(-i\omega-\frac{\gamma_a}{2}-\lambda\right)\qquad\left(-i\omega-\frac{\gamma_b}{2}-\lambda\right)$$
$$=\lambda^2+\lambda(\gamma_a+\gamma_b)/2+\left(i\omega(\gamma_a+\gamma_b)+\frac{\gamma_a\gamma_b}{4}\right)+\omega^2. \tag{9.180}$$

Thus, the eigenvalues are

$$\lambda_\pm=-\frac{1}{4}(\gamma_a+\gamma_b)-\frac{i}{2}(2\omega)\pm\frac{1}{2}\sqrt{g^2-\left(\frac{\gamma_a-\gamma_b}{2}\right)^2}. \tag{9.181}$$

The eigenvectors of M determine the weights $A_\pm$. The eigenvectors corresponding to $\lambda_\pm$ satisfy

$$\begin{bmatrix} i\omega-\frac{\gamma_a}{2}-\lambda_\pm & -ig \\ ig & -i\omega-\frac{\gamma_b}{2}-\lambda_\pm \end{bmatrix}\begin{bmatrix} v_1 \\ v_2 \end{bmatrix}=0. \tag{9.182}$$

By solving this system, we obtain

$$\begin{aligned} A_+ &= \frac{\gamma_a-\gamma_b+2\sqrt{4g^2-(\gamma_a-\gamma_b)^2}}{4g} \\ A_- &= \frac{\gamma_a-\gamma_b-2\sqrt{4g^2-(\gamma_a-\gamma_b)^2}}{4g}. \end{aligned} \tag{9.183}$$

These determine the relative contributions of the two modes. The asymmetry between A_+ and A_- shows that the system has non-reciprocal dynamics unless $\gamma_a=\gamma_b$. The presence of a drive on oscillator B breaks the detailed balance, leading to different weights in the forward and backward correlation functions. The correlation functions $C(\tau)$ and $C'(-\tau)$ for the driven two-oscillator system are

$$\begin{aligned} C(\tau)=\frac{1}{4g}\Big[&(\gamma_a-\gamma_b-2\sqrt{4g^2-(\gamma_a-\gamma_b)^2})e^{\frac{\tau}{4}(\gamma_a+\gamma_b+4i\omega-\sqrt{4g^2-(\gamma_a-\gamma_b)^2})} \\ &+(\gamma_a-\gamma_b+2\sqrt{4g^2-(\gamma_a-\gamma_b)^2})e^{\frac{\tau}{4}(\gamma_a+\gamma_b+4i\omega+\sqrt{4g^2-(\gamma_a-\gamma_b)^2})}\Big]e^{-\frac{\tau}{4}(2\gamma_a+2\gamma_b+8i\omega)} \end{aligned} \tag{9.184}$$

and

$$\begin{aligned} C'(-\tau)=\frac{1}{4g}\Big[&(\gamma_a-\gamma_b-2\sqrt{4g^2-(\gamma_a-\gamma_b)^2})e^{\frac{\tau}{4}(\gamma_a+\gamma_b+4i\omega+\sqrt{4g^2-(\gamma_a-\gamma_b)^2})} \\ &+(\gamma_a-\gamma_b+2\sqrt{4g^2-(\gamma_a-\gamma_b)^2})e^{\frac{\tau}{4}(\gamma_a+\gamma_b+4i\omega-\sqrt{4g^2-(\gamma_a-\gamma_b)^2})}\Big]. \end{aligned} \tag{9.185}$$

We observe that

$$C(\tau)\neq C'(-\tau). \tag{9.186}$$

This asymmetry arises because the prefactor in $C(\tau)$ includes an additional exponential term

$$e^{-\frac{\tau}{4}(2\gamma_a + 2\gamma_b + 8i\omega)}, \tag{9.187}$$

which does not appear in $C'(-\tau)$. This term is a direct consequence of the driven non-equilibrium dynamics, leading to a broken time-reversal symmetry and violation of detailed balance. When the damping rates of both oscillators are equal, the correlation functions are simplified to

$$C(\tau) = (e^{\tau g} - 1)e^{-\frac{\tau}{2}(\gamma+2i\omega+g)} \tag{9.188}$$

and

$$C'(-\tau) = \left(e^{\frac{\tau}{2}(\gamma+2i\omega-g)} - e^{\frac{\tau}{2}(\gamma+2i\omega+g)}\right). \tag{9.189}$$

Even when the damping rates are equal, the time-reversal symmetry remains broken due to the coherent coupling g and the presence of a steady-state drive. Detailed balance is broken as the external drive introduces an energy flow into the system and breaks the time-reversal symmetry in the steady state.

Correlation functions such as

$$G^{(2)}(\tau) = \langle a^\dagger(0)a^\dagger(\tau)a(\tau)a(0)\rangle \tag{9.190}$$

are time symmetric by construction. Using stationarity we have and enforcing time ordering we have

$$\begin{aligned} G^{(2)}(-\tau) &= \langle :a^\dagger(0)a^\dagger(-\tau)a(-\tau)a(0): \rangle \\ &= \langle :a^\dagger(\tau)a^\dagger(0)a(0)a(\tau): \rangle \\ &= \langle a^\dagger(0)a^\dagger(\tau)a(\tau)a(0)\rangle \\ &= G^{(2)}(\tau). \end{aligned} \tag{9.191}$$

This makes sense as both physically refer to the relative probability of detecting a photon at τ given a detection at $\tau = 0$, a photon detection conditioned by a previous one. But for

$$H = \langle a^\dagger(0)a_\theta(\tau)a(0)\rangle, \tag{9.192}$$

we are looking at a quadrature measurement conditioned by a photodetection. The time-reversed one would be the probability of a photodetection conditioned by a previous quadrature measurement, two very different things. This will be more clear once we learn about the quantum trajectories in chapter 11.

References

[1] Geiger H and Müller W 1928 Elektronenzählrohr zur Messung schwächster Aktivitäten (Electron counting tube for the measurement of the weakest radioactivities) *Naturwissenschaften* **16** 617

[2] Mandel L 1959 Fluctuations of photon beams: the distributions of the photoelectrons *Proc. Phys. Soc.* **74** 233

[3] Einstein A 1923 Nobel Prize lecture: Fundamental ideas and problems of the theory of relativity The *Nobel Prize* https://www.nobelprize.org/prizes/physics/1921/einstein/lecture/

[4] Fano U 1947 Ionization yield of radiations. II. The fluctuations of the number of ions *Phys. Rev.* **72** 26

[5] Mandel L 1979 Sub-Poissonian photon statistics in resonance fluorescence *Opt. Lett.* **4** 205

[6] Kimble H J, Dagenais M and Mandel L 1977 Photon antibunching in resonance fluorescence *Phys. Rev. Lett.* **39** 691

[7] Nogueira W A T, Walborn S P, Pádua S and Monken C H 2004 Generation of a two-photon singlet beam *Phys. Rev. Lett.* **92** 043602

[8] Carmichael H, Castro-Beltran H, Denisov A, Foster G and Orozco L 2000 Measuring nonclassical fluctuations of light through wave-particle correlations *Phys. Rev. Lett.* **5** 1855–8

[9] Carmichael H J, Foster G T, Orozco L A, Reiner J E and Rice P R 2004 *Intensity-Field Correlations of Non-Classical Light* Progress in Optics vol 46 (Amsterdam: North Holland)

[10] Reiner J, Smith W P, Orozco L A, Carmichael H and Rice P 2000 Time evolution and squeezing of the field amplitude in cavity QED *J. Opt. Soc. Am.* B **18** 12

[11] Carmichael H, Castro-Beltran H, Foster G and Orozco L A 2000 Giant violations of classical inequalities through conditional homodyne detection of the quadrature amplitudes of light *Phys. Rev. Lett.* **85** 1855–8

[12] Foster G, Orozco L A, Castro-Beltran H and Carmichael H 2000 Quantum state reduction and conditional time evolution of wave-particle correlations in cavity QED *Phys. Rev. Lett.* **85** 3149–52

[13] Leach J R, Mumba M and Rice P R 2011 Quantum fluctuations in a cavity-QED system with quantized center-of-mass motion *Phys. Rev.* A **83** 063844

[14] Leach J 2003 Photon statistics and field-intensity correlation of a cavity QED system with external potentials *PhD Thesis* Miami University http://rave.ohiolink.edu/etdc/view?acc_num=miami1058466870

IOP Publishing

An Introduction to Quantum Optics (Second Edition)
An open systems approach
Perry Rice

Chapter 10

The density matrix and the master equation or wave functions: the big lie

10.1 Open quantum systems

In the 'real world', quantum systems are not isolated from their environment. This is not a lack, as there are systems that we want to interact with by putting energy and information in and getting energy and information out. This is true in classical mechanics as well, but we can add driving forces and treat dissipation in an ad hoc phenomenological way.

In classical mechanics, the addition of a velocity-dependent force $\overrightarrow{F} = -\gamma\overrightarrow{v} = -\gamma\overrightarrow{p}/m$ can account for dissipation to the environment, as seen in a ballistics problem that accounts for air resistance. Let us consider a similar treatment in quantum mechanics for adding a damping mechanism to a harmonic oscillator:

$$H = \frac{p^2}{2m} + \frac{1}{2}m\omega^2 q^2. \tag{10.1}$$

The equations of motion associated with equation (10.1) after a damping force of $-\gamma p$ is added are

$$\dot{q} = \frac{p}{m}, \qquad \dot{p} = -\gamma p - m\omega^2 q. \tag{10.2}$$

If we look at the time evolution of the commutator $[q, p]$, we see

$$\frac{d}{dt}[q, p] = \dot{q}p + q\dot{p} - \dot{p}q - p\dot{q} \tag{10.3}$$

$$= -\gamma[q, p], \tag{10.4}$$

which has the solution

$$[q(t), p(t)] = e^{-\gamma t}[q(0), p(0)] = i\hbar e^{-\gamma t}. \tag{10.5}$$

This obviously cannot be the correct quantum mechanical treatment of dissipation, as the commutator of q and p would decay with time, making the lower bound of the uncertainty principle decay to 0:

$$\begin{aligned}\Delta q\Delta p &\geq |\frac{i}{2}\langle[q, p]\rangle| \\ &\geq \frac{\hbar}{2}e^{-\gamma t}.\end{aligned} \tag{10.6}$$

Because of problems such as these in the quantum mechanical treatment of dissipation, the description of a typical quantum optics system that incorporates losses must be fundamentally different from the classical description.

10.2 Density matrix and reduced density matrix

In elementary quantum mechanics, we deal with isolated, closed systems. The total energy of the system is conserved, and the system is described by a wavefunction or state vector that contains information about the pure state of the system, including any superpositions or entanglements. A Hermitian Hamiltonian describes the time evolution of the state vector as well as the energy eigenstates of the system. This evolution preserves the norm of the state vector. Quantum systems that interact with their environment cannot generally be described by a wave function or state vector; we must use a quantity known as a density matrix that characterizes our state of knowledge of the system.

Quantum optics, as a field, began with the invention of the laser, an open system by its nature. Treating open systems complicates the formalism that we use for modeling the system. In quantum optics, we define a system about whose dynamics we wish to know in detail, and an environment about whose dynamics we do not care in detail. We just need the gross properties of the environment such as temperature, pressure, etc. However, when we allow the system (say a two level atom) to couple with the environment (all fields modes surrounding the atom), then the system is no longer in a pure state but in a statistical mixture of states. We wish to follow the state of the atom as it absorbs and emits light, but we do not care with which particular mode the atom is interacting. The density operator is the object that describes the statistical mixture of states of an open quantum system. The density operator exists as an object unto itself in a mathematical sense, just as the state vector does. It is defined as the sum of the density operators of pure states that span the basis of possible states

$$\rho = \sum_i P_i|\psi_i\rangle\langle\psi_i|. \tag{10.7}$$

The P_i signify our classical ignorance; they are probabilities for finding the state in $|i\rangle$ if we make a projective measurement on that basis

$$\text{Proj} = |i\rangle\langle i|. \tag{10.8}$$

However, we can find a matrix representation for the density operator in any basis, which we will call the density matrix. As with any operator, we find a matrix representation by allowing the operator to act on the basis vectors. Note that this is not a unique expansion, as any basis may be used. For example, the density matrix in the energy eigenbasis of a two-level atom is

$$\rho \rightarrow \begin{pmatrix} \langle e|\rho|e\rangle & \langle e|\rho|g\rangle \\ \langle g|\rho|e\rangle & \langle g|\rho|g\rangle \end{pmatrix}, \tag{10.9}$$

where $|e\rangle$ denotes the excited state and $|g\rangle$ denotes the ground state. To learn about the properties of a density matrix, let us consider the density matrix for a two-level atom in an arbitrary superposition of the excited state and the ground state

$$|\psi_S\rangle = C_e|e\rangle + C_g|g\rangle. \tag{10.10}$$

For a pure state, the density matrix is

$$\begin{pmatrix} \langle e|\psi_S\rangle\langle\psi_S|e\rangle & \langle e|\psi_S\rangle\langle\psi_S|g\rangle \\ \langle g|\psi_S\rangle\langle\psi_S|e\rangle & \langle g|\psi_S\rangle\langle\psi_S|g\rangle \end{pmatrix} = \begin{pmatrix} C_eC_e^* & C_eC_g^* \\ C_gC_e^* & C_gC_g^* \end{pmatrix}. \tag{10.11}$$

We see that the diagonal elements of the density matrix tell us the populations of states $P_g = |C_g|^2$, $P_e = |C_g|^2$, and the off-diagonal elements tell us about the coherence between states. Nonzero off diagonal elements tell us that in the basis we represent the density matrix we have a superposition, which can lead to interference terms in certain measurements. In the state above, we have more than just statistical uncertainty in terms of which state the system is in, excited or ground. The system is in both states simultaneously, which results in quantum interference.

The trace of ρ is just the sum of the probabilities for being in the various states, so normalization of the quantum state results in

$$\mathrm{Tr}(\rho) = 1. \tag{10.12}$$

In the pure state case, as the density operator is really a projection operator in the state of the system, leaving it in that state, we have

$$\begin{aligned} \rho^2 &= |\psi\rangle\langle\psi|\psi\rangle\langle\psi| \\ &= |\psi\rangle\langle\psi| \\ &= \rho. \end{aligned} \tag{10.13}$$

We can also have mixed states more generally, $\rho = \sum_i P_i|\psi_i\rangle\langle\psi_i|$. Here we have some classical uncertainty in terms of our knowledge of the quantum state of the system is. For a mixed state that is in the excited state with some probability P_e and the ground state with some probability P_g, the density matrix is

$$\begin{pmatrix} \langle e|(P_e|e\rangle\langle e| + P_g|g\rangle\langle g|)|e\rangle & \langle e|(P_e|e\rangle\langle e| + P_g|g\rangle\langle g|)|g\rangle \\ \langle g|(P_e|e\rangle\langle e| + P_g|g\rangle\langle g|)|e\rangle & \langle g|(P_e|e\rangle\langle e| + P_g|g\rangle\langle g|)|g\rangle \end{pmatrix}, \tag{10.14}$$

which reduces to

$$\begin{pmatrix} P_e & 0 \\ 0 & P_g \end{pmatrix}, \tag{10.15}$$

after we averaged over a uniform distribution of the excited and ground state phases.

Here we *have* a case where either the atom is in the excited or ground state, with some probability, with no interference; just classical statistical uncertainty. For a mixed state, we still have $\mathrm{Tr}(\rho) = 1$ as the probabilities sum to one, but

$$\begin{aligned} \rho^2 &= \sum_{i,j} P_i P_j |\psi_i\rangle \langle\psi_i|\psi_j\rangle \langle|\psi_j \\ &= \sum_j P_j^2 |\psi_j\rangle \langle\psi_j| \\ &\neq \rho. \end{aligned} \tag{10.16}$$

One can also show that for a mixed state, we have

$$\mathrm{Tr}(\rho^2) = \sum_j P_j^2 \leq \sum_j P_j \tag{10.17}$$

and so $\mathrm{Tr}(\rho^2) \leq 1$, where the equality holds for a pure state only. One can use $1 - \mathrm{Tr}(\rho^2)$ as a measure of how far your state is from a pure state; in fact, this is referred to as the impurity of the state.

The decomposition of a mixed-state density matrix into a weighted sum of pure-state density matrices is not unique. Consider the state

$$\rho = \begin{pmatrix} 1/2 & 1/2 \\ 1/2 & 1/2 \end{pmatrix}. \tag{10.18}$$

This can be written as both

$$\begin{aligned} \rho &= 1/2(|e\rangle + |g\rangle)(\langle e| + \langle g|) \\ &= 1/2(|+\rangle + |-\rangle)(\langle +| + \langle -|), \end{aligned} \tag{10.19}$$

where the dressed states are $|\pm\rangle = (1/\sqrt{2})(|e\rangle \pm |g\rangle)$. In both cases, there is a probability of 1/2 for the two basis states.

We can use the density operator to gain information about the system of interest by calculating the expectation values and correlation functions of various orders. The expected values of the operators are calculated by taking the trace of both the operator and the density matrix over the system states

$$\langle O\rangle = \mathrm{Tr}_S(\rho O), \tag{10.20}$$

which reduces to the standard form

$$\begin{aligned} \langle O\rangle &= \mathrm{Tr}_S(|\psi\rangle\langle\psi|O) \\ &= \mathrm{Tr}_S(\langle\psi|O|\psi\rangle) \\ &= \langle\psi|O|\psi\rangle, \end{aligned} \tag{10.21}$$

when $\rho = |\psi\rangle\langle\psi|$ describes a pure state.

In the mixed-state case, we also obtain something that makes sense

$$\begin{aligned}\langle O\rangle &= \mathrm{Tr}_S\left(\sum\nolimits_i P_i|\psi_i\rangle\langle\psi_i|O\right)\\ &= \sum\nolimits_i \mathrm{Tr}_S(P_i\langle\psi_i|O|\psi_i\rangle)\\ &= \sum\nolimits_i P_i\langle\psi_i|O|\psi\rangle\\ &= \sum\nolimits_i P_i\langle O\rangle_i.\end{aligned} \tag{10.22}$$

The time evolution of the density matrix under interactions that are described by a Hamiltonian is found by differentiating equation (10.7) and using the Schrödinger equation:

$$\dot{\rho} = \frac{d}{dt}\sum\nolimits_i P_i|\psi_i\rangle\langle\psi_i| \tag{10.23}$$

$$= \sum\nolimits_i P_i(|\psi_i\rangle\langle\dot{\psi}_i| + |\dot{\psi}_i\rangle\langle\psi_i|) \tag{10.24}$$

$$= -\frac{i}{\hbar}\sum\nolimits_i P_i(H|\psi_i\rangle\langle\psi_i| - |\psi_i\rangle\langle\psi_i|H) \tag{10.25}$$

$$= -\frac{i}{\hbar}[H, \rho]. \tag{10.26}$$

We can now treat systems in mixed states under Hamiltonian evolution using this formalism. However, we wish to treat open quantum systems with dissipation into an environment using the density matrix. For this, we use a master equation to describe the time evolution of the density matrix [1].

Although quantum optics is generally not concerned with the specific dynamics of the full environment (we may not be interested in which field mode the atom has radiated into, we care only that the atom has radiated), it is instructive to first characterize a general density operator that represents the state of both the system we are concerned with and the environment. As we have seen, the density operator for a mixed state is the sum of the density operators for the possible pure states of the full system (atom + field) with a weight factor of P_i, the probability of being projected onto state i:

$$\chi = \sum\nolimits_i P_i|\psi_i\rangle\langle\psi_i|. \tag{10.27}$$

For a pure state:

$$\chi = |\psi\rangle\langle\psi|. \tag{10.28}$$

This density operator has a matrix representation using the basis vectors of the full system, we call this the density matrix. If our system S (the atom) and the reservoir R (the EM field modes) are coupled together, they form a direct product space $S\bigotimes R$. Because the number of possible states for the system and the reservoir are different, we designate S as an $N \times N$ space, and R as a $M \times M$ space, with

$N \ll M$. The density matrix for $S \otimes R$, $\chi(t)$, takes the form of an $NM \times NM$ matrix,

$$\chi(t) = \begin{pmatrix} R_{11}\rho(t) & R_{12}\rho(t) & R_{13}\rho(t) & \dots & R_{1M}\rho(t) \\ R_{21}\rho(t) & R_{22}\rho(t) & R_{23}\rho(t) & \dots & R_{2M}\rho(t) \\ R_{31}\rho(t) & R_{32}\rho(t) & R_{33}\rho(t) & \dots & R_{3M}\rho(t) \\ \vdots & \vdots & \vdots & \vdots & \vdots \\ R_{M1}\rho(t) & R_{M2}\rho(t) & R_{M3}\rho(t) & \dots & R_{MM}\rho(t) \end{pmatrix}, \tag{10.29}$$

where R_{ii} is the probability of the reservoir being in the state i, and R_{ij} is a measure of the coherence between the reservoir states i and j. This form is true at $t = 0$, but at later times the reservoir and system will generally become entangled.

$\rho(t)$ is the reduced density matrix that is a representation of the statistical mixture of states in the system S is in. Following the general treatment, $\rho(t)$ takes the form

$$\rho(t) = \begin{pmatrix} \rho_{11} & \rho_{12} & \rho_{13} & \cdots & \rho_{1N} \\ \rho_{21} & \rho_{22} & \rho_{23} & \cdots & \rho_{2N} \\ \rho_{31} & \rho_{32} & \rho_{33} & \cdots & \rho_{3N} \\ \vdots & \vdots & \vdots & \vdots & \vdots \\ \rho_{N1} & \rho_{N2} & \rho_{N3} & \cdots & \rho_{NN} \end{pmatrix}, \tag{10.30}$$

where, similar to $\chi(t)$, ρ_{ii} is the probability of the system S being in state i, and ρ_{ij} is a measure of the coherence between system states i and j. From equation (10.29), we see that the reduced density matrix is really just the trace of the density matrix over the reservoir states

$$\rho(t) = \mathrm{Tr}_R[\chi(t)] = \left[\sum\nolimits_i R_{ii}\right]\rho(t) = 1\ \rho(t), \tag{10.31}$$

and that the trace of the reduced density matrix is evaluated as

$$\mathrm{Tr}_S[\rho(t)] = \sum\nolimits_i \rho_{ii}(t) = 1. \tag{10.32}$$

To calculate the expectation values for the operators within the basis of S, we take the trace of the product of $\chi(t)$ and the operator O:

$$\langle O \rangle = \mathrm{Tr}_{S \otimes R}[O\chi(t)]. \tag{10.33}$$

This reduces to

$$\begin{aligned} \mathrm{Tr}_S[\mathrm{Tr}_R[O\chi(t)]] &= \mathrm{Tr}_S[O\mathrm{Tr}_R[\chi(t)]] = \mathrm{Tr}_S[O\rho(t)] \\ &= \mathrm{Tr}[O\rho(t)] \end{aligned} \tag{10.34}$$

because O does not act on the reservoir basis.

Although we know that the reduced density matrix $\rho(t)$ (from here on referred to as the density matrix, as $\chi(t)$ serves us no computational purpose) describes the state of the system, we do not yet know anything about the time evolution of the density matrix, and hence know nothing about the evolution of the system. Using the definition of the density operator (equation (10.28)) and the Schrödinger equation, we obtain

$$\begin{aligned}\dot{\rho} &= \sum_i p_i [|\dot{\psi}_i\rangle\langle\psi_i| + |\psi_i\rangle\langle\dot{\psi}_i|] \\ &= \sum_i p_i \left[-\frac{i}{\hbar} H_S |\psi_i\rangle\langle\psi_i| + \frac{i}{\hbar}|\psi_i\rangle\langle\psi_i| H_S \right] \\ &= -\frac{i}{\hbar}[H_S, \rho],\end{aligned} \tag{10.35}$$

where H_S is the Hamiltonian for the system only, containing no terms that act on the reservoir, nor any interaction terms. Thus, if the system Hamiltonian is known to us, we can solve for the time evolution of the system at hand. However, we still have not added dissipation into our formalism; we have simply set up a framework into which dissipative terms may be added in a clear, consistent manner. For the consideration of dissipation, we must generalize the form of equation (10.35).

10.3 The master equation with dissipation

An open quantum system typically consists of several atoms and field modes coupled with an environment with many degrees of freedom. While in principle the evolution of all the environmental degrees of freedom could be followed, it is computationally expensive, and detailed information about the environment is not of interest. We need a formalism that properly models the action of the environment on the system but does not require us to keep detailed information about the environment. (For example, we wish to know that an atom has emitted a photon, but not which direction the photon went.) One way to do this is to derive a master equation where the time evolution of the density operator is given by a Liouvillian operator

$$\dot{\rho} = \mathcal{L}\rho \tag{10.36}$$

with an reversible Hamiltonian evolution and irreversible, dissipative processes. A brief outline of the derivation follows, and then we will present a more detailed derivation. The approach we follow is inspired by the works of Carmichael [2–4], Bruer and Pettrucione [5], and Gardiner and Zoller [6].

In a typical system, we are concerned with the dynamics of one or many two-level atoms and the quantized electromagnetic field within a lossy optical cavity. Energy is being provided to this system via the driving field, a laser treated classically (in this work) coming through the cavity's input mirror. Dissipation arises in the system through the coupling of the atoms and the cavity field modes to an environment with many degrees of freedom. In general, we are not concerned with the details of this reservoir's evolution, we only care about the reservoir's effect on the atom–cavity system, for it is the source of dissipation. To formally treat the system's evolution, we use the master equation,

$$\dot{\rho} = \mathcal{L}\rho, \tag{10.37}$$

where $\mathcal{L}$ is a generalized Liouvillian superoperator (it operates on other operators, not states) [2]. The Liouvillian contains within it the same Hamiltonian evolution given by equation (10.35), but also contains the contributions of dissipative processes. Sometimes people will define

$$\mathcal{L}\rho = -\frac{i}{\hbar}[H, \rho] + \mathcal{L}'\rho, \tag{10.38}$$

where $\mathcal{L}'$ just includes the environmental terms

The reservoir to which our system couples is modeled as a collection of quantized harmonic oscillator modes in thermal equilibrium at some temperature. This allows us to sufficiently model different environments. In this work, the environment is considered to be a reservoir of electromagnetic field modes. To arrive at the final Liouvillian evolution of our system, we essentially do the following:

1. Couple the system to the reservoir.
2. Use the Born and Markov approximations to deal with the system/reservoir's coupling strength and the response time of the reservoir.
3. Trace over the reservoir states to extract the system dynamics.

One type of dissipation we will examine is the decay of the cavity field mode into the vacuum radiation field outside the optical cavity, the coupling occurring via the mirror. This system is characterized by the Hamiltonian

$$H_S = \hbar\omega_0 a^\dagger a. \tag{10.39}$$

The reservoir Hamiltonian is given by

$$H_R = \sum\nolimits_j \hbar\omega_j r_j^\dagger r_j, \tag{10.40}$$

where a and $a^\dagger$ (r_j and $r^\dagger{}_j$) are the annihilation and creation operators for the cavity field mode, respectively. (The jth reservoir field mode.) The interaction Hamiltonian between the cavity field mode and the reservoir in the rotating wave approximation (RWA) is given by

$$H_{SR} = \sum\nolimits_j \hbar\left(\kappa_j^* a r_j^\dagger + \kappa_j a^\dagger r_j\right), \tag{10.41}$$

where κ_j is the coupling strength between the cavity field mode and the jth mode of the reservoir field. The first approximation we make in solving for the system's evolution is the Born approximation. The Born approximation enters the formal derivation as a statement that the full density matrix can be factored as

$$\tilde{\chi}(t) = \tilde{\rho}(t)R_0, \tag{10.42}$$

where the tildes above $\rho(t)$ and $\chi(t)$ denote that this equation occurs in the interaction picture and not the Schrödinger picture that we have worked in thus far.

Equation (10.42) is equivalent to saying that the reservoir's properties are independent of time. In this work, this means that due to the reservoir's many degrees of freedom, it is assumed that the system couples only very weakly to any one mode of the vacuum field ($\kappa_j \langle < 1$ for all j). Thus, the system has very little effect on the reservoir, while the reservoir modes collectively have a large effect on the system. A simple analogy might involve dropping an ice cube into the Atlantic Ocean. The ocean affects the ice cube profoundly, but the ice cube disturbs the ocean to a very small degree. The next approximation made is the Markov approximation; a Markoffian system's future dynamics only depend on the system's present state. This approximation enters the formal derivation in the form of a qualitative comparison between the reservoir's time scale and that of the system. If the reservoir is able to return to its steady-state very quickly as compared with the time scale of the system's dynamics, then the reservoir will 'have no memory' of past interactions. This irreversibility is then in line with our intuition of the dissipative processes. After these approximations are made in the formal derivation, one can then trace over the reservoir's states to obtain the Liouvillian evolution of the system's density matrix in the form of the master equation. A more formal discussion of the Born–Markov approximation and a derivation of the master equation are presented in a later section.

The Liouvillian evolution of a single cavity field mode coupled to the vacuum radiation field given by equation (10.40) is

$$\begin{aligned}\dot{\rho}(t) = &- i\omega_0[a^\dagger a, \rho] + \kappa(\bar{n} + 1)[2a\rho(t)a^\dagger - \rho(t)a^\dagger a - a^\dagger a\rho(t)] \\ &+ \kappa\bar{n}[2a\rho(t)a^\dagger - aa^\dagger\rho(t) - \rho(t)aa^\dagger],\end{aligned} \tag{10.43}$$

where κ and $\bar{n}$ are the cavity field decay rate and the thermal photon number (the mean number of photons in the reservoir at thermal equilibrium), respectively. We can summarize this by defining a Liouvillian super-operator (an operator that acts on operators)

$$\begin{aligned}\mathcal{L} = &- i\omega_0[a^\dagger a, \rho] + \kappa(\bar{n} + 1)[2a\rho(t)a^\dagger - \rho(t)a^\dagger a - a^\dagger a\rho(t)] \\ &+ \kappa\bar{n}[2a\rho(t)a^\dagger - aa^\dagger\rho(t) - \rho(t)aa^\dagger].\end{aligned} \tag{10.44}$$

To see that equation (10.113) gives results that are consistent with our intuition of a decaying field, we calculate the time-dependent photon number of the cavities. Using equation (10.33) to calculate the time derivative of the photon number's expectation value, we find

$$\begin{aligned}\langle \dot{n} \rangle &= \mathrm{Tr}[a^\dagger a\dot{\rho}(t)] \\ &= - i\omega_0 \,\mathrm{Tr}(a^\dagger aa^\dagger a\rho - a^\dagger a\rho a^\dagger a) + \kappa \,\mathrm{Tr}\,(2a^\dagger a^2\rho a^\dagger - a^\dagger aa^\dagger a\rho \\ &\quad - a^\dagger a\rho a^\dagger a) + 2\kappa\bar{n}\,\mathrm{Tr}(a^\dagger a^2\rho a^\dagger + a^\dagger aa^\dagger\rho a - a^\dagger aa^\dagger a\rho - a^\dagger a\rho aa^\dagger) \\ &= 2\kappa\,\mathrm{Tr}((a^\dagger)^2a^2\rho - (a^\dagger a)^2\rho) \\ &\quad + 2\kappa\bar{n}\,\mathrm{Tr}((a^\dagger)^2a^2\rho + (a^\dagger a)^2\rho - (a^\dagger a)^2\rho a(a^\dagger)^2 a\rho) \\ &= - 2\kappa[\langle n\rangle - \bar{n}].\end{aligned} \tag{10.45}$$

This differential equation has the solution

$$\langle n(t)\rangle = \langle n(0)\rangle e^{-2\kappa t} + \bar{n}(1 - e^{-2\kappa t}). \tag{10.46}$$

From this example, we see that our method provides us with a photon number that starts at some initial value $\langle n(0)\rangle$ and evolves to its thermal steady-state value $\bar{n}$; thus, our method is in congruence with our intuition.

Another type of dissipation that plays an important role in the evolution of a cavity QED system is the decay of the atomic inversion, that is, the radiative decay of the two-level atom. Here, our system Hamiltonian is given by

$$H_S = \frac{1}{2}\hbar\omega_0\sigma_z, \tag{10.47}$$

where σ_z is the Pauli pseudo-spin operator that describes the atomic inversion:

$$\sigma_z = \begin{pmatrix} & 0 \\ 0 & -1 \end{pmatrix}. \tag{10.48}$$

The environment that we will couple the system to is a reservoir of cavity field modes, which has the same Hamiltonian given by equation (10.40); the interaction Hamiltonian is

$$H_{SR} = \sum\nolimits_{\vec{k},p}\hbar(\kappa^*_{\vec{k},p} r^\dagger_{\vec{k},p}\sigma_- + \kappa_{\vec{k},p} r_{\vec{k},p}\sigma_+), \tag{10.49}$$

where $\vec{k}$ denotes the wavevector of the field mode, and p denotes the mode's polarization. σ_+ and σ_- are the Pauli pseudo-spin operators that take a two-level atom from the excited state to the ground state and the atom from the ground state to the excited state, respectively,

$$\sigma_+ = \begin{pmatrix} 0 & 1 \\ 0 & 0 \end{pmatrix}, \tag{10.50}$$

$$\sigma_- = \begin{pmatrix} 0 & 0 \\ & 0 \end{pmatrix}. \tag{10.51}$$

Following a derivation analogous to that for the damped cavity field and making the same approximations, we find that the master equation for a radiatively damped two-level atom is given by

$$\begin{aligned} \dot{\rho} = &-\frac{i}{2}\omega_0[\sigma_z, \rho] + \frac{\gamma}{2}(\bar{n}_a + 1)(2\sigma_-\rho\sigma_+ - \sigma_+\sigma_-\rho - \rho\sigma_+\sigma_-) \\ &+ \frac{\gamma}{2}\bar{n}_a(2\sigma_+\rho\sigma_- - \sigma_-\sigma_+\rho - \rho\sigma_-\sigma_+), \end{aligned} \tag{10.52}$$

where $\bar{n}_a$ is now the thermal photon number of the reservoir to which the atom couples, and γ now denotes the rate of spontaneous emission.

At the optical wavelengths often considered in cavity QED experiments, the thermal photon number $\bar{n}$ is often very small and can be ignored. At $\omega_0 = 3 \times 10^{15}$, $T = 300$ K and $\bar{n} \approx 7 \times 10^{-34} \approx 0$. Taking this approximation, we will show the simple calculation of the spontaneous emission dynamics of the two-level atom. By the definition of the reduced density matrix $\rho(t)$, the probability of finding the two-level atom in its excited state at time t is given by the population

$$\rho_{ee}(t) = \langle e|\rho(t)|e\rangle . \tag{10.53}$$

Thus, the time derivative of this population is given by

$$\begin{aligned}\dot{\rho}_{ee} &= \langle e|\dot{\rho}_{ee}|e\rangle \\ &= -\frac{i}{2}\omega_0\langle e|(\sigma_z\rho - \rho\sigma_z)|e\rangle + \frac{\gamma}{2}\langle e|(2\sigma_-\rho\sigma_+ - \sigma_+\sigma_-\rho - \rho\sigma_+\sigma_-)|e\rangle \\ &= -\gamma\rho_{ee}.\end{aligned} \tag{10.54}$$

This is a familiar differential equation with the familiar solution

$$\rho_{ee}(t) = \rho_{ee}(0)e^{-\gamma t}. \tag{10.55}$$

This result matches the classical result and is consistent with our intuition of radiative decay.

10.4 Quantum regression theorem

We have seen earlier that aspects of the optical spectra and photon statistics are described by two-time correlation functions. For example, the power spectrum of light is given by

$$\begin{aligned}I(\omega) &= \frac{T}{2\pi}\int_{-\infty}^{\infty} d\tau e^{-i\omega\tau}\langle a^\dagger(t+\tau)a(t)\rangle \\ &= \frac{\langle a^\dagger a\rangle T}{2\pi}\int_{-\infty}^{\infty} d\tau e^{-i\omega\tau}g^{(1)}(\tau).\end{aligned} \tag{10.56}$$

In addition, the probability of detecting a photon at time τ given that one was detected at $t = 0$, scaled by the probability of detecting two independent photons (randomly generated via a Poisson process, as would result from a coherent state) is

$$g^{(2)}(\tau) = (G^{(2)}\tau)/G^{(2)}(\infty) \tag{10.57}$$

$$= \frac{\langle a^\dagger(0)a^\dagger(\tau)a(\tau)a0\rangle}{\langle a^\dagger a\rangle_{ss}^2}. \tag{10.58}$$

The key to calculating the correlation functions from the master equation is the quantum regression theorem. As we have seen, the correlation functions take the form

$$\langle O_1(t)\ O_2(t+\tau)\rangle = \mathrm{Tr}\{O_2(0)e^{\mathcal{L}\tau}[\rho(t)O_1(0)]\} \tag{10.59}$$

$$\langle O_1(t)\ O_2(t+\tau)\ O_3(t)\rangle = \mathrm{Tr}\{O_2(0)e^{\mathcal{L}\tau}[O_3(0)\rho(t)O_1(0)]\}. \tag{10.60}$$

To calculate equation (10.59) for a given time t, we define an operator

$$A(\tau) = e^{\mathcal{L}\tau}\rho(t)O_1(0). \tag{10.61}$$

Here, we have assumed that the statistics are stationary; A depends only on the delay time τ. We are *not* saying that everything is stationary, that nothing changes. Merely that after the system has been going for a while (defined by some number of the longest lifetime in the problem), we can pick any time interval we wish, beginning at any time $\tau = 0$ we wish, and the resulting averages, variances, etc, would not change. Then, we can write for a complete set of operators $\overrightarrow{A}$

$$\overrightarrow{A}(\tau) = e^{\mathcal{L}\tau}\overrightarrow{A}(0) \tag{10.62}$$

$$\dot{\overrightarrow{A}} = \mathcal{L}\overrightarrow{A}. \tag{10.63}$$

Note that $\overrightarrow{A}(\tau)$ is governed by the same master equation as $\rho(t)$ although with initial conditions that depend on $\rho(t)$. All that is needed then is to solve for $\overrightarrow{A}(\tau)$ by one of the methods described at the beginning of this section, using the result in equation (10.59). The solution of equation (10.60) follows from a similar method. In other words, the quantum regression theorem states that two-time correlation functions obey the same equations of motion with respect to τ as expectation values obey with respect to t. Often in quantum optics, we are interested in the spectrum or photon statistics of the field after the system has settled down to some steady state. In this case, we let $t \to \infty$ so that $\rho(t)$ is replaced by ρ_{ss} and then correlation functions depend only on τ.

The quantum regression theorem provides a method to formally represent and solve two-time correlation functions, that is, an expectation value of the product of two operators evaluated at two different times. In the Heisenberg picture, this expectation value is given by

$$\langle O_1(t)O_2(t')\rangle = \mathrm{Tr}_{S\oplus R}[\chi(0)O_1(t)O_2(t')]. \tag{10.64}$$

Defining $\tau = t' - t$, using the cyclic property of the trace, and tracing over the reservoir states eventually leads to the result

$$\langle O_1(t)O_2(t+\tau)\rangle = \mathrm{Tr}_S[O_2(0)e^{\mathcal{L}\tau}[\rho(t)O_1(0)]]. \tag{10.65}$$

If we consider time t to be the system steady-state and exploit statistical stationarity, we may write equation (10.65) as

$$\langle O_1(0)O_2(\tau)\rangle = \mathrm{Tr}_S\Big[O_2(0)e^{\mathcal{L}\tau}\big[\rho_{ss}O_1(0)\big]\Big]. \tag{10.66}$$

Next, we define a set of operator,

$$\overrightarrow{A}(\tau) = e^{\mathcal{L}\tau}\rho_{ss}O_1(0), \tag{10.67}$$

where we only see the evolution of A in τ, with the initial condition

$$A(0) = \rho_{ss}O_1(0). \tag{10.68}$$

Now we can write

$$A(\tau) = e^{\mathcal{L}\tau}A(0), \tag{10.69}$$

$$\dot{A} = \mathcal{L}A. \tag{10.70}$$

Equation (10.70) shows that $A(\tau)$ obeys the same master equation as $\rho(t)$, with $A(0)$ having initial conditions dependent upon ρ_{ss}. The familiar result of this is that the two-time correlation function in equation (10.66) satisfies the same equation of motion with respect to τ as expectation values do with respect to t:

$$\langle O_1(0)O_2(\tau)\rangle = \mathrm{Tr}_S[O_2(0)A(\tau)]. \tag{10.71}$$

How does one calculate the correlation functions with negative delays? With

$$\frac{d\rho}{dt} = \mathcal{L}\rho, \tag{10.72}$$

then the two-time correlation function for $\tau\rangle 0$

$$\langle A^{\dagger}(0)B(\tau)A(0)\rangle \tag{10.73}$$

can be computed by 'propagating' the operator B forward in time using the same generator $\mathcal{L}$. When dealing with a negative delay, say

$$\langle A^{\dagger}(0)B(-\tau)A(0)\rangle \quad (\tau\rangle 0), \tag{10.74}$$

if the system is stationary, you can use the time-translation invariance to 'flip' the time argument. One can show that

$$\langle A^{\dagger}(0)B(-\tau)A(0)\rangle = \langle A^{\dagger}(\tau)B(0)A(\tau)\rangle. \tag{10.75}$$

In this picture, the correlation with a negative delay is recast as a correlation at a positive delay, albeit with the time argument shifted from the operators A rather than B. Alternatively, one can view the negative delay as evolving the operator B 'backward in time'. Instead of using the Liouvillian $\mathcal{L}$ (which generates forward evolution), one uses its adjoint $\mathcal{L}^{\dagger}$ to propagate the operators backward

$$B(-\tau) = e^{\mathcal{L}^{\dagger}\tau}B(0), \quad (\tau\rangle 0). \tag{10.76}$$

Then, the correlation function is obtained as

$$\langle A^{\dagger}(0)B(-\tau)A(0)\rangle = \langle A^{\dagger}(0)[e^{\mathcal{L}^{\dagger}\tau}B(0)]A(0)\rangle.$$

This approach explicitly recognizes that evolving 'backward' requires the adjoint dynamics. Both perspectives are equivalent when the system is in a stationary state and the dynamics are Markovian. In practice, one often uses the time-translation invariance property to relate negative-delay correlations to the corresponding positive-delay ones, thereby sidestepping the need to directly implement a backward time evolution.

10.5 Derivation of the master equation in the Born–Markov approximation

To begin the formal derivation of the For the master equation in the Born–Markoff approximation, we start with a generalized form of equation (10.35) that applies to the full density operator, $\chi(t)$,

$$\dot{\tilde{\chi}}(t) = \frac{i}{\hbar}[\tilde{\chi}(t), H_i], \tag{10.77}$$

where, as before, the tilde denotes that we are working in the interaction picture and H_i is the interaction Hamiltonian for the system and the reservoir(s). We now formally integrate equation (10.77) and solve for $\tilde{\chi}(t)$

$$\tilde{\chi}(t) = \tilde{\chi}(0) + \frac{i}{\hbar}\int_0^t dt'[\tilde{\chi}(t'), H_i(t')]. \tag{10.78}$$

We now substitute this back into the right-hand side of equation (10.77) to obtain

$$\dot{\tilde{\chi}}(t) = \frac{i}{\hbar}[\tilde{\chi}(0), H_i] - \frac{1}{\hbar^2}\int_0^t dt'[[\tilde{\chi}(t'), H_i(t')], H_i(t)]. \tag{10.79}$$

We now assume that at $t = 0$, the full density operator factorizes as

$$\tilde{\chi}(0) = \tilde{\rho}(0)R_0, \tag{10.80}$$

where $\tilde{\rho}$ is the reduced density operator in the interaction picture and R_0 is the density operator of the reservoir(s) at $t = 0$. This assumption states that at $t = 0$, we have chosen that there is no interaction between the system and the reservoir(s). We now trace over the reservoir(s) states to obtain an equation of motion for $\tilde{\rho}$ alone:

$$\dot{\tilde{\rho}} = \frac{i}{\hbar}\mathrm{Tr}_R[\tilde{\rho}(0)R_0(r), H_i(t)]\frac{1}{\hbar^2}\mathrm{Tr}_R\int_0^t dt'[[\tilde{\chi}(t'), H_i(t')], H_i(t)]. \tag{10.81}$$

For simplicity, we assume $\mathrm{Tr}_R[R_0(r), H_i(t)] = 0$. This means that the mean value of the interaction The Hamiltonian with respect to the reservoir is zero. This leaves us with

$$\dot{\tilde{\rho}} = -\frac{1}{\hbar^2}\mathrm{Tr}_R\int_0^t dt'[[\tilde{\chi}(t'), H_i(t')], H_i(t)]. \tag{10.82}$$

We now state the Born approximation, that to the lowest order in H_i The dynamics of the full density operator only depend upon the dynamics of the system's reduced density operator:

$$\tilde{\chi}(t) = \tilde{\chi}(t)R_0 + \mathcal{O}(H_i). \tag{10.83}$$

We may make this assumption because we consider the coupling between the system and the reservoir to be small. With this weak coupling, the reservoir is considered to be unaffected by the system due to its vast size. Our system, however, because of its small size compared to the reservoir, is greatly affected by the weak coupling to many reservoir modes. Using the Born approximation, to the lowest order in H_i, equation (10.82) becomes

$$\dot{\tilde{\rho}} = -\frac{1}{\hbar^2}\mathrm{Tr}_R\int_0^t dt'[[\tilde{\rho}(t')R_0, H_i(t')], H_i(t)]. \tag{10.84}$$

It is instructive to complete the derivation with an example such that the form of the Liouvillian superoperator (an operator that acts on operators) is justified in the equation $\dot{\rho} = \mathcal{L}\rho$. To do this, we must first further specify the interaction

Hamiltonian H_i to be a collection of products between the system and reservoir operators:

$$H_i = \hbar \sum_{j,k} S_k \Gamma_{j,k}, \tag{10.85}$$

where S_k are the system operators, and $\Gamma_{j,k}$ are the operators of the reservoir only. Notice here that the term we neglected in equation (10.81) is identically zero, as it would be the average of the linear reservoir system operators. If these are the field modes, we are saying that the average field is zero. Quantum mechanics forbids us from having the variance of the field be zero. Equation (10.84) now becomes

$$\dot{\tilde{\rho}} = -\frac{1}{\hbar^2}\mathrm{Tr}_R \int_0^t dt' \left[\left[\tilde{\rho}(t')R_0, \hbar \sum_{j,k} S_k(t')\Gamma_{j,k}(t')\right], \hbar \sum_{j,k} S_k(t)\Gamma_{j,k}(t)\right]. \tag{10.86}$$

Expanding equation (10.86), and accounting for the sum indexing, we have

$$\begin{aligned}\dot{\tilde{\rho}} = -\sum_{j,k,l,m} \int_0^t dt' [\; &\mathrm{Tr}_R[\tilde{\rho}(t')R_0 S_k(t')\Gamma_{j,k}(t')S_m(t)\Gamma_{l,m}(t)] \\ + \; &\mathrm{Tr}_R[S_k(t')\Gamma_{j,k}(t')\tilde{\rho}(t')R_0 S_m(t)\Gamma_{l,m}(t)] \\ + \; &\mathrm{Tr}_R[S_m(t)\Gamma_{l,m}(t)\tilde{\rho}(t')R_0 S_k(t')\Gamma_{j,k}(t')] \\ + \; &\mathrm{Tr}_R[S_m(t)\Gamma_{l,m}(t)S_k(t')\Gamma_{j,k}(t')\tilde{\rho}(t')R_0]].\end{aligned} \tag{10.87}$$

After performing the trace over the reservoir states, equation (10.87) simplifies to

$$\begin{aligned}\dot{\tilde{\rho}} = -\sum_{j,k,l,m} \int_0^t dt'[&[S_k(t')\tilde{\rho}(t')S_m(t) - \tilde{\rho}(t')S_k(t')S_m(t)]\langle \Gamma_{j,k}(t')\Gamma_{l,m}(t)\rangle_R \\ &+ [S_m(t)\tilde{\rho}(t')S_k(t') - S_m(t)S_k(t')\tilde{\rho}(t')]\langle \Gamma_{l,m}(t)\Gamma_{j,k}(t')\rangle_R].\end{aligned} \tag{10.88}$$

We are now in a position to specify the system and reservoir operators in equation (10.85) to work the example of a single cavity field mode coupled to a reservoir of harmonic oscillator modes. We define the creation and annihilation operators for the cavity field mode in the interaction picture

$$\begin{aligned}\tilde{S}_1(t) &= e^{i\omega_0 a^\dagger a t} a^\dagger e^{-i\omega_0 a^\dagger a t} = a^\dagger e^{-i\omega_0 t}, \\ \tilde{S}_2(t) &= e^{i\omega_0 a^\dagger a t} a e^{-i\omega_0 a^\dagger a t} = a e^{+i\omega_0 t}.\end{aligned} \tag{10.89}$$

Furthermore, we define the reservoir oscillator mode operators to be

$$\begin{aligned}\tilde{\Gamma}_{1,j}(t) &= e^{i\omega_j c_j^\dagger c_j t}\kappa_j c_j e^{-i\omega_j c_j^\dagger c t} = \kappa_j c_j e^{-i\omega_j t}, \\ \tilde{\Gamma}_{2,j}(t) &= e^{i\omega_j c_j^\dagger c_j t}\kappa_j^* c_j^\dagger e^{-i\omega_j c_j^\dagger c t} = \kappa_j^* c_j^\dagger e^{+i\omega_j t}.\end{aligned} \tag{10.90}$$

Using these operator definitions, we may now bring the sum through equation (10.88) to arrive at

$$\begin{aligned}\dot{\tilde{\rho}} = \int_0^t dt'(\; &[a\tilde{\rho}(t')a - aa\tilde{\rho}(t')]e^{-i\omega_0(t+t')}\sum_{j,l}\langle \Gamma_{2,j}(t')\Gamma_{2,l}(t)\rangle_R + \text{h.c.} \\ + \; &[a^\dagger\tilde{\rho}(t')a^\dagger - a^\dagger a^\dagger\tilde{\rho}(t')]e^{+i\omega_0(t+t')}\sum_{j,l}\langle \Gamma_{1,j}(t')\Gamma_{1,l}(t)\rangle_R + \text{h.c.} \\ + \; &[a^\dagger\tilde{\rho}(t')a - aa^\dagger\tilde{\rho}(t')]e^{-i\omega_0(t-t')}\sum_{j,l}\langle \Gamma_{2,j}(t)\Gamma_{1,l}(t')\rangle_R + \text{h.c.} \\ + \; &[a\tilde{\rho}(t')a^\dagger - a^\dagger a\tilde{\rho}(t')]e^{-i\omega_0(t+t')}\sum_{j,l}\langle \Gamma_{1,j}(t)\Gamma_{2,l}(t')\rangle_R + \text{h.c.}),\end{aligned} \tag{10.91}$$

where h.c. denotes the Hermitian conjugate of the term immediately preceding. The specific forms of the reservoir correlations needed to evaluate equation (10.91) are

$$\begin{aligned}
\sum_{j,l}\langle\widetilde{\Gamma}_{2,j}(t')\widetilde{\Gamma}_{2,l}(t)\rangle &= \sum_{j,l}\kappa_j^*\kappa_l^* e^{i\omega_j t'}e^{i\omega_l t}\mathrm{Tr}_R\left(R_0 c_j^\dagger c_l^\dagger\right)\\
&=0,\\
\sum_{j,l}\langle\widetilde{\Gamma}_{1,j}(t)\widetilde{\Gamma}_{1,l}(t')\rangle &= \sum_{j,l}\kappa_j\kappa_l e^{-i\omega_j t'}e^{-i\omega_l t}\mathrm{Tr}_R\left(R_0 c_j c_l\right)\\
&=0,\\
\sum_{j,l}\langle\widetilde{\Gamma}_{2,j}(t)\widetilde{\Gamma}_{1,l}(t')\rangle &= \sum_{j,l}\kappa_j^*\kappa_l e^{+i\omega_j t}e^{-i\omega_l t'}\mathrm{Tr}_R\left(R_0 c_j^\dagger c_l\right)\\
&= \sum_j|\kappa_j|^2 e^{+i\omega_j(t-t')}\bar{n}\left(\omega_j,\, T\right),\\
\sum_{j,l}\langle\widetilde{\Gamma}_{1,j}(t)\widetilde{\Gamma}_{2,l}(t')\rangle &= \sum_{j,l}\kappa_j\kappa_l^* e^{-i\omega_j t}e^{+i\omega_l t'}Tr_R\left(R_0 c_j c_l^\dagger\right)\\
&= \sum_j|\kappa_j|^2 e^{-i\omega_j(t-t')}\left(\bar{n}\left(\omega_j,\, T\right)+1\right),
\end{aligned} \tag{10.92}$$

where $\bar{n}(\omega_j,\, T)$ is the average thermal photon number for a reservoir mode j in thermal equilibrium at temperature T, as discussed in chapter 2. More specifically,

$$\begin{aligned}
\bar{n}\left(\omega_j,\, T\right) &= \mathrm{Tr}_R\left(R_0 c_j^\dagger c_j\right)\\
&= \langle c_j^\dagger c_j\rangle\\
&= \frac{e^{-\hbar\omega_j/k_B T}}{1-e^{-\hbar\omega_j/k_B T}}.
\end{aligned} \tag{10.93}$$

To continue, it is helpful if we make a change in variables and an approximation about the nature of the reservoir's mode spacing. First, we define the delay time $\tau = t - t'$. Next, we state that the reservoir's modes are very tightly spaced, warranting the conversion of the sum in equations (10.92) to integrals. Defining the mode density $g(\omega)$ such that $g(\omega)d\omega$ is the number of reservoir modes of frequencies between ω and $\omega + d\omega$ and using τ in place of $t - t'$, we have

$$\sum_j|\kappa_j|^2 e^{+i\omega_j(t-t')}\bar{n}\left(\omega_j,\, T\right) \Rightarrow \int_0^\infty d\omega e^{+i\omega\tau}g(\omega)|\kappa(\omega)|^2\bar{n}(\omega,\, T), \tag{10.94}$$

$$\sum_j|\kappa_j|^2 e^{-i\omega_j(t-t')}(\bar{n}(\omega_j,\, T)+1) \Rightarrow \int_0^\infty d\omega e^{-i\omega\tau}g(\omega)|\kappa(\omega)|^2(\bar{n}(\omega,\, T)+1). \tag{10.95}$$

Equation (10.91) may now be written as

$$\dot{\tilde{\rho}} = \int_0^t d\tau\Bigg([a^\dagger\tilde{\rho}(t-\tau)a - aa^\dagger\tilde{\rho}(t-\tau)]\int_0^\infty d\omega e^{+i(\omega-\omega_0)\tau}g(\omega)|\kappa(\omega)|^2\bar{n}(\omega,\, T) + \text{h.c.} \tag{10.96}$$

$$+[a\tilde{\rho}(t-\tau)a^\dagger - a^\dagger a\tilde{\rho}(t-\tau)]\int_0^\infty d\omega e^{-i(\omega-\omega_0)\tau}g(\omega)|\kappa(\omega)|^2(\bar{n}(\omega,\, T)+1) + \text{h.c.}\Bigg). \tag{10.97}$$

Now we can make a formal statement of the Markoff approximation. The integrals of equation (10.94) are essentially delta functions, as any sizable τ will average the 'slower varying' functions $g(\omega)$, $|\kappa(\omega)|^2$ and $\bar{n}(\omega, T)$ to zero. This means that the reservoir correlations 'die off' on a time scale much shorter than the natural time scale of the cavity field mode system (and hence $\tilde{\rho}$). We may therefore replace $\tilde{\rho}(t-\tau)$ with $\tilde{\rho}(t)$ and extend the upper limit on the time integral of equation (10.96) to infinity. We may now write

$$\lim_{t\to\infty}\int_0^t d\tau e^{-i(\omega-\omega_0)\tau} \approx \pi\delta(\omega-\omega_0), \tag{10.98}$$

where we have ignored a small, imaginary offset (this offset is equivalent to the Lamb shift for our cavity field mode 'oscillator'). This delta function reduces the remaining integral over ω, and we can now write (10.96) as

$$\dot{\tilde{\rho}} = \frac{\kappa'}{2}(2a\tilde{\rho}a^\dagger - a^\dagger a\tilde{\rho} - \tilde{\rho}a^\dagger a) + \kappa'\bar{n}(a\tilde{\rho}a^\dagger + a^\dagger\tilde{\rho}a - a^\dagger a\tilde{\rho} - \tilde{\rho}aa^\dagger), \tag{10.99}$$

where

$$\kappa' \equiv 2\pi g(\omega_0)|\kappa(\omega_0)|^2, \tag{10.100}$$

$$\bar{n} \equiv \bar{n}(\omega_0, T). \tag{10.101}$$

We may transform this equation back into the Schrödinger picture by the usual prescription,

$$\dot{\rho} = \frac{-i}{\hbar}[H_S, \rho] + e^{-(i/\hbar)H_S t}\dot{\tilde{\rho}}e^{+(i/\hbar)H_S t}, \tag{10.102}$$

to obtain a familiar result,

$$\begin{aligned}\dot{\rho} = &- i\omega_0[a^\dagger a, \rho] + \frac{\kappa'}{2}(\bar{n}+1)(2a\rho a^\dagger - a^\dagger a\rho - \rho a^\dagger a)\\ &+ \frac{\kappa'}{2}\bar{n}(2a^\dagger\rho a - aa^\dagger\rho - \rho aa^\dagger).\end{aligned} \tag{10.103}$$

10.6 Other types of reservoirs

We can also have the atomic inversion coupled to a heat bath. Here, the interaction is given by

$$H_{SB} = \sum\nolimits_k (a_k + a_k^\dagger)\sigma_z. \tag{10.104}$$

This results in the master equation addition

$$\dot{\rho} = -\Gamma(\sigma_z\rho\sigma_z - \rho). \tag{10.105}$$

This does not appear in Lindblad form, but it can be written as

$$\dot{\rho} = -\frac{\Gamma}{2}(\sigma_z\rho\sigma_z - \sigma_z^2\rho - \rho\sigma_z^2), \tag{10.106}$$

which is identical, once we realize that $\sigma_z^2 = 1$.

Also, we can have phase-dependent reservoirs where

$$\langle a_k a_k \rangle \tag{10.107}$$

is not zero. For a bath of squeezed vacuum states, we have

$$\mathrm{Tr}_R\left(Rc_j^\dagger c_j\right) = N \tag{10.108}$$

$$\mathrm{Tr}_R\left(Rc_j c_j^\dagger\right) = N + 1 \tag{10.109}$$

$$\mathrm{Tr}_R\left(Rc_j c_j\right) = M \tag{10.110}$$

$$\mathrm{Tr}_R\left(Rc_j^\dagger c_j^\dagger\right) = M^*, \tag{10.111}$$

where we must have

$$|M|^2 \leq N(N + 1). \tag{10.112}$$

For a squeezed vacuum, we have $N = \sinh^2(r)$, and $M = \sinh(r)\cosh(r)e^{\phi}$, where ϕ is the phase of the squeezed vacuum, which is related to the direction in which the fluctuations are reduced. This produces a master equation of the form

$$\begin{aligned}\dot{\rho}(t) = &-i\omega_0[a^\dagger a, \rho] + \kappa(N + 1)[2a\rho(t)a^\dagger - \rho(t)a^\dagger a - a^\dagger a\rho(t)]\\ &+ \kappa N[2a\rho(t)a^\dagger - aa^\dagger\rho(t) - \rho(t)aa^\dagger]\end{aligned} \tag{10.113}$$

$$+\kappa M[2a\rho(t)a - a^2\rho(t) - \rho(t)a^2] \tag{10.114}$$

$$+\kappa M^*[2a^\dagger\rho(t)a^\dagger - a^{\dagger 2}\rho(t) - \rho(t)a^{\dagger 2}]. \tag{10.115}$$

The corresponding equation for atomic decay into a squeezed reservoir is simpler than $\sigma_\pm^2 = 0$

$$\begin{aligned}\dot{\rho} = &-\frac{i}{2}\omega_0[\sigma_z, \rho] + \frac{\gamma}{2}(N + 1)(2\sigma_-\rho\sigma_+ - \sigma_+\sigma_-\rho - \rho\sigma_+\sigma_-)\\ &+ \frac{\gamma}{2}N_a(2\sigma_+\rho\sigma_- - \sigma_-\sigma_+\rho - \rho\sigma_-\sigma_+)\end{aligned} \tag{10.116}$$

$$+\frac{\gamma}{2}M(2\sigma_\rho\sigma_-) \tag{10.117}$$

$$+\frac{\gamma}{2}M^*(2\sigma_+\rho\sigma_+). \tag{10.118}$$

10.7 Alternative derivation of the Lindblad equation

A completely positive, trace-preserving (CPTP) map $\mathcal{E}$ acting on a density operator ρ can always be written in the Kraus form

$$\mathcal{E}(\rho) = \sum\nolimits_\alpha E_\alpha \rho E_\alpha^\dagger, \tag{10.119}$$

where the Kraus operators E_α satisfy the completeness relation

$$\sum_\alpha E_\alpha^\dagger E_\alpha = I. \tag{10.120}$$

When describing the continuous time evolution, we break the time evolution from t to $t + \delta t$ into an infinitesimal completely positive map,

$$\rho(t + \delta t) = \sum_\alpha E_\alpha(\delta t)\rho(t)E_\alpha^\dagger(\delta t). \tag{10.121}$$

We assume a Markovian scenario, which means that over small times δt, the map's behavior depends only on the current state $\rho(t)$ (memoryless property). We would like to take the limit $\delta t \to 0$ in such a way that leads to a first-order differential equation. For a Markovian process, one typically requires that the family of CPTP maps $\{\mathcal{E}(t)\}_{t\geqslant 0}$ forms a semigroup, with the following properties, i.e.

$$\mathcal{E}(t)\mathcal{E}(s) = \mathcal{E}(t + s), \quad \mathcal{E}(0) = \mathrm{Id}. \tag{10.122}$$

Semigroups are essentially groups without an inverse, which is appropriate to describe no unitary evolution. This semigroup property implies that the time evolution is generated by a Lindbladian superoperator $\mathcal{L}$. In other words,

$$\frac{d}{dt}\rho(t) = \mathcal{L}[\rho(t)], \quad \rho(t) = e^{t\mathcal{L}}\rho(0). \tag{10.123}$$

These two properties (composition and resolution of the identity) imply that the evolution from time 0 to time $t + s$ can be decomposed into evolution from 0 to s and then s to $t + s$. Physically, this is a mathematical statement of Markov memorylessness; the only thing that matters for the next step is the current state, not how it was reached. In the semigroup theory (both in classical probability and operator semigroups), one defines the generator $\mathcal{L}$ by looking at the short-time behavior:

$$\mathcal{L} = \lim_{t\to 0}\frac{\mathcal{E}(t) - I}{t}. \tag{10.124}$$

This $\mathcal{L}$ is a linear map of the density operators (an example of a Liouvillian or superoperator). The assumption that this limit exists is essentially the assumption of the continuity of the semigroup at $t = 0$. If that limit exists, you can rewrite the semigroup $\mathcal{E}(t)$ by formally expanding in infinitesimal steps

$$\mathcal{E}(t) = \lim_{n\to\infty}\left(\mathcal{E}\left(\frac{t}{n}\right)\right)^n. \tag{10.125}$$

If

$$\mathcal{E}\left(\frac{t}{n}\right) = \mathrm{Id} + \frac{t}{n}\mathcal{L} + O\left(\left(\frac{t}{n}\right)^2\right) \tag{10.126}$$

then as $n \to \infty$ this limit becomes

$$\mathcal{E}(t) = \exp(t\mathcal{L}). \tag{10.127}$$

Here, $\exp(t\mathcal{L})$ is the usual exponential of a (super)operator:

$$\exp(t\mathcal{L}) = \mathrm{Id} + t\mathcal{L} + \frac{1}{2!}(t\mathcal{L})^2 + \cdots. \tag{10.128}$$

If $\rho(t)$ evolves under $\mathcal{E}(t)$ from an initial state $\rho(0)$, then

$$\rho(t) = \mathcal{E}(t)[\rho(0)]. \tag{10.129}$$

However, because $\mathcal{E}(t) = e^{t\mathcal{L}}$, we simply have

$$\rho(t) = e^{t\mathcal{L}}\rho(0). \tag{10.130}$$

Taking the time derivative,

$$\frac{d}{dt}\rho(t) = \frac{d}{dt}(e^{t\mathcal{L}})\rho(0) = \mathcal{L}e^{t\mathcal{L}}\rho(0) = \mathcal{L}\rho(t). \tag{10.131}$$

Thus

$$\frac{d}{dt}\rho(t) = \mathcal{L}[\rho(t)]. \tag{10.132}$$

One subtlety is that if the system's evolution is truly time-homogeneous, then $\mathcal{E}(t)$ depends only on the length of the time interval (i.e. from 0 to t is the same map as from s to $s + t$). Physically, this means that the environment does not change in time, and the rates of interaction are constant. This leads to a time-independent generator $\mathcal{L}$. If the rates (or environment parameters) vary in time, one would still have $\frac{d}{dt}\rho(t) = \mathcal{L}(t)[\rho(t)]$, but $\mathcal{L}(t)$ would be time-dependent, and the semigroup property in its strict form would not hold. Instead, one obtains a more general divisible or completely positive (CP)-divisible map scenario. In that case, formally,

$$\rho(t) = \mathcal{T}\exp\left[\int_0^t \mathcal{L}(\tau)d\tau\right]\rho(0), \tag{10.133}$$

where $\mathcal{T}$ denotes time-ordering. To find $\mathcal{L}$ from Kraus operators, suppose that for a small δt, we have the Kraus operators $E_\alpha(\delta t)$. We expand each to the first order in δt. A common way to parameterize them is as follows. Define a single 'no-jump' or 'free evolution' Kraus operator that is close to the identity plus some Hamiltonian-like term. One writes

$$E_0(\delta t) = I - iH_{\mathrm{eff}}\delta t + O((\delta t)^2), \tag{10.134}$$

where H_{eff} is (in general) not a Hermitian but can be thought of as an 'effective Hamiltonian' for the non-jump part of the evolution. Then, we define one or more 'jump' Kraus operators $E_j(\delta t)$, each being first order in $\sqrt{\delta t}$. For example, one might write something like

$$E_j(\delta t) = \sqrt{\gamma_j \delta t}\, L_j, \tag{10.135}$$

where L_j is some operator (often called a jump operator), and γ_j is a positive rate. These forms automatically produce a complete positivity condition to first order in δt and incorporate the typical scaling that if a jump occurs, it happens with probability $\mathcal{O}(\delta t)$. Consider the map from t to $t + \delta t$:

$$\rho(t+\delta t) = \sum_\alpha E_\alpha(\delta t)\rho(t)E_\alpha^\dagger(\delta t). \tag{10.136}$$

For the no-jump operator $E_0(\delta t)$

$$E_0(\delta t) = I - iH_{\text{eff}}\delta t + \mathcal{O}((\delta t)^2). \tag{10.137}$$

Then

$$E_0(\delta t)\rho(t)E_0^\dagger(\delta t) = (I - iH_{\text{eff}}\delta t)\rho(t)\left(I + iH_{\text{eff}}^\dagger\delta t\right) + \mathcal{O}((\delta t)^2). \tag{10.138}$$

Expanding to the first order in δt,

$$E_0(\delta t)\rho(t)E_0^\dagger(\delta t) = \rho(t) - i\delta t\left(H_{\text{eff}}\rho(t) - \rho(t)H_{\text{eff}}^\dagger\right) + \mathcal{O}((\delta t)^2). \tag{10.139}$$

Now for the jump operators $E_j(\delta t) = \sqrt{\gamma_j\delta t}\,L_j$

$$E_j(\delta t)\rho(t)E_j^\dagger(\delta t) = \gamma_j\delta t L_j\rho(t)L_j^\dagger + \mathcal{O}((\delta t)^2). \tag{10.140}$$

Hence, summing over α, the increment $\rho(t+\delta t) - \rho(t)$ is, to first order in δt,

$$\rho(t+\delta t) - \rho(t) = -i\delta t\left(H_{\text{eff}}\rho(t) - \rho(t)H_{\text{eff}}^\dagger\right) + \sum_j \gamma_j\delta t L_j\rho(t)L_j^\dagger + \mathcal{O}((\delta t)^2). \tag{10.141}$$

Recall that for CPTP completeness, the Kraus operators must satisfy

$$\sum_\alpha E_\alpha^\dagger(\delta t)E_\alpha(\delta t) = I. \tag{10.142}$$

Expanding this to first order in δt forces

$$H_{\text{eff}}^\dagger - H_{\text{eff}} = -i\sum_j \gamma_j L_j^\dagger L_j, \tag{10.143}$$

which implies that the non-Hermitian part of H_{eff} is related to the jump operators' contributions. One then rewrites $H_{\text{eff}} = H - \frac{i}{2}\sum_j \gamma_j L_j^\dagger L_j$, where H is the system's Hamiltonian. Putting it all together and dividing by δt, and finally taking the limit $\delta t \to 0$, we obtain the following differential form

$$\frac{d}{dt}\rho(t) = -i[H, \rho(t)] + \sum_j \gamma_j\left[L_j\rho(t)L_j^\dagger - \frac{1}{2}\left\{L_j^\dagger L_j, \rho(t)\right\}\right]. \tag{10.144}$$

10.8 Resonance fluorescence

Here, we consider resonance fluorescence (figure 10.1) in some detail, coherently driving a two-level atom in free space. The Hamiltonian is

$$H = \frac{\hbar\omega_A\sigma_z}{2} + i\hbar\Omega(\sigma_+e^{-i\omega_L t} + \sigma_- e^{i\omega_L t}). \tag{10.145}$$

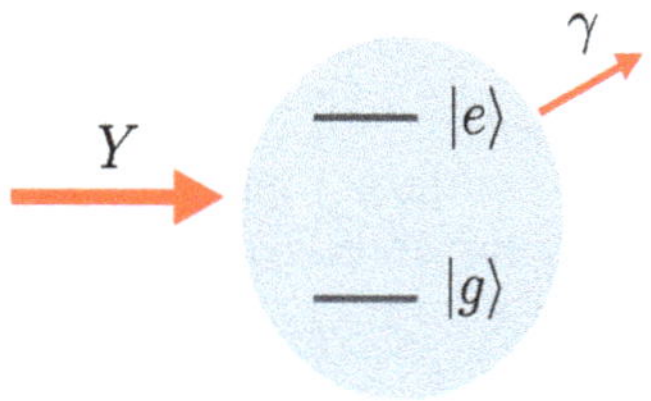

Figure 10.1. Resonance fluorescence.

Here ω_L is the frequency of the driving laser, ω_A is the resonant frequency of the atomic transition, and

$$\Omega = -\frac{\mu E_0}{\hbar}, \tag{10.146}$$

where

$$\mu = -\langle \epsilon \cdot e\vec{r} \rangle \tag{10.147}$$

is the (real) induced dipole moment. The master equation is then

$$\dot{\rho} = -i\frac{1}{2}\omega_A[\sigma_z, \rho] + i(\Omega/2)[e^{-i\omega_L t}\sigma_+ + e^{i\omega_L t}\sigma_-, \rho] \tag{10.148}$$

$$+\frac{\gamma}{2}(2\sigma_-\rho\sigma_+ - \sigma_+\sigma_-\rho - \rho\sigma_+\sigma_-). \tag{10.149}$$

From this we can derive the optical Bloch equations

$$\langle\dot{\sigma}_-\rangle = -\left[\frac{\gamma}{2} + i\omega_A\right]\langle\sigma_-\rangle - i(\Omega/2)e^{-i\omega_L t}\langle\sigma_z\rangle \tag{10.150}$$

$$\langle\dot{\sigma}_+\rangle = -\left[\frac{\gamma}{2} - i\omega_A\right]\langle\sigma_+\rangle + i(\Omega/2)e^{i\omega_L t}\langle\sigma_z\rangle \tag{10.151}$$

$$\langle\dot{\sigma}_z\rangle = -\gamma[\langle\sigma_z\rangle + 1] + i\Omega e^{-i\omega_L t}\langle\sigma_+\rangle - i\Omega e^{i\omega_L t}\langle\sigma_-\rangle. \tag{10.152}$$

We will only consider the case when the electromagnetic vacuum is at zero temperature, $\bar{n} = 0$. If we neglect spontaneous emission, ($\gamma = 0$). We then have

$$\langle\dot{\sigma}_-\rangle = -i\omega_A\langle\sigma_-\rangle - i(\Omega/2)e^{-i\omega_L t}\langle\sigma_z\rangle \tag{10.153}$$

$$\langle\dot{\sigma}_+\rangle = i\omega_A\langle\sigma_+\rangle + i(\Omega/2)e^{i\omega_L t}\langle\sigma_z\rangle \tag{10.154}$$

$$\langle\dot{\sigma}_z\rangle = i\Omega e^{-i\omega_L t}\langle\sigma_+\rangle - i\Omega e^{i\omega_L t}\langle\sigma_-\rangle. \tag{10.155}$$

These can be written in the more compact form

$$\frac{d}{d\tau}\langle\vec{\sigma}\rangle = \vec{B} \times \langle\vec{\sigma}\rangle \tag{10.156}$$

$$\langle \vec{\sigma} \rangle = \langle \sigma_x \rangle x + \langle \sigma_y \rangle y + \langle \sigma_z \rangle z \tag{10.157}$$

and

$$\vec{B} = -(\Omega \cos \omega_0 t)x - (\Omega \sin \omega_0 t)y + \omega_0 z, \tag{10.158}$$

where we see that the Bloch vector $\langle \vec{\sigma}$ precesses around the applied field on the Bloch sphere. In a frame that rotates at the resonant frequency of the atom

$$\sigma_+ \to \sigma_+ e^{i\omega_A t} \tag{10.159}$$

$$\sigma_- \to \sigma_+ e^{-i\omega_A t} \tag{10.160}$$

$$\sigma_z \to \sigma_z \tag{10.161}$$

we have the Bloch equations in the following form, which just depends on the detuning of the atomic transition from the incident field, $\Delta = (\omega_A - \omega_L)/\gamma$,

$$\frac{d}{dt}\langle \sigma_+ \rangle = -\frac{\gamma}{2}(1 - \mathrm{i}\Delta)\langle \sigma_+ \rangle - \mathrm{i}\Omega\langle \sigma_z \rangle, \tag{10.162}$$

$$\frac{d}{dt}\langle \sigma_- \rangle = -\frac{\gamma}{2}(1 + \mathrm{i}\Delta)\langle \sigma_- \rangle + \mathrm{i}\Omega\langle \sigma_z \rangle, \tag{10.163}$$

$$\frac{d}{dt}\langle \sigma_z \rangle = -\left(\langle \sigma_z \rangle + \frac{\gamma}{2}\right) - \mathrm{i}\frac{\Omega}{2}(\langle \sigma_+ \rangle - \langle \sigma_- \rangle). \tag{10.164}$$

These have steady state solutions

$$\langle \sigma_z \rangle_{\mathrm{ss}} = -\frac{1}{2}\frac{1 + \Delta^2}{1 + \Delta^2 + Y^2} \tag{10.165}$$

$$\langle \sigma_+ \rangle_{\mathrm{ss}} = \frac{i}{\sqrt{2}}\frac{Y(1 + i\Delta)}{1 + \Delta^2 + Y^2} \tag{10.166}$$

where $Y = \Omega/\Omega_{\mathrm{sat}}$ with $\Omega_{\mathrm{sat}} = \sqrt{2}/\gamma^2$ is the saturation Rabi frequency, a flux of one photon every two lifetimes over the cross section of the atom (resonant):

$$I_s = \frac{\hbar\omega_0}{2\tau_0\sigma_0} = \frac{\pi}{3}\frac{\gamma_0\hbar\omega_0}{\lambda_0^2}. \tag{10.167}$$

We also scaled the time to the spontaneous emission rate, $\tau = \gamma t$. If $I = I_0$ the rate of the stimulated emission is equal to the rate of the spontaneous emission (Rabi frequency Ω) and the population in the excited state 1/4:

$$\Omega = \frac{\vec{d}\cdot\vec{E}}{\hbar} = \gamma\sqrt{I/I_s}. \tag{10.168}$$

This is the Rabi frequency at which the probability of the excited state of the atom being 1/4 (on resonance) and nonlinear effects begin to manifest. Below the saturation frequency $\Omega \ll \gamma$, all oscillations in the response will be damped, as the photon flux is small compared to the dissipation rate. At resonance, we expand

$$\langle \sigma_+ \rangle = Y - Y^3 + \cdots \tag{10.169}$$

and we see that we have a $\chi^{(3)}$ nonlinearity, plus higher order terms.

We may then cast the Bloch equations in the form

$$\frac{d}{dt}\langle \vec{\sigma} \rangle = \mathbf{M}\langle \vec{\sigma} \rangle + \vec{B}, \tag{10.170}$$

where

$$\langle \vec{\sigma}(\tau) \rangle = \mathbf{S}e^{\Lambda t}\mathbf{S}^{-1}\langle \vec{\sigma}(0) \rangle + (\mathbf{I} + \mathbf{S}e^{\Lambda \tau}\mathbf{S}^{-1})\mathbf{M}^{-1}\vec{B} \tag{10.171}$$

and

$$\mathbf{M} = \begin{pmatrix} -(1 + i\Delta) & 0 & iY/\sqrt{2} \\ -(1 - i\Delta) & 1 & -iY/\sqrt{2} \\ i\sqrt{2}\,Y & -i\sqrt{2}\,Y & -2 \end{pmatrix} \tag{10.172}$$

and

$$\vec{B} = \begin{pmatrix} 0 \\ 0 \\ 1 \end{pmatrix}. \tag{10.173}$$

Further $\mathbf{S}$ is the matrix of the column eigenvectors or $\mathbf{M}$, and

$$\Lambda = \mathrm{diag}(\lambda_1, \lambda_2, \lambda_3), \tag{10.174}$$

where the λ_i are the eigenvalues of $\mathbf{M}$. On resonance, we have

$$\lambda_1 = -\gamma/2 \tag{10.175}$$

$$\lambda_2 = -3\gamma/4 + \sqrt{(\gamma^2/4 - \Omega^2)} \tag{10.176}$$

$$\lambda_2 = -3\gamma/4 - \sqrt{(\gamma^2/4 - \Omega^2)}. \tag{10.177}$$

For weak driving fields, ($\Omega \ll \gamma$) or $\Omega \ll \Omega_{\mathrm{sat}}$ these are

$$\lambda_1 = -\gamma/2 \tag{10.178}$$

$$\lambda_2 = -\gamma/2 \tag{10.179}$$

$$\lambda_2 = -\gamma. \tag{10.180}$$

Please note the degeneracy as this will be important in weak field calculations. In this regime, we will find spectral components at atomic resonance, and the polarization scales linearly with the applied field. A linear system in the dynamics sense responds at the driven frequency, and the response is linearly proportional to the drive. For strong driving fields, we have

$$\lambda_1 = -\gamma/2 \tag{10.181}$$

$$\lambda_2 = -3\gamma/4 - i\Omega \tag{10.182}$$

$$\lambda_2 = -3\gamma/4 + i\Omega. \tag{10.183}$$

Here we are in the nonlinear regime. The atomic response saturates, and we have sidebands at $\pm\Omega$, which leaves us with a triplet structure in the spectrum, known as the Mollow triplet. We now consider the emission spectra. This will entail the Fourier transform of the correlation function

$$G^{(1)} = \langle \sigma_+(\tau)\sigma_-(0) \rangle_{\text{ss}} \tag{10.184}$$

$$= \langle\sigma_+\rangle_{\text{ss}}\langle\sigma_-\rangle_{\text{ss}} + \langle \Delta\sigma_+(\tau)\Delta\sigma_-(0) \rangle_{\text{ss}}, \tag{10.185}$$

where we have defined the fluctuation operators

$$\Delta\sigma_- = \sigma_- - \langle\sigma_-\rangle_{\text{ss}} \tag{10.186}$$

$$\Delta\sigma_+ = \sigma_+ - \langle\sigma_+\rangle_{\text{ss}}. \tag{10.187}$$

The first term in $G^{(2)}$ is a constant, and its Fourier transform will yield a δ function at the line center, which we will refer to as the coherent spectrum

$$I_{\text{coh}}(\omega) = \langle\sigma_+\sigma_-\rangle_{\text{ss}}\,\delta(0). \tag{10.188}$$

This is due to Rayleigh scattering, as depicted in the one-photon scattering process To calculate the other term, which results from multi-photon scattering, we invoke the quantum regression theorem. We first write the equations of motion for the fluctuations

$$\frac{d}{dt}\langle\Delta\tilde{\sigma}_-\rangle = -i(\Omega/2)\langle\Delta\sigma_z\rangle - \frac{\gamma}{2}\langle\Delta\tilde{\sigma}_-\rangle \tag{10.189}$$

$$\frac{d}{dt}\langle\Delta\tilde{\sigma}_+\rangle = i(\Omega/2)\langle\Delta\sigma_z\rangle - \frac{\gamma}{2}\langle\Delta\tilde{\sigma}_+\rangle \tag{10.190}$$

$$\frac{d}{dt}\langle\Delta\sigma_z\rangle = i\Omega\langle\Delta\tilde{\sigma}_+\rangle - i\Omega\langle\Delta\tilde{\sigma}_-\rangle - \gamma\langle\Delta\sigma_z\rangle. \tag{10.191}$$

The quantum regression theorem yields

$$\frac{d}{d\tau}\langle\Delta\tilde{\sigma}_+(0)\Delta\vec{\sigma}(\tau)\rangle_{\text{ss}} = \mathbf{M}\langle\Delta\sigma_+\vec{\Delta}(\tau)\rangle_{\text{ss}} \tag{10.192}$$

with the initial conditions

$$\langle \Delta\tilde{\sigma}_+ \Delta\vec{\sigma} \rangle_{ss} = \frac{1}{2}\frac{Y^2}{(1+Y^2)^2}\begin{pmatrix} Y^2 \\ 0 \\ i\sqrt{2}\,Y \end{pmatrix}, \tag{10.193}$$

where

$$\Delta\vec{\sigma} = \begin{pmatrix} \Delta\tilde{\sigma}_- \\ \Delta\tilde{\sigma}_+ \\ \Delta\sigma_z \end{pmatrix}. \tag{10.194}$$

Solving for $G^{(1)}$ we find

$$\begin{aligned}\langle \Delta\tilde{\sigma}_+(0)\Delta\tilde{\sigma}_-(\tau)\rangle_{ss} &= \frac{1}{4}\frac{Y^2}{1+Y^2}e^{-(\gamma/2)\tau} \\ &\quad - \frac{1}{8}\frac{Y^2}{(1+Y^2)^2}\left[1 - Y^2 + (1-5Y^2)\frac{(\gamma/4)}{\delta}\right]e^{-[(3\gamma/4)-\delta]\tau} \\ &\quad - \frac{1}{8}\frac{Y^2}{(1+Y^2)^2}\left[1 - Y^2 - (1-5Y^2)\frac{(\gamma/4)}{\delta}\right]e^{-[(3\gamma/4)+\delta]\tau}.\end{aligned}$$

In figure 10.2, we plot $g^{(1)}(\tau)$ for a variety of driving strengths. These all begin above one, more so as the drive is increased. At $Y = 0.80$, we see the beginnings of Rabi oscillations, which are in full bloom for larger values of drive.

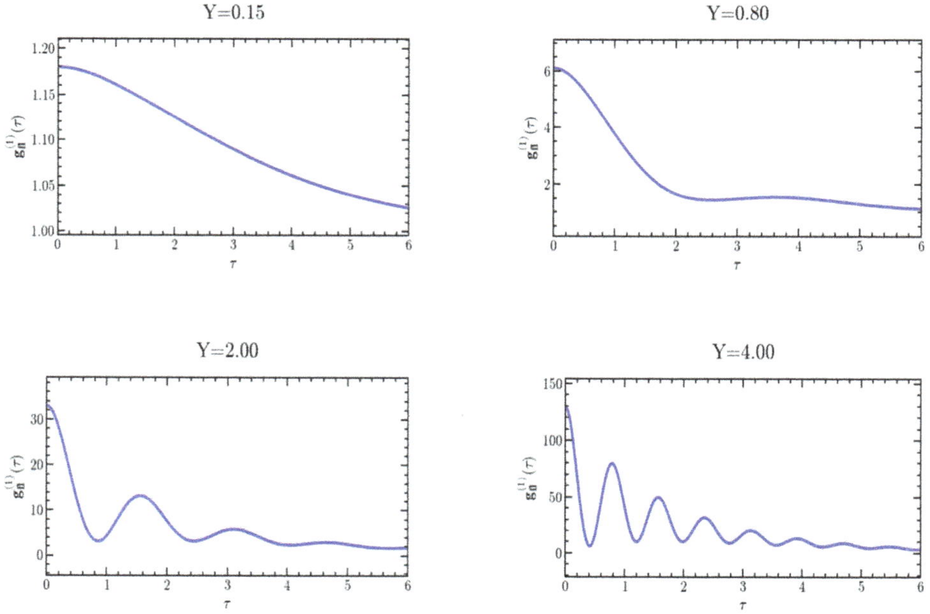

Figure 10.2. $g^{(1)}(\tau)$ for resonance fluorescence.

Now we are after the Fourier transform of this, where we have normalized the area under the spectrum to unity and with $\delta = (\gamma/2)\sqrt{1 - 8Y^2}$. As they say, 'after some algebra',

$$\begin{aligned} S_{\text{inc}}(\omega) = & \frac{1}{4}\frac{Y^2}{1 + Y^2}e^{-(\gamma/2)\tau} \\ & - \frac{1}{8}\frac{Y^2}{(1 + Y^2)^2}\left[1 - Y^2 + (1 - 5Y^2)\frac{(\gamma/4)}{\delta}\right]e^{-[(3\gamma/4)-\delta]\tau} \\ & - \frac{1}{8}\frac{Y^2}{(1 + Y^2)^2}\left[1 - Y^2 - (1 - 5Y^2)\frac{(\gamma/4)}{\delta}\right]e^{-[(3\gamma/4)+\delta]\tau}. \end{aligned} \tag{10.195}$$

We notice a central peak with full width γ, and two Rabi sidebands at $\omega \sim \Omega$ with a width $\sim 3\gamma/2$. In the weak field driving limit $Y^2 \ll 1$ [7], these simplify to

$$S_{\text{inc}}(\bar{\omega}) = \frac{2}{[1 + (\bar{\omega} - \bar{\omega}_0)^2]^2}. \tag{10.196}$$

Note that the linewidth (half-width at half-maximum) for the spectrum given by equation (10.196) is less than the natural width $\gamma/2$; it is

$$\begin{aligned} \Delta\omega &= (\sqrt{2} - 1)^{1/2}(\gamma/2) \\ &\approx 0.64(\gamma/2). \end{aligned} \tag{10.197}$$

The essence of this behavior was noticed by Mollow. This narrowing is a result of the degeneracy of the eigenvalues for the weak fields. If we have a differential equation with two distinct eigenvalues a and b, $(D - a)(D - b)y = 0$, the solution is the sum of the exponentials

$$y(t) = Ae^{at} + Be^{bt}, \tag{10.198}$$

where A and B are determined from the initial conditions. If $a = b$, we do not have two linearly independent solutions; we must use

$$y(t) = Ae^{at} + Bt\, e^{at}. \tag{10.199}$$

The second term will cause y to persist longer in time (assuming negative real parts of the eigenvalues), and hence a reduced linewidth in the frequency space, a width about 60% of the natural width.

In figure 10.3, we plot the incoherent spectrum for various driving fields.

In figure 10.3, we plot the incoherent spectrum for various drive strengths. We do see a sub-natural linewidth for weak driving, and then the beginnings of Rabi oscillations for larger drives, the famous Mollow triplet.

The incoherent optical spectrum and the spectrum of squeezing are both defined in terms of the first-order correlation functions of the observed field. The formal definition of the incoherent spectrum is

$$\pi S_{\text{inc}}(\bar{\omega}) = (\langle \Delta\tilde{\sigma}_+ \Delta\tilde{\sigma}_- \rangle_{ss})^{-1} \tag{10.200}$$

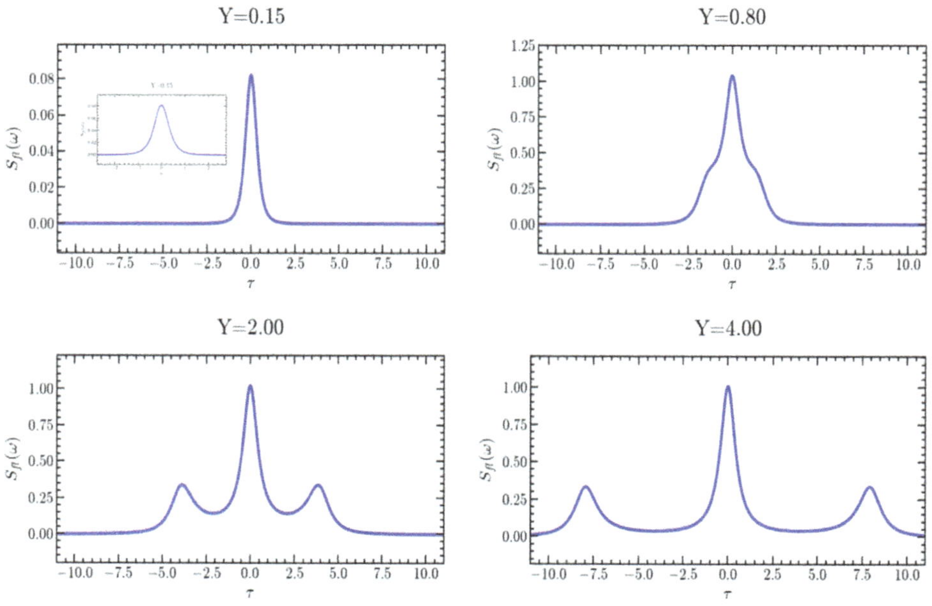

Figure 10.3. Incoherent spectra for resonance fluorescence.

$$\left\{\int_0^\infty d\bar{\tau}\, \exp\left[i(\bar{\omega} - \bar{\omega}_0)\bar{\tau}\right]\langle\Delta\tilde{\sigma}_+(0)\Delta\tilde{\sigma}_-(\bar{\tau})\rangle_{ss}\right\} \tag{10.201}$$

The spectrum of squeezing is defined by

$$\begin{aligned} S(\bar{\omega},\, \theta) = 8\eta \int_0^\infty d\bar{\tau}\, \cos\bar{\omega}\bar{\tau}\ \mathrm{Re}\left[\langle\Delta\tilde{\sigma}_+(0)\Delta\tilde{\sigma}_-(\bar{\tau})\rangle_{ss}\right. \\ \left. + e^{2i\theta}\langle\Delta\tilde{\sigma}_+(0)\Delta\tilde{\sigma}_+(\bar{\tau})\rangle_{ss}\right] \end{aligned} \tag{10.202}$$

The incoherent spectrum will be symmetric if the first-order field correlation function is real and will be asymmetric if it is not. In the former case, the incoherent spectrum can be divided into two independent contributions, each describing the fluctuations in one of the two quadrature phase amplitudes of the observed field. In resonance fluorescence $\langle\Delta\tilde{\sigma}_+(0)\Delta\tilde{\sigma}_-(\bar{\tau})\rangle_{ss}$ is real, and we can decompose equation (10.202) into separate contributions from the fluctuations in any two quadrature phase amplitudes of the induced atomic dipole. This is expressed in the following relationship between the incoherent optical spectrum and the spectrum of squeezing:

$$\pi S_{\text{inc}}(\bar{\omega} + \bar{\omega}_0) = (16\eta\langle\Delta\tilde{\sigma}_+\Delta\tilde{\sigma}_-\rangle_{ss})^{-1}[S(\bar{\omega},\, \theta) + S(\bar{\omega},\, \theta + \pi/2)]. \tag{10.203}$$

We set the phase reference $\theta = 0$ so that $S(\bar{\omega},\, \theta)$ measures atomic dipole fluctuations in phase with the mean induced dipole moment. At weak excitation intensities, these fluctuations are squeezed, with

$$S(\bar{\omega},\, 0) = -4\eta\frac{Y^2}{1 + Y^2}\frac{\bar{\omega}^2 + 4(1 - 2Y^2)}{\bar{\omega}^4 + \bar{\omega}^2(5 - 4Y^2) + 4(1 + Y^2)^2}. \tag{10.204}$$

Fluctuations in the quadrature to the mean induced dipole in phase with the mean dipole moment give

$$S(\bar{\omega}, \pi/2) = 4\eta \frac{Y^2}{1 + Y^2} \frac{1}{1 + \bar{\omega}^2}. \tag{10.205}$$

This is a subtraction of the Lorentzians of slightly different widths for weak fields. The incoherent part of the optical spectrum for resonance fluorescence has a subnatural linewidth at weak excitation intensities because of the squeezing of atomic dipole fluctuations in phase with the mean induced dipole moment. This has been called squeezing-induced linewidth narrowing.

In figure 10.4, we plot the spectrum of squeezing for various drives. There is not much squeezing, but as we have seen, it is responsible for squeezing-induced linewidth narrowing.

What about the photon statistics? Well, we have

$$\begin{aligned} G_{ss}^{(2)}(\tau) &= \langle \sigma_+(0)\sigma_+(\tau)\sigma_-(\tau)\sigma_-(0)\rangle_{ss} \\ &= \frac{1}{2}[\langle \sigma_+\sigma_-\rangle_{ss} + \langle \sigma_+(0)\sigma_z(\tau)\sigma_-(0)\rangle_{ss}] \end{aligned} \tag{10.206}$$

Again, we calculate the correlation function $\langle \sigma_+(0)\sigma_z(\tau)\sigma_-(0)\rangle_{ss}$ using the quantum regression theorem as the third component of the vector $\langle \sigma_+ \boldsymbol{s}(\tau)\sigma_-\rangle_{ss}$

$$\frac{d}{d\tau}\langle \sigma_+(0)\boldsymbol{s}(\tau)\sigma_-(0)\rangle_{ss} = \boldsymbol{M}[\langle \sigma_+(0)\boldsymbol{s}(\tau)\sigma_-(0)\rangle_{ss} + \langle \sigma_+\sigma_-\rangle_{ss}\boldsymbol{M}^{-1}\boldsymbol{b}] \tag{10.207}$$

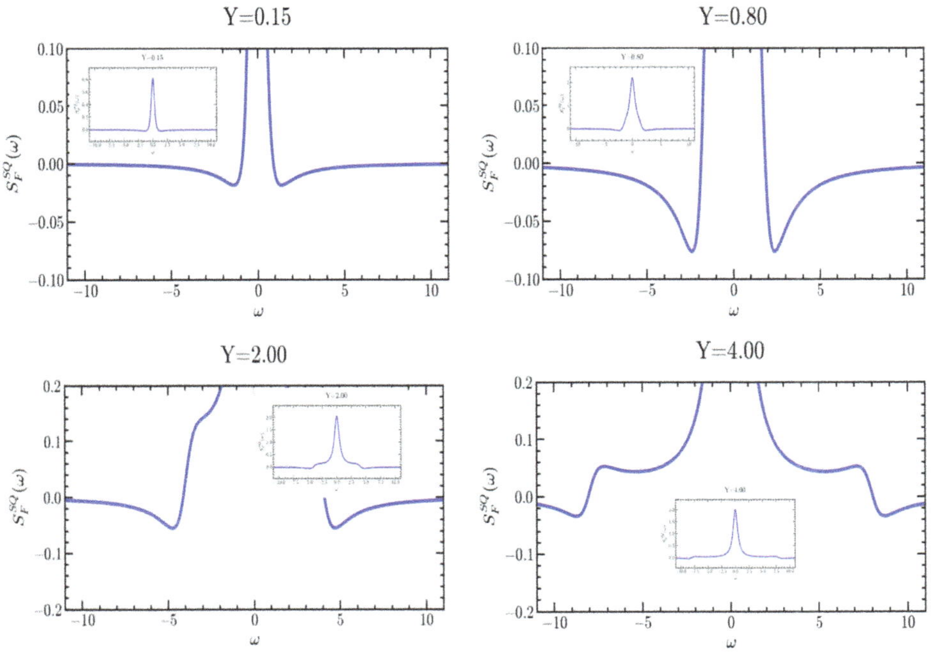

Figure 10.4. Spectrum of squeezing resonance fluorescence.

The formal solution to this equation is

$$\langle\sigma_+(0)\boldsymbol{s}(\tau)\sigma_-(0)\rangle_{ss} = -\langle\sigma_+\sigma_-\rangle_{ss}\boldsymbol{M}^{-1}\boldsymbol{b} + \exp(\boldsymbol{M}\tau)[\langle\sigma_+\boldsymbol{s}\sigma_-\rangle_{ss} + \langle\sigma_+\sigma_-\rangle_{ss}\boldsymbol{M}^{-1}\boldsymbol{b}] \tag{10.208}$$

with the initial conditions

$$\langle\sigma_+\vec{s}\rangle_{ss} = \begin{pmatrix}\langle\sigma_+\tilde{\sigma}_-\sigma_-\rangle_{ss}\\ \langle\sigma_+\tilde{\sigma}_+\sigma_-\rangle_{ss}\\ \langle\sigma_+\sigma_z\sigma_-\rangle_{ss}\end{pmatrix} \tag{10.209}$$

$$= \langle\sigma_+\sigma_-\rangle_{ss}\begin{pmatrix}0\\0\\1\end{pmatrix}. \tag{10.210}$$

Equation refsol provides the solution for $\langle\sigma_z(t)\rangle$ from which

$$g_{ss}^{(2)}(\tau) = 1 - e^{-(3\gamma/4)\tau}\left(\cosh\delta\tau + \frac{3\gamma/4}{\delta}\sinh\delta\tau\right). \tag{10.211}$$

We notice that $g^{(2)}(0) = 0$, physically because a single atom cannot emit two photons at once. From the math it is because

$$g^{(2)}(0) = \langle(\sigma_+)^2(\sigma_-)^2\rangle = 0 \tag{10.212}$$

due to the properties of the Pauli operators, $(\sigma_+)^2 = 0$. For many atoms, the result, which we will not derive here, is

$$g^{(2)}(0) = 1 - \frac{1}{N}. \tag{10.213}$$

In figure 10.5, we plot the second-order intensity correlation function. It always starts at zero as required by the Pauli algebra and the fact a single atom cannot emit two photons at once, and as the drive is increased, we see the emergence of Rabi oscillations. The facts that this starts below one (at zero) and has an initial rise are *both* nonclassical effects. We can also find

$$h(|\tau|) = 1 - e^{-\frac{3}{4}\gamma|\tau|}\left[\cosh\delta|\tau| + \frac{1-2Y^2}{4\delta/\gamma}\sinh\delta|\tau|\right]. \tag{10.214}$$

In figure 10.6, we plot the intensity–field correlation function. These all start at zero, and Rabi oscillations appear at larger drives. Starting at zero, the maximum value allowed classically is 2, which is not the case in the last two panels, as they violate

$$0 \le h_\theta(0) - 1 \le 1. \tag{10.215}$$

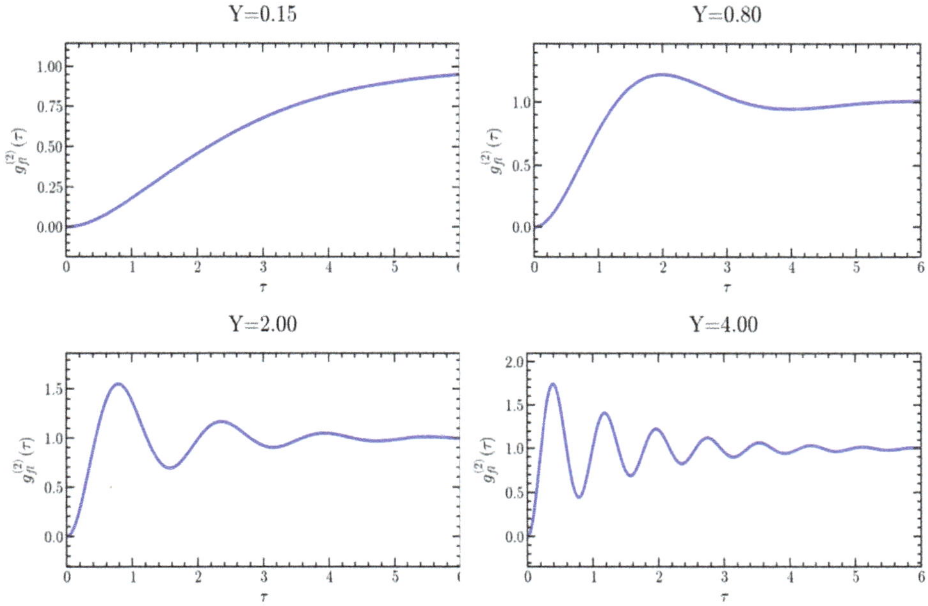

Figure 10.5. $g^{(2)}\tau$ for resonance fluorescence.

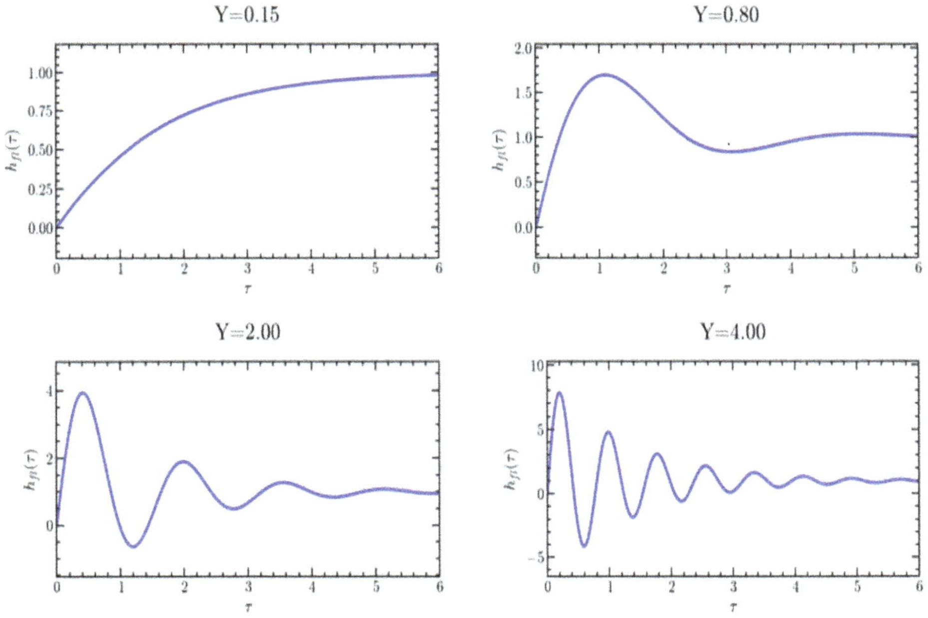

Figure 10.6. $h(\tau)$ for resonance fluorescence.

10.9 Cavity quantum electrodynamics

Let us consider a two-level atom in a cavity with a classical driving field [8]. We have

$$\begin{aligned}\dot{\rho} &= \mathcal{E}[a^\dagger - a, \rho] + g[a^\dagger\sigma_- - a\sigma_+, \rho] \\ &+ \frac{\gamma}{2}(2\sigma_-\rho\sigma_+ - \sigma_+\sigma_-\rho - \rho\sigma_+\sigma_-) \\ &+ \kappa(2a\rho a^\dagger - a^\dagger a\rho - \rho a^\dagger a).\end{aligned} \tag{10.216}$$

The equations of motion for the expectation values of the atomic operators are

$$\langle\dot{\sigma}_-\rangle = 2g\langle\sigma_z a\rangle - (\gamma/2)\langle\sigma_-\rangle \tag{10.217}$$

$$\langle\dot{\sigma}_+\rangle = 2g\langle a^\dagger\sigma_z\rangle - (\gamma/2)\langle\sigma_+\rangle \tag{10.218}$$

$$\langle\dot{\sigma}_z\rangle = -g(\langle a^\dagger\sigma_-\rangle + \langle\sigma_+ a\rangle) - \gamma\left(\langle\sigma_z\rangle + \frac{1}{2}\right). \tag{10.219}$$

Seeing terms such as $\langle a\sigma_+\rangle$, we must find equations of motion for that and others. In the end, we will end up generating an infinite hierarchy of equations. For now, let us restrict ourselves to the bad cavity limit (figure 10.7), where the largest rate is κ. We write Langevin equations for the field operator and the bath

$$\dot{a} = g\sigma_- + \mathcal{E} - \kappa a + \xi, \tag{10.220}$$

where we have the typical bath contribution

$$\xi = -i\sum_j \kappa_j r_j(0)e^{-i(\omega_j - \omega_0)t}. \tag{10.221}$$

This can be integrated formally to give

$$\begin{aligned}a(t) &= a(0)e^{-\kappa t} + (\varepsilon/\kappa)(1 - e^{-\kappa t}) \\ &+ g\int_0^t dt'\sigma_-(t - t')e^{-\kappa t'} \\ &+ \int_0^t dt'\xi(t - t')e^{-\kappa t'}.\end{aligned} \tag{10.222}$$

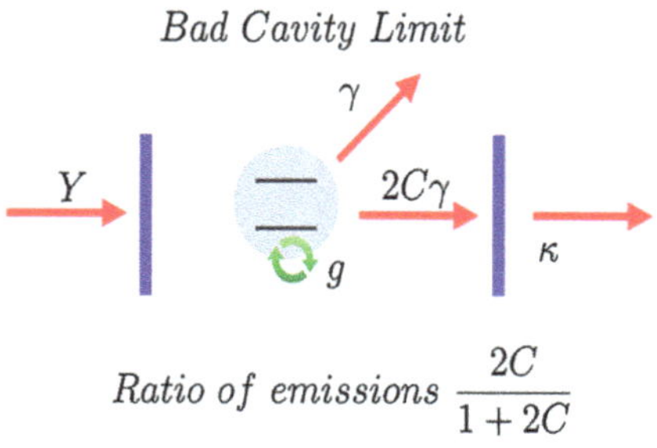

Figure 10.7. Cavity QED set-up.

We define some useful parameters as

$$C = \frac{2g^2}{\kappa\gamma} \tag{10.223}$$

$$n_s = \frac{\gamma^2}{8g^2} \tag{10.224}$$

$$Y = \frac{\mathcal{E}}{\kappa\sqrt{n_s}} \tag{10.225}$$

$$\mu = \frac{2\kappa}{\gamma}. \tag{10.226}$$

The bad cavity limit is $\kappa \gg \gamma$ ($\mu \to \infty$) with C and Y arbitrary. This requires a large $\mathcal{E}$, and we can have $g\rangle\kappa$ In this limit, we have

$$a(t) = \sqrt{n_s}[Y + 2\sqrt{2}\,C\sigma_-(t)] + \kappa^{-1}\xi(t). \tag{10.227}$$

Now, in the original equations of motion, we have written a to the right and $a^\dagger$ to the left of the atomic operators in all the operator products. In fact, the operators for the cavity mode and the atom commute at equal times, and the ordering in is immaterial. However, when we substitute $a(t)$ from equation (10.227), the chosen ordering brings considerable simplification: ξ always acts to the right, and $\xi^\dagger$ always acts to the left, in the reservoir vacuum state; thus, every term involving the reservoir operators is zero. Then, the equations of motion for the atomic variables become

$$\langle\dot{\sigma}_-\rangle = (Y/\sqrt{2})\langle\sigma_z\rangle - \frac{1}{2}(1 + 2C)\langle\sigma_-\rangle \tag{10.228}$$

$$\langle\dot{\sigma}_+\rangle = (Y/\sqrt{2})\langle\sigma_z\rangle - \frac{1}{2}(1 + 2C)\langle\sigma_+\rangle \tag{10.229}$$

$$\langle\dot{\sigma}_z\rangle = -(Y/2\sqrt{2})(\langle\sigma_+\rangle + \langle\sigma_-\rangle) \tag{10.230}$$

$$-(1 + 2C)\left(\langle\sigma_z\rangle + \frac{1}{2}\right), \tag{10.231}$$

where time is defined in terms of $t \to \gamma t$. These are the equations for the resonance fluorescence in free space, but with an enhanced decay rate,

$$\gamma \to \gamma(1 + 2C). \tag{10.232}$$

The term $2C\gamma = 2g^2/\kappa$ represents emission into the cavity mode, the γ term is the fluorescence out the side of the cavity. We take γ to be the free-space decay rate, as for most experimental situations it is well approximated by that. We already know the spectra and photon statistics of the field observed from the side of the cavity if we

Figure 10.8. Two emission pathways cavity QED, bad cavity limit.

use the cavity-enhanced decay rate, but we will see new physics in the output of the cavity. As for light leaking out of the cavity, there are two contributions, the coherent drive passing through the cavity and the emission from the atom (figure 10.8).

The average cavity field can be calculated using equation (10.227). We find for example

$$\langle a \rangle_{ss} = \sqrt{n_s}\left(Y + 2\sqrt{2}\,\langle \sigma_- \rangle_{ss}\right). \tag{10.233}$$

Likewise, the average photon number is

$$\langle a^\dagger a \rangle_{ss} = n_s(Y^2 + 2\sqrt{2}\,Y\,\langle(\sigma_+ + \sigma_-)\rangle + 8\langle \sigma_+\sigma_- \rangle). \tag{10.234}$$

As a last example, we find

$$\begin{aligned} \langle a^2 \rangle &= n_s(Y^2 + 2\sqrt{2}\langle \sigma_- + \sigma_+ \rangle + \langle \sigma_+^2 \rangle) \\ &= n_s(Y^2 + 2\sqrt{2}\langle \sigma_- + \sigma_+ \rangle). \end{aligned} \tag{10.235}$$

In terms of the system parameters, we have

$$\sqrt{n_s}X \equiv \langle a \rangle_{ss} = \sqrt{n}_s Y \frac{1 + 2C + Y^2}{(1 + 2C)^2 + Y^2}. \tag{10.236}$$

The average photon number is

$$\langle a^\dagger a \rangle_{ss} = n_s Y^2 \frac{1 + Y^2}{(1 + 2C)^2 + Y^2}, \tag{10.237}$$

which can be divided into coherent and incoherent components,

$$\langle a^\dagger a \rangle_{ss} = n_s X^2 + n_s 4C^2 \left[\frac{Y^2}{(1 + 2C)^2 + Y^2}\right]^2. \tag{10.238}$$

Using equation (10.227), we can show that the intensity correlation function for zero time delay is given by

$$\begin{aligned} g_{ss}^{(2)}(0) &= \langle (a^\dagger)^2 a^2 \rangle_{ss} / (\langle a^\dagger a \rangle)^2 \\ &= (1 + Y^2)^{-2}[(1 - 4C)^2 + 2Y^2(1 + 4C^2) + Y^4]. \end{aligned} \tag{10.239}$$

It is straightforward to calculate the time-dependent intensity correlation function. We have

$$\begin{aligned} g_{\text{ss}}^{(2)}(\tau') = 1 &+ \left(\frac{2\sqrt{2}C}{1+Y^2}\right)^2 e^{-(3/4)\tau'} \\ &\times \{(Y^2 + 2C^2 - 1)\cosh(\Omega\tau') \\ &+ (4\Omega)^{-1}[Y^2(5-2C)/(1+2C) \\ &- 1 - 2C^2]\sinh(\Omega\tau')\}, \end{aligned} \tag{10.240}$$

where

$$\Omega = (1/4)[1 - 8Y^2/(1+2C)^2]^{1/2} \tag{10.241}$$

$$\tau' = \gamma(1+2C)\tau. \tag{10.242}$$

For weak driving fields ($Y^2 \ll 1$), equation (10.240) takes the simple form

$$g_{\text{ss}}^{(2)}(\tau') = (1 - 4C^2 e^{-\tau'/2})^2. \tag{10.243}$$

We have $g^{(2)}(0) = 0$ for $C = 1/2$, but it is nonzero for other values of C. There are two ways to detect two photons in transmission, two photons from emission from the cavity, or one cavity field photon and one emitted from the atom that escapes, and there can be interference between the two.

Let us examine the origin of the nonclassical effects. This means the usual trick of making something messier and then rearranging terms into meaningful bits. Here we have

$$\begin{aligned} g_{\text{ss}}^{(2)}(\tau) &= \langle a^\dagger(0)a^\dagger(\tau)a(\tau)a(0)\rangle_{\text{ss}}/(\langle a^\dagger a\rangle_{\text{ss}})^2 \\ &= 1 + (\langle a^\dagger a\rangle_{\text{ss}}/n_s)^{-2} 8C^2 \cdot \{4X^2\langle :\Delta\sigma_1(0)\Delta\sigma_1(\tau):\rangle_{\text{ss}} \\ &\quad + 2\sqrt{2}CX[\langle\Delta\sigma_+(0)\Delta\sigma_+(\tau)(\Delta\sigma_-(\tau) \\ &\quad + \Delta\sigma_-(0))\rangle_{\text{ss}} + h.c.] \\ &\quad + 8C^2[\langle\Delta\sigma_+(0)\Delta\sigma_+(\tau)\Delta\sigma_-(\tau)\Delta\sigma_-(0)\rangle_{\text{ss}} \\ &\quad - (\langle\Delta\sigma_+\Delta\sigma_-\rangle_{\text{ss}})^2]\} \end{aligned} \tag{10.244}$$

where $\Delta\sigma_1 = (1/2)(\Delta\sigma_- + \Delta\sigma_+)$ and $\langle :\Delta\sigma_1(0)\ \Delta\sigma_1(\tau):\rangle_{\text{ss}}$ are the normally ordered, time-ordered correlation functions that enter the calculation of the spectrum of squeezing. Equation (10.244) describes a self-homodyning of the light from the fluctuating dipole source $\Delta\sigma_\pm$ with the local oscillator amplitude X; since the local oscillator is not an independent field, its amplitude cannot be arbitrarily increased to justify dropping the last two terms on the right-hand side of equation (10.240), as in the homodyne detection of squeezing. The atomic correlation functions appearing in equation (10.240) can be evaluated using the quantum regression theorem, the Bloch equations, and their steady-state solutions.

For weak driving fields, the dominant terms in equation (10.244) are given by

$$\langle :\Delta\sigma_1(0)\Delta\sigma_1(\tau'):\rangle_{\text{ss}} = -(X^2/4)e^{-\tau'/2} \tag{10.245}$$

$$\langle \Delta\sigma_+(0)\Delta\sigma_+(\tau')\Delta\sigma_-(\tau')\Delta\sigma_-(0)\rangle_{ss} \quad - (\langle \Delta\sigma_+\Delta\sigma_-\rangle_{ss})^2 = (X^4/4)e^{-\tau'}, \tag{10.246}$$

and $\langle a^\dagger a\rangle_{ss}/n_s = [Y/(1+2C)]^2 = X^2$. The other atomic correlation functions can be neglected as they add only second-order corrections. These correlation functions contribute the two exponentially decaying terms in the weak-field form for the intensity correlation function,

$$g_{ss}^{(2)}(\tau') = 1 - 8C^2e^{-\tau'/2} + 16C^4e^{-\tau'}. \tag{10.247}$$

as shown in figure 10.9.

The important feature is the minus sign before the first exponential. This arises from the squeezed fluctuations of the radiation from the in-phase quadrature of the atomic dipole $\Delta\sigma_\pm$ [10.244]. It is not possible for a classical correlation function to have a negative sign here. However, the self-homodyning interpretation of $g_{ss}^{(2)}(\tau)$ is not perfect. the local oscillator is not an independent field. Thus, two sources of fluctuations contribute to $g_{ss}^{(2)}(\tau)$ in competition: squeezed amplitude fluctuations and 'classical' intensity fluctuations.

It is the tradeoff between the two sources of fluctuations that brings the C dependence to the condition for photon antibunching $-g_{ss}^{(2)}(0)\langle 1$ only for $C\langle 1/\sqrt{2}$; the largest effect, $g_{ss}^{(2)}(0) = 0$, is observed only for $C = 0.5$. However, with the time dependence included, we see that the squeezed fluctuations make their presence known for all $C\rangle 0.5$. This occurs because the contributions to $g_{ss}^{(2)}(\tau')$ from amplitude and intensity fluctuations regress at different rates. For $C\rangle 1/\sqrt{2}$, the squeezed fluctuations do not dominate at $\tau' = 0$, but they do for all $\tau'\rangle 2\ln(2C^2)$, and for $\tau' = \tau'_m = 2\ln(4C^2)$ they cancel the intensity fluctuations and cause the vanishing of $g_{ss}^{(2)}(\tau')$ noted above.

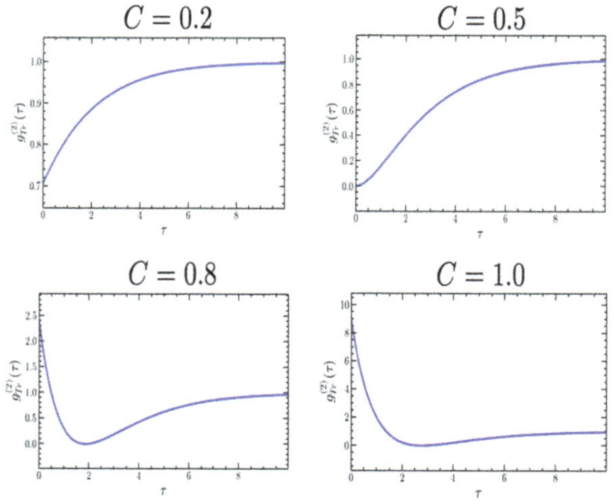

Figure 10.9. Cavity QED: $g^{(2)}(\tau)$ in the bad cavity limit.

10.10 Time asymmetric correlations

As we discussed in chapter 9, a time asymmetric correlation function implies a breakdown of detailed balance. In the cavity QED system, we can ask for the relative probability of detecting a fluorescent photon given a previous detection of a transmitted on, or vice versa. These need not be time symmetric, and indeed they are not as shown in figures 10.10 and 10.11.

The same thing occurs with a cross correlation involving photodetection in fluorescence and a quadrature measurement of the field or the inverse ordering,

$$h_\theta^{TF} = \frac{\langle a^\dagger(0)\sigma_\theta(\tau)a(0)\rangle}{\langle\sigma_\theta\rangle_{ss}\langle a^\dagger a\rangle_{ss}} \tag{10.248}$$

$$h_\theta^{FT} = \frac{\langle \sigma_+(0)a_\theta(\tau)\sigma_-(0)\rangle}{\langle\sigma_\theta\rangle_{ss}\langle a^\dagger a\rangle_{ss}} \tag{10.249}$$

as shown in figures 10.12 and 10.13.

The cross correlations do agree at $\tau = 0$ as they must.

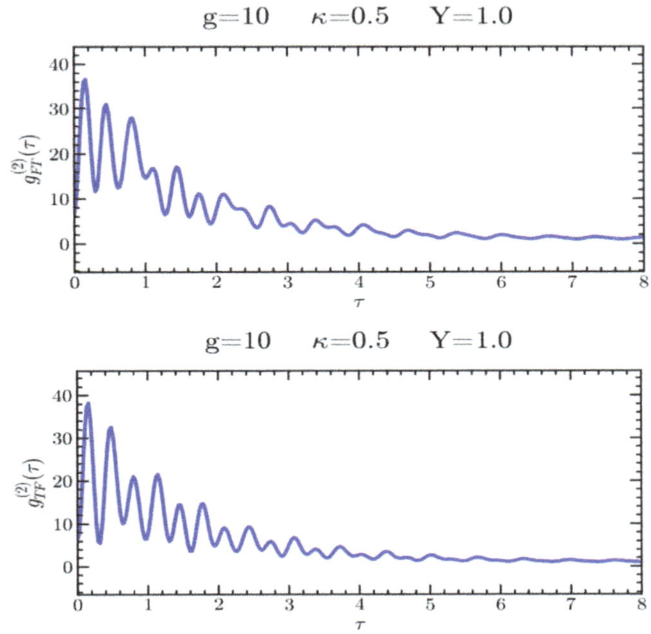

Figure 10.10. Asymmetric cross correlations in CQED in the bad cavity limit.

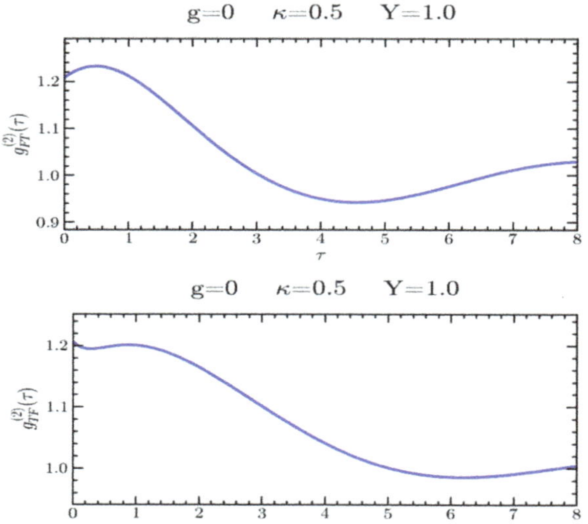

Figure 10.11. Asymmetric cross correlations in CQED in the bad cavity limit.

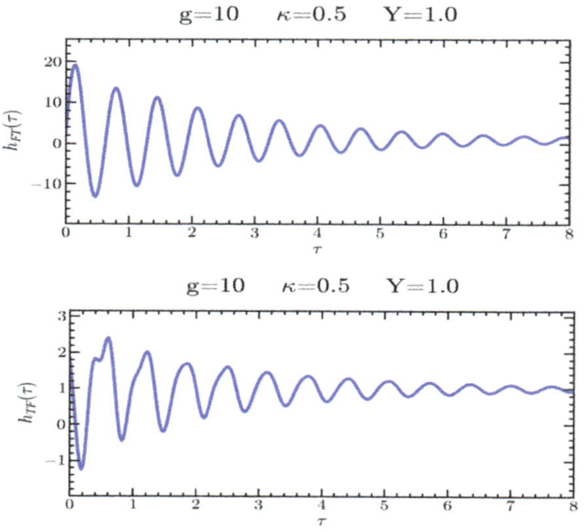

Figure 10.12. Asymmetric cross correlations in CQED in the bad cavity limit.

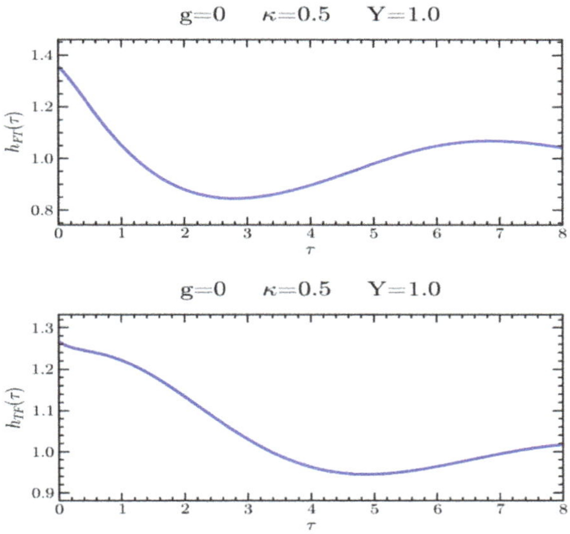

Figure 10.13. Asymmetric cross correlations in CQED in the bad cavity limit.

References

[1] Leach J 2003 Photon statistics and field–intensity correlation of a cavity QED system with external potentials *PhD Thesis* Miami University http://rave.ohiolink.edu/etdc/view?acc_num=miami1058466870

[2] Carmichael H J 1993 *An Open Systems Approach to Quantum Optics* (Berlin: Springer)

[3] Carmichael H J 1999 *Statistical Methods in Quantum Optics 1: Master Equations and Fokker–Planck Equations* (Berlin: Springer)

[4] Carmichael H J 2007 *Statistical Methods in Quantum Optics 2: Non-Classical Fields* Theoretical and Mathematical Physics (Berlin: Springer)

[5] Breuer H-P and Petruccione F 2002 *The Theory of Open Quantum Systems* (Oxford: Oxford University Press)

[6] Gardiner C and Zoller P 2004 *Quantum Noise* (Berlin: Springer)

[7] Rice P R and Carmichael H J 1988 Nonclassical effects in optical spectra *J. Opt. Soc. Am.* **B 5** 1661–8

[8] Rice P R and Carmichael H J 1988 Single-atom cavity-enhanced absorption. I. Photon statistics in the bad-cavity limit *IEEE J. Quantum Electron.* **24** 1351–66

IOP Publishing

An Introduction to Quantum Optics (Second Edition)
An open systems approach
Perry Rice

Chapter 11

Quantum trajectory theory

11.1 A new way

We have seen that to formally treat the open quantum system's evolution, we can use the master equation,

$$\dot{\rho} = \mathcal{L}\rho, \tag{11.1}$$

where $\mathcal{L}$ is a generalized Liouvillian superoperator (it operates on other operators, not states) [1]. We can use it directly to calculate the expectation values and, with the quantum regression theorem, we can calculate the two-time averages or correlation functions. The diagonal elements of the density matrix tell us about the populations of the various states of the system, and the appearance of nonzero off-diagonal density-matrix elements tells us that there is coherence of some type or another. For given initial states, we obtain perfectly correct answers for what is to appear on the meters in our lab; quantum mechanics actually tells us that we cannot determine *how* a system gets from the initial to the final state. Recently, there has been a new method developed by several workers that results in describing open systems by wave functions with backaction on the system based on measurements made by detectors (real or imagined). For example, one might imagine an atom sitting in its excited state, and at some point a detector goes click, a photon has been detected, and then we know the atom is in its ground state. For an individual atom, one can only describe the likelihood that an atom will decay at some time. This so-called quantum jump seems quite mysterious, and the ensemble nature of the density-matrix theory actually hides it at first glance. In this chapter, we learn about a method that is *derived* from the general theory of open systems (not just thrown together ad hoc) that allows us to say *what we can* about the process. This is akin in spirit to Feynman's sum-over-histories approach.

We begin with the master equation and unravel it in a particular way based on a simulated photon counting experiment. Here we follow the formalism of Carmichael [2, 3] although there are other equivalent formulations [4]. There are also different

doi:10.1088/978-0-7503-6014-2ch11

unravelings of the master equation. One is quantum state diffusion, which is based on balanced homodyne detection [5]. Nice reviews are available, by Wiseman [6], Jacobs and Steck [7], and Dorsselaer and Nienhuis [8]. In the quantum trajectory formalism (QTF), we abandon the density-matrix, mixed-state representation of the system in favor of an ensemble of systems represented by pure-state wave functions. Each of these pure-state systems will undergo continuous, non-Hermitian Hamiltonian evolution interrupted at random times by some form of discontinuous state change. These state changes are the realizations of the system decay processes, including the spontaneous emission and cavity decay. The pure-state representation for each member of the ensemble will take the form governed by its history of discontinuous state changes and is thus called the system's *measurement record.* This can always be done for a master equation in the Lindblad form and is particularly useful if we wish to examine the statistics of photodetection. For other types of measurements, notably those including homodyne or heterodyne measurements, another approach will be discussed later in this chapter.

The master equation that results in most if not all of the quantum optics, where one has treated the interaction with the bath/environment by the Born–Markov approximation, is the so-called Lindblad form

$$\dot{\rho} = -i[H_{\text{sys}}, \rho] + \Gamma(2C\rho C^{\dagger} - C^{\dagger}C\rho - \rho C^{\dagger}C), \tag{11.2}$$

where C is a linear combination of the creation and annihilation operators for the source. The part of the master equation $-\Gamma(C^{\dagger}C\rho + \rho C^{\dagger}C)$ is an anticommutator. By considering the evolution of a quantum state with a non-Hermitian Hamiltonian, we can accommodate these terms into a quasi-Hamiltonian evolution:

$$\begin{aligned}
\dot{\rho} &= |\dot{\psi}\rangle\langle\psi| + |\psi\rangle\langle\dot{\psi}| \\
&= -\frac{i}{\hbar}H|\psi\rangle\langle\psi| + \frac{i}{\hbar}|\psi\rangle\langle\psi|H^{\dagger} \\
&= -\frac{i}{\hbar}(H_S + i\hbar H_D)|\psi\rangle\langle\psi| + \frac{i}{\hbar}|\psi\rangle\langle\psi|(H_S - i\hbar H_D) \\
&= -\frac{i}{\hbar}[H_S, \rho] + H_D\rho + \rho H_D \\
&= -\frac{i}{\hbar}[H_S, \rho] + [H_D, \rho]_{+},
\end{aligned} \tag{11.3}$$

with $H_D = \hbar\Gamma C^{\dagger}C$. We can put all this together in one non-Hermitian Hamiltonian

$$H_{\text{eff}} = H_S - i\hbar\Gamma C^{\dagger}C, \tag{11.4}$$

where we have

$$|\dot{\psi}\rangle = (-i/\hbar)H_{\text{eff}}|\psi\rangle \tag{11.5}$$

$$\langle\dot{\psi}| = (i/\hbar)\langle\psi|H_{\text{eff}}^{\dagger}, \tag{11.6}$$

for evolution via this non-Hermitian Hamiltonian. Note that this evolution does not preserve the norm of the state vector, a point to which we shall return to.

Now, the master equation is pulled apart into two pieces

$$\dot{\rho} = (\mathcal{L} - \mathcal{S})\rho + \mathcal{S}\rho, \tag{11.7}$$

where we (for now) choose $\mathcal{S}$ to contain all the terms of the form $C\rho C^\dagger$. Each operator C in $\mathcal{S}$ becomes a collapse operator. The remaining terms are written

$$(\mathcal{L} - \mathcal{S})\rho = -\frac{i}{\hbar}[H_S, \rho] + [H_D, \rho]_+, \tag{11.8}$$

where $[O, \rho]_+$ denotes the anticommutator and H_D contains the dissipation terms that are excluded in $\mathcal{S}$. The non-Hermitian Hamiltonian that governs the trajectory evolution is

$$H = H_S + i\hbar H_D. \tag{11.9}$$

How do we handle the evolution from the $\mathcal{S}$ part of the master equation? Again continuing with the idea that there is some type of wave function type object underlying ρ, we have

$$\begin{aligned} \mathcal{S}\rho &= 2\Gamma C\rho C^\dagger \\ &= \sqrt{2\Gamma}\, C|\psi\rangle\, \langle\psi|C^\dagger\sqrt{2\Gamma}, \end{aligned} \tag{11.10}$$

where we can see the emergence of a collapse operator, $\sqrt{2\Gamma}\,C$. If, for example, this is the annihilation operator for the field, a, this has the effect of detecting a photon out of the field at a given time step. There will be some cases where we have 'collapses' where, say, the photon state is increased! These collapse processes need not correspond solely with the energy leaving the system [9].

When do these collapses occur? The probability of photodetection in a time step dt is the rate of photons leaving times the time step,

$$P_{\text{coll}} = 2\kappa\,\langle\psi_c(t)|a^\dagger a|\psi_c(t)\rangle\, dt. \tag{11.11}$$

In general, we will have

$$P_{\text{coll}} = \Gamma\,\langle\psi_c(t)|C^\dagger C|\psi_c(t)\rangle\, dt. \tag{11.12}$$

We see that the field evolves continuously according to this Hamiltonian, but punctuated by photon emission events that occur with a probability proportional to the photon number.

As an example, for the single cavity field mode, the master equation is

$$\dot{\rho} = -i\omega_0[a^\dagger a, \rho] + \kappa(2a\rho a^\dagger - a^\dagger a\rho - \rho a^\dagger a). \tag{11.13}$$

We identify

$$(\mathcal{L} - S)\rho = -i\omega_0[a^\dagger a, \rho] - \kappa[a^\dagger a, \rho]_+ \tag{11.14}$$

$$\mathcal{S}\rho = 2\kappa a \rho a^{\dagger}, \tag{11.15}$$

with the non-Hermitian Hamiltonian

$$H_{\text{eff}} = \hbar\omega a^{\dagger}a - i\hbar\kappa a^{\dagger}a. \tag{11.16}$$

At this point, everything we have done is ad hoc; we will formalize it more later. A schematic of the implementation is shown in figure 11.1.

The quantum trajectories are evolved in the simulations as follows, as shown in figure 11.1. Starting from some initial state, the wave function evolves according to the Schrödinger equation

$$\frac{d}{dt}|\psi_{\text{REC}}(t)\rangle = -\frac{i}{\hbar}H_{\text{eff}}|\psi_{\text{REC}}(t)\rangle. \tag{11.17}$$

This evolution is punctuated by collapses occurring at random times with a probability determined by

$$P_{\text{REC}} = \frac{\langle\psi_{\text{REC}}(t)|C^{\dagger}C|\psi_{\text{REC}}(t)\rangle}{\langle\psi_{\text{REC}}(t)|\psi_{\text{REC}}(t)\rangle}. \tag{11.18}$$

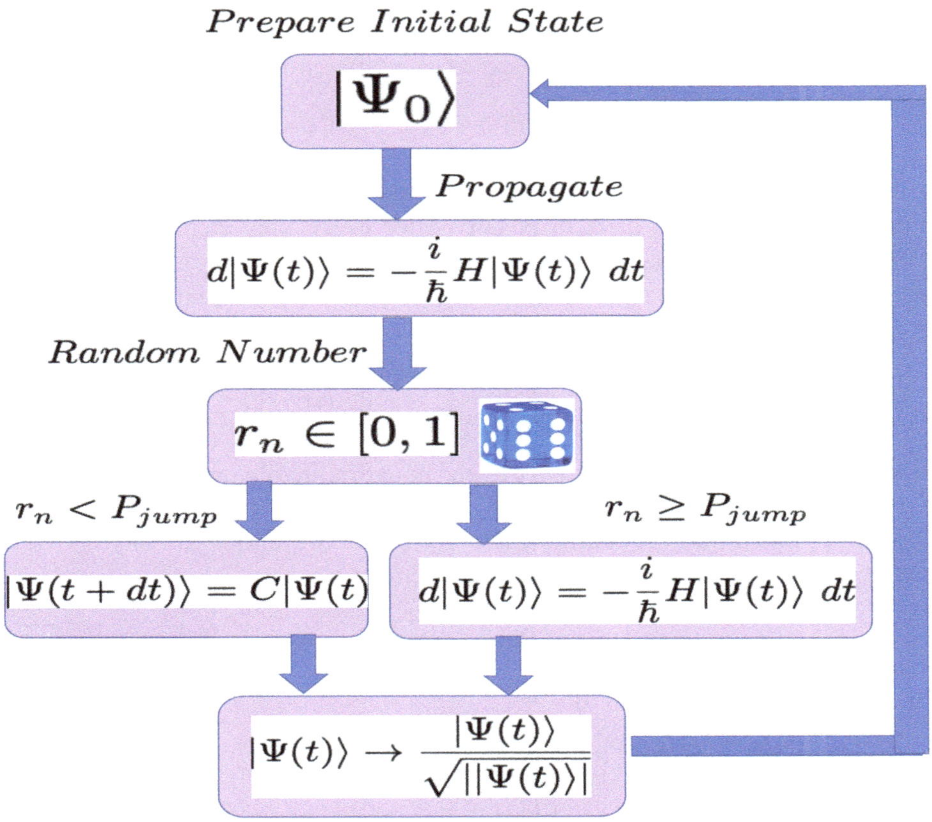

Figure 11.1. Flow chart for the quantum trajectory simulations.

At any given time. The denominator is necessary because the non-Hermitian Hamiltonian does not preserve the norm of the wave function. In practice, the wave function is numerically integrated and normalized at each time step. The wave function $|\psi_c(t)\rangle$ is conditioned on a particular series of collapses that have occurred.

Obviously, the state vector changes dramatically upon a detection event, which corresponds to a change in our state of knowledge of the state. The length of the state vector can be changed drastically, and we must normalize it. If there is *not* a click, we also obtain information, but not as much. Here too, we must then normalize the wave function every time step. Mathematically, this results from the fact that a non-Hermitian Hamiltonian leads to a non-unitary propagation, and the norm of the state vector is not preserved.

The ensemble average of the conditioned wave functions is formally equivalent to the density matrix because

$$\rho = \overline{|\psi_{\mathrm{REC}}(t)\rangle\langle\psi_{\mathrm{REC}}(t)|}. \tag{11.19}$$

Expectation values are calculated from the wave function in the usual way and then averaged,

$$\langle O\rangle = \overline{\langle\psi_{\mathrm{REC}}(t)|O|\psi_{\mathrm{REC}}(t)\rangle}. \tag{11.20}$$

One can also write the whole process as one differential equation

$$d|\psi_{\mathrm{REC}}\rangle = \frac{-1}{\hbar}H_{\mathrm{eff}}|\psi\rangle\, dt + (\sqrt{\Gamma}\, C - 1)|\psi_{\mathrm{REC}}\rangle\, dN, \tag{11.21}$$

where the increment dN satisfies $dN^2 = dN$, which is satisfied by $dN = 0, 1$; that is, either the jump happens or it does not. After every time step, one still has to normalize the wave function. We can interpret the normalization process into our equations directly, which results in a nonlinear Schrödinger equation

$$d|\psi_{\mathrm{REC}}\rangle = \frac{-i}{\hbar}(H_{\mathrm{eff}} + \imath\hbar\Gamma\langle C^\dagger C\rangle)|\psi\rangle\, dt + \left(\frac{2\sqrt{\Gamma}\, C}{\sqrt{2\Gamma\langle C^\dagger C\rangle}} - 1\right)|\psi\rangle\, dN.$$

The nonlinearity stems from the part of the equation involving $\langle C^\dagger C\rangle$ multiplying $d|\psi\rangle$.

It can be shown that the ensemble average of the conditional density matrices is equivalent to the system's real density matrix:

$$\rho(t) = \overline{|\psi_c(t)\rangle\langle\psi_c(t)|}. \tag{11.22}$$

If we unravel this equation as

$$S\rho = 2\kappa a\rho\, a^\dagger, \tag{11.23}$$

$$(L - S)\rho = -i\omega_0[a^\dagger a, \rho] - \kappa[a^\dagger a, \rho]_+. \tag{11.24}$$

We adopted a formalism based on a simulated photon counting experiment, where the detector is modeled as an absorptive device. Thus, to calculate the system's photon statistics, we simply employ the same QTF simulation presented earlier and keep a record of the collapse event times. By making a histogram of the photon absorption delay times, we can characterize the transmitted field's dynamics in precisely the same manner as is done experimentally.

What if there is more than one collapse operator? Let us say there are two, with probabilities P_1 and P_2, both small numbers due mainly to dt. Then, we perform collapse 1 if the random number falls between 0 and P_1, collapse 2 if the random number falls between P_1 and P_2, and no collapse if the random number is larger than $P_1 + P_2$. There are other protocols that can be used.

To mathematically formalize this approach, we first unravel the master equation by writing equation (11.1) as

$$\dot{\rho} = (\mathcal{L} - \mathcal{S})\rho + \mathcal{S}\rho, \tag{11.25}$$

where $\mathcal{S}\rho$ contains all terms of form $O^{\dagger}\rho O$. These O operators are those that model collapse events for the system, although in actuality we can choose *any* operator for $\mathcal{S}$. We express the remaining terms as

$$(L - S)\rho = -\frac{i}{\hbar}[H_S, \rho] + [H_D, \rho]_+, \tag{11.26}$$

where $[O, \rho]_+ = O\rho + \rho O$, and H_D contains the dissipative terms not included in $S\rho$. The The Hamiltonian that determines the continuous evolution of the pure states is not a Hermitian and is given by

$$H = H_s + i\hbar H_D, \tag{11.27}$$

where H_S is the Hermitian Hamiltonian for the system, not accounting for the dissipation processes.

In the next section, we will see how this approach allows us to gain more intuition about open systems; going beyond noticing that off-diagonal density-matrix elements are nonzero and that there is some type of quantum coherence. Beyond that, what are the advantages of this approach? The density matrix involves $N(N - 1)/2 \sim N^2$ elements, after considering the hermeticity of the density matrix and the condition that $\text{Tr}(\rho) = 1$. Therefore, as our system size increases, the number of equations that must be solved increases quadratically. In the quantum trajectory approach, one has to evolve N equations, a significant savings in space requirements. Of course, the downside is that we must solve these N equations many times to perform an ensemble average; the density-matrix/quantum regression approach is generally more efficient for crunching numbers.

The unraveling we have examined here involves the choice $\mathcal{S} = \Gamma C$ where C is an annihilation operator of the system, and the unraveling is usually referred to as the direct detection version of quantum trajectories.

11.2 Some examples

Let us consider the example of the damped two-level atom, for which the master equation is

$$\begin{aligned}\dot{\rho} = &-\frac{i}{2}\omega_A[\sigma_z, \rho] + \frac{\gamma}{2}(\bar{n}+1)(2\sigma_-\rho\sigma_+ - \sigma_+\sigma_-\rho - \rho\sigma_+\sigma_-) \\ &+ \frac{\gamma}{2}\bar{n}(2\sigma_+\rho\sigma_- - \sigma_-\sigma_+\rho - \rho\sigma_-\sigma_+),\end{aligned} \tag{11.28}$$

where γ is the rate of spontaneous emission. To simplify our example, let us assume that the thermal photon number $\bar{n}$ is equal to zero; this is a valid approximation in many experimental regimes, particularly for optical transitions, and will not qualitatively change the form of our example. Therefore, the master equation for our damped two-level atom is now written as

$$\dot{\rho} = -\frac{i}{2}\omega_A[\sigma_z, \rho] + \frac{\gamma}{2}(2\sigma_-\rho\sigma_+ - \sigma_+\sigma_-\rho - \rho\sigma_+\sigma_-). \tag{11.29}$$

We can immediately recognize $\frac{\gamma}{2}(2\sigma_-\rho\sigma_+)$ as a collapse term of the form $O^\dagger\rho O$, and thus we define $S\rho$ to be

$$\begin{aligned}S\rho &= \gamma\sigma_-\rho\sigma_+ \\ &= \sqrt{\gamma}\sigma_-|\psi_{\text{REC}}\rangle\langle\psi_{\text{REC}}|\sqrt{\gamma}\sigma_+.\end{aligned} \tag{11.30}$$

Thus,

$$(L - S)\rho = -\frac{i}{2}\omega_A[\sigma_z, \rho] - \frac{\gamma}{2}[\sigma_+\sigma_-, \rho]_+, \tag{11.31}$$

with an effective non-Hermitian Hamiltonian

$$H_{\text{eff}} = \frac{\hbar\omega}{2}\sigma_z - \frac{i\hbar\gamma}{2}\sigma_+\sigma_-. \tag{11.32}$$

Equation (11.30) identifies $\sqrt{\gamma}\sigma_-$ as the collapse operator for the system; this follows from our model of radiative damping for the two-level atom, as a transition from the excited state to the ground state will radiate a photon into the environment. With the identification of the appropriate collapse operator, we can now evaluate the probability of a collapse event occurring $P_{\text{coll}}(t)$ during any random time interval $t + dt$

$$\begin{aligned}P_{\text{coll}}(t) &= \text{Tr}\left[S\rho_{\text{REC}}(t)\right]dt \\ &= \gamma\frac{\langle\psi_{\text{REC}}(t)|\sigma_+\sigma_-|\psi_{\text{REC}}(t)\rangle}{\langle\psi_{\text{REC}}(t)|\psi_{\text{REC}}(t)\rangle}dt,\end{aligned} \tag{11.33}$$

where the subscript c denotes the use of the conditional wave function and density matrix for the system. The denominator of equation (11.33) is needed because the non-Hermitian Hamiltonian does not generally preserve the normalization of the conditional wave function. With these fundamental procedures and details explained, we can now work through the typical operation of a QTF simulation carried out on the computer. First, we consider an ensemble of systems each in some common initial state; for the purpose of this explanation, let us follow one such system and its conditional wave function $|\psi_c(t)\rangle$, through the simulation. The systems undergo evolution as governed by the Schrödinger equation:

$$i\hbar|\dot{\psi}_c(t)\rangle = H|\psi_c(t)\rangle, \tag{11.34}$$

where H is the Hamiltonian of equation (11.27). At each time step (t_n) of the program, a random number (r_n) In the interval [0, 1] is generated and compared to the collapse probability as evaluated in equation (11.33). One of the two conditions can be satisfied at this point, determining the evolution of the system for the time step. If $r_n \geqslant P_{\text{coll}}(t_n)$, then the system will experience a collapse and the evolved conditional wave function at time t_{n+1} will be given by

$$|\psi_{\text{REC}}(t_{n+1})\rangle = \frac{\sigma_-|\psi_{\text{REC}}(t_n)\rangle}{\sqrt{\langle\psi_{\text{REC}}(t_n)|\sigma_+\sigma_-||\psi_{\text{REC}}(t_n)\rangle}}. \tag{11.35}$$

If $r_n < p_c(t_n)$, then the system will undergo continuous Hamiltonian evolution during the time step, giving the conditional wave function to be

$$|\psi_{\text{REC}}(t_{n+1})\rangle = \frac{e^{-(i/\hbar)H\Delta t}|\psi_{\text{REC}}(t_n)\rangle}{\sqrt{\langle\psi_{\text{REC}}(t_n)|e^{(i/\hbar)(H^\dagger - H)\Delta t}|\psi_{\text{REC}}(t_n)\rangle}}. \tag{11.36}$$

Note that the evolved conditional wave functions are to be normalized at each time step.

For our simulation, we have

$$H_{\text{eff}} = \hbar\omega_0\sigma_z - 1\hbar\sigma_+\sigma_- \tag{11.37}$$

$$|\psi\rangle = C_g|g\rangle + C_e|e\rangle \tag{11.38}$$

with the evolution equations

$$\dot{C}_g = i\omega_0 C_g \tag{11.39}$$

$$\dot{C}_e = -i\omega_0 C_e - \frac{\gamma}{2}C_e. \tag{11.40}$$

A simulation for an atom initially in the excited state is shown below in figure 11.2. The excited-state probability stays at 1 until there is a jump, and then the excited state is 0. There is no evolution between jumps here. The excited-state probability amplitude decays via the non-Hermitian $-i\hbar\gamma$ term, but then when we normalize the wave function, this is wiped out.

As one increases the number of trajectories averaged over (figure 11.3), we see convergence to the full master equation result. We see N jumps in general for N trajectories, although they may occur simultaneously. The master equation result is obtained in the $N \to \infty$ limit, but sometimes 11–100 is sufficient for simple systems.

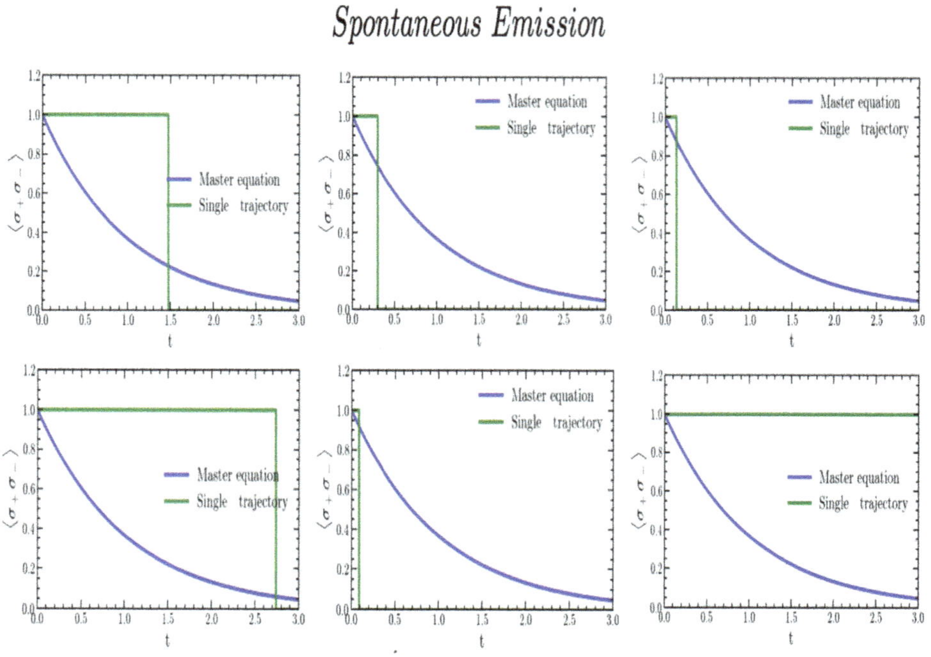

Figure 11.2. Single trajectories and full master equation solution for an atom in the excited state.

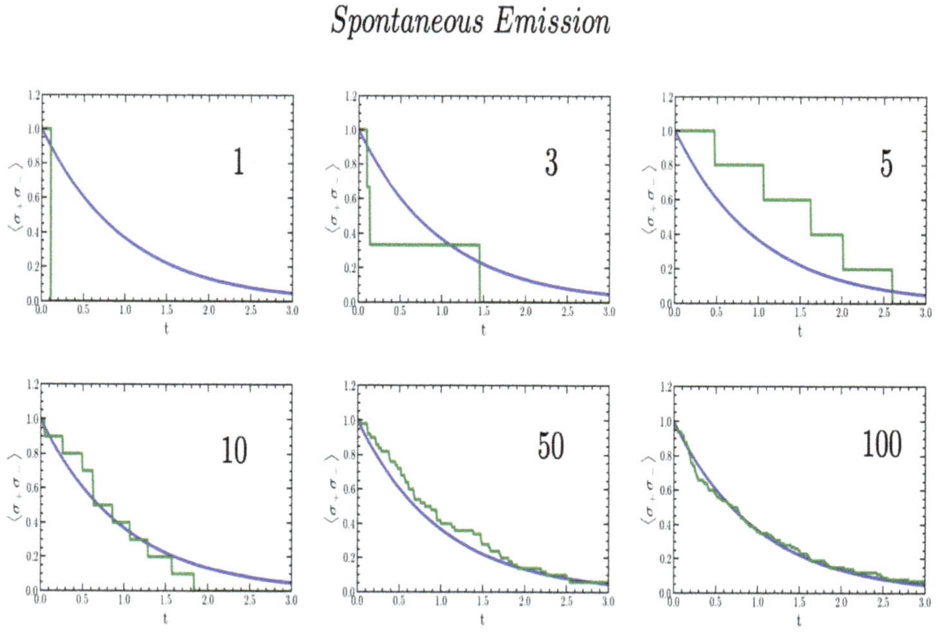

Figure 11.3. Spontaneous emission simulations for an increasing number of trajectories averaged over.

Now consider the single-mode cavity, described by

$$|\psi\rangle = \sum_n C_n|n\rangle \tag{11.41}$$

$$H_{\text{eff}} = \hbar\omega_0 a^\dagger a - \imath\hbar\kappa a^\dagger a. \tag{11.42}$$

The probability for a jump at any time step is $P_c = 2\kappa\langle\psi|a^\dagger a|\psi\rangle dt$, and the probability amplitudes have a long jump evolution described by

$$\dot{C}_n = -(\imath\omega_0 + 2\kappa)nC_n \tag{11.43}$$

with solution

$$C_n(t) = \exp\left(-(\imath\omega_0 + 2\kappa)n\right)C_n. \tag{11.44}$$

There is an exponential decay of these amplitudes, reflecting the fact that the *lack* of a detection event tells us something about the relative probability of the various states. However, as in the spontaneous emission example, this is countered by the decrease in the norm of the state. A sample trajectory is shown in figure 11.4, where we start with four photons in the cavity and plot $\langle a^\dagger(t)a(t)\rangle$. We see four jumps at random times, eventually emptying the cavity.

Next, we examine the effect of increasing the number of trajectories and observe a fairly rapid convergence to the full master equation result (figure 11.5).

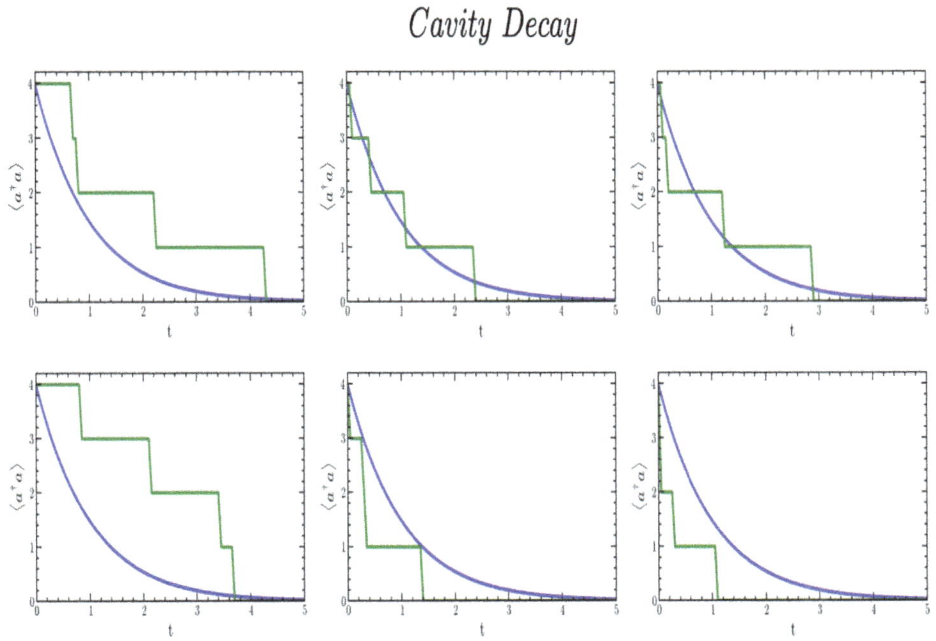

Figure 11.4. Single trajectories and master equation solution for a cavity initially in a Fock state with four photons.

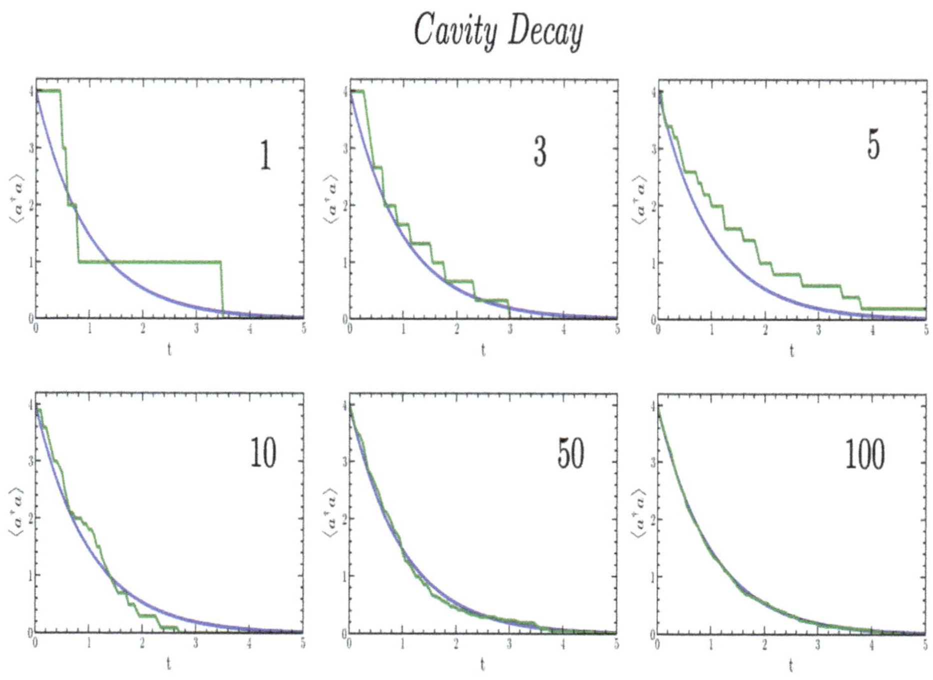

Figure 11.5. Cavity decay simulations for an increasing number of trajectories averaged over.

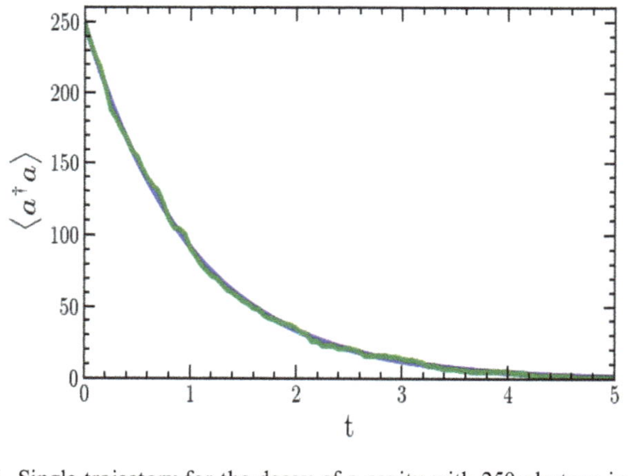

Figure 11.6. Single trajectory for the decay of a cavity with 250 photons in it initially.

In the case of a highly excited cavity, the effect of the jumps becomes less noticeable, as the size of the jump is small for a large number of photons, and a single trajectory can well approximate the master equation solution, as shown in figure 11.6.

A particularly interesting case is that of a cavity prepared in a coherent state. There is *no* collapse in the sense that the collapse operator does *not* change the state of the system, $a|\alpha\rangle = \alpha|\alpha\rangle \rightarrow |\alpha\rangle$, where the latter relation holds after normalization. The non-jump evolution gives us

$$\begin{aligned} |\psi(t)\rangle &= e^{-2\kappa \alpha t}|\psi(0)\rangle \\ &= |\alpha_0 e^{-\kappa t}\rangle . \end{aligned} \tag{11.45}$$

This is another indication of the classical nature of the coherent state! See figure 11.7 for the case of a coherent state with one photon on average.

How about if we start with a superposition state $(1/\sqrt{2})(|2\rangle + |0\rangle)$? An example trajectory and ensemble average is shown in figure 11.8.

Here, we see exponential decay and upward collapses. If a photon is detected, it must have come from the $n = 2$ part of the state, leaving the cavity in $n = 1$, which is higher than the master equation solution; the collapse operator annihilates the vacuum. Occasionally, there are no collapses as the trajectory decays faster than the master equation solution. Of course, when averaged over many trajectories, this balances out to give the master equation solution as it must, as shown in figure 11.9.

Another case of interest is a cavity with no photons coupled to a thermal field with $\langle n\rangle = 5$. We see a mixture of upward jumps, with some downward jumps, with lots of noise for single trajectories, as in figure 11.10.

As we expect, as we average over a set of trajectories, the master equation result eventually obtains as the number of trajectories n_{traj} is increased, as in figure 11.11.

Let us now consider an optical parametric oscillator. This is a device with a $\chi^{(2)}$ crystal that is pumped externally and results in pairs of photons being generated in the cavity. The Hamiltonian is

$$H = \Omega(a^2 + a^{\dagger 2}). \tag{11.46}$$

We assume no depletion of the pump and assume that we are pumping at a small enough level to avoid gain saturation. For $\Omega = 1$, we see in figure 11.12 that the

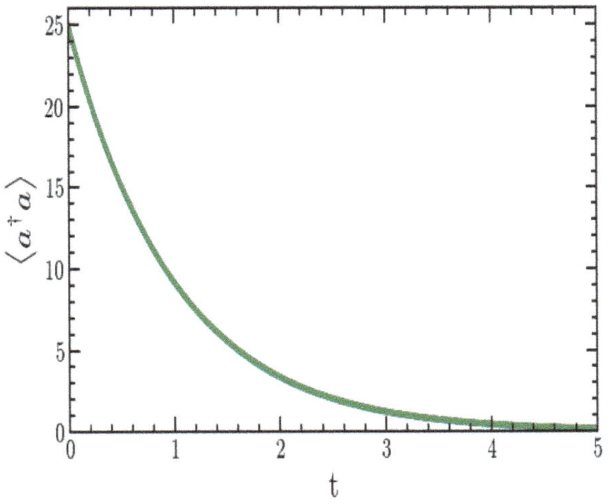

Figure 11.7. Single trajectory for a cavity in a coherent state with $\alpha = 1$, with and without jump operators.

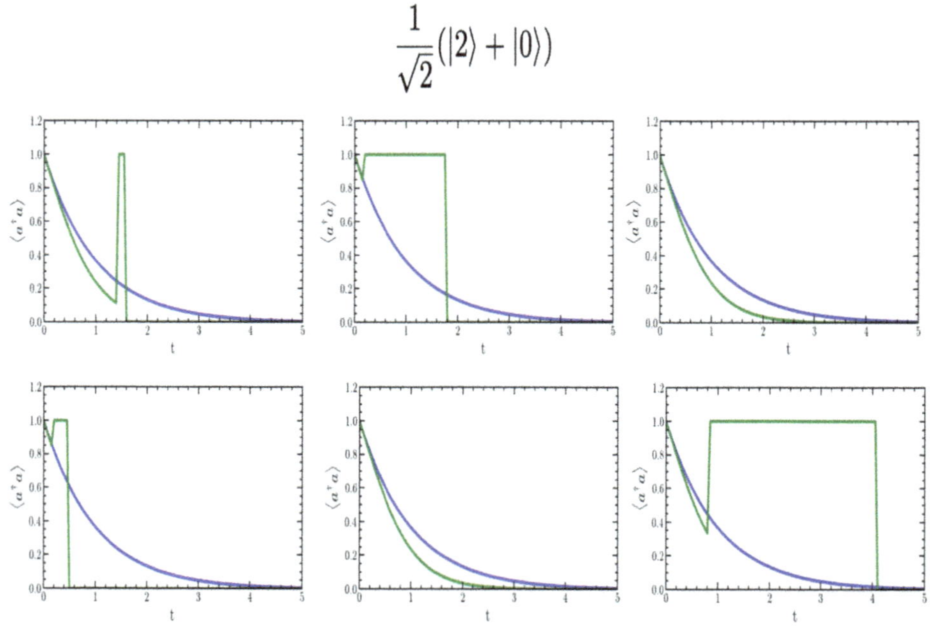

Figure 11.8. Single trajectories for the cavity decay of a superposition state $(1/\sqrt{2})(|2\rangle + |0\rangle)$.

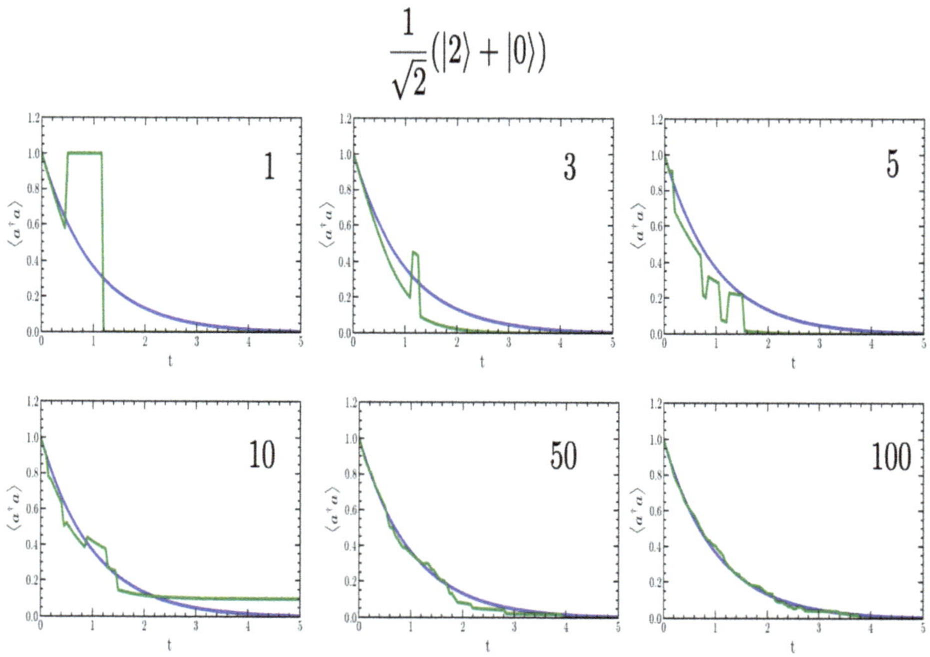

Figure 11.9. Increasing number of trajectories for the cavity decay of a superposition state $(1/\sqrt{2})(|2\rangle + |0\rangle)$.

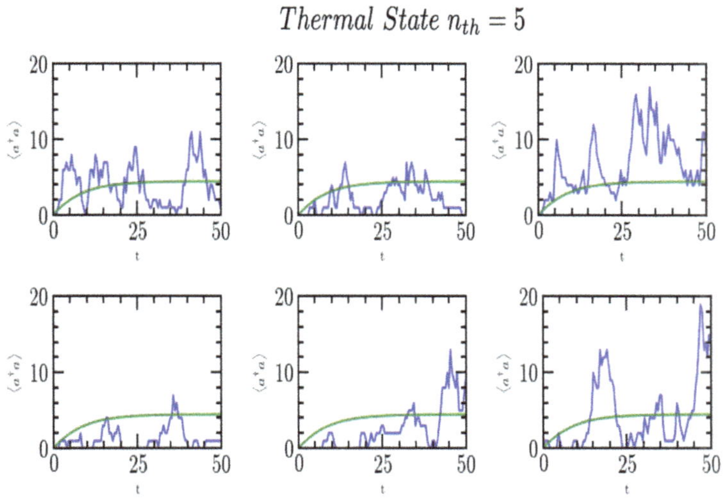

Figure 11.10. Single trajectories for the cavity decay of a thermal state.

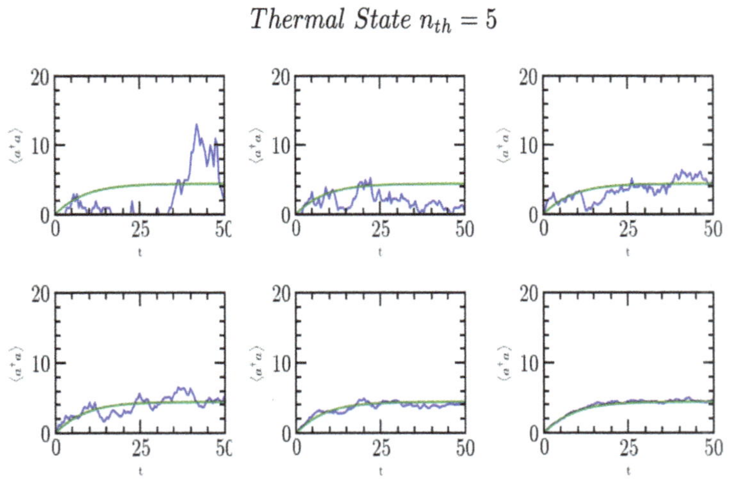

Figure 11.11. Increasing number of trajectories for the cavity decay of a thermal state.

average photon number rises to a level ϵ below the master equation solution, and then there is an upward jump of $1 + \epsilon$. Then, there is a decay back to ϵ. For weak pumps, the wave function is dominated by the vacuum, with single and dual photon state amplitudes on the order of ϵ^2 and then higher order terms. Upon collapse, the vacuum and single-photon states are annihilated, and after normalization, the average photon number is greater than 1, reflecting the now dominant contributions from $a|2\rangle$, and less from $a|3\rangle$, etc.

When one turns up the pump some, we notice a very noisy trace for single trajectories as in figure 11.13. As we will see later in chapter 14, the OPO has a threshold, and we see hints of the large noise near that phase transition.

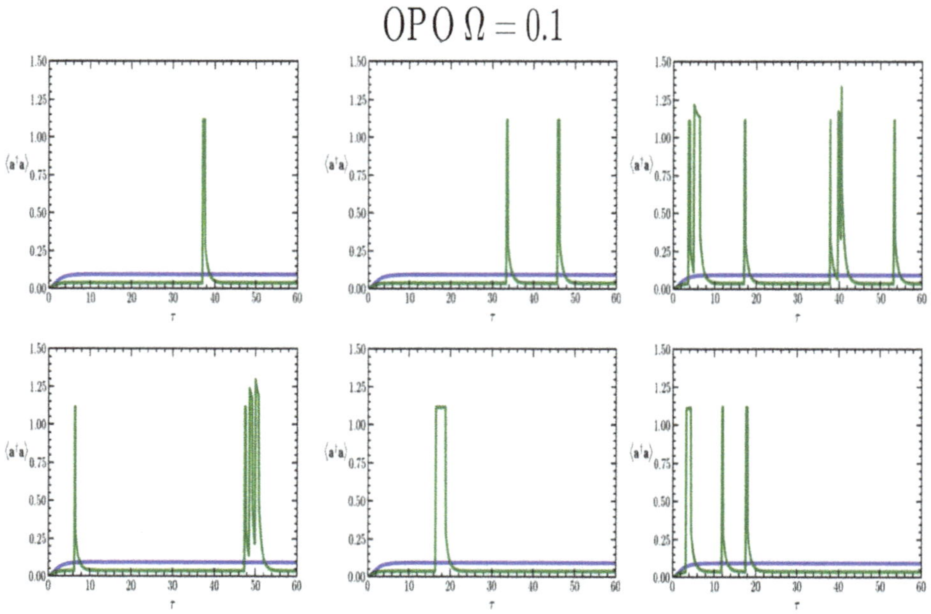

Figure 11.12. Single trajectories for a weakly pumped OPO.

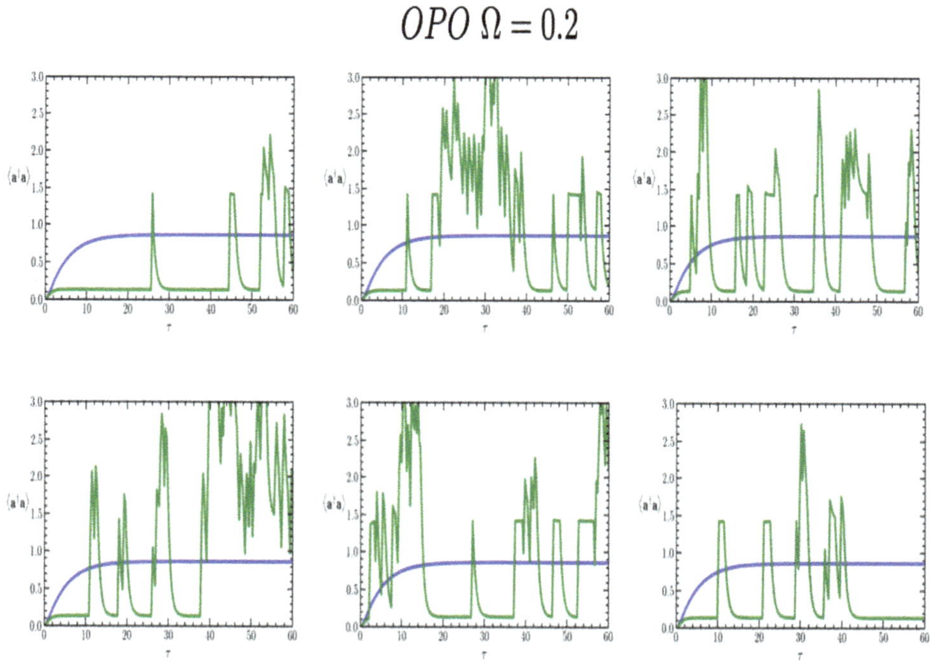

Figure 11.13. Single trajectories for an OPO slightly below threshold.

Note that the mean photon number rises with increasing pump, as does the noise. Further increase in the pump takes us to a phase transition, where the fluctuations are quite large. Returning to the weak pumping case, as we increase the number of trajectories, we see many upward jumps initially, but they decrease in magnitude and the master equation result of course obtains in the large N limit, averaging over many trajectories (figure 11.14).

As our last example, we consider a two-level atom driven by a classical field, resonance fluorescence. For Rabi frequencies well below the spontaneous emission rate, we see in figure 11.15 for a single trajectory an initial increase in the excited-state population, and then a jump back to the ground state, and the process starts anew.

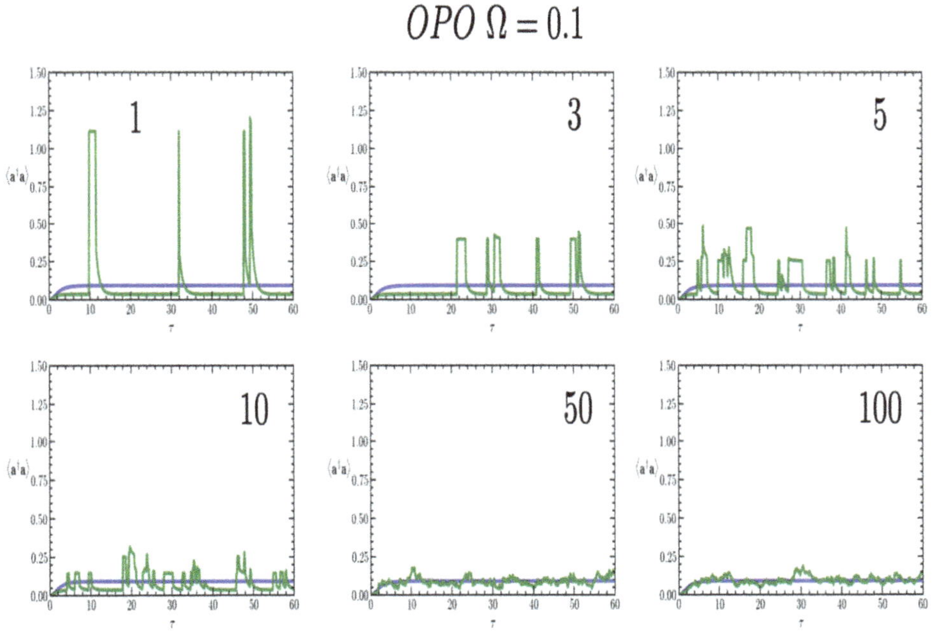

Figure 11.14. The effect of increasing the number of trajectories for a weakly driven OPO.

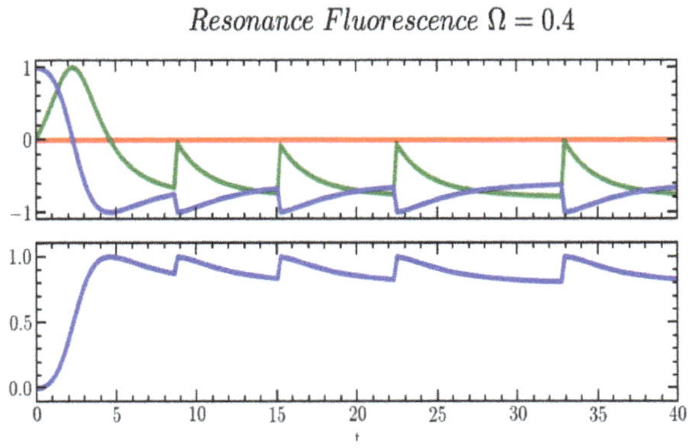

Figure 11.15. Single trajectories for the weakly driven resonance fluorescence.

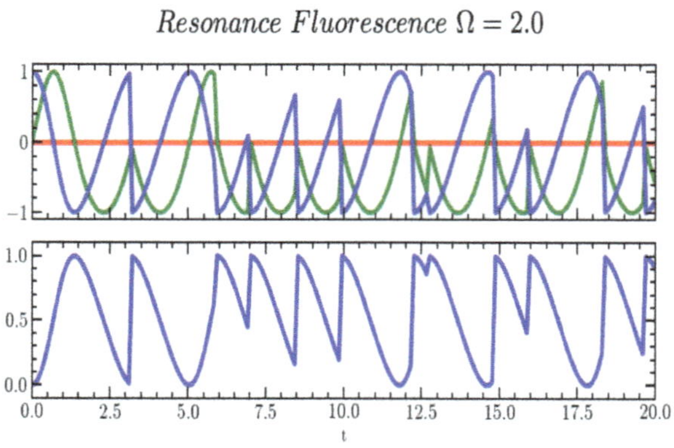

Figure 11.16. Single trajectories for strongly driven resonance fluorescence.

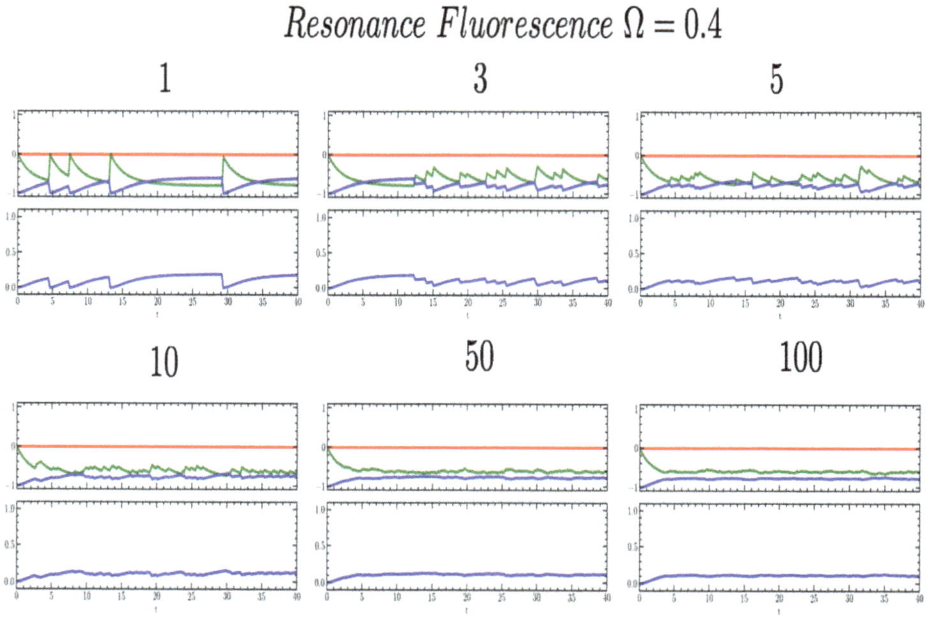

Figure 11.17. Effect of increasing the number of trajectories for weakly driven resonance fluorescence.

In figure 11.16, we see that increasing the Rabi frequency to twice the spontaneous emission rate results in visible oscillations, followed by jumps. In figure 11.17, we see that the master equation result obtains as the number of trajectories is increased.

11.3 Now for a little formality

So how does this dual-evolution stem from the master equation in a more formal way. The formal solution to the master equation is

$$\rho(t) = e^{\mathcal{L}t}\rho(0). \tag{11.47}$$

We use the following identity, known as the Dyson expansion,

$$\begin{aligned} e^{\mathcal{L}t} &= e^{[(\mathcal{L}-\mathcal{S})+S]t} \\ &= \sum_{m=0}^{\infty}\int_0^t dt_m \int_0^{t_m} dt_{m-1} \ldots \int_0^{t_2} dt_1 e^{(\mathcal{L}-\mathcal{S})(t-t_m)}\mathcal{S} \\ &\times e^{(\mathcal{L}-\mathcal{S})(t_m - t_{m-1})}\mathcal{S} \ldots \mathcal{S}e^{(\mathcal{L}-\mathcal{S})t_1}\rho(0) \end{aligned} \tag{11.48}$$

We now define an operator

$$\bar{\rho}_{\mathrm{REC}} = e^{(\mathcal{L}-\mathcal{S})(t-t_m)}\mathcal{S} \times e^{(\mathcal{L}-\mathcal{S})(t_m - t_{m-1})}\mathcal{S} \ldots \mathcal{S}e^{(\mathcal{L}-\mathcal{S})t_1}\rho(0) \tag{11.49}$$

The subscript REC stems from the notion that this is a density-matrix-like object *conditioned* on a given set of *detection* events. If we use $\mathcal{S} = \sqrt{2\kappa}\, a$, these detections would be actual photodetector clicks. We are free to use any choice for $\mathcal{S}$; one can optimize the choice depending on what one wishes to measure/calculate. This can be used to define the density operator

$$\rho_{\mathrm{REC}}(t) = \frac{\bar{\rho}_{\mathrm{REC}}}{\mathrm{Tr}\left[\bar{\rho}_{\mathrm{REC}}\right]}, \tag{11.50}$$

which is properly normalized. We can then write

$$\rho(t) = \sum_{m=0}^{\infty}\int_0^t dt_m \int_0^{t_m} dt_{m-1} \ldots \int_0^{t_2} dt_1 \overline{\rho_{\mathrm{REC}}}. \tag{11.51}$$

Thus, we can see that the 'real' density matrix is the sum of the conditioned density matrices that correspond to different detection records. This is sort of a Feynman diagrammatic version of the density-matrix theory. It is most useful to make a choice for $\mathcal{S}$ for which the conditioned density operator can be factorized, as in the examples we have seen already,

$$\rho_{\mathrm{REC}}(t) = |\psi_{\mathrm{REC}}(t)\rangle\langle\psi_{\mathrm{REC}}(t)|, \tag{11.52}$$

where we can use this decomposition for either the normalized or un-normalized version (the overbarred version or not). Each of these trajectories, or wave functions $|\psi_{\mathrm{REC}}\rangle$ evolve in time via a evolution via $e^{(\mathcal{L}-S)t}$ punctuated by applications of $\mathcal{S}$ at the times t_1, t_2, $\ldots t_m$, where m is the total number of detection events for a particular trajectory. We use the phrase detection event, but some would use collapse. The collapse term can cause long arguments; it is a collapse in the Bayesian sense of a dramatic change in our knowledge of the system, and not necessarily in the Bohr sense of an actual collapse of a physical system. How do we calculate the expectation values?

$$\langle O\rangle = \mathrm{Tr}(\rho O). \tag{11.53}$$

In terms of the trajectory decomposition, we have

$$\begin{aligned} \langle O\rangle &= \mathrm{Tr}(\Sigma|\psi_{\mathrm{REC}}(t)\rangle\langle\psi_{\mathrm{REC}}(t)|O) \\ &= \Sigma\mathrm{Tr}(\langle\psi_{\mathrm{REC}}(t)|O|\psi_{\mathrm{REC}}(t)\rangle) \\ &= \Sigma\langle\psi_{\mathrm{REC}}(t)|O|\psi_{\mathrm{REC}}(t)\rangle, \end{aligned} \tag{11.54}$$

where we have used the shorthand $\Sigma \equiv \sum_{m=0}^{\infty} \int_0^t dt_m \int_0^{t_m} dt_{m-1} \ldots \int_0^{t_2} dt_1$, which is essentially an ensemble average over all realizations of the trajectories. This is a generalized sum over the time and total number of detection events. The method usually used to calculate these averages is to use an ensemble of trajectories, normalizing the trajectories at each time step, and calculating

$$\frac{1}{N} \sum_{\text{REC}} \langle \psi_{\text{REC}}(t) | O | \psi_{\text{REC}}(t) \rangle, \tag{11.55}$$

where N is the number of trajectories simulated in the manner discussed earlier in this chapter. To calculate the steady-state values of the expectation values, one can usually replace the ensemble average with a time average while mumbling about ergodicity.

11.4 Homodyne detection and quantum state diffusion

Recall that balanced homodyne detection measures the quadrature of the output field. The cavity output field is mixed on a 50:50 beam splitter (the 'balanced' part) with a strong local oscillator at a chosen phase. The difference in the photocurrent at the two photodiodes is proportional to a particular quadrature X_ϕ of the field leaving the cavity. If we choose the local oscillator phase to measure the quadrature $X = \frac{1}{\sqrt{2}}(b + b^\dagger)$, this is often written as 'homodyne detection of the amplitude quadrature'. In the absence of observation, the system evolves via the unconditional master equation

With observation, each measurement outcome slightly updates the conditional state of the system in real time. In the continuum limit, these updates become a stochastic differential equation for the system's wave function $|\psi_t\rangle$. Suppose we discretize time into steps δt. Over each short interval $[t, t + \delta t]$ we have an unconditional evolution due to H and damping κ and a measurement of the (outgoing) field quadrature once in that interval, giving some outcome X_{meas}. We want to see how the system's conditional state $\rho_c(t)$ changes due to both (i) the unitary + Lindblad part and (ii) the measurement result. In the pure-state 'quantum trajectory' picture, we track $|\psi_c(t)\rangle$ instead of $\rho_c(t)$. First, ignoring the measurement, the system evolves from $|\psi(t)\rangle$ to

$$|\psi'(t + \delta t)\rangle = \left[I - \frac{i}{\hbar} H \delta t - \frac{\kappa}{2} a^\dagger a \delta t \right] |\psi(t)\rangle \tag{11.56}$$

to the first order in δt. (This is the non-jump 'no-photon' evolution part, or more precisely, the non-collapse part of the unraveling. In a more formal treatment, you might see the Kraus-operator expansions, etc.) During the same interval δt, the leaking photon field is measured by the homodyne detector. For the balanced homodyne detection of the amplitude quadrature, the measurement outcome $I(t)dt$ (the photocurrent) is proportional to

$$I(t)dt = \langle X_{\text{out}} \rangle_t \delta t + \sqrt{\frac{1}{2\delta t}} \xi(t), \tag{11.57}$$

where $\xi(t)$ is (in the $\delta t \to 0$ limit) white noise. In the discrete model, ξ would be Gaussian-distributed with mean zero and variance $\sim 1/\delta t$. We now want to make a Bayesian update (or 'collapse') associated with the observed quadrature. Furthermore, we must perform normalization to ensure $\langle \psi_c | \psi_c \rangle = 1$. Over the interval $[t, t + \delta t]$, we combine the short 'free' evolution (including damping), and the projector/measurement associated with the particular quadrature outcome, and then renormalize. Note that most of the detected photons come from the local oscillator, but the difference current reveals the squeezing. By doing this carefully, these updates turn into an Itô stochastic differential equation for $|\psi_c(t)\rangle$. n the continuum limit, we denote the infinitesimal Wiener increment by $dW(t)$, satisfying $\langle dW(t) \rangle = 0$ and $(dW)^2 = dt$. The homodyne photocurrent can be written as

$$I(t) = \sqrt{\kappa}\,\langle X \rangle_t + \frac{1}{\sqrt{2dt}}\xi(t) \tag{11.58}$$

or equivalently, the conditional update has a noise term dW. One finds a standard result for the conditional state vector:

$$\begin{aligned} d|\psi_c(t)\rangle = &-\frac{i}{\hbar}H|\psi_c(t)\rangle dt - \frac{\kappa}{2}a^\dagger a|\psi_c(t)\rangle dt \\ &+ \sqrt{\kappa}(a - \langle a \rangle_t)|\psi_c(t)\rangle dW(t). \end{aligned} \tag{11.59}$$

Here, $\langle a \rangle_t = \langle \psi_c(t)|a|\psi_c(t)\rangle$ is the instantaneous expectation value (in the conditional state). Let us look a little deeper at the source of this equation.

In the homodyne detection scheme, a photodetector would generate a current proportional to the intensity, as follows:

$$\begin{aligned} i(t) &= \left(rE_{\text{LO}} + tE_{\text{signal}}\right)^2 \\ &= X^2 + 2XtE_{\text{signal}} + t^2E_{\text{signal}}^2, \end{aligned} \tag{11.60}$$

where we have defined $rE_{\text{LO}} \equiv X$. In such a scheme, one does not want to lose much signal, so $t \to 1$, and one would like to have a strong local oscillator such that $X = rE_{\text{LO}} \gg tE_{\text{signal}}$. In such a case, the photodetector will fire almost continuously. Each detection tells us something about the combined local oscillator/signal field, but very little about the signal itself. In fact, in a time increment dt that is small compared to $1/f_{\text{max}}$, where f_{max} is the largest frequency in the signal field that is of interest, there will be a large number of detection events. To utilize the quantum trajectory approach as we have developed it, we would have to go to a time step small compared to the inverse count rate due to the local oscillator, $1/X^2$, which decreases with the intensity of the local oscillator. We do not wish to track every detection anymore; we just examine the dynamics of the system on a time scale relevant to it. Also, the collapse operator is no longer one that takes a photon away from the source; that is not the action of the detector in the homodyne detection scheme. We can address these problems by a new choice for the unraveling of our master equation, with

$$\mathcal{S} = x + \sqrt{2\kappa}\, a_{\text{signal}}, \tag{11.61}$$

where x is the field of the local oscillator multiplied by r in photon flux units,and we set $t = 1$ and used the signal field in photon flux units $E_{\text{signal}} \rightarrow \sqrt{2\kappa}\,a$. This choice considers that collapses do *not* reflect a photon coming from the source. The other problem, that we would have to use a minute time step, can be avoided by a type of coarse graining. We look at time increments over which the system does not change very much, but during which the number of photocounts is large enough to consider the number of counts to be Poisson distributed. This results in

$$\begin{aligned}(\mathcal{L} - S)\rho &= (-i/\hbar)[H_S, \rho] - \kappa a^\dagger a\rho - \kappa\rho a^\dagger a - x^2\rho - x\sqrt{2\kappa}(\rho a^\dagger + a\rho) \\ &= (-i/\hbar)[H_{\text{eff}}, \rho],\end{aligned} \tag{11.62}$$

where we have an effective Hamiltonian

$$H_{\text{eff}} = H_S - i(\sqrt{\kappa}x(a - a^\dagger) + (x + \sqrt{2\kappa}a)(x + \sqrt{2\kappa}a^\dagger)). \tag{11.63}$$

This is then the Hamiltonian used in the no-jump evolution. What about the jumps? The probability of a collapse in time dt is given by

$$P_{\text{coll}} = 2\kappa\left\langle(x^2 + \sqrt{2\kappa}x(a + a^\dagger) + 2\kappa a^\dagger a)\right\rangle dt. \tag{11.64}$$

As we can see, to resolve the jumps, one must have $dt \ll 1/x^2$, but we do not wish to resolve all of these clicks! To determine how to proceed, we return to the nonlinear Schrödinger equation that contains both jumps and jump-free evolution

$$d|\psi\rangle = \frac{-i}{\hbar}(H_S - i\hbar\Gamma C^\dagger C - i\hbar\Gamma\langle C^\dagger C\rangle)|\psi\rangle dt + \left(\frac{C}{\langle C^\dagger C\rangle - 1}\right)|\psi\rangle dN, \tag{11.65}$$

where we must use $C = x + \sqrt{2\kappa}\,a$. We now wish to do coarse graining, and we have no intention of keeping track of all the detector clicks, but how many happen over a short time interval. We begin by considering the number of jumps in a time step dt; we no longer wish to look at a case where the jump happens or not, with small probability but consider a situation where there are many detection events in time interval dt. We start with

$$\overline{dN} = (x^2 + x\sqrt{2\kappa}\langle a - a^\dagger\rangle + 2\kappa\langle a^\dagger a\rangle)dt, \tag{11.66}$$

where we will drop the last term, as we are only interested in keeping the leading term in the expression that involves the source, and take the limit $x \gg \langle a^\dagger a\rangle$. Now, in any time interval dt, we will not always obtain the same number of counts; there will be fluctuations. If the number of counts is large enough, they will be Poisson distributed about the mean. This is not just an appeal to the mean value theorem, it is the strict result of applying the quantum theory of detection, according to the work of Mandel, Kelly, and Kleiner. In this case, we may write down a 'noisy' count operator

$$dN = (x^2 + \sqrt{2\kappa}x(a + a^\dagger))dt + xdW, \tag{11.67}$$

where we have retained the leading noise term and dW is a Weiner process defined by

$$\langle dW \rangle = 0 \tag{11.68}$$

$$\langle dW^2 \rangle = dt. \tag{11.69}$$

In this, we have a state vector that evolves continuously, albeit with a stochastic element. If we preserve the normalization in the equation for $d|\psi\rangle$ we also have a nonlinearity as before. Physically, this does correspond to our state of knowledge of the system *if* we monitored it with a homodyne detector; the unraveling can also be used to calculate whatever we need for this system, although for photon statistics it is much more difficult than using the direct detection unraveling.

Another way to think about this is to use the Kraus operators. The unconditional master equation is

$$\frac{d\rho}{dt} = -\frac{i}{\hbar}[H, \rho] + \kappa\mathcal{D}[a]\rho, \quad \text{where} \quad \mathcal{D}[a]\rho \equiv a\rho a^\dagger - \frac{1}{2}\{a^\dagger a, \rho\}, \tag{11.70}$$

In the balanced homodyne detection, the light leaking out of the cavity is mixed with a strong local oscillator at a 50:50 beam splitter, and the difference photocurrent is measured. Adjusting the local oscillator phase corresponds to measuring a particular quadrature of the outgoing field. First, we ignore the measurements and focus on the short-time propagator from t to $t + \delta t$. To the first order in δt,

$$U_{\text{eff}}(\delta t) \approx \exp\left(-\frac{i}{\hbar}H\delta t\right)\exp\left(-\frac{\kappa}{2}a^\dagger a\delta t\right) \approx \mathbb{I} - \frac{i}{\hbar}H\delta t - \frac{\kappa}{2}a^\dagger a\delta t. \tag{11.71}$$

In a Kraus-operator picture, we find

$$E_0 = \mathbb{I} - \frac{i}{\hbar}H\delta t - \frac{\kappa}{2}a^\dagger a\delta t, \tag{11.72}$$

$$E_1 = \sqrt{\kappa\delta t}\, a. \tag{11.73}$$

For a small δt, applying U_{eff} to a pure state $|\psi(t)\rangle$ gives

$$|\psi'(t + \delta t)\rangle = \left(\mathbb{I} - \frac{i}{\hbar}H\delta t - \frac{\kappa}{2}a^\dagger a\delta t\right)|\psi(t)\rangle, \tag{11.74}$$

up to order δt. The Kraus operator (to first order in δt) can be written as

$$E_I = \left(\mathbb{I} - \frac{i}{\hbar}H\delta t - \frac{\kappa}{2}a^\dagger a\delta t\right) \times \exp\left(+\sqrt{\kappa\delta t}\, Ia^\dagger - \frac{1}{2}\kappa\delta t a^\dagger a\right), \tag{11.75}$$

where the real number I is the measurement outcome (proportional to the integrated photocurrent in δt). This expression can be simplified and expanded in the powers of $\sqrt{\delta t}$ and δt. We now expand E_I to first order in δt and the first order in $\sqrt{\delta t}$. Let us define a small real fluctuation

$$I = \sqrt{\frac{1}{2\delta t}}\xi + \ldots \quad (\xi \sim \mathcal{N}(0, 1)). \tag{11.76}$$

The factor $1/\sqrt{\delta t}$ ensures that $\delta t \to 0$, the distribution of I remains finite and becomes 'white-noise'-like. In typical homodyne settings, the measured current has the form

$$I_{\delta t} \approx \sqrt{\kappa}\langle X \rangle_t + \frac{1}{\sqrt{2\delta t}}\xi, \tag{11.77}$$

where $X = \frac{1}{2}(a + a^\dagger)$ is the real quadrature. Let $|\psi_c(t)\rangle$ be the conditioned state at time t. Applying E_I yields

$$|\psi_c(t + \delta t)\rangle = \frac{E_I|\psi_c(t)\rangle}{\|E_I|\psi_c(t)\rangle\|}. \tag{11.78}$$

Expanding carefully, we find a Hamiltonian part $-\frac{i}{\hbar}H\delta t$, a damping part $-\frac{\kappa}{2}a^\dagger a\delta t$, and a measurement coupling $\sqrt{\kappa\delta t}\, Ia^\dagger$ + h.c.. Then, we must normalize, which requires dividing by the norm, which also must be expanded in δt. After a bit of algebra (and using that ξ is a small random increment with $\langle\xi\rangle = 0$ and $\langle\xi^2\rangle = 1$), we obtain the form

$$|\psi_c(t + \delta t)\rangle \approx \left[\mathbb{I} - \frac{i}{\hbar}H\delta t - \frac{\kappa}{2}a^\dagger a\delta t + \sqrt{\kappa}(a - \langle a\rangle_t)\delta W\right]|\psi_c(t)\rangle, \tag{11.79}$$

where δW is defined via

$$\delta W \equiv \frac{\xi}{\sqrt{2}}\sqrt{\delta t}, \quad (\text{so that}\,(\delta W)^2 = \delta t). \tag{11.80}$$

The subtraction of $\langle a\rangle_t$ in the last term is precisely the piece that enforces proper normalization to the first order in δt. Physically, it is the part that 'centers' the measurement so that the outcome has zero mean noise in the quantum sense. Now, we take the continuous limit $\delta t \to 0$. Define $dW(t) \equiv \delta W$. The increment $\langle dW(t)\rangle = 0$ and $(dW)^2 = dt$. Then

$$|\psi_c(t + dt)\rangle - |\psi_c(t)\rangle = d|\psi_c(t)\rangle, \tag{11.81}$$

becomes

$$d|\psi_c\rangle = -\frac{i}{\hbar}H|\psi_c\rangle dt - \frac{\kappa}{2}a^\dagger a|\psi_c\rangle dt + \sqrt{\kappa}(a - \langle a\rangle_t)|\psi_c\rangle dW(t). \tag{11.82}$$

This is the Itô form of the stochastic Schrödinger equation for homodyne detection (measuring the 'real' quadrature of the output field). The first two terms are exactly the Hamiltonian and damping (no measurement). and the last term is the 'measurement innovation', which depends on the noise increment $dW(t)$. The $\langle a\rangle_t$ subtraction ensures correct normalization in Itô calculus, which we discuss in detail in chapter 13. The photocurrent observed in the balanced homodyne can be written (in the continuum limit) as

$$I(t)dt = \sqrt{\kappa}\langle X\rangle_t dt + dW(t). \tag{11.83}$$

11.5 Two-time averages

How do we use the quantum trajectory approach to calculate the two-time averages needed for the correlation functions? In the density-matrix formalism, we were able to use the quantum regression theorem, where we solved for steady-state density-matrix elements. These were then used as the initial conditions to evolve the two-time averages using the same Liouvillian. How do we proceed here? Let us start with the example of

$$\eta = \langle a^{\dagger}(0)a^{\dagger}(\tau)a(\tau)a(0)\rangle . \tag{11.84}$$

This is the numerator of $g^{(2)}(\tau)$, the denominator being $\langle n\rangle^2 = \langle a^{\dagger}a\rangle^2_{\rm SS}$. In this case, we have a symmetric two-time average, in that we have an equal number of creation and annihilation operators acting at a similar time. We can define a quantum state $|\psi_C\rangle = a|\psi(t)\rangle$, and then write

$$\eta = \langle \psi_C|a^{\dagger}(\tau)a(\tau)|\psi_C\rangle . \tag{11.85}$$

Therefore, our answer is just the inner product of a state, where that state is evolved for a time τ after a collapse. If the system has a steady state, one finds the steady-state wave function, applies the collapse operator and evolves. The further evolution can contain further jumps of course. If there is no steady state, but the system is stationary, the choice of initial time will change the trajectories, but *not* the ensemble average.

For

$$\langle a^{\dagger}(\tau)a(0)\rangle \tag{11.86}$$

we must evolve *two* stochastic wave functions.

$$|A(t)\rangle = U(t)a|\psi(0)\rangle \tag{11.87}$$

$$|B(t)\rangle = aU(t)|\psi(0)\rangle . \tag{11.88}$$

The wave function $|A\rangle$ is formed from a collapse at $t = 0$ and then propagated to $t = \tau$. The wave function $|B\rangle$ is evolved from $t = 0$ to $t = \tau$, and then a collapse was applied. An ensemble average of the inner product of these two leaves us with the desired result

$$\langle a^{\dagger}(\tau)a(0)\rangle = (\langle \psi(0)|a^{\dagger}U^{\dagger}\tau)a(0)U(\tau)|\psi(0)\rangle . \tag{11.89}$$

11.6 Quantum jumps

Cook and Kimble [10] proposed a method to observe quantum jumps directly. In a three-level system, such as a V-configuration (figure 11.18), the transitions between the states are $|2\rangle \rightarrow |1\rangle$ which is strongly driven, and a weak transition.

The strong driving field applied to the transition between levels $|2\rangle$ and $|1\rangle$ causes a Rabi oscillation between these two states. This means that the atom will oscillate

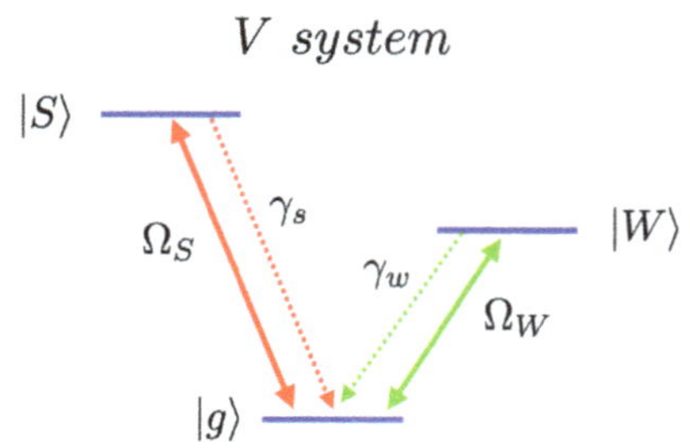

Figure 11.18. Three-level atom structure to see quantum jumps.

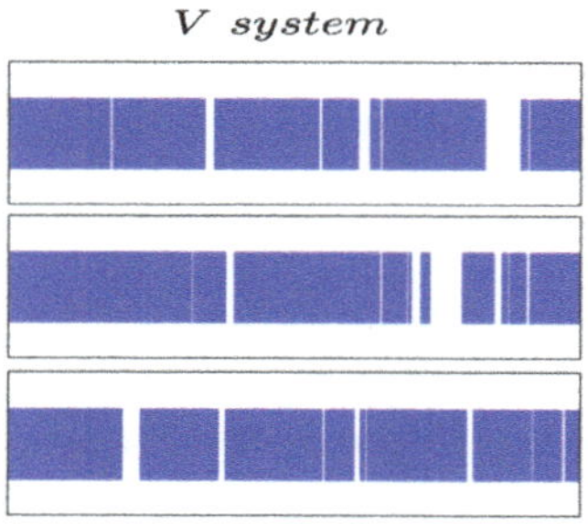

Figure 11.19. A plot of the detection clicks on the strongly driven transition, showing intermittent fluorescence.

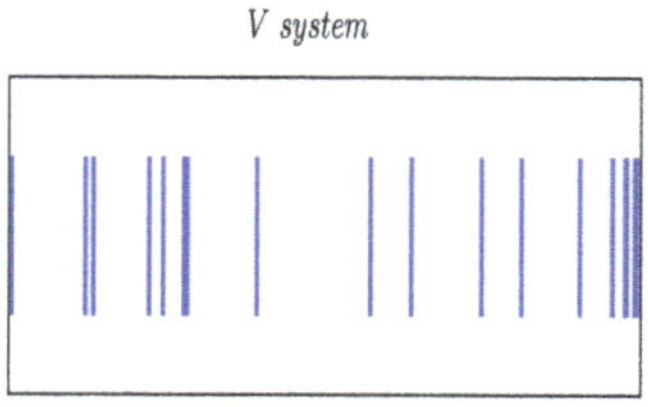

Figure 11.20. A blowup of a bright section, showing individual clicks.

between the states $|2\rangle$ and $|1\rangle$ in time depending on the strength of the laser field. A weak driving field, typically associated with a weak coupling or decay process, drives the transition between states $|2\rangle$ and $|3\rangle$. This coupling might cause the atom to go to state $|3\rangle$ with a lower probability. With a weak driving field, there is occasionally a $|2\rangle \rightarrow |3\rangle$ transition. This puts the atom in $|3\rangle$, and the atom would stay in that state until it decays back. This will interrupt the fluorescence observed due to the Rabi flopping between $|1\rangle$ and $|2\rangle$. Figure 11.19 shows a plot of the telegraphic fluorescence. The blue lines denote an emission event on the strong transition, and it shows obvious gaps (figure 11.20). During these gaps, the excitation is 'shelved' in the upper state of the weak transition. This can be seen in the quantum trajectory simulation.

For a ladder-style system, the separation of timescales is not very different (figures 11.21 and 11.22).

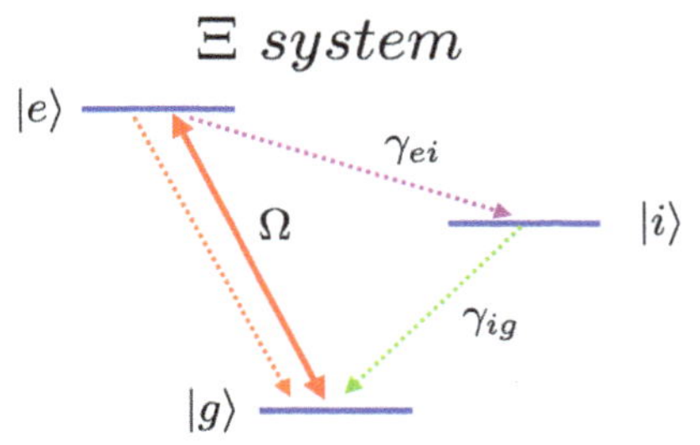

Figure 11.21. Three-level ladder system for intermittent fluorescence.

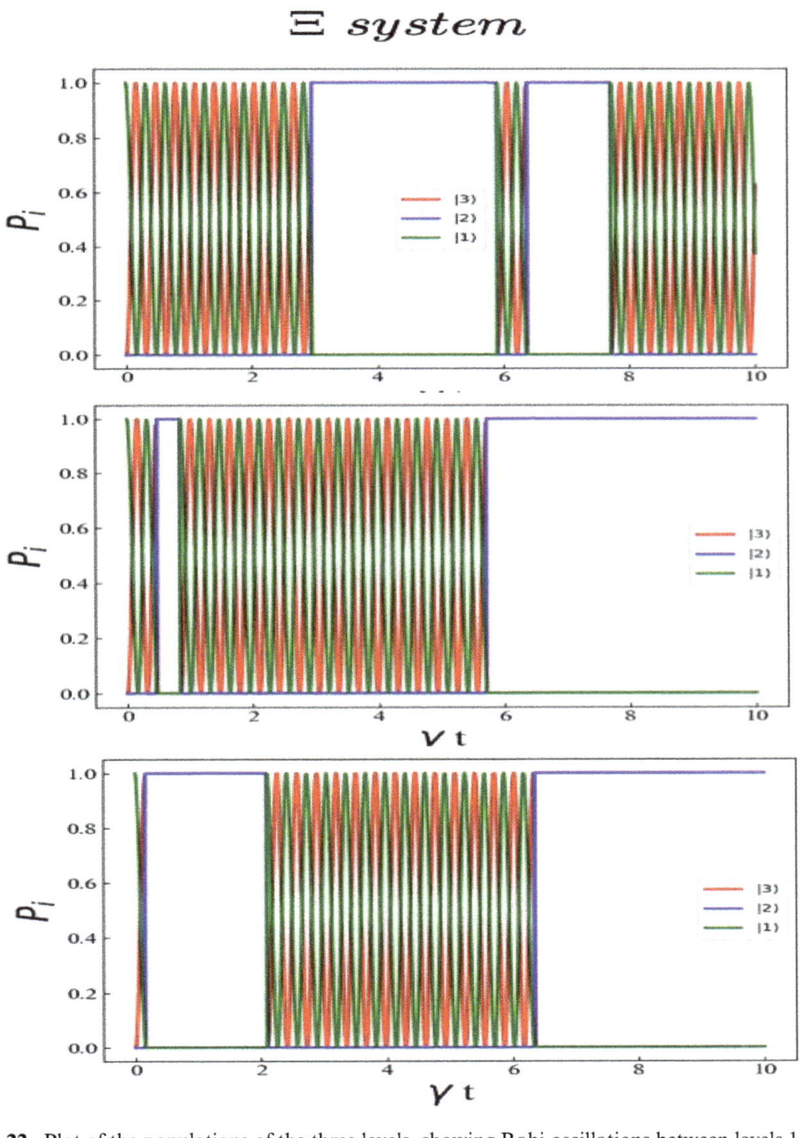

Figure 11.22. Plot of the populations of the three levels, showing Rabi oscillations between levels 1 and 3, and dark periods where the atom is shelved in level 2.

The driven transition has a very weak decay rate, hence a small induced dipole moment. Therefore, to obtain a large Rabi frequency requires a large amplitude laser. Occasionally, the excited state decays to an intermediate state, which then cascades down to the ground state.

11.7 Optical bistability

Consider a ring cavity with an injected field and intracavity atom(s) acting as a saturable absorber [11],

$$\varepsilon_{n+1}(0) = \sqrt{T}E_I + Re^{-\alpha l}e^{iKL}\varepsilon_n(0), \tag{11.90}$$

with boundary condition $\varepsilon_{n+1}(0) = \varepsilon_n(0) = \varepsilon_0$. Applying the boundary condition results in

$$\varepsilon_0 = \sqrt{T}E_\mathrm{I} + Re^{-\alpha l}e^{iKL}\varepsilon_0. \tag{11.91}$$

The transmitted field is then

$$E_T = \sqrt{T}\varepsilon(l) = \sqrt{T}\varepsilon_o e^{(-\alpha+iK)l}. \tag{11.92}$$

Expressing this in terms of the input field, we find that

$$\frac{E_T}{E_\mathrm{I}} = \frac{Te^{iK(l-L)}}{e^{\alpha l - iKL} - R}, \tag{11.93}$$

where we take the atomic response as a saturable absorber

$$\alpha = \frac{\alpha_o}{1 + I/I_s}. \tag{11.94}$$

In terms of this, we have

$$\frac{E_T}{E_I} = \frac{1}{1 + \alpha l/T} \tag{11.95}$$

if $\alpha L \ll 1$, we find

$$E_I = E_T\left[1 + \frac{\alpha_o l/T}{1 + I_T/I_s T}\right] \tag{11.96}$$

with

$$I = \frac{I_T}{T}. \tag{11.97}$$

The ratio of losses—atomic losses per round trip (αl) to cavity losses per round trip (T)—is the cooperativity

$$C = \frac{\alpha_o l}{T} = \frac{\sigma_0 \rho l}{T} = \frac{\sigma_0 N}{\mathrm{Area}_{\mathrm{mode}}}\frac{1}{T}$$

and we define $y = E_I/\sqrt{I_s}$ and $x = E_T/\sqrt{TI_s}$ leading to

$$y = x\left[1 + \frac{2C}{1 + x^2}\right]. \tag{11.98}$$

When we add detunings, the result is

$$y = x\left(1 + \frac{2C}{1 + \Delta^2 + |x|^2}\right) + ix\left(\theta - \frac{2C\Delta}{1 + \Delta^2 + |x|^2}\right) \tag{11.99}$$

with

$$\Theta = \frac{\omega_c - \omega_l}{\kappa} \tag{11.100}$$

$$\Delta = \frac{\omega_a - \omega_l}{\gamma/2}. \tag{11.101}$$

These results also follow from the Maxwell–Bloch equations of chapter 4. At intermediate intensities, there can be saturation (denominator of $1 + X$) when $I = I_s$, occurs in this simple model for the case of C > 4. The input–output curve shows hysteresis for $C > 4$ (figure 11.23), where the output jumps upward (to the upper branch) as the drive is increased when the atom(s) saturate, but then on decreasing the drive the system stays on the upper branch. The switchback of the S-shaped curve is unstable. This is known as optical bistability. Quantum mechanics is linear, so there cannot be hysteresis in the plot of an expectation value. What survives this behavior at the single-atom level? What survives in the upper and lower branches? We examine a single-atom system with parameters chosen such that there is optical bistability in the semiclassical model. For low drives, the $\langle a + a^{\dagger}\rangle$ is mainly a low value with occasional excursions to a larger value in the first two panels

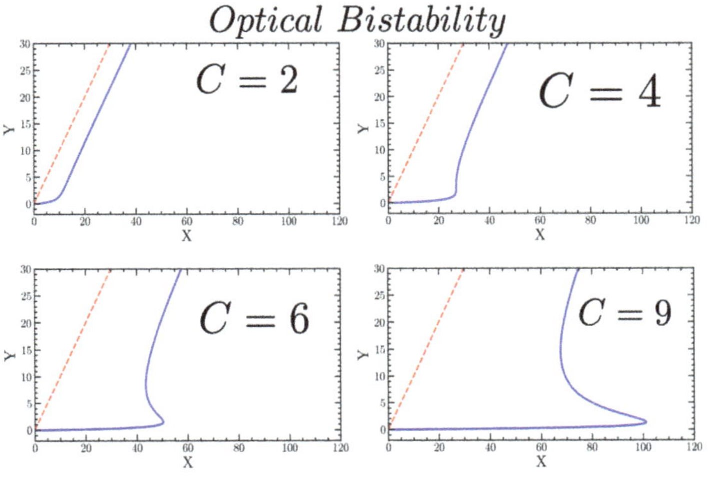

Figure 11.23. Semiclassical optical bistability with hysteresis for $C > 4$. The dashed line is $Y = X$.

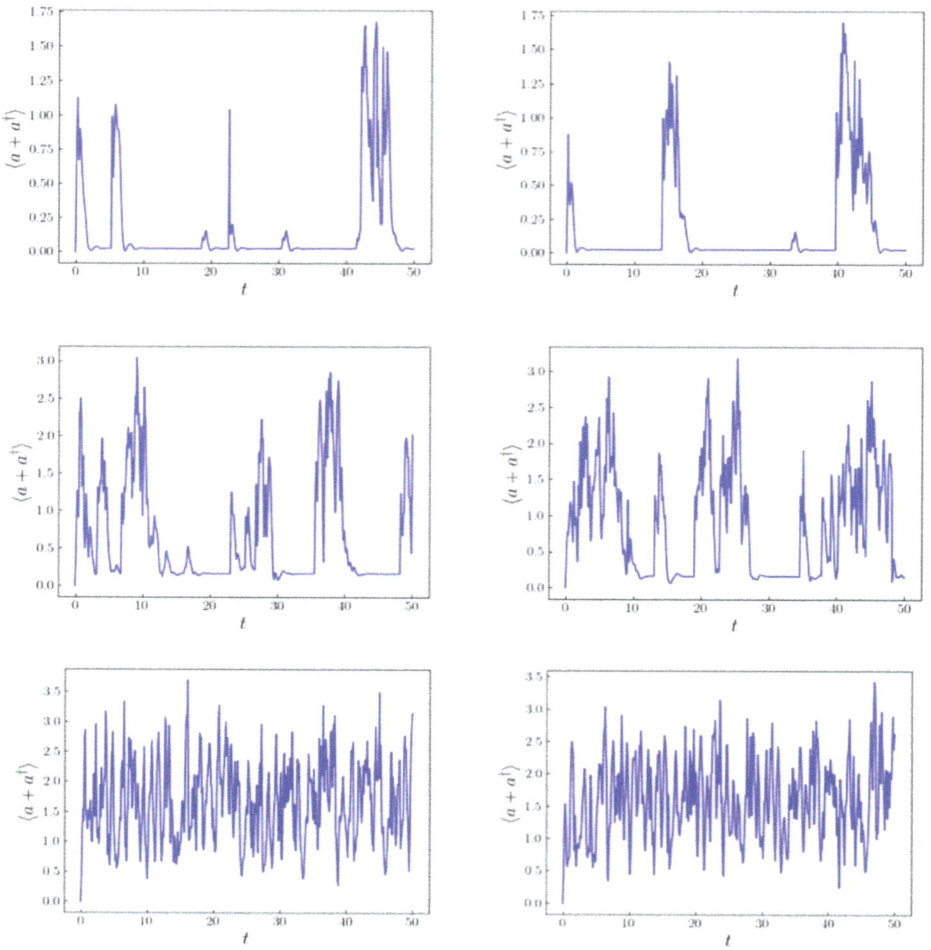

Figure 11.24. Stochastic switching.

of figure 11.24. At intermediate drives, the system stochastically switches between the remnants of the upper and lower branches, as shown in the middle two panels of figure 11.24. At higher drives, the system oscillates with large fluctuations about the remnants of the upper branch.

11.8 Cavity quantum electrodynamics

This section has been reproduced with permission from [12].

The single-atom system (figure 11.25) can show a rich variety of behaviors, such as squeezing and antibunching. Some of this was seen in chapter 10, where we considered the system in the bad cavity limit $\kappa \gg \gamma$, g. Here, we will examine the weak driving limit, where only one or two excitations are present [12].

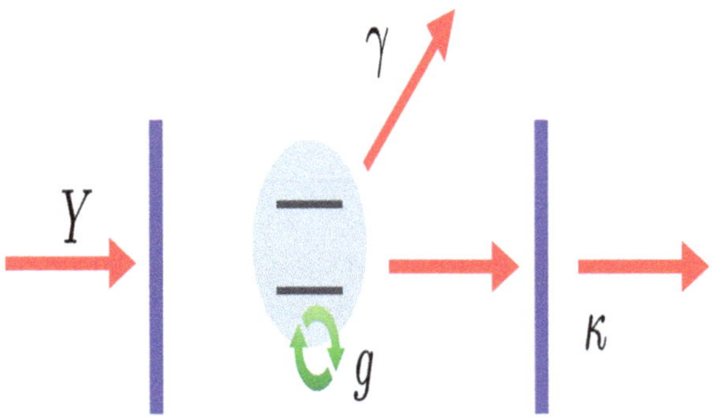

Figure 11.25. Single atom cavity quantum electrodynamic set-up. Y is the classical drive, κ is the cavity decay rate, γ is the spontaneous emission rate, and g is the atom field coupling constant.

This regime should exhibit strong quantum fluctuations. The quantum trajectory wave function and non-Hermitian Hamiltonian are

$$|\Psi_c(t)\rangle = \sum_n^{\infty}(C_{g,n}(t)e^{-iE_{g,n}t}|g, n\rangle + C_{e,n}(t)e^{-iE_{e,n}t}|e, n\rangle)$$

$$H = \hbar g(a^\dagger J_- + aJ_+) - iKa^\dagger a - i\frac{Y}{2}\sigma_+\sigma_- + i\hbar\varepsilon(a^\dagger - a)$$

with the collapse operators

$$C_{\text{cav}} = \sqrt{2\kappa}\, a$$
$$C^i_{\text{atom}} = \sqrt{\frac{\gamma}{2}}\sigma^i_-,$$

associated with photons exiting the output mirror and spontaneous emission out the side of the cavity. The latter could be collective as well; here we focus on the transmitted light. As a reminder, the indices $e(g)$ indicate the atom in the excited (ground) state, while n is the number of photons in the mode. We are particularly interested in $g^{(2)}(\tau)$ in particular, so we will need to include up to two excitations in our truncated basis. We will then use

$$|g0\rangle, |e0\rangle, |g1\rangle, |e1\rangle, |2e\rangle, |g2\rangle.$$

We use the Schrödinger equation to find the equations of motion for the As, Here, we used the scaling $|n \text{ excitations} = m\rangle \sim Y^m$ and dropped higher order terms in Y

$$\dot{A}_1 = -\kappa A_1 + \sqrt{N}gA_2 + Y$$
$$\dot{A}_2 = -(\gamma/2)A_2 - \sqrt{N}gA_1,$$

while A_3, A_4, and A_5 satisfy the equations

$$\dot{A}_3 = -2\kappa A_3 + \sqrt{2}\sqrt{N}gA_4 + \sqrt{2}\, YA_1 \tag{11.102}$$

$$\dot{A}_4 = -(\kappa + \gamma/2)A_4 - \sqrt{2}\sqrt{N}gA_3 \tag{11.103}$$

$$+\sqrt{2}\sqrt{N-1}gA_5 + YA_2 \tag{11.104}$$

$$\dot{A}_5 = -\gamma A_5 - \sqrt{2}\sqrt{N-1}gA_4. \tag{11.105}$$

One can also take the master equation and show that the density operator factorizes in the form $\rho(t) = |\Psi(t)\rangle\langle\Psi(t)|$, to the leading order in Y, with

$$|\Psi\rangle = |0g\rangle + YA_{1,g}|1g\rangle + YA_{0,e}|0e\rangle + Y^2A_{2,g}|2g\rangle + Y^2A_{1,e}|1e\rangle + Y^2A_{0,2e}|0, 2e\rangle.$$

In the weak driving limit, the system reaches a quasi-steady-state wave function. Because the probability of an excited state is small, so is the probability of a jump downward, so the system persists in this state for some time, where in the steady state the A_{ij} are

$$A_{1,g} = \alpha \tag{11.106}$$

$$A_{0,e} = \beta \tag{11.107}$$

$$A_{1,e} = \alpha\beta q \tag{11.108}$$

$$A_{2,g} = \alpha^2 pq/\sqrt{2} \tag{11.109}$$

$$A_{0,2e} = \beta^2 qr/\sqrt{2} \tag{11.110}$$

with

$$\alpha = \frac{\varepsilon}{\kappa(1 + 2C_1)} \tag{11.111}$$

$$\beta = \frac{-2g}{\gamma}\alpha \tag{11.112}$$

$$p = 1 - 2C_1' \tag{11.113}$$

$$q = \frac{(1 + 2C_1)}{\left(1 + 2C_1 - 2C_1'\right)} \tag{11.114}$$

$$C_1 = \frac{g^2}{\kappa\gamma} \tag{11.115}$$

$$C_1' = C_1\frac{2\kappa}{(2\kappa + \gamma)}. \tag{11.116}$$

The quantities p and q would be one for coupled harmonic oscillators. We are primarily interested in $g^{(2)}(\tau)$ here. We note that we can write the numerator as

$$\begin{aligned} G^{(2)}(\tau) &= \langle |a^{\dagger}(0)a^{\dagger}(\tau)a(\tau)a(0)\rangle \\ &= \langle \psi_c|a^{\dagger}(\tau)a(\tau)|\psi_c\rangle, \end{aligned} \tag{11.117}$$

where the state

$$|\psi_c\rangle = a|\Psi\rangle_{ss}$$

is the collapsed state after the detection of a photon in transmission from the cavity. In full, we have

$$|\psi_c\rangle = YA_{1g}|0, g\rangle + Y^2A_{2,g}|1, g\rangle + Y^2A_{1,e}|0, e\rangle.$$

After normalization (to first order in Y), we obtain

$$\begin{aligned} |\psi_c\rangle &= |0, g\rangle + Y\frac{A_{2,g}}{A_{1g}}|1, g\rangle + Y\frac{A_{1,e}}{A_{1g}}|0, e\rangle \\ &= |0, g\rangle + \bar{A}_{1g}(t)|1, g\rangle + \bar{A}_{0e}|0, e\rangle. \end{aligned} \tag{11.118}$$

We then find

$$g^{(2)}(\tau) = |\alpha|^2\left|\bar{A}_{1g}(\tau)\right|^2/|\alpha|^4,$$

where $\bar{A}_{1g}(\tau)$ is the solution of the one excitation amplitude equation with the initial conditions $\bar{A}_{1g}(0) = \beta q$ and $\bar{A}_{0e}(0) = \alpha pq$ determined by the collapsed state. We find

$$\begin{aligned} g(\tau) = \Bigg\{1 &+ (\Delta\alpha/\alpha)\exp\left[-\frac{1}{2}(\kappa + \gamma/2)\tau\right] \\ &\times\left[\cosh\Omega\tau + \frac{1}{2}(\kappa + \gamma/2)\Omega^{-1}\sinh\Omega\tau\right]\Bigg\}^2, \end{aligned} \tag{11.119}$$

where

$$\Delta\alpha/\alpha = -2C_1'\left[2C/\left(1 + 2C - 2C_1'\right)\right]$$

and

$$\Omega = \left[\frac{1}{4}(\kappa - \gamma/2)^2 - Ng^2\right]^{1/2}.$$

Two examples are shown in figure 11.26 which exhibit nonclassical behavior. The first panel starts above 1 and initially decreases, but it exhibits an 'undershoot' which violates

$$|g^{(2)}(\tau) - 1| \leqslant |g^{(2)}(0) - 1|,$$

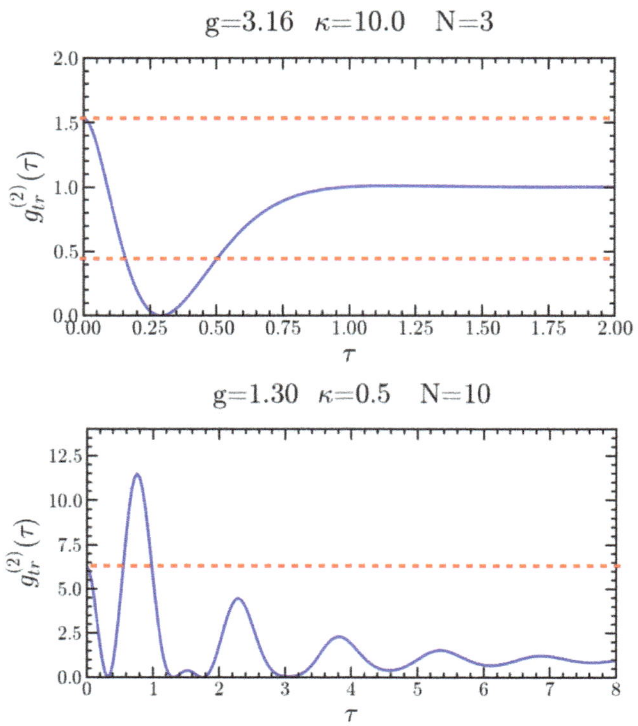

Figure 11.26. $g^{(2)}(\tau)$ in the weak field limit. The red dashed line indicates the classically allowed regions.

which was derived in chapter 9. We have used quantum trajectory theory to find an analytic solution in the weak field limit, and we now use those trajectories to provide some insight into the behavior. The physical explanation of the quantum fluctuations and associated nonclassical correlations illustrates the roles of the interference of the probability amplitudes and the collapse of the wave function in quantum mechanics. We discuss the interference of the probability amplitudes first. For this purpose, we consider the bad cavity limit from back in chapter 10, where we found that there were two contributions to the output field; one two cavity photons leak out, or an existing cavity photon plus one radiated from the atom leak out. For N atoms, we have a third possibility, that two photons emitted by atoms leak out. This picture only holds in the bad cavity limit, where it is likely that photons hitting the output mirror for the first time leak out.

We find that once there is a detection event, the wave function collapses, which changes the cavity field. This is due to a change in the field radiated by the induced atomic dipole. The absolute magnitude of the induced dipole increases by the factor q when the wave function collapses. Before the collapse, we have

$$\begin{aligned}\langle \Psi_{ss}|\sigma_-|\Psi_{ss}\rangle &= YA_{0e}\langle 0, g|\sigma_-|1, g\rangle \\ &= Y\beta.\end{aligned} \tag{11.120}$$

After the collapse, we find

$$\begin{aligned}\langle\psi_c|\sigma_-|\psi_c\rangle &= \frac{YA_{1e}}{A_{1g}}\langle 0, g|\sigma_-|1, g\rangle \\ &= Y\beta q,\end{aligned} \tag{11.121}$$

which gives us the ratio of q for the relative size of the dipole before and after detection.

Since the dipole field is out-of-phase with the driving field, this decreases the amplitude of the cavity field, as shown in figure 11.27. The increase in the amplitude of the induced atomic dipole can be explained in terms of the 'turning off' of one of the atoms after the emission of the first transmitted photon. As the system relaxes back to the steady state (with negligible chance of a second detection beforehand), the intracavity field passes through zero, perhaps multiple times if vacuum Rabi oscillations are present. The magnitude of the induced dipole is not simply rescaled by the ratio $(N-1)/N$ as it would be if one of the atoms had been returned to its

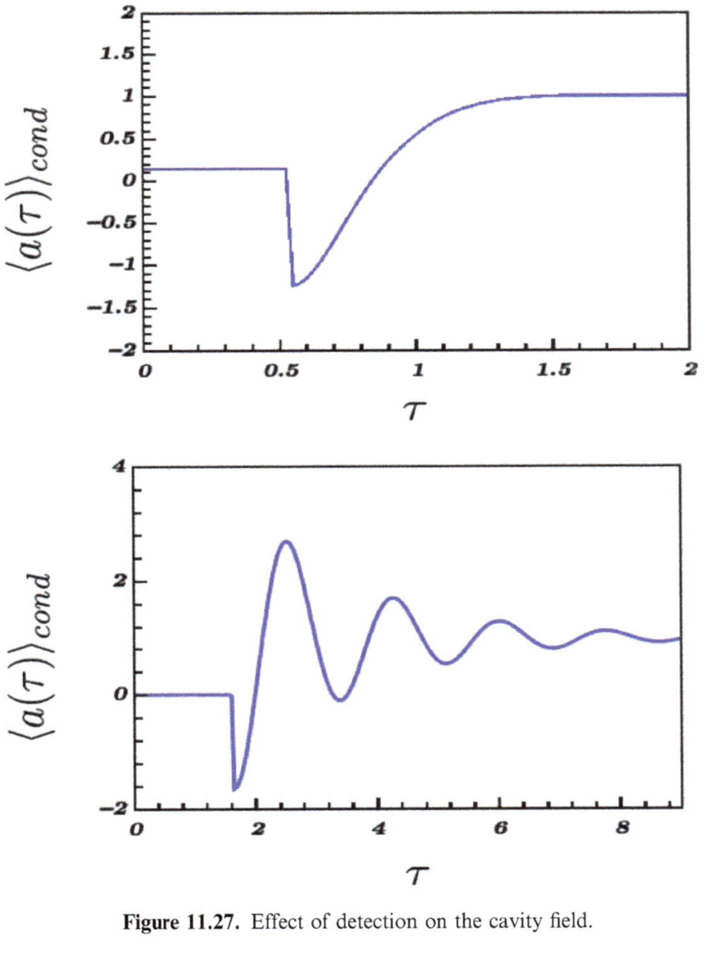

Figure 11.27. Effect of detection on the cavity field.

ground state. The absolute magnitude of the induced dipole increases, while rescaling by $(N-1)N$ results in a decrease. It would be correct to rescale by $(N-1)/N$ if we knew that the collapse was due to the transmission of a photon emitted by one of the atoms. But we do not know this. The photon responsible for the collapse may have traversed the cavity without ever being absorbed. The correct interpretation must be drawn from the rescaling by q. From this, we conclude that after the wave function collapses, the induced dipole field is still radiated by N atoms, but it now has a magnitude determined by the cavity-enhanced emission rate $\gamma(1+2C-2C_1)$ rather than $\gamma(1+2C)$. In effect, the cavity enhancement due to the reaction field radiated by one of the N atoms is turned off.

If N is large, this single-atom effect changes the magnitude of the induced dipole by something negligible ($\sim 1/N$). However, the size of the fluctuations is determined by the ratio $\Delta\alpha/\alpha$, not by the relative change in the magnitude of the dipole, and for large N, $\Delta\alpha/\alpha$ is independent of N. The cavity field amplitude α is formed by the (out-of-phase) interference of the radiated dipole field and the empty cavity excitation Y/κ. The magnitude of the cavity field is therefore much less than the magnitude of the dipole field alone. Thus, when N is large, $\Delta\alpha$ and α are both very small; however, when measured with respect to α, the change $\Delta\alpha$ can be large. Thus, we can have nonclassical behavior for any number of atoms. As we remarked earlier, for the resonance fluorescence from N atoms one has

$$g^{(0)}_{\mathrm{NARF}}(0) = 1 - \frac{1}{N}. \tag{11.122}$$

11.9 Quantum Zeno effect

Does a watched quantum pot boil? [13]. Well, it usually does not, but it may boil immediately. What do I mean? Consider a two-level atom. In quantum mechanics, an unobserved system evolves according to the Schrödinger equation, typically transitioning between different possible states over time. When a measurement is made, the wave function 'collapses' to a specific outcome. If you measure the system repeatedly and rapidly—faster than it would naturally transition to a new state—you effectively reset its wave function each time. This repeated collapse keeps the system in (or very near) its initial state. This is known as the quantum Zeno effect (figure 11.28). Similarly, in the quantum realm, continuous (or extremely frequent) measurement can prevent a system from 'moving' (i.e. transitioning) away from its initial state. In practice, the quantum Zeno effect has implications for fields such as quantum control, quantum computing, and fundamental quantum mechanics experiments. It highlights how measurement—an act that might naively seem passive—actually plays a profound, active role in the behavior of quantum systems.

Suppose the atom is driven by a classical field at frequency ω_0 (on resonance). In the interaction picture, the effective Hamiltonian becomes

$$H_{\mathrm{I}} = \frac{\hbar\Omega}{2}\sigma_x,$$

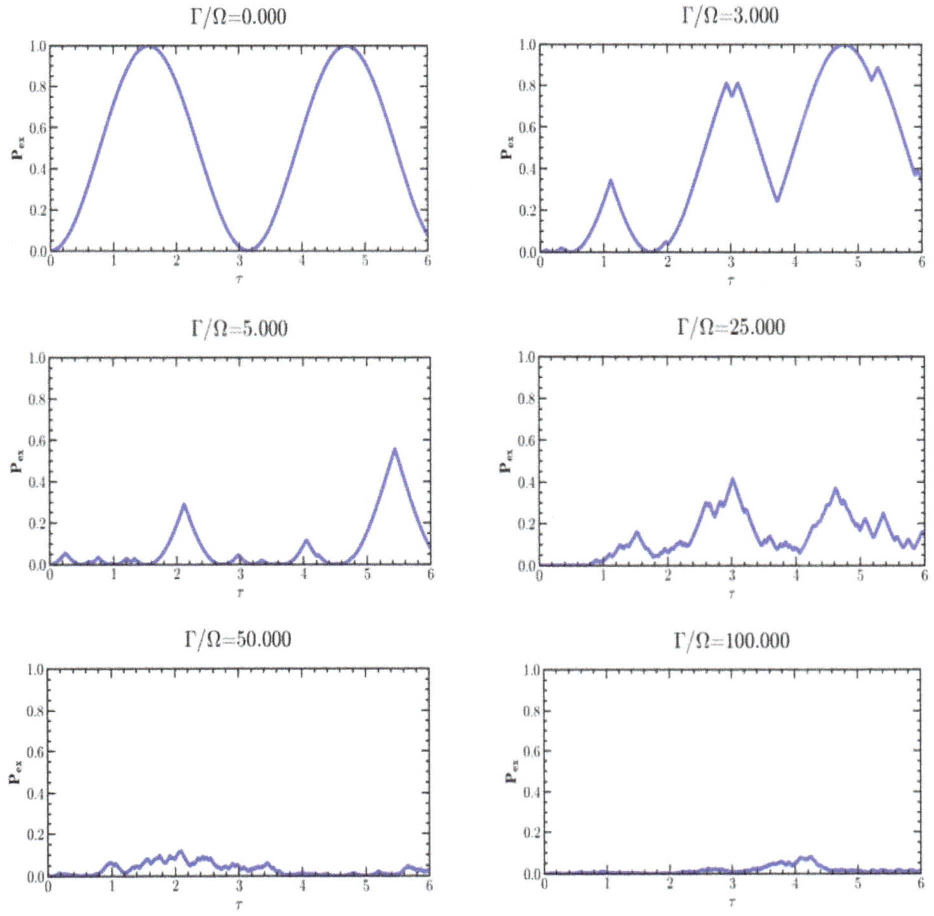

Figure 11.28. Quantum Zeno effect.

where Ω is the Rabi frequency (the strength of the coupling to the driving field), and σ_x is the Pauli X operator. Due to the drive, if you start in $|g\rangle$, there is a nonzero amplitude to transition to $|e\rangle$. The time evolution operator is

$$U(t) = e^{-\frac{i}{\hbar}H_1 t} = e^{-\frac{i}{\hbar}\frac{\hbar\Omega}{2}\sigma_x t} = e^{-\frac{iQt}{2}\sigma_x}.$$

A well-known Pauli-matrix identity gives:

$$e^{-i\theta\sigma_x} = \cos\theta I - i\sin\theta\sigma_x.$$

Hence,

$$U(t) = \cos\left(\frac{\Omega t}{2}\right)I - i\sin\left(\frac{\Omega t}{2}\right)\sigma_x.$$

Assume that the atom is initially in the ground state $|g\rangle$ at $t = 0$. Then

$$|\psi(t)\rangle = U(t)|g\rangle = \cos\left(\frac{\Omega t}{2}\right)|g\rangle - i\sin\left(\frac{\Omega t}{2}\right)\sigma_x|g\rangle.$$

But $\sigma_x|g\rangle = |e\rangle$. So

$$|\psi(t)\rangle = \cos\left(\frac{\Omega t}{2}\right)|g\rangle - i\sin\left(\frac{\Omega t}{2}\right)|e\rangle.$$

The amplitude remaining in $|g\rangle$ after time t is $\cos\left(\frac{\Omega t}{2}\right)$. Therefore, the probability is

$$P_{g\to g}(t) = |\langle g|\psi(t)\rangle|^2 = \cos^2\left(\frac{\Omega t}{2}\right).$$

Thus, if no measurements are performed, the system Rabi-oscillates between $|g\rangle$ and $|e\rangle$. Now, we incorporate the idea of projective measurements. After each short interval δt, you measure in the basis $\{|g\rangle, |e\rangle\}$ to check if the system is still in the ground state $|g\rangle$. Let the total evolution time be T. We slice it into N short intervals, each $\delta t = T/N$. In each short interval, the atom evolves from $|g\rangle$ under H_I for time δt. The probability that it is found in $|g\rangle$ at the end of that interval is

$$p(\delta t) = \cos^2\left(\frac{\Omega\delta t}{2}\right).$$

If the measurement outcome is 'still in $|g\rangle$', the state collapses back to the pure state $|g\rangle$. This 'resets' the next short interval so that evolution again starts from $|g\rangle$. The measurements are projective and repeated N times. Assuming that each time we find 'still in $|g\rangle$' (i.e. we never find it in $|e\rangle$), the overall probability is the product

$$P_\text{survival}(T) = [p(\delta t)]^N = \left[\cos^2\left(\frac{\Omega T}{2N}\right)\right]^N.$$

For large N, $\Omega T/2N$ is negligible. A small-angle expansion gives $\cos(x) \approx 1 - x^2/2$. Thus

$$\cos^2\left(\frac{\Omega T}{2N}\right) \approx 1 - \left(\frac{\Omega T}{2N}\right)^2.$$

Hence

$$\left[\cos^2\left(\frac{\Omega T}{2N}\right)\right]^N \approx \left[1 - \frac{\Omega^2 T^2}{4N^2}\right]^N.$$

However, we may find the atom in the excited state after one small time step, albeit quite rarely.

$$\ln\left[\left(1 - \frac{\Omega^2T^2}{4N^2}\right)^N\right] = N\ln\left(1 - \frac{\Omega^2T^2}{4N^2}\right) \approx N\left(-\frac{\Omega^2T^2}{4N^2}\right) = -\frac{\Omega^2T^2}{4N}.$$

As $N \to \infty$, $1/N \to 0$, we find

$$\lim_{N\to\infty} P_{\text{survival}}(T) = \lim_{N\to\infty} \exp\left(-\frac{\Omega^2 T^2}{4N}\right) = 1$$

and so a continuous measurement would of course leave the atom in the ground state, no matter the drive, but $\Omega\delta t$ is not zero in most situations.

For the measurement, how often is it needed to freeze the state? We can examine this with a quantum trajectory simulation We consider a driven two-level atom

$$H = \Omega(\sigma_+ + \sigma_-)$$

with a jump operator

$$S = \sqrt{\gamma}\sigma_z.$$

This simulates an elastic collision process, where the interrogation is provided by the background gas. The essential part of the measurement is killing the coherence between the two levels.

We find that 'often' is γ around 100 Ω. A different scheme was proposed by Cook [14], where the measurement was performed by excitation out of the ground state to a third level and returning. This is essentially the opposite of the intermittent fluorescence of the three-level system mentioned earlier.

11.10 Catching and reversing a jump

In a brilliant experiment, Minev and coworkers monitored a system, and upon seeing a precursor to a jump, they would reverse it, with an 83% efficiency [15]. Traditionally, these jumps are considered fundamentally random, but the authors ask can quantum jumps be predicted in the short term? The experiment used a superconducting artificial atom (a transmon qubit) inside a cavity. The system was designed to have three relevant energy levels as in the section above. A ground state ($|G\rangle$), a dark state ($|D\rangle$), which is hidden from direct observation, and a bright state ($|B\rangle$), which is continuously monitored. A strong Rabi drive was applied between $|G\rangle$ and $|B\rangle$, and a weaker drive was used between $|G\rangle$ and $|D\rangle$. The population of the Bright state ($|B\rangle$) was continuously monitored using a readout cavity. The authors found that before a quantum jump occurred, there was a 'lull' period, during which the system briefly stopped switching between $|G\rangle$ and $|B\rangle$. This lull acted as a warning signal, indicating an imminent jump to $|D\rangle$. They confirmed that this precursor always occurs before the jump, proving that the jump is not entirely random but follows a predictable pattern. The fact that fluorescence stops before the system fully reaches $|D\rangle$ (while in a superposition) allows for early warning to allow them to apply a feedback pulse to reverse the jump, effectively returning the system to $|G\rangle$ before the transition to $|D\rangle$ completes.

The state of the superconducting artificial atom is represented by a three-level Hilbert space:

$$|\psi(t)\rangle = c_G(t)|G\rangle + c_D(t)|D\rangle + c_B(t)|B\rangle.$$

The evolution of the quantum state under continuous measurement is described by the stochastic master equation:

$$d\rho = -\frac{i}{\hbar}[H_{\text{eff}}, \rho]dt + \sum_i \mathcal{D}[C_i]\rho dt + \sqrt{\eta}\sum_i \Big(C_i\rho + \rho C_i^\dagger - Tr\big[(C_i + C_i^\dagger)\rho\big]\rho\Big)dW_i, \tag{11.123}$$

where H_{eff} is the effective non-Hermitian Hamiltonian,

$$\mathcal{D}[C_i]\rho = C_i\rho C_i^\dagger - \frac{1}{2}\{C_i^\dagger C_i, \rho\} \tag{11.124}$$

which describes Lindblad (dissipative) evolution. dW_i are stochastic Wiener noise terms from measurement backaction and η is the quantum efficiency of the detector, and the C_i remain measurement operators corresponding to the photon emission events. The system oscillates between $|G\rangle$ and $|B\rangle$, producing random clicks from the fluorescence of $|B\rangle$. The absence of clicks (a prolonged lull in fluorescence) signals the approach of a quantum jump to $|D\rangle$. Once the fluorescence stops, the system enters a smooth deterministic evolution toward the superposition of $|G\rangle$ and $|D\rangle$ and then gradually collapses into $|D\rangle$ due to weak driving. The tomography shows that during the transition, the system is in a coherent superposition of $|G\rangle$ and $|D\rangle$. The time taken to reach a 50–50 superposition ($Z_{GD} = 0$) defines the mid-flight time Δt_{mid} and is considered when a control pulse is applied. The measured current for monitoring is

$$I(t)dt = \langle X\rangle dt + \frac{dW}{\sqrt{8\eta k}},$$

where $\langle X\rangle = \text{Tr}(X\rho)$ is the expectation value of the observable monitored, k is the measurement strength (rate at which information is extracted), η is the measurement efficiency ($0 \leqslant \eta \leqslant 1$, with $\eta = 1$ for a perfect measurement), and dW is a Wiener process (white noise with mean zero and variance dt). We have seen this before in our discussion of the stochastic master equation for the balanced homodyne spectra. Here we must include this with the jump dynamics already introduced, and so this term is added to the master equation

11.11 Some final thoughts on the trajectories

We mention that there are alternative algorithms available. One common method is to use the waiting time distribution for a jump

$$W(t) = e^{-\gamma t} \tag{11.125}$$

and then select a random number in the interval [0, 1] and evolve until the norm of the state equals that random number, and then apply a jump. In cases where the waiting time distribution is known, analytic answers are sometimes possible.

Finally, what element of reality, if any, can we assign to these trajectories? The best we can say is that with a complete measurement record, we can *retrodict* what one member of an ensemble would have been doing. Different unraveling/measurement schemes yield different results for that of course. So perhaps we can say that we play dice with the world, but those that are loaded based on our measurement record. An individual system is of course not deterministic.

References

[1] Carmichael H J 1999 *Statistical Methods in Quantum Optics 1: Master Equations and Fokker–Planck Equations* (Berlin: Springer)
[2] Carmichael H J 1993 *An Open Systems Approach to Quantum Optics* (Berlin: Springer)
[3] Carmichael H J 2007 *Statistical Methods in Quantum Optics 2: Non-Classical Fields* Theoretical and Mathematical Physics (Berlin: Springer)
[4] Mølmer K, Castin Y and Dalibard J 1993 Monte Carlo wave-function method in quantum optics *J. Opt. Soc. Am.* B **10** 524
[5] Gisin N and Percival I C 1992 The quantum-state diffusion model applied to open systems *J. Phys. A: Math. Gen.* **25** 5677
[6] Wiseman H 1996 Quantum trajectories and quantum measurement theory *Quantum Semiclass. Opt.: J. Eur. Opt. Soc.* B **8** 205
[7] Jacobs K and Steck D 2006 A straightforward introduction to continuous quantum measurement *Contemp. Phys.* **47** 279
[8] van Dorsselaer F E and Nienhuis G 2000 Quantum trajectories *J. Opt.* B **2** R25
[9] Leach J 2003 Photon statistics and field–intensity correlation of a cavity QED system with external potentials *PhD Thesis* Miami University http://rave.ohiolink.edu/etdc/view?acc_num=miami1058466870
[10] Cook R J and Kimble H J 1985 Possibility of direct observation of quantum jumps *Phys. Rev. Lett.* **54** 1023–6
[11] Grant D E and Kimble H J 1983 Optical bistability for two-level atoms in a standing-wave cavity: errata *Opt. Lett.* **8** 66
[12] Carmichael H J, Brecha R J and Rice P R 1991 Quantum interference and collapse of the wavefunction in cavity QED *Opt. Commun.* **82** 73–9
[13] Degasperis A, Fonda L and Ghirardi G C 1974 Does the lifetime of an unstable system depend on the measuring apparatus? *Nuovo Cimento* A **21** 471–84
[14] Cook R J 1988 What are quantum jumps? *Phys. Scr.* **1988** 49
[15] Minev Z K, Mundhada S O, Shankar S, Reinhold P, Gutiérrez-Jáuregui R, Schoelkopf R J, Mirrahimi M, Carmichael H J and Devoret M H 2019 To catch and reverse a quantum jump mid-flight *Nature* **570** 200–4

IOP Publishing

An Introduction to Quantum Optics (Second Edition)

An open systems approach

Perry Rice

Chapter 12

Quasiprobability distributions

12.1 Glauber–Sudarshan *P* representation

As the coherent states form a basis set, albeit overcomplete, we may write the density matrix as [2, 7]

$$\rho = \int P(\alpha)|\alpha\rangle\,\langle\alpha|d^2\alpha, \tag{12.1}$$

where $P(\alpha)$ is a probability distribution. This means that we can represent the density matrix as a weighted sum over coherent state density matrices. We anticipate where the quantum weirdness is manifest here; the coherent states are essentially classical fields with quantum noise. It would seem that we will not see quantum effects, but we know there are systems that display behavior that can only be described by a quantum field. The rub will be that $P(\alpha)$ cannot have the properties of a classical distribution function, positive definiteness for one. How can we have squeezed fluctuations in the sum of vacuum fluctuations? Only if there is destructive interference, which would be described by the negative values of $P(\alpha)$ for some α. We do require that

$$\int P(\alpha)d^2\alpha = 1. \tag{12.2}$$

How can we use this distribution to calculate the expectation values? Consider

$$\begin{aligned}\langle a^\dagger a\rangle &= \mathrm{Tr}(\rho a^\dagger a)\\ &= \mathrm{Tr}\Big(\int P(\alpha)a^\dagger a|\alpha\rangle\langle\alpha|d^2\alpha\Big)\\ &= \int(\mathrm{Tr}(P(\alpha)|\alpha|^2|\alpha\rangle\langle\alpha|)d^2\alpha)\\ &= \int(P(\alpha)|\alpha|^2\,\mathrm{Tr}(\langle\alpha|\alpha\rangle)d^2\alpha)\\ &= \int P(\alpha)|\alpha|^2 d^2\alpha,\end{aligned} \tag{12.3}$$

doi:10.1088/978-0-7503-6014-2ch12

where we have used $\langle\alpha|\alpha\rangle = 1$, $a|\alpha\rangle = \alpha|\alpha\rangle$, and the cyclic property of the trace. This can be generalized to

$$\langle a^{\dagger n} a^m \rangle = \int P(\alpha)\alpha^{*n}\alpha^m d^2\alpha. \tag{12.4}$$

The averages in this form are referred to as *normal ordering*, where all the annihilation operators are to the left acting on the bra, and the creation operators act to the right on the ket. It is possible to express averages in this form by using the proper commutation relations, for example

$$\langle aa^\dagger \rangle = \langle a^\dagger a \rangle + 1. \tag{12.5}$$

This is only possible because the coherent states are overcomplete. Glauber initially investigated

$$\begin{aligned} \rho &= \frac{1}{\pi^2}\int d^2\alpha \int d^2\beta |\alpha\rangle\langle\alpha|\rho|\beta\rangle\langle\beta| \\ &= \int d^2\alpha \int d^2\beta R(\alpha, \beta)\rho_{\alpha,\beta}. \end{aligned} \tag{12.6}$$

Note that the P distribution only relies on the diagonal matrix elements of ρ in the coherent state basis.

Before turning to the examples, let us consider the normally ordered characteristic function

$$\chi_N = \mathrm{Tr}(\rho e^{i\beta a^\dagger} e^{i\beta * a}). \tag{12.7}$$

One can use this to obtain all normally ordered averages via

$$\langle a^{\dagger n} a^m \rangle = \left. \frac{\partial^{m+n}\chi_N}{\partial^m(i\beta)\partial^n(i\beta^*)} \right|_{\beta=\beta*=0}. \tag{12.8}$$

Hence, there must be some relationship between this characteristic function and $P(\alpha)$; they are in fact Fourier pairs

$$P(\alpha) = \frac{1}{\pi^2}\int \chi_N(\beta, \beta^*) e^{i\beta a^\dagger} e^{i\beta * a} \tag{12.9}$$

and

$$\chi_N(\alpha) = \frac{1}{\pi^2}\int P(\beta, \beta^*) e^{-i\beta a^\dagger} e^{-i\beta * a}. \tag{12.10}$$

We can treat α as a complex number with real and imaginary parts, or α and α^* as two independent variables, and we will use that again later. The easiest state to find the P distribution for is a coherent state; it is merely

$$P_{\alpha_0}(\alpha) = \delta(\alpha - \alpha_0). \tag{12.11}$$

Another simple case is that of a thermal state with

$$\bar{n} = \langle n \rangle = \frac{1}{e^{\hbar\omega/kT} - 1}. \tag{12.12}$$

There are only diagonal elements of the density matrix for a thermal system, and one finds

$$P_{\text{thermal}}(\alpha) = \frac{1}{\pi\bar{n}} e^{-|\alpha|^2/\bar{n}}, \tag{12.13}$$

a simple Gaussian. The density operator for the thermal state can be written as

$$\rho_{\text{thermal}} = \frac{1}{Z} e^{-\beta a^\dagger a\hbar\omega} \tag{12.14}$$

with $Z = \text{Tr}(e^{-\beta a^\dagger a\hbar\omega})$. Inserting a coherent state identity, we find

$$\begin{aligned}\rho_{\text{thermal}} &= \frac{1}{\pi Z}\int d^2\alpha e^{-\beta a^\dagger a\hbar\omega/2}|\alpha\rangle\langle\alpha|e^{-\beta a^\dagger a\hbar\omega/2} \\ &= \frac{1}{\pi Z}\int d^2\alpha e^{-\beta a^\dagger a\hbar\omega/2}|\alpha\rangle\langle\alpha|e^{-|\alpha|^2/\bar{n}}.\end{aligned} \tag{12.15}$$

The P distribution for a Fock state is a highly singular function. In fact, nonclassical states exhibit this type of behavior. In fact, negative values of $P(\alpha)$ are associated with non-classical states. Defining $\Delta X = X - \langle X\rangle$ for the X quadrature, we obtain

$$|\Delta X|^2 = \frac{1}{4}\int P(\alpha)|\Delta\alpha^* + \Delta\alpha|^2 d^2\alpha. \tag{12.16}$$

For this to be below 1/4, $P(\alpha)$ must be negative for some values of α. We will not derive the formula, but we have

$$P_{\text{Fock}}(\alpha) = \left(\frac{e^{|\alpha|^2}}{n!}\frac{\partial^{2n}}{\partial^n\alpha\partial^n\alpha^*}\right)\delta^2(\alpha), \tag{12.17}$$

an nth order derivative of a two-dimensional δ function. These are called generalized functions. As with a δ function, they are not really functions in the mathematical sense, but make sense under an integral where they play the role of a distribution. A physicist, Dirac, invented them, so we reserve the right to be lazy and refer to them as functions, while keeping their true nature in our minds. It makes conversations easier really, just a convention. It is often more convenient (although still not *convenient*) to use this in a polar form

$$P_{\text{Fock}=l}(\alpha) = \frac{1}{2\pi r}\frac{l!}{(2l)!}e^{r^2}\frac{\partial^{2l}}{\partial^{2l}r}. \tag{12.18}$$

We can use these forms for some calculations, for example, the average photon number. Consider the case $|l = 2\rangle$:

$$\langle a^\dagger a\rangle = \frac{1}{2}\int r^2 e^{r^2}\frac{\partial^2}{\partial^2 r}\delta(r). \tag{12.19}$$

Therefore, only two derivatives of a δ function are needed. We use the identity

$$\int f(x)\delta'(x)dx = -f'(0) \tag{12.20}$$

twice, as well as

$$r^2 e^{r^2} = \frac{1}{4}\left(\frac{\partial^2}{\partial^2 r} - 2\right)e^{r^2} \tag{12.21}$$

and find

$$\langle a^\dagger a \rangle_1 = \frac{1}{2}\int r^2 e^{r^2} \frac{\partial^2}{\partial^2 r}\delta(r). \tag{12.22}$$

Using

$$\frac{\partial^2}{\partial^2 r} e^{r^2} = (2 + 4r^2)e^{r^2} \tag{12.23}$$

we find that

$$\langle a^\dagger a \rangle_1 = (\frac{1}{4}(2 + 4r^2 - 2)e^{r^2})|_r = 0. \tag{12.24}$$

Not too bad. How about for $|l = \rangle$?
We use the previous result for

$$\frac{\partial^2}{\partial^2 r} e^{r^2} \tag{12.25}$$

and find

$$\frac{\partial^4}{\partial^4 r} e^{r^2} = (8 + 40r^2 + 16r^4)e^{r^2} \tag{12.26}$$

to yield

$$\langle a^\dagger a \rangle_1 = \frac{1}{4}(8 + 40r^2 + 16r^4 - 4 - rr^2)e^{r^2}\,|_{r=0} = 1. \tag{12.27}$$

Due to the singular nature of the P distribution for interesting states, we will not be presenting any. Most of the work with the P distribution has been done with a linearization of the dynamics, where small nonclassicalities 'ride' along on a Gaussian distribution.

12.2 Wigner distribution

Another quasi-probability of utility is one used for calculating symmetrically ordered operators. We start with the characteristic function [8]

$$\chi_S = \mathrm{tr}(\rho e^{i(\beta a^\dagger + i\beta * a)}). \tag{12.28}$$

This will be useful for calculating the expectation values of symmetrically ordered operators such as

$$(a^{\dagger}a)_S = \frac{1}{2}(aa^{\dagger} + a^{\dagger}a) \tag{12.29}$$

$$(a^{\dagger 2}a)_S = \frac{1}{2}(a^{\dagger 2}a + 2a^{\dagger}aa^{\dagger} + a^{\dagger 2}a). \tag{12.30}$$

The Wigner quasi-probability distribution is the Fourier transform of χ_S,

$$W(\alpha, \alpha^*) = \frac{1}{\pi}\int d^2z\chi_S e^{-iz\alpha}e^{-iz*\alpha*}. \tag{12.31}$$

We can calculate our expectation values via several equivalent methods:

$$\begin{aligned}\langle (a^{\dagger p}a^q)\rangle_S &= \mathrm{tr}(\rho a^{\dagger p}a^q)\\ &= \frac{\partial^{q+p}}{\partial^p(iz^*)\partial^p(iz)}\chi_S|_{(z=z*=0)}\\ &= \int d^2\alpha W(\alpha, \alpha^*)\alpha^{*p}\alpha^q.\end{aligned} \tag{12.32}$$

The relation between the Wigner function and the matrix elements of the density operator is

$$W(x_\alpha, y_\alpha) = \frac{1}{\pi}\int dy\langle X + y|\rho|X - y\rangle e^{-2iYy}, \tag{12.33}$$

where X and Y are our quadrature operators, $x_\alpha = \mathrm{Re}(\alpha)$ and $y_\alpha = \mathrm{Im}(\alpha)$. It was originally written in terms of the position and momentum operators,

$$W(x, p) = \frac{1}{\pi\hbar}\int dy\langle x + y|\rho|x - y\rangle e^{-2ipy}. \tag{12.34}$$

We also have the relation

$$\rho = \frac{1}{\pi}\int d^2\alpha W(\alpha)T(\alpha), \tag{12.35}$$

where the operator $T(\alpha)$ is given by

$$T(\alpha) = \frac{1}{\pi}\int d^2\beta D(\beta)e^{i(z*a^{\dagger} + za)}, \tag{12.36}$$

where $D(\beta)$ is our old friend the displacement operator. The P distribution gives us a weighted sum of the coherent state density matrices. The Wigner function seems more unwieldy, but it has one major utility

$$\int dp \quad W(x, p) = P(x) \tag{12.37}$$

$$\int dx \quad W(x, p) = P(p), \tag{12.38}$$

where we then have the probability densities for x and y. We can also measure the Wigner function using balanced homodyne detection. For a CAT scan in medical situations, one measures the x-ray absorption along a set of radial directions. From knowing the total absorption in all directions, one can reconstruct a 2D image of x-ray absorption via a Radon transform. The same method applies for the Wigner function, roughly. One measures the balanced homodyne signal for all choices of the local oscillator phase. The set of signals is then Radon transformed to yield the Wigner distribution. As in medical applications, there are many new improvements beyond this technique for image reconstruction, but the basic method is the same. This was first demonstrated by Smithey *et al* [6].

What do the Wigner functions look like? Let us start with the Fock states. First, we show some relations between the Wigner function and the P distribution

$$W(\alpha) = \frac{2}{\pi} \int d^2\beta P(\beta) e^{-2|\alpha-\beta|^2}. \tag{12.39}$$

$W(\alpha)$ is a convolution of the P distribution with a Gaussian, which acts as a smoothing filter. Upon integration of the highly singular P_{Fock} with the Gaussian, and again using the identity for integration of a function against the derivative of a delta function, leading to evaluation of the derivative of the function at the location of the delta function, we find

$$W_n(\alpha) = \frac{2}{\pi}(-1)^n L_n(4|\alpha|^2) e^{-2|\alpha|^2}. \tag{12.40}$$

This is easy to see by examining the Rodrigues formula for the Laguerre polynomials, L_n,

$$L_n(x) = \frac{e^x}{n!} \frac{d^n}{dx^n}(x^n e^x). \tag{12.41}$$

In figure 12.1 we show the Wigner distribution for the vacuum state, which is a rather harmless looking Gaussian. It is a coherent state with zero amplitude.

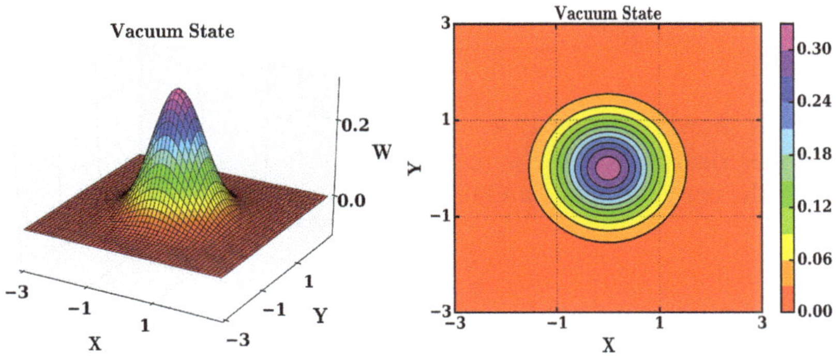

Figure 12.1. Wigner function for the vacuum state.

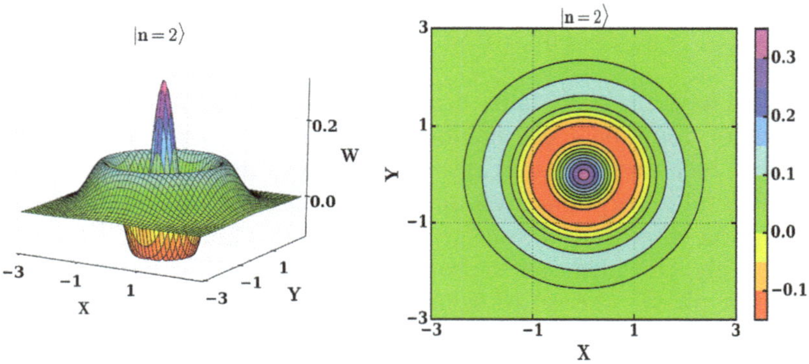

Figure 12.2. Wigner function for $|n = 2\rangle$.

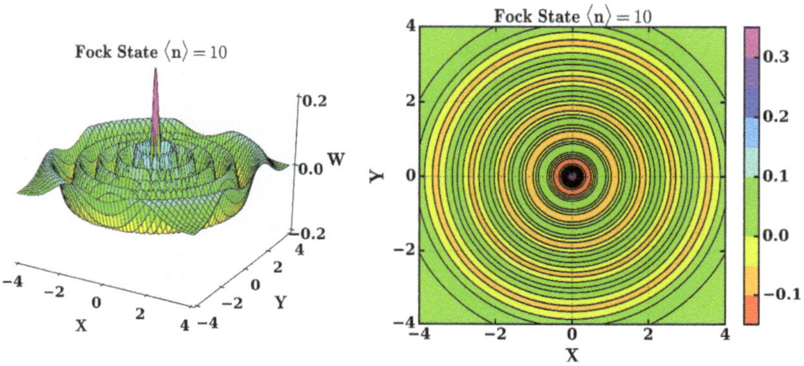

Figure 12.3. Wigner function for $|n = 10\rangle$.

In figure 12.2 we plot the Wigner function for $|n = 2\rangle$. You can see that it takes on negative values around $\alpha = \sqrt{2}$. This negativity indicates a nonclassical state.

Figure 12.3 shows nonclassicality and characteristic ripples of the Wigner function of a Fock state, this being $|n = 10\rangle$.

For a thermal state, the P distribution is a Gaussian, and the convolution with another Gaussian to find the Wigner function of course yields another Gaussian

$$W_{\text{th}} = \frac{1}{\pi(\bar{n} + 1/2)} e^{-|\alpha|^2/(\bar{n}+1/2)}. \tag{12.42}$$

An example is shown in figure 12.4.

The Wigner distribution for a coherent state $|\alpha_0\rangle$ is the easiest to calculate because it is the convolution of a Gaussian with a δ function

$$W_{\text{coh}}(\alpha) = \frac{2}{\pi} e^{-|\alpha-\alpha_0|^2}. \tag{12.43}$$

An example for $\alpha = 2$ is shown in figure 12.5.

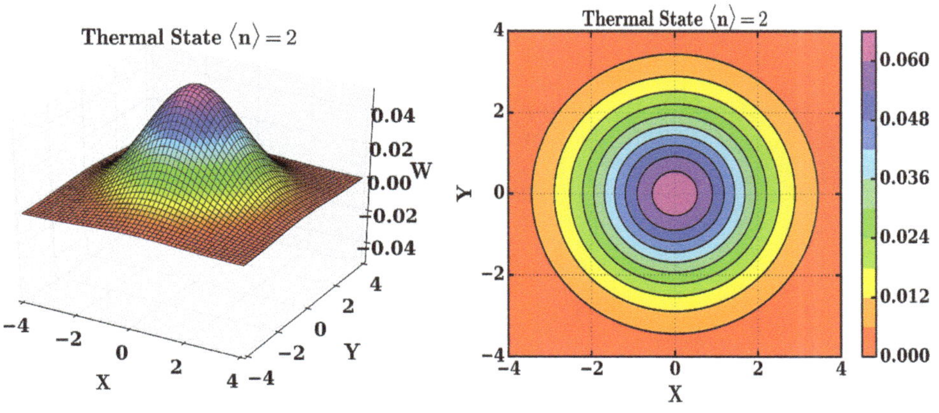

Figure 12.4. Wigner function for a thermal state with $\bar{n} = 10$.

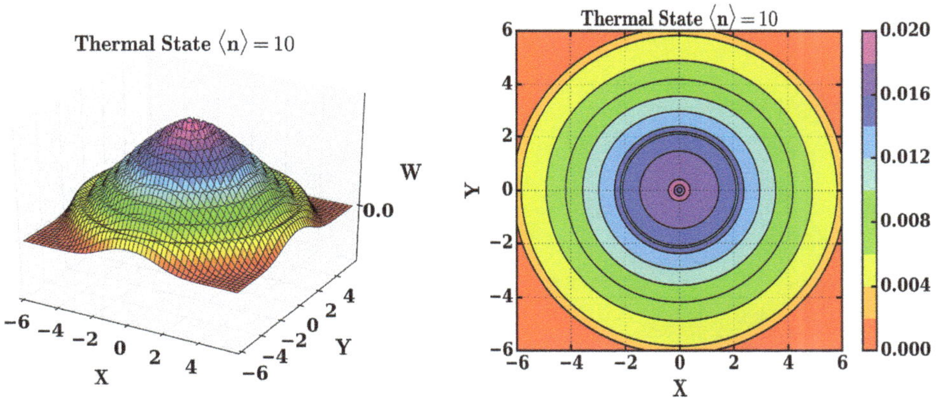

Figure 12.5. Wigner function for a coherent state with $|\alpha|^2 = 2$.

To find the Wigner function for a squeezed vacuum state, it is again easier to obtain it from the Gaussian convolution with the corresponding P distribution

$$W_{\mathrm{sv}}(x, y) = \frac{2}{\pi} e^{-2(e^{2r}x^2 - e^{-2r}p^2)}, \tag{12.44}$$

where the Y quadrature is squeezed. An example for $|r = 10\langle$ is shown in figure 12.6. There are negative values that indicate its nonclassicality, beyond a width for the distribution of x values below 1/4. Squeezing in the X quadrature is shown in figure 12.7.

Increasing the amount of squeezing leads to more prominent ripples and negativity, as shown in figure 12.8.

Finally, we examine the cat states of the form

$$|\mathrm{cat}\rangle = \frac{1}{\sqrt{2}}(|\alpha\rangle + |-\alpha\rangle). \tag{12.45}$$

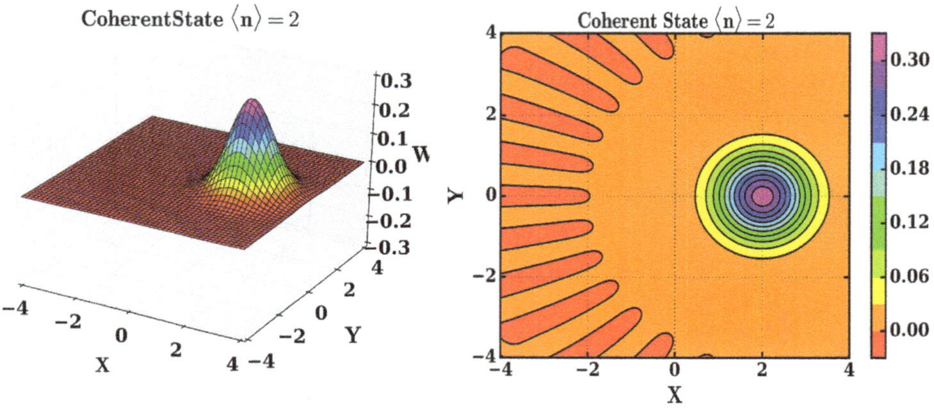

Figure 12.6. Wigner function for a squeezed vacuum state with $|\sinh(r)|^2 = 10$ with squeezing in the Y quadrature.

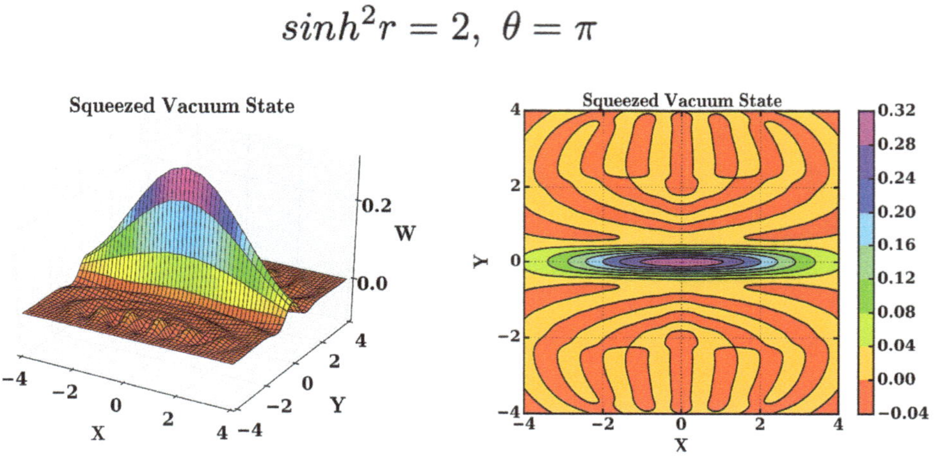

Figure 12.7. Wigner function for a squeezed vacuum state with $|\sinh(r)|^2 = 2$ with squeezing in the X quadrature.

The resulting Wigner function is the sum of two coherent state Wigner functions centered at α_0 and $-\alpha_0$, with an interference term given by

$$W_{\text{int}} = \frac{2}{\pi} e^{-2(e^{2r}x^2 - e^{-2r}p^2)} \cos(xp/4). \tag{12.46}$$

This interference term is sometimes negative and indicates that this sum of two essentially classical states is indeed nonclassical. This is shown for $\alpha = 2$ in figure 12.9, and for $\alpha = 10$ in figure 12.10.

Two other examples with the anti-cat state

$$|\text{acat}\rangle = \frac{1}{\sqrt{2}}(|\alpha\rangle | -\alpha\rangle) \tag{12.47}$$

are shown in figures 12.11 and 12.12.

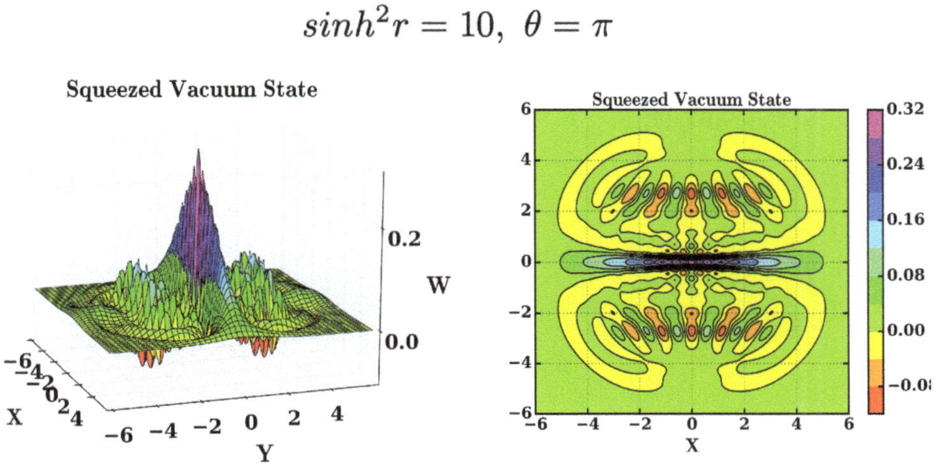

Figure 12.8. Wigner function for a squeezed vacuum state with $|\sinh(r)|^2 = 10$ with squeezing in the X quadrature.

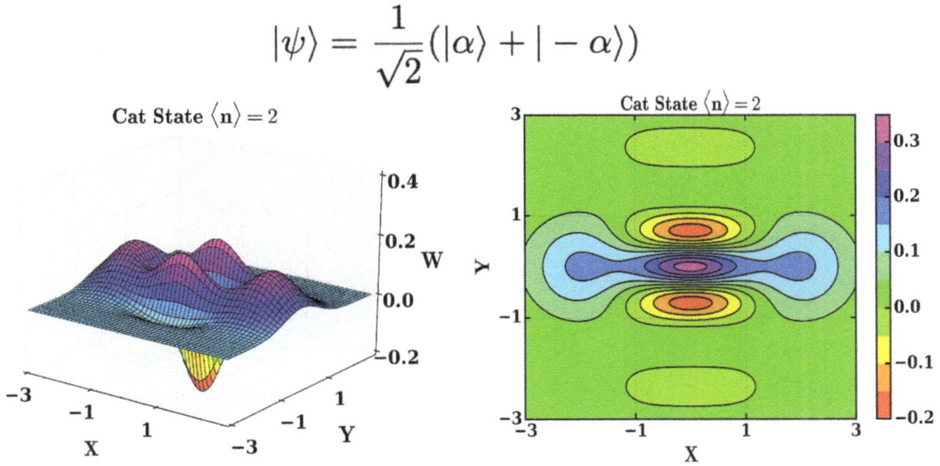

Figure 12.9. Wigner distribution for the cat state with $\alpha = 2$.

12.3 Husimi *Q* function

The Husimi, or *Q*, function [3] is useful for calculating expectation values of antinormally ordered operators such as

$$(a^\dagger a)_A = aa^\dagger. \tag{12.48}$$

The characteristic function for this distribution is

$$\chi_A = \mathrm{tr}(\rho e^{i\alpha * a} e^{i\alpha a^\dagger}), \tag{12.49}$$

$$|\psi\rangle = \frac{1}{\sqrt{2}}(|\alpha\rangle + |-\alpha\rangle)$$

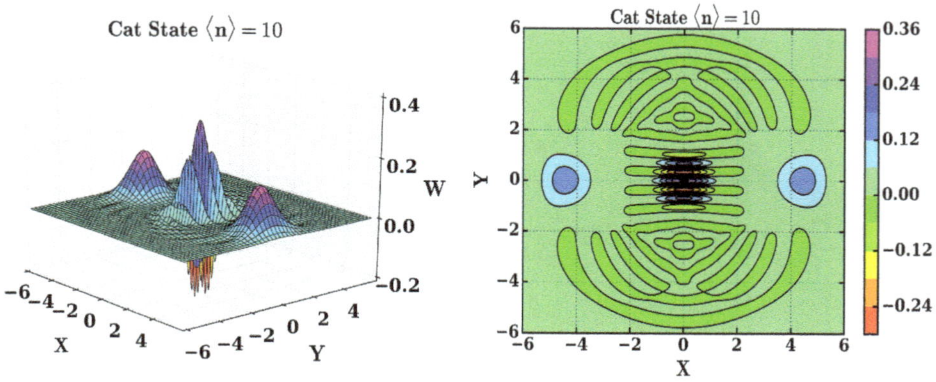

Figure 12.10. Wigner distribution for the cat state with $\alpha = 10$.

$$|\psi\rangle = \frac{1}{\sqrt{2}}(|\alpha\rangle - |-\alpha\rangle)$$

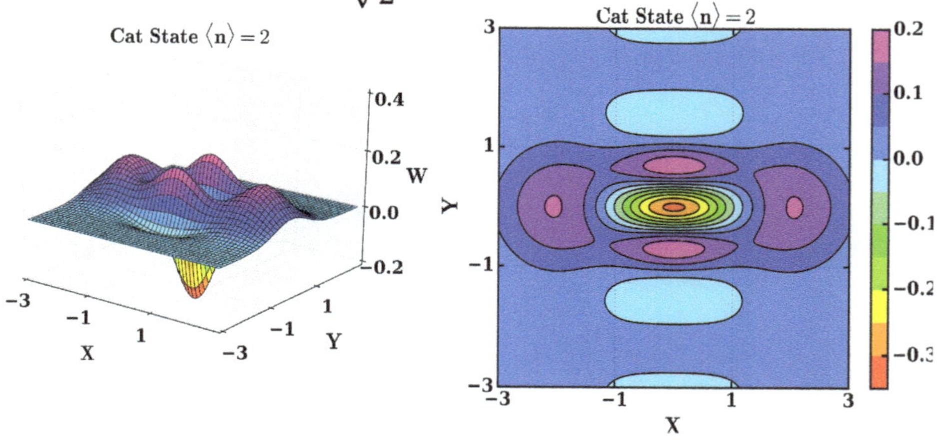

Figure 12.11. Wigner distribution for the anti-cat state with $\alpha = 2$.

where we find

$$\langle a^n a^{\dagger m} \rangle = \frac{\partial^{m+n} \chi_A}{\partial^n (i\beta) \partial^m (i\beta^*)}_{\beta = \beta * = 0}. \tag{12.50}$$

The Q function is given by the diagonal matrix elements of the density matrix in the coherent state basis

$$Q(\alpha) = \langle \alpha | \rho | \alpha \rangle. \tag{12.51}$$

$$|\psi\rangle = \frac{1}{\sqrt{2}}(|\alpha\rangle - |-\alpha\rangle)$$

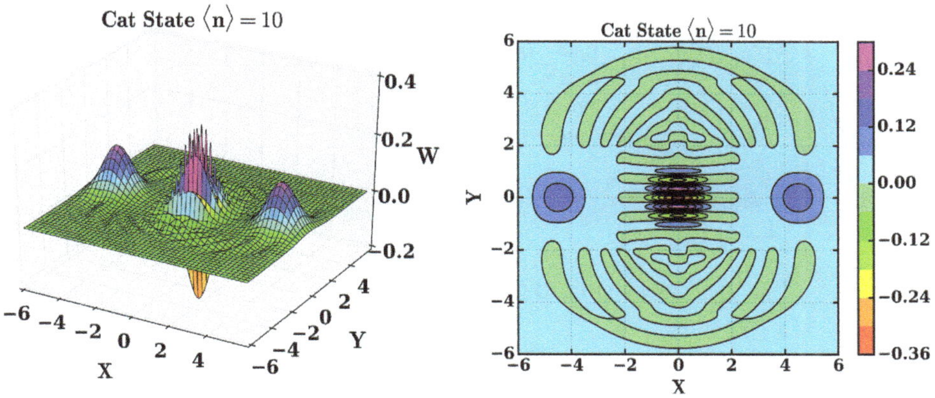

Figure 12.12. Wigner distribution for the anti-cat state with $\alpha = 10$.

As such, the Q distribution is positive definite and can be regarded as a true probability distribution.

We can calculate the antinormally ordered expectation values via

$$\langle a^n a^{\dagger m}\rangle = \int Q(\alpha)\alpha^{*m}\alpha^n d^2\alpha. \tag{12.52}$$

The Q distribution can also be written as a convolution of Gaussians with P or W distributions,

$$\begin{aligned} Q(\alpha) &= \frac{1}{\pi}\int d^2\beta P(\beta)e^{-|\alpha-\beta|^2} \\ &= \frac{2}{\pi}\int d^2\beta W(\beta)e^{-2|\alpha-\beta|^2}. \end{aligned} \tag{12.53}$$

Armed with these, we now look at some examples. For a Fock state, we find

$$Q_n(\alpha) = \frac{-|\alpha|^{2n}}{\pi n!}e^{-|\alpha|^2}. \tag{12.54}$$

We see something interesting for the vacuum state; the Q function has support at $|\alpha| = 1$, as shown in figure 12.13. You would think that for the vacuum, the distribution would peak at $\alpha = 0$. However, recall it is used for the expectation of antinormally ordered operators, so

$$\langle((a^\dagger a)_A)\rangle = \langle aa^\dagger\rangle = 1. \tag{12.55}$$

We see that $Q_1(\alpha)$ has similar support at $\alpha = 2$ as shown in figure 12.14.

For a thermal state, we find

$$Q_{\text{th}} = \frac{1}{\pi(\bar{n}+1)}e^{-|\alpha|^2/(\bar{n}+1)}. \tag{12.56}$$

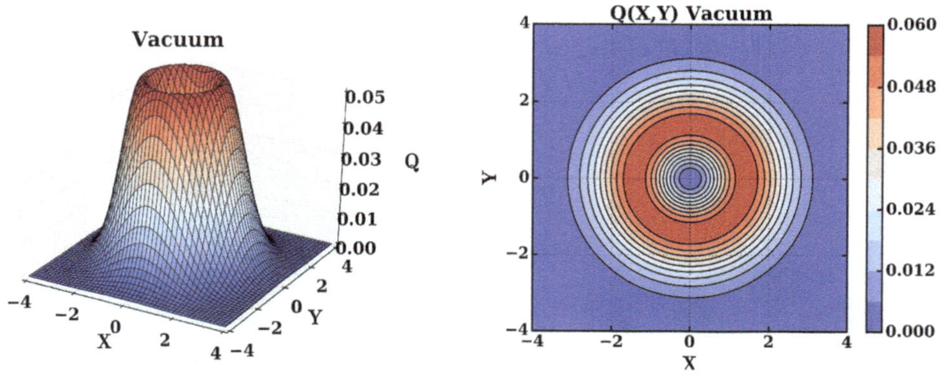

Figure 12.13. Q function for the vacuum state.

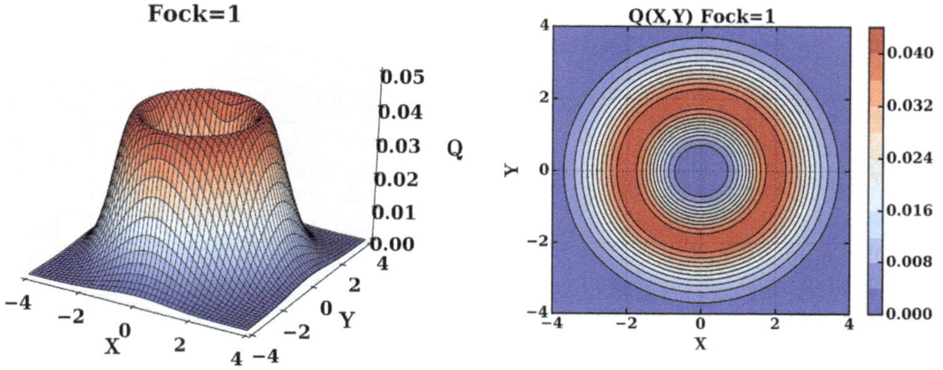

Figure 12.14. Q function for the single-photon state.

We plot an example with $\langle n \rangle = 2$ in figure 12.15.

For a coherent state, we merely obtain a Gaussian of the form

$$Q_{\text{coh}} = e^{-|\alpha|^2}, \tag{12.57}$$

which we exhibit for a coherent state with $\langle n \rangle = 2$ in figure 12.16. It is rather indistinguishable from the thermal state.

For a squeezed vacuum state, we find no pleasant closed form. We show the Q function for the squeezed vacuum states in figure 12.17, and in figure 12.18 for squeezing in the other quadrature. One can indeed notice the decrease in fluctuations in one quadrature at the expense of the other.

For our last exploration of the Q function, we examine the cat states of the form

$$|\text{cat}\rangle = \frac{1}{\sqrt{2}}(|\alpha\rangle + |-\alpha\rangle). \tag{12.58}$$

We see essentially a double-peaked distribution in figure 12.19 that has constructive interference in the center; this is where the nonclassicality is hidden.

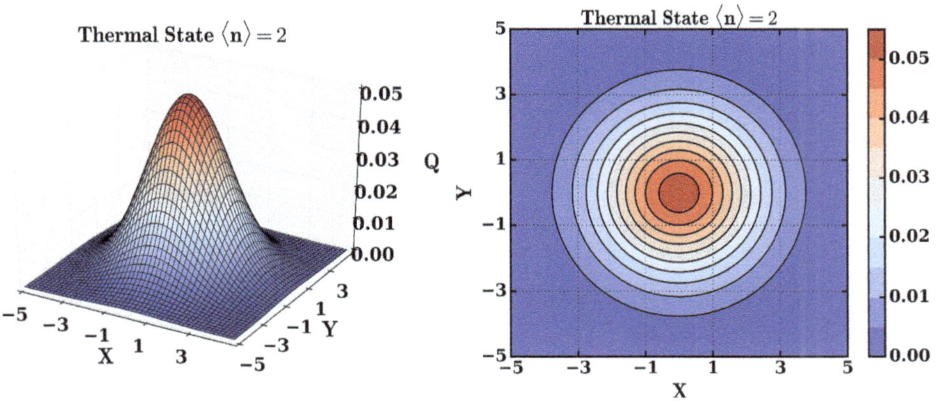

Figure 12.15. Q function for a thermal state with $\langle n \rangle = 2$.

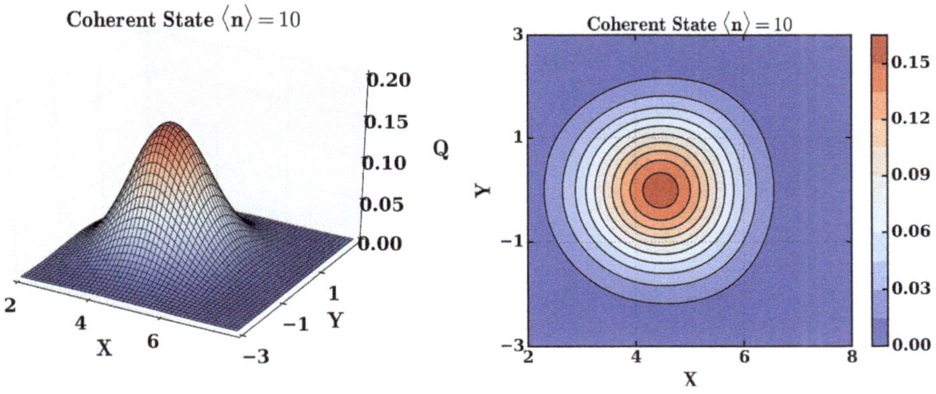

Figure 12.16. Q function for a coherent state with $\langle n \rangle = 2$.

$$sinh^2 r = 10$$

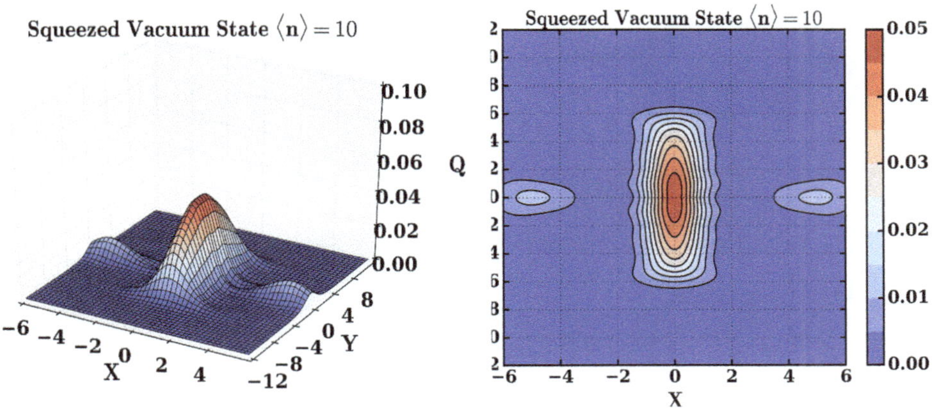

Figure 12.17. Q function for a squeezed vacuum state with $\langle n \rangle = 2$.

$$sinh^2 r = 10, \ \theta = \pi$$

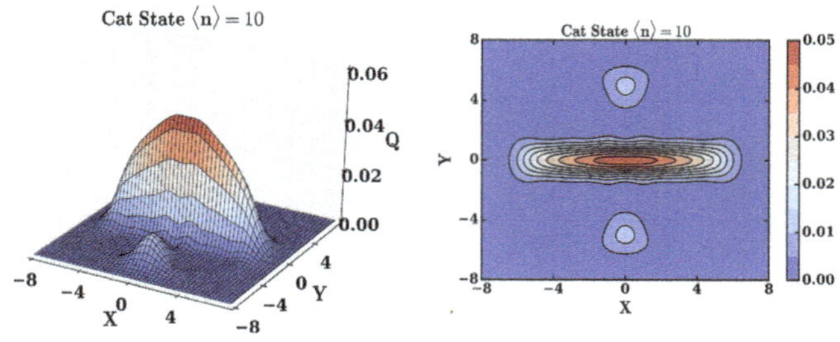

Figure 12.18. Q function for a squeezed vacuum state with $\langle n \rangle = 2$ with squeezing in the other quadrature.

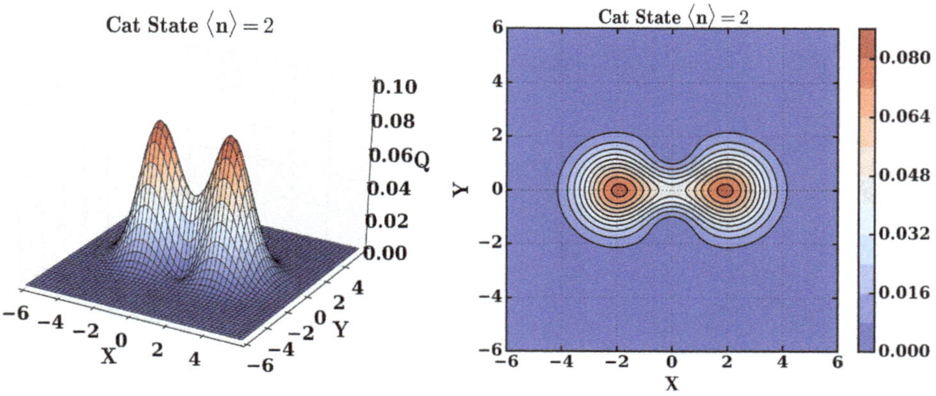

Figure 12.19. Q distribution for the cat state with $\alpha = 2$.

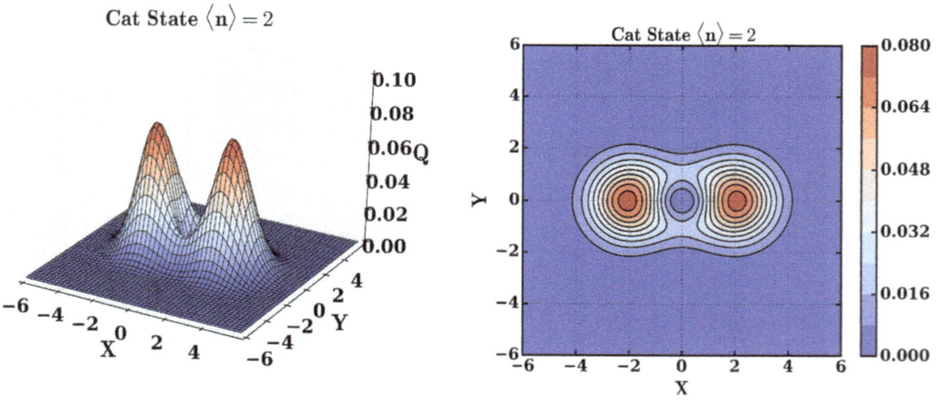

Figure 12.20. Q distribution for the anti-cat state with $\alpha = 2$.

For the anti-cat state

$$|\text{acat}\rangle = \frac{1}{\sqrt{2}}(|\alpha\rangle + |-\alpha\rangle). \tag{12.59}$$

We again see a two-peaked structure but this time with a negative interference clear in the center (figure 12.20).

12.4 Fokker–Planck equations

What about the time evolution of a quasi-probability distribution? We know we have the master equation for the density operator, say for a field mode coupled to a thermal reservoir:

$$\begin{aligned}\dot{\rho}(t) = &- i\omega_0[a^\dagger a, \rho] + \kappa(\bar{n} + 1)[2a\rho(t)a^\dagger - \rho(t)a^\dagger a - a^\dagger a\rho(t)] \\ &+ \kappa\bar{n}[2a\rho(t)a^\dagger - aa^\dagger\rho(t) - \rho(t)aa^\dagger].\end{aligned} \tag{12.60}$$

If we replace ρ with its P representation, we see how the left side of the master equation yields

$$\dot{\rho} = \int d^2\alpha \dot{P}|\alpha\rangle\langle\alpha|. \tag{12.61}$$

What happens to the right-hand side? We know if we have terms with $a|\alpha\rangle = \alpha|\alpha\rangle$ or the conjugate, they are easy to deal with. But what about $a^\dagger|\alpha\rangle$?

$$a^\dagger|\alpha\rangle = a^\dagger e^{-\alpha\alpha^*}e^{\alpha a^\dagger}|0\rangle\langle 0|e^{\alpha * a}. \tag{12.62}$$

We can produce an $a^\dagger$ in front of $|0\rangle$ via a derivative with respect to α, and in fact

$$\frac{\partial}{\partial\alpha}e^{-\alpha\alpha^*}e^{\alpha a^\dagger}|0\rangle\langle 0|e^{\alpha * a} = (a^\dagger - \alpha^*)e^{-\alpha\alpha^*}e^{\alpha a^\dagger}|0\rangle\langle 0|e^{\alpha * a}. \tag{12.63}$$

So we have the following correspondences:

$$a\rho \to \alpha P(\alpha) \tag{12.64}$$

$$\rho a^\dagger \to \alpha^* P(\alpha) \tag{12.65}$$

$$\rho a \to (\alpha - \frac{\partial}{\partial\alpha^*}P(\alpha)) \tag{12.66}$$

$$a^\dagger\rho \to (\alpha^* - \frac{\partial}{\partial\alpha}P(\alpha)). \tag{12.67}$$

Plugging these into the master equation, we find

$$\begin{aligned}\frac{\partial P}{\partial t} = &(\gamma + i\omega_0)\frac{\partial}{\partial\alpha}\alpha P + (\gamma - i\omega_0)\frac{\partial}{\partial\alpha^*}\alpha^* P \\ &+ \frac{\gamma}{2}\frac{\partial^2}{\partial\alpha^*\partial\alpha^*}P.\end{aligned} \tag{12.68}$$

We can perform the same procedure with the W and Q distributions, using

$$a\rho \to (\alpha + \frac{1}{2}\frac{\partial}{\partial\alpha^*})W(\alpha) \tag{12.69}$$

$$\rho a^\dagger \to (\alpha^* + \frac{1}{2}\frac{\partial}{\partial\alpha})W(\alpha) \tag{12.70}$$

$$\rho a \to (\alpha - \frac{1}{2}\frac{\partial}{\partial \alpha^*})W(\alpha) \tag{12.71}$$

$$a^\dagger \rho \to (\alpha^* - \frac{1}{2}\frac{\partial}{\partial \alpha})W(\alpha) \tag{12.72}$$

and

$$a^\dagger \rho \to \alpha^* Q(\alpha) \tag{12.73}$$

$$\rho a \to \alpha Q(\alpha) \tag{12.74}$$

$$a\rho \to (\alpha + \frac{\partial}{\partial \alpha^*}Q(\alpha)) \tag{12.75}$$

$$\rho a^\dagger \to (\alpha^* + \frac{\partial}{\partial \alpha}Q(\alpha)). \tag{12.76}$$

The resulting equations for Q and W time evolution are the same as for P with the replacements

$$\langle n \rangle \to \langle n \rangle + 1/2 \tag{12.77}$$

for W, and

$$\langle n \rangle \to \langle n \rangle + 1 \tag{12.78}$$

for Q. The solution for these with an initial coherent state $|\alpha_0\rangle$ is given by

$$\begin{aligned} P/W/Q = &\frac{1}{\pi}\frac{1}{K + \langle n \rangle(1 - e^{-2\gamma t})} \\ &\times \exp\left(-\frac{|\alpha - \alpha_0(\exp-(\gamma + i\omega_0))t|^2}{K + \langle n \rangle(1 - e^{-2\gamma t})}\right) \end{aligned} \tag{12.79}$$

with $K = 0$ for P, 1/2 for W, and 1 for Q.

The time evolution of a quasi-probability can often take the following form

$$\frac{\partial P}{\partial t} = -(\frac{\partial}{\partial x}A(x)P(x) + \frac{1}{2} - \frac{\partial^2}{\partial^2 x}D(x)P(x)), \tag{12.80}$$

which is of the form of a Fokker–Planck equation (FPE) used in statistical mechanics. See the excellent review by Risken [5]. So it looks classical except for the quantum weirdness hidden in the quasi-probability distribution itself. Often to obtain an equation in this form it is necessary to make a small noise approximation, where we have a classical distribution with correction terms of the order of an inverse system size, say atom number or saturation photon number, to some power. We can solve for the expectation value and variance of x,

$$\langle \dot{x} \rangle = \langle A(x) \rangle \tag{12.81}$$

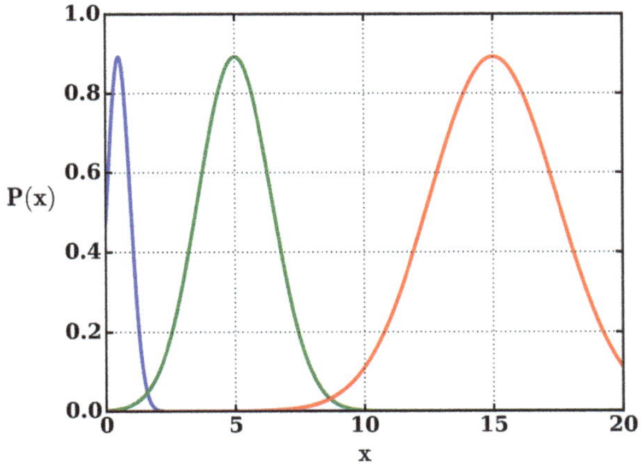

Figure 12.21. Time development of $P(x)$ for $t = 0.\ 2,\ 1,\ 5$.

$$\langle \dot{x}^2 \rangle = 2\langle A(x)x \rangle + \langle D(x) \rangle \tag{12.82}$$

leading to

$$\Delta\dot{x}^2 = 2\langle A(x)x\rangle - s\langle x\rangle \langle A(x)\rangle + \langle D(x)\rangle . \tag{12.83}$$

In the case of a linear FPE,

$$A(x) = Ax \tag{12.84}$$

$$D(x) = D \tag{12.85}$$

the solution is

$$P(x,\ t) = \frac{1}{4\pi Dt}\exp\left(-\frac{(x - At)^2}{4Dt} \right). \tag{12.86}$$

Here we see that the resulting distribution is a Gaussian distribution where the mean is centered at $x = At$, with a width $2Dt$ as in figure 12.21.

So the mean moves at a drift rate A with a diffusion of the width of $2Dt$. Hence A (or set of As in multiple variables) are known as drift coefficients, and the set of D_{ij} are known as diffusion coefficients.

12.5 Two-level atoms

So far, we have only discussed the quasi-probability distributions of bosons, what about the fermions? We can certainly construct a characteristic function

$$\chi_N = \mathrm{tr}(\rho e^{i\mu * \sigma_+} e^{iz\sigma_z} e^{i\mu\sigma_-}). \tag{12.87}$$

Here the N refers to normal ordering, lowering operators to the right and raising operators to the left, just as for bosons.

We can calculate the expectation values via

$$\begin{aligned}\langle\sigma_+^p\sigma_z^q\sigma_-^r\rangle &= \mathrm{tr}\left(\rho\sigma_+^p\sigma_z^q\sigma_-^r\right)\\ &= \frac{\partial^{p+q+r}}{\partial^p(i\mu^*)\partial^q(iz)\partial^r(i\mu)}\chi_N|_{(\mu=\mu*=z=0)}\end{aligned} \tag{12.88}$$

and write down a quasi-probability function

$$P(\mu,\, \mu^*,\, z) = \frac{1}{2\pi^3}\int d^2wd\eta\chi_N(w,\, w^*,\, \eta)e^{-iw\mu-w*\mu*-\eta z}. \tag{12.89}$$

The difficulty arises when we convert this into an FPE, where we have terms in the diffusion part of the form

$$e^{\frac{\partial}{\partial z}}, \tag{12.90}$$

which has an infinite number of derivatives, which brings on enormous difficulties. We can find atomic coherent states that are useful, however. They are states of minimum and symmetric noise. Consider the $|j,\, -j\rangle$ state of a spin. We know that $J_z = -j$ and that $J^2 = j(j+1)$, suppressing the $\hbar$s. Thus, this is a state where

$$\Delta J_z = 0 \tag{12.91}$$

$$\Delta J_x \Delta J_y = \frac{j}{2} \tag{12.92}$$

with $\Delta J_x = \Delta J_y$. The atomic coherent states are just this state with symmetric, and minimum noise rotated to some other location on the Bloch sphere. This is shown schematically in figure 12.22.

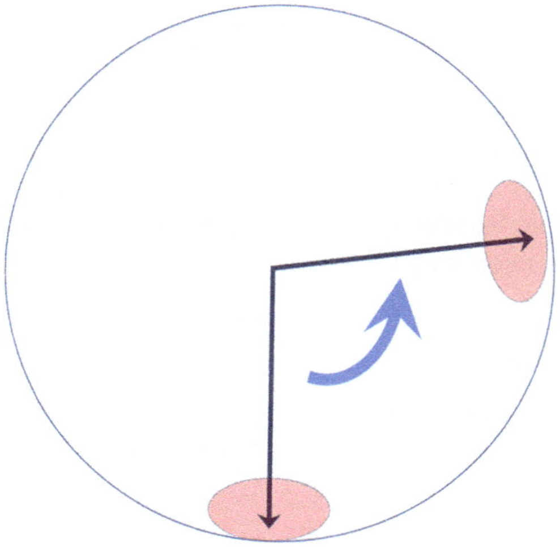

Figure 12.22. Schematic of atomic coherent states.

The phase uncertainty of these states is given by

$$\Delta\phi = \Delta J \langle J_z \rangle \approx 1/\sqrt{N}, \tag{12.93}$$

where N is the number of atoms. We can formalize this by defining [1]

$$|\mathrm{ACS}\rangle = |\theta, \phi\rangle = (\cos(\theta/2)|-\rangle + e^{i\phi}\sin(\theta/2)|+\rangle) \tag{12.94}$$

in the single-atom case. As with the bosonic coherent states, we can generate these states with a type of displacement operator or rotation operator

$$|\theta, \phi\rangle = \exp(i(zJ_+ + z^* J_-))|j, -j\rangle \equiv R(z)|j, -j\rangle, \tag{12.95}$$

where $|j, -j\rangle$ is the state with

$$J_z|j, -j\rangle = -j|j, -j\rangle \tag{12.96}$$

$$J^2|j, -j\rangle = j(j+1)|j, -j\rangle, \tag{12.97}$$

the state of N atoms with the lowest z component of angular momentum. The rotation parameter z is given by

$$z = -\frac{\theta}{2}e^{-i\phi}. \tag{12.98}$$

The atomic squeezed states can be defined above, with minimum noise $\Delta J_x \Delta J_y = \frac{j}{2}$ but with

$$\Delta J_x = e^{-2r}\sqrt{j/2} \tag{12.99}$$

$$\Delta J_y = e^{2r}\sqrt{j/2}. \tag{12.100}$$

A more formal definition of a squeezing parameter, by analogy with bosonic squeezed states, is [4]

$$s = 2(\Delta J_i)^2/\langle J_k\rangle \tag{12.101}$$

with $j \neq k$. Unfortunately, this parameter can be less than unity, implying squeezing, for an atomic coherent state. The analogy with bosonic squeezed states breaks down due to a preferred direction, given by $\langle \vec{J} \rangle$. This stems from the fact that these states live on a Bloch sphere instead of a plane. Wineland defined a useful squeezing parameter for metrology as [9]

$$s_W = N(\Delta J_{\vec{n}\perp})^2/\langle \vec{J} \rangle^2, \tag{12.102}$$

where $\vec{n_\perp} \cdot \langle \vec{J} \rangle = 0$. This always yields the ratio

$$s_W = \frac{\Delta\phi^2}{\Delta\phi^2_{\mathrm{ACS}}}. \tag{12.103}$$

Spin squeezing is shown schematically in figure 12.23.

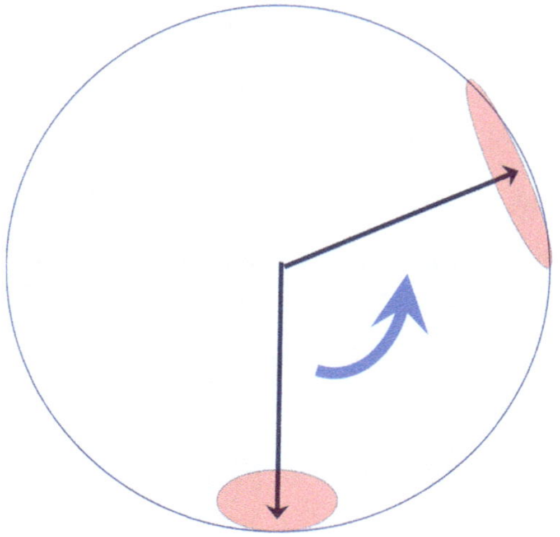

Figure 12.23. Schematic of the spin squeezing.

Due to the interactions of the spin with the magnetic fields, these latter states can be used for quantum enhanced magnetic field measurements.

The states of definite angular momentum are given by

$$|j, m\rangle = \frac{a^{\dagger k} b^{\dagger n}}{k!n!}|0\rangle \tag{12.104}$$

with $j = (k + n)/2$ and $m = (k - n)/2$, which is reminiscent of generating photonic states from the ground state. We have the transformation

$$J_x = (a^\dagger b + b^\dagger a)/2 \tag{12.105}$$

$$J_y = (a^\dagger b - b^\dagger a)/2 \tag{12.106}$$

$$J_z = (a^\dagger a + b^\dagger b)/2. \tag{12.107}$$

An interaction of the form J_x^2 or J_y^2 will contain terms of the form

$$a^{\dagger 2} b^2 + \text{h.c.}, \tag{12.108}$$

which are analogous to the bosonic squeezing operators. These are known as twisting interactions.

12.6 System size expansion

Most systems are not analytically tractable without some approximations. One common approach is to find the average value of system variables and then ask for the behavior of the fluctuations around that average. In this case, the fluctuations are generally assumed to be Gaussian in nature, but this is not always true. Why do we like Gaussians so much? They have one parameter (the standard deviation), and we can do the integrals. For a simple example, suppose the coherent state amplitude α of a cavity mode obeys a classical equation of motion such as

$$\frac{d\alpha}{dt} = -\frac{\gamma}{2}\alpha + E, \tag{12.109}$$

where γ is the damping rate and E is the external driving field. In a quantum description, the state of the field can be described by a Glauber P function, $P(\alpha, \alpha^*, t)$, that obeys an FPE with drift and diffusion terms For example, a simplified FPE might look like

$$\frac{\partial P}{\partial t} = \left\{-\frac{\partial}{\partial \alpha}A(\alpha) - \frac{\partial}{\partial \alpha^*}A^*(\alpha) + \frac{1}{2}\frac{\partial^2}{\partial \alpha \partial \alpha^*}D(\alpha)\right\}P, \tag{12.110}$$

where $A(\alpha)$ represents the drift (incorporating damping and driving) and $D(\alpha)$ the diffusion (accounting for quantum noise). in the regime where the mean number of photons N is large, one writes

$$\alpha = \alpha_0 + \frac{1}{\sqrt{N}}\eta, \tag{12.111}$$

where α_0 satisfies the classical (mean-field) equation

$$\frac{d\alpha_0}{dt} = -\frac{\gamma}{2}\alpha_0 + E \tag{12.112}$$

and η represents the quantum fluctuations around α_0. Substituting $\alpha = \alpha_0 + \frac{1}{\sqrt{N}}\eta$ into the FPE and expanding in powers of $1/\sqrt{N}$. The zeroth order returns the classical dynamics for α_0. To first order in $(1/\sqrt{N})$ the equation for the fluctuations η becomes linear. In this linearized regime, the FPE for the fluctuation variable η takes the form

$$\frac{\partial P(\eta, t)}{\partial t} = -\sum_i \frac{\partial}{\partial \eta_i}[L_{ij}\eta_j P(\eta, t)] + \frac{1}{2}\sum_{ij} D_{ij}\frac{\partial^2 P(\eta, t)}{\partial \eta_i \partial \eta_j}, \tag{12.113}$$

where L_{ij} and D_{ij} are the elements of the linearized drift and diffusion matrices, respectively. The linearized dynamics of the fluctuation η come from expanding the drift $A(\alpha)$ about α_0. We have

$$A(\alpha) = A\left(\alpha_0 + \frac{1}{\sqrt{N}}\eta\right) \approx A(\alpha_0) + \frac{1}{\sqrt{N}}\frac{\partial A}{\partial \alpha}\bigg|_{\alpha_0}\eta. \tag{12.114}$$

Since α_0 already satisfies the classical (zeroth-order) dynamics, the leading correction for η is given by

$$\frac{d\eta}{dt} = L\eta, \tag{12.115}$$

where the drift matrix (or simply drift coefficient in this one-mode case) is

$$L \equiv \left. \frac{\partial A}{\partial \alpha} \right|_{\alpha_0}. \tag{12.116}$$

For our example, since

$$A(\alpha) = -\frac{\gamma}{2}\alpha + E \tag{12.117}$$

we have a derivative

$$\frac{\partial A}{\partial \alpha} = -\frac{\gamma}{2}. \tag{12.118}$$

Thus,

$$L = -\frac{\gamma}{2}. \tag{12.119}$$

If we also consider the complex conjugate variable (i.e. writing equations for η and η^*), the linearization yields the same damping rate:

$$L_{\eta*} = -\frac{\gamma}{2}. \tag{12.120}$$

In our simplified FPE, the diffusion term is given by

$$\frac{1}{2}\frac{\partial^2}{\partial \alpha \partial \alpha^*}[D(\alpha)P]. \tag{12.121}$$

When we perform the system size expansion, the diffusion term is already of the order required to balance the fluctuations. In the linearized FPE for the fluctuation variable η (and its complex conjugate), the diffusion matrix appears in the mixed derivative term. In other words, we write the fluctuation FPE as

$$\frac{\partial P(\eta, t)}{\partial t} = -\frac{\partial}{\partial \eta}[L\eta P] - \frac{\partial}{\partial \eta^*}[L^*\eta^* P] + \frac{1}{2}D\frac{\partial^2 P}{\partial \eta \partial \eta^*}, \tag{12.122}$$

with

$$D = \gamma. \tag{12.123}$$

Thus, the elements of the diffusion 'matrix' (or coefficient in this one-mode example) are given by

$$D_{\eta\eta*} = \gamma. \tag{12.124}$$

When many atoms interact with a cavity field, one common approach is to treat the collective atomic degrees of freedom via a Holstein–Primakoff transformation and then perform a system size expansion. For N atoms, the idea is that the collective atomic polarization scales as $\sqrt{N}$, and you can write the operators as a large mean-field piece plus small fluctuations. The set of mean-field equations for a driven, damped cavity coupled to atoms can be written as follows:

$$\dot{\alpha}_0 = -\left(\frac{\kappa}{2} + i\Delta_c\right)\alpha_0 - ig\sqrt{N}\beta_0 + E, \tag{12.125}$$

$$\dot{\beta}_0 = -\left(\frac{\gamma}{2} + i\Delta_a\right)\beta_0 - ig\sqrt{N}\alpha_0, \tag{12.126}$$

where α_0 is the classical amplitude of the cavity mode, β_0 is the collective (scaled) atomic polarization, κ is the cavity decay rate, γ is the atomic decay rate, and Δ_c and Δ_a are the detunings of the cavity and the atoms. As usual g is the single-atom coupling strength, and E is the coherent drive amplitude. Because the atoms form a large collective spin, we write

$$\alpha = \alpha_0 + \frac{1}{\sqrt{N}}\delta\alpha, \tag{12.127}$$

$$\beta = \beta_0 + \frac{1}{\sqrt{N}}\delta\beta. \tag{12.128}$$

Here, $\delta\alpha$ and $\delta\beta$ represent the quantum fluctuations around the mean-field values. In the large N limit, the expansion parameter $1/\sqrt{N}$ makes these fluctuations small compared to the leading terms. To obtain the linearized dynamics of the fluctuations, we take the Jacobian of the mean-field equations with respect to α and β. In our case, the linearized (fluctuation) equations take the form

$$\frac{d}{dt}(\delta\alpha \quad \delta\beta) = L \quad \delta\alpha \quad \delta\beta + \text{noise} \tag{12.129}$$

with the drift matrix L given by

$$L = \begin{pmatrix} -\left(\frac{\kappa}{2} + i\Delta_c\right) & -ig \\ ig & -\left(\frac{\gamma}{2} + i\Delta_a\right) \end{pmatrix}. \tag{12.130}$$

The off-diagonal entries $-ig$ come from the coupling terms after the $\sqrt{N}$ factors have been absorbed into the definition of the mean fields. This matrix is the Jacobian evaluated at the steady state (α_0, β_0).

In the full Fokker–Planck description for the Glauber P function, the noise terms arise from the damping of both the cavity and the atoms. In a simplified model, one can often approximate the diffusion matrix as being diagonal, with the elements determined by the corresponding decay rates. That is, the noise acting on the cavity and the atomic polarization can be written as

$$D = \begin{pmatrix} \kappa & 0 \\ 0 & \gamma \end{pmatrix}. \tag{12.131}$$

More precisely, if the FPE (for the fluctuations) is written in terms of the complex variables (and their conjugates), one typically finds

$$\frac{1}{2}D_{\alpha\alpha*}\frac{\partial^2 P}{\partial\delta\alpha \quad \partial\delta\alpha^*} + \frac{1}{2}D_{\beta\beta*}\frac{\partial^2 P}{\partial\delta\beta \quad \partial\delta\beta^*}, \tag{12.132}$$

with

$$D_{\alpha\alpha*} = \kappa, \quad D_{\beta\beta*} = \gamma. \tag{12.133}$$

For a system with a cavity field (complex variable α), atomic polarization (complex variable β), and inversion (real variable w), the full FPE for the phase-space distribution $P(\alpha, \alpha^*, \beta, \beta^*, w; t)$ as

$$\frac{\partial P}{\partial t} = -\sum_i \frac{\partial}{\partial x_i}[A_i(\mathbf{x})P] + \frac{1}{2}\sum_{i,j}\frac{\partial^2}{\partial x_i \partial x_j}[D_{ij}(\mathbf{x})P], \tag{12.134}$$

where the vector of the phase-space variables is

$$\mathbf{x} = (\alpha, \alpha^*, \beta, \beta^*, w). \tag{12.135}$$

Here $A_i(\mathbf{x})$ represents the drift terms derived from the semiclassical (or mean-field) equations of motion and $D_{ij}(\mathbf{x})$ is the diffusion matrix that encodes the noise arising from the damping (for both the cavity and the atoms):

$$\dot{\alpha}_0 = -\left(\frac{\kappa}{2} + i\Delta_c\right)\alpha_0 - ig\sqrt{N}\beta_0 + E, \tag{12.136}$$

$$\dot{\beta}_0 = -(\gamma_\perp + i\Delta_a)\beta_0 - ig\sqrt{N}\alpha_0 w_0, \tag{12.137}$$

$$\dot{w}_0 = -\gamma_\parallel(w_0 - w_{eq}) + 2ig\sqrt{N}\left(\alpha_0^*\beta_0 - \alpha_0\beta_0^*\right). \tag{12.138}$$

In a full treatment of a collection of N atoms (each modeled as a two-level system), the inversion w and polarization β are actually collective variables. Their scales and fluctuations depend on N. Because the system is large (i.e. N is large), one assumes that the overall dynamics are dominated by the mean fields and that the fluctuations scale as $1/\sqrt{N}$. We decompose each variable into its mean-field part plus a fluctuation as

$$\alpha = \alpha_0 + \frac{1}{\sqrt{N}}\delta\alpha \tag{12.139}$$

$$\beta = \beta_0 + \frac{1}{\sqrt{N}}\delta\beta \tag{12.140}$$

$$w = w_0 + \frac{1}{\sqrt{N}}\delta w. \tag{12.141}$$

Here, α_0, β_0, and w_0 satisfy the mean-field equations (deterministic part). The fluctuations $\delta\alpha$, $\delta\beta$ (each complex, so two real components each), and δw (real) describe the noise about the mean. We also use the traditional notation that $\gamma_\parallel$ is the spontaneous emission rate and $\gamma_\perp = \gamma_\parallel + \gamma_{\text{dephase}}$ is the total decay rate of the dipole. The drift matrix is defined by

$$L_{ij} = \left.\frac{\partial A_i}{\partial x_j}\right|_{\mathbf{x}=\mathbf{x}_0}, \tag{12.142}$$

with $\mathbf{x}_0 = (\alpha_0, \alpha_0^*, \beta_0, \beta_0^*, w_0)$ the mean-field solution. For example, choosing the ordering

$$x_1 = \alpha,\ x_2 = \alpha^*,\ x_3 = \beta,\ x_4 = \beta^*,\ x_5 = w, \tag{12.143}$$

the nonzero elements are (with evaluation at $\mathbf{x}_0$):

For the cavity field we have

$$L_{11} = \frac{\partial A_\alpha}{\partial \alpha} = -\left(\frac{\kappa}{2} + i\Delta_c\right), \tag{12.144}$$

$$L_{13} = \frac{\partial A_\alpha}{\partial \beta} = -ig\sqrt{N}, \tag{12.145}$$

$$L_{22} = \frac{\partial A_{\alpha*}}{\partial \alpha^*} = -\left(\frac{\kappa}{2} - i\Delta_c\right), \tag{12.146}$$

$$L_{24} = \frac{\partial A_{\alpha*}}{\partial \beta^*} = ig\sqrt{N}. \tag{12.147}$$

For the atomic polarization we find

$$L_{31} = \frac{\partial A_\beta}{\partial \alpha} = -ig\sqrt{N}w_0, \tag{12.148}$$

$$L_{33} = \frac{\partial A_\beta}{\partial \beta} = -(\gamma_\perp + i\Delta_a), \tag{12.149}$$

$$L_{35} = \frac{\partial A_\beta}{\partial w} = -ig\sqrt{N}\alpha_0, \tag{12.150}$$

$$L_{42} = \frac{\partial A_{\beta*}}{\partial \alpha^*} = ig\sqrt{N}w_0, \tag{12.151}$$

$$L_{44} = \frac{\partial A_{\beta*}}{\partial \beta^*} = -(\gamma_\perp - i\Delta_a), \tag{12.152}$$

$$L_{45} = \frac{\partial A_{\beta*}}{\partial w} = ig\sqrt{N}\alpha_0^*, \tag{12.153}$$

and finally for the inversion

$$L_{51} = \frac{\partial A_w}{\partial \alpha} = -2ig\sqrt{N}\beta_0^*, \tag{12.154}$$

$$L_{52} = \frac{\partial A_w}{\partial \alpha^*} = 2ig\sqrt{N}\beta_0, \tag{12.155}$$

$$L_{53} = \frac{\partial A_w}{\partial \beta} = 2ig\sqrt{N}\alpha_0^*, \tag{12.156}$$

$$L_{54} = \frac{\partial A_w}{\partial \beta^*} = -2ig\sqrt{N}\alpha_0, \tag{12.157}$$

$$L_{55} = \frac{\partial A_w}{\partial w} = -\gamma_{\|}. \tag{12.158}$$

These derivatives form the 5×5 drift matrix L. The diffusion matrix elements D_{ij} come from the noise terms in the FPE. In general, if the master equation is written in the Lindblad form with jump operators L_k, one can show that the Fokker–Planck diffusion coefficients are given by

$$D_{ij} = \sum_k \frac{1}{2}\left\langle L_k^\dagger \{\Delta x_i, \Delta x_j\} L_k \right\rangle, \tag{12.159}$$

where $\Delta x_i = x_i - \langle x_i \rangle$. In practice, one obtains phenomenological forms for the noise associated with cavity decay and atomic dissipation.

For the cavity field

$$D_{\alpha\alpha*} = \kappa, \quad D_{\alpha\alpha} = D_{\alpha*\alpha*} = 0. \tag{12.160}$$

For the atomic polarization

$$D_{\beta\beta*} = \gamma_\perp F(w_0), \tag{12.161}$$

where $F(w_0)$ is a function that depends on the inversion (often, $F(w_0)$ is proportional to a factor like $(1 - w_0)/2$ in a two-level system). The off-diagonal elements $D_{\beta\beta}$ and $D_{\beta*\beta*}$ are typically zero. Finally for the inversion

$$D_{ww} = 2\gamma_{\|}(1 - w_0^2). \tag{12.162}$$

12.7 Correlation functions

If the FPE for the fluctuations is

$$\frac{\partial P(\delta\mathbf{x}, t)}{\partial t} = -\sum_i \frac{\partial}{\partial(\delta x_i)}\left[\sum_j M_{ij}\delta x_j P(\delta\mathbf{x}, t)\right] + \frac{1}{2}\sum_{ij}\frac{\partial^2}{\partial(\delta x_i)\partial(\delta x_j)}[D_{ij}P(\delta\mathbf{x}, t)]. \tag{12.163}$$

In the steady state, the covariance matrix

$$\mathbf{V} = \langle \delta \mathbf{x}(t) \delta \mathbf{x}^{\dagger}(t) \rangle \tag{12.164}$$

satisfies the Lyapunov equation

$$\mathbf{MV} + \mathbf{VM}^{\dagger} = -\mathbf{D}. \tag{12.165}$$

This equation tells us how the balance between the deterministic drift (which tends to damp fluctuations) and the noise (described by **D**) sets the magnitude of the steady-state fluctuations. The formal solution of the linearized Langevin equations is

$$\delta \mathbf{x}(t + \tau) = e^{\mathbf{M}\tau} \delta \mathbf{x}(t) + \int_{t}^{t+\tau} e^{\mathbf{M}(t+\tau-t')} \boldsymbol{\eta}(t') dt'. \tag{12.166}$$

Assuming that the system has reached a stationary state, the two-time correlation function between fluctuations at times t and $t + \tau$ is given by

$$\langle \delta \mathbf{x}(t + \tau) \delta \mathbf{x}^{\dagger}(t) \rangle = e^{\mathbf{M}\tau} \langle \delta \mathbf{x}(t) \delta \mathbf{x}^{\dagger}(t) \rangle, \tag{12.167}$$

which is essentially the quantum regression theorem. Thus, each element of this matrix gives the two-time correlation functions for the corresponding fluctuation variables. For example, if you are interested in the correlation of the first variable, it is given by

$$\left\langle \delta x_1(t + \tau) \delta x_1^{\dagger}(t) \right\rangle = [e^{\mathbf{M}\tau} \mathbf{V}]_{11}. \tag{12.168}$$

References

[1] Arecchi F T, Courtens E, Gilmore R and Thomas H 1972 Atomic coherent states in quantum optics *Phys. Rev.* A **6** 2211

[2] Glauber R J 1963 Coherent and incoherent states of the radiation field *Phys. Rev.* **131** 2766

[3] Husimi K 1940 Some formal properties of the density matrix *Proc. Phys. Math. Soc. Jpn.* **22** 264

[4] Ma J, Wang X, Sun C P and Nori F 2011 Quantum spin squeezing *Phys. Rep.* **509** 89

[5] Risken H 1996 *The Fokker–Planck Equation: Methods of Solutions and Applications* (Springer Series in Synergetics) (Berlin: Springer)

[6] Smithey D T, Beck M, Raymer M G and Faridani A 1993 Measurement of the Wigner distribution and the density matrix of a light mode using optical homodyne tomography: application to squeezed states and the vacuum *Phys. Rev. Lett.* **70** 1244–7

[7] Sudarshan E C G 1983 Equivalence of semiclassical and quantum mechanical descriptions of statistical light beams *Phys. Rev. Lett.* **10** 277

[8] Wigner E P 1932 On the quantum correction for thermodynamic equilibrium *Phys. Rev.* **40** 749

[9] Wineland D J, Bollinger J J, Itano W M, Moore F L and Heinzen D J 1992 Spin squeezing and reduced quantum noise in spectroscopy *Phys. Rev.* A **46** R6797

IOP Publishing

An Introduction to Quantum Optics (Second Edition)

An open systems approach

Perry Rice

Chapter 13

Stochastic differential equations

13.1 Langevin equations

For systems coupled to reservoirs with zero mean values for their fields, it is possible to model the fluctuations via a stochastic process. As we now know, it must be done differently for quantum systems [1–6] For our purposes, a Langevin equation is a Heisenberg equation of motion for a set of operators, with a noise term representing interactions with a bath, as well as a decay/gain term arising from those same interactions, more properly called a Heisenberg–Langevin equation. As a simple example, let us consider a single field mode interacting with a continuum of field modes

$$H = \hbar\,\omega_0\, a^{\dagger}a + \sum_k \hbar\omega_k b_k^{\dagger} b_k + \sum_k g_k(ab_k^{\dagger} + a^{\dagger}b_k). \tag{13.1}$$

The Heisenberg equations of motion are given by

$$\dot{a} = -i\omega_0 a - i\sum_k g_k b_k \tag{13.2}$$

$$\dot{b}_k = -i\omega_k b_k - ig_k a. \tag{13.3}$$

Substituting the equations for $\dot{b}_k$ into the equation for $\dot{a}$ we have

$$\dot{a} = -i\omega_0 a - i\sum_k |g_k|^2 a(t')e^{-i\omega_k(t-t')} + f_a(t), \tag{13.4}$$

where we have defined the noise operator

$$f_a(t) = -i\sum_k g_k b^k(0)e^{-i\omega_k t}. \tag{13.5}$$

Here, we have replaced $b_k(t)$ with $b_k(0)e^{-i\omega_k t}$, keeping only the lowest order in the (assumed small) coupling constant g_k. Similarly, we can replace

$$a(t') \to a(t)e^{-i\omega_0(t'-t)}. \tag{13.6}$$

doi:10.1088/978-0-7503-6014-2ch13

In the field mode equation. This is exactly what we did back in chapter 7 when we discussed vacuum fluctuations and the radiation reaction, as well as deriving the Wigner–Weiskopff formula. This results in

$$\dot{a} = -i\omega_0 a - \Gamma(t')e^{-i\omega_k(t-t')} + F_a(t), \tag{13.7}$$

where we have

$$\Gamma = 2\pi \sum_k |g_k|^2 \tag{13.8}$$

and

$$F_a = -i\sum_k b_k(0)e^{-i(\omega_k - \omega_0)t}. \tag{13.9}$$

If we ignore the fluctuating noise operator, F_a, the solution is given by

$$a(t) = a(0)e^{-\Gamma t/2}. \tag{13.10}$$

The difficulty is that the commutator between a and $a^\dagger$ will decay to zero! To have decay/gain, one must have fluctuations. This is the heart of the celebrated fluctuation–dissipation theorem. The noise term has a zero mean, but a non-zero intensity

$$\langle F_a \rangle = 0 \tag{13.11}$$

$$\langle F_a^\dagger(t)F_a(t') \rangle = \Gamma\delta(t - t'). \tag{13.12}$$

As the noise is completely different from one time to another, even with negligible separation, unless $t = t'$, the correlation is zero. To simulate this behavior, we write

$$da = -(i\omega_0 + \Gamma/2)da + \sqrt{\Gamma}dW. \tag{13.13}$$

Here dW is what is known as a Weiner process, and it is simulated by drawing a random number from a Gaussian distribution of unit mean and zero average, and multiplying by the square root of the time step

$$dW_{\text{sim}} = r\sqrt{dt}, \tag{13.14}$$

where r is the random number drawn from the Gaussian distribution. The reason for the square root is that this is essentially a diffusive process, where the intensity of the noise grows linearly with time. It is an odd process. Each time step it jumps randomly, and as such it has no meaningful derivative. This is why we (should) never write dW/dt. A simulation for a non-zero initial amplitude will decay noisily to zero, and an average of such simulations will yield the result for the expected value, as shown in figure 13.1,

$$\langle a(t) \rangle = \langle a(0) \rangle e^{-\Gamma t/2}. \tag{13.15}$$

Let us now examine the Weiner process in more detail, following the elegant treatment of Steck [7]. If we have a sum of random numbers X with average $\langle X \rangle = 0$ and variance $\Delta X \equiv \langle X^2 \rangle - \langle X \rangle^2 = \sigma$, what is the behavior or that sum?

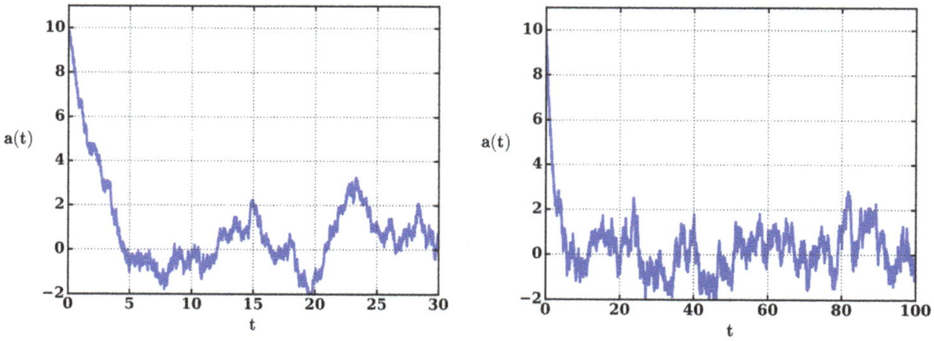

Figure 13.1. Solution of a damped harmonic oscillator for short and long times.

This is essentially the random walker who takes a random step left or right at each time step. You might expect that after N steps the walker is at $\langle X \rangle = 0$ but they are not. The central limit theorem states that the probability distribution for the position after N steps is given by a Gaussian with zero mean and variance $\sqrt{N}\sigma$. If we have N steps of Δt, with $N\Delta t = t$, we have a standard deviation, the square root of the variance, given by

$$\sigma(t) = \sigma\sqrt{t/\Delta t}. \tag{13.16}$$

This is a diffusive process with a root mean square displacement of

$$\Delta x(t) = D\sqrt{t}. \tag{13.17}$$

In this example, the diffusion coefficient depends on the time step, $D = \sigma/\sqrt{\Delta t}$. A key fact is that for random steps, even if they are not identically distributed, the variances add and not the average. If $X = X_1 + X_2$, then

$$(\Delta X)^2 = (\Delta X_1)^2 + (\Delta X_2)^2. \tag{13.18}$$

If the distributions are identical, with variance σ, we obtain

$$\sigma_N = N\sigma_0. \tag{13.19}$$

What if our steps are continuous and not discrete? This is the Weiner process, $W(t)$, which satisfies the distribution

$$P(W(t)) = \frac{1}{2\pi t}e^{-W^2/2t}. \tag{13.20}$$

With the choice $W(0) = 0$. we have

$$P(W(t)) = \delta(0). \tag{13.21}$$

We can then write the variance

$$\langle(\Delta W)^2\rangle = \Delta t. \tag{13.22}$$

Therefore, we can say that the Weiner increment

$$W(t + dt) - W(t) = \sqrt{\Delta t}. \tag{13.23}$$

This all suggests that for stochastic differential equations, we must keep second order in dW to be consistent with keeping terms linear in dt In particular

$$\Delta W/dt = 1/\sqrt{\Delta t}, \tag{13.24}$$

which tends to infinity in the limit of $\Delta t \to 0$. Consider now

$$dy = \alpha\, dt + \beta\, dW, \tag{13.25}$$

a variable that changes in time and has a random component. A function of y, say $z = e^y$, has a derivative

$$\frac{d}{dt}z = e^{y(t+dt)} - y(t) \tag{13.26}$$

$$= (z(dy+)dy)^2/2, \tag{13.27}$$

where we keep the dy^2 due to the need to keep dW^2 to keep terms of order dt. The result is

$$dz = (\alpha + \beta^2/2)dt + z\beta dW. \tag{13.28}$$

We can see that for a decay/gain term γ, we must have noise on the order of $\sqrt{\gamma}$. If there is gain/dissipation, there will be noise. This form is known as the Itô form of a stochastic differential equation. There are other forms. In general, we have

$$y(t + dt) = y(t) + \alpha(Y(t), t)dt + \beta(Y(t + \tau, \tau))dW \tag{13.29}$$

with τ in the interval $t \to t + dt$. The Itô choice takes $\tau = t$. Another choice is the Stratonovich approach which takes $\tau = t + dt/2$ [8].

One major difference between the two approaches is the chain rule for stochastic differentiation. For the Itô calculus, we have

$$dz = df(y) = f'dy + (1/2)f''dy^2 \tag{13.30}$$

$$= [f'\alpha + (1/2)f''\beta]dt + f'\beta dW, \tag{13.31}$$

which is not the usual chain rule. An appeal of the Stratonovich approach is that

$$df = f'\alpha dt + f'\beta dW, \tag{13.32}$$

which is the usual chain rule. However, because things are evaluated at the middle of the interval between points, the strangeness is hidden. One must be careful to specify at the beginning whether one is using the Itô or Stratonovich approach, and proceed consistently

13.2 Coupled oscillators

The standard quantum Langevin equations (in the Heisenberg picture) for such a system, in the frame rotating at ω_a, are given by

$$\frac{d\tilde{a}}{dt} = -\frac{\gamma_a}{2}\tilde{a} + g\tilde{b} + \sqrt{\gamma_a}\tilde{\xi}_a. \tag{13.33}$$

$$\frac{d\tilde{b}}{dt} = -\frac{\gamma_b}{2}\tilde{b} - g\tilde{a} + \sqrt{\gamma_b}\tilde{b}_b. \tag{13.34}$$

If $\gamma_b \gg \gamma_a$, we set $\frac{d\tilde{b}}{dt} \approx 0$, giving

$$\tilde{b} \approx \frac{g\tilde{a} - \sqrt{\gamma_b}\tilde{\xi}_b}{\frac{\gamma_b}{2}}. \tag{13.35}$$

Essentially, we are saying that the b oscillator damps so fast, it is a 'steady state' that is dependent on A. This is like a rapidly rotating magnetic dipole, which performs a Larmor precession around the magnetic field. If you change the magnetic field slowly enough, the rotation will generally be precession around the axis of the magnetic field. Substituting into the equation for $\tilde{a}$

$$\frac{d\tilde{a}}{dt} = -\frac{\gamma_a}{2}\tilde{a} + g\left(\frac{g\tilde{a} - \sqrt{\gamma_b}\tilde{\xi}_b}{\frac{\gamma_b}{2}}\right) + \sqrt{\gamma_a}\tilde{\xi}_a. \tag{13.36}$$

Simplifying

$$\frac{d\tilde{a}}{dt} = \left(-\frac{\gamma_a}{2} - \frac{2g^2}{\gamma_b}\right)\tilde{a} + \sqrt{\gamma_a}\tilde{\xi}_a + \frac{2g\sqrt{\gamma_b}\tilde{\xi}_b}{\gamma_b}. \tag{13.37}$$

We can write this as

$$\frac{d\tilde{a}}{dt} = -\frac{\gamma_a}{2}\tilde{a} - \frac{2g^2}{\gamma_b}\tilde{a} + \sqrt{\gamma_a}\tilde{\xi}_a - \frac{2g}{\sqrt{\gamma_b}}\tilde{\xi}_b \tag{13.38}$$

with an effective damping of $\tilde{a}$

$$\gamma_{\text{eff}} = \gamma_a + \frac{4g^2}{\gamma_b} \tag{13.39}$$

and a noise term $\tilde{\xi}_b$ which introduces additional fluctuations into $\tilde{a}$. If the oscillators are coupled to the same heat bath, we find it convenient to define

$$\epsilon = \frac{4g^2}{\gamma_a\gamma_b} \tag{13.40}$$

and then

$$\frac{da}{dt} = -\frac{\gamma_a}{2}(1 + \epsilon)a + \sqrt{\gamma_a(1 + \epsilon)}\xi, \tag{13.41}$$

where we have the enhanced loss reflecting the loss through the b channel. The effective damping rate increases by a factor of $1 + \epsilon$. The effective noise strength also increases by $\sqrt{1 + \epsilon}$, meaning that the fluctuations in a become stronger due to coupling to noise in b. The noise correlations must satisfy

$$\left\langle \xi_{\text{eff}}(t)\xi^{\dagger}_{\text{eff}}(t')\right\rangle = \gamma_{\text{eff}}\delta(t - t'). \tag{13.42}$$

13.3 Relation to the Fokker–Planck equation

For a Fokker–Planck equation in the form

$$\frac{\partial P(\mathbf{x}, t)}{\partial t} = -\sum_i \frac{\partial}{\partial x_i}[A_i(\mathbf{x})P(\mathbf{x}, t)] + \frac{1}{2}\sum_{i,j}\frac{\partial^2}{\partial x_i \partial x_j}[D_{ij}(\mathbf{x})P(\mathbf{x}, t)], \tag{13.43}$$

the corresponding Itô Langevin equations are

$$dx_i = A_i(\mathbf{x})dt + \sum_j B_{ij}(\mathbf{x})dW_j(t), \tag{13.44}$$

where $B(\mathbf{x})$ is chosen such that

$$D_{ij}(\mathbf{x}) = \sum_k B_{ik}(\mathbf{x})B_{jk}(\mathbf{x}), \tag{13.45}$$

and $dW_j(t)$ are independent Wiener increments with

$$\left\langle dW_j(t)\right\rangle = 0, \quad \left\langle dW_j(t)dW_k(t)\right\rangle = \delta_{jk}dt. \tag{13.46}$$

For constant D (or equivalently constant B), there is no additional Itô correction. For an Itô stochastic differential equation (SDE)

$$dx_i = A_i(x)dt + \sum_j B_{ij}(x)dW_j(t). \tag{13.47}$$

The equivalent Stratonovich SDE is given by

$$dx_i = \left[A_i(x) - \frac{1}{2}\sum_j\sum_k B_{kj}(x)\frac{\partial B_{ij}(x)}{\partial x_k}\right]dt + \sum_j B_{ij}(x)\circ dW_j(t), \tag{13.48}$$

where the '0' indicates that the stochastic integral is interpreted in the Stratonovich sense. For constant diffusion coefficients (i.e. $B_{ij}(x) = \text{constant}$), the derivative term $\partial B_{ij}(x)/\partial x_k$ vanishes, and hence there is no additional drift term.

13.4 Input–output theory

In the preceding chapters, we used a master-equation treatment to calculate the photon statistics inside an optical cavity when the internal field is damped. This approach is based on treating the field external to the cavity, to which the system is coupled, as a heat bath. The heat bath is simply a passive system with which the system gradually comes into equilibrium. In this chapter, we will explicitly treat the heat bath as the external cavity field, as in figure 13.2. Our objective is to determine the effect of the intracavity dynamics on the quantum statistics of the output field. Within this perspective, we will also treat the field input to the cavity explicitly. This approach is necessary in the case of the squeezed state generation due to the interference effects at the interface between the intracavity field and the output field. An input–output formulation is also required if the input field state is specified as other than simply a vacuum or thermal state. We know that an optical parametric oscillator generates squeezed light because the interaction Hamiltonian is essentially the squeezing operator. But how much squeezing? What is the squeezing in the output field? It was found that the intracavity squeezing was limited to half of the maximum. As the Hamiltonian is a squeezing operator, the damping of the cavity must play a role, and so a theory of the output was needed. It turns out the squeezing is 1/2 the max outside as well, for a vacuum input the cavity field just leaks out. However, when one looks at the frequency decomposition of the squeezing, we will see that you find perfect squeezing at zero frequency. This will follow the seminal treatment of Collett and Gardiner, and the treatment presented by Walls and Milburn [2, 9].

A single-ended optical cavity has one partially transmitting mirror (coupled to the external field) and one perfectly reflecting mirror. The Hamiltonian of the system can be written as

$$H = \hbar\omega_c a^\dagger a + \hbar\int_{-\infty}^{\infty} \omega b_\omega^\dagger b_\omega d\omega + \hbar\int_{-\infty}^{\infty} i\sqrt{\kappa}\left(b_\omega^\dagger a - a^\dagger b_\omega\right)d\omega. \tag{13.49}$$

The third term is the interaction between the cavity mode (a) and the bath modes (b_i). Using the Heisenberg equations, we can derive the equations of motion for the system operators. For the cavity mode a we find

$$\dot{a} = -i\omega_c a - \kappa\int_{-\infty}^{\infty} b_\omega d\omega. \tag{13.50}$$

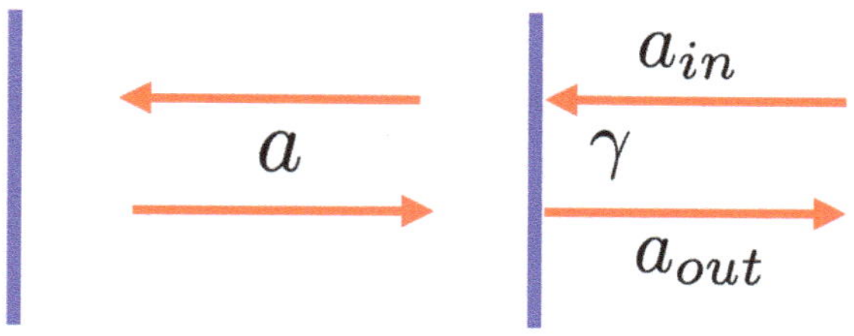

Figure 13.2. Fields involved in the input–output theory.

For an external field mode b_ω we have

$$\dot{b}_\omega = -i\omega b_\omega - \sqrt{\kappa}a. \tag{13.51}$$

We now define the input and output fields. The input field $b_{\text{in}}(t)$ represents the incoming external field, and the output field $b_{\text{out}}(t)$ represents the outgoing field. These are connected to the external field modes b_ω by

$$b_{\text{in}}(t) = \frac{1}{\sqrt{2\pi}} \int_{-\infty}^{\infty} b_\omega(0)e^{-i\omega t}d\omega \tag{13.52}$$

$$b_{\text{out}}(t) = \frac{1}{\sqrt{2\pi}} \int_{-\infty}^{\infty} b_\omega(t)e^{-i\omega t}d\omega. \tag{13.53}$$

By solving the equation of motion for b_ω, we obtain

$$b_\omega(t) = b_\omega(0)e^{-i\omega t} - \sqrt{\kappa} \int_0^t a(\tau)e^{-i\omega(t-\tau)}d\tau, \tag{13.54}$$

which is the input–output relation we desire. It is essentially a boundary condition imposed at the output mirror. Substituting this into the definition of $b_{\text{out}}(t)$, we find that

$$b_{\text{out}}(t) = b_{\text{in}}(t) + \sqrt{\kappa}a(t). \tag{13.55}$$

Returning to the cavity mode equation of motion and using the definition of $b_{\text{in}}(t)$, we find

$$\dot{a} = -\left(i\omega_c + \frac{\kappa}{2}\right)a - \sqrt{\kappa}b_{\text{in}}(t). \tag{13.56}$$

Often the equations of motion are linear and may be written in the form

$$\frac{d}{dt}\boldsymbol{a}(t) = \boldsymbol{A}\boldsymbol{a}(t) - \frac{\gamma}{2}\boldsymbol{a}(t) + \sqrt{\kappa}\boldsymbol{a}_{\text{in}}(t), \tag{13.57}$$

where

$$\boldsymbol{a}(t) = \begin{pmatrix} a(t) \\ a^\dagger(t) \end{pmatrix}, \tag{13.58}$$

$$\boldsymbol{a}_{\text{i}}(t) = \begin{pmatrix} a_{\text{in}}(t) \\ a_{\text{in}}^\dagger(t) \end{pmatrix}, \tag{13.59}$$

and we now replace b with a_{in}. The Fourier components of the intracavity field is

$$\boldsymbol{a}(t) = \frac{1}{\sqrt{2\pi}} \int_{-x}^{x} e^{-i\omega(t-t_0)}\mathbf{a}(\omega)d\omega \tag{13.60}$$

and a frequency component vector

$$\boldsymbol{a}(\omega) = \begin{pmatrix} a(\omega) \\ a^{\dagger}(\omega) \end{pmatrix}, \tag{13.61}$$

where $a^{\dagger}(\omega)$ is the Fourier transform of $a^{\dagger}(t)$. The equations of motion become

$$[\boldsymbol{A} + (i\omega - \kappa/2)\boldsymbol{I}]\boldsymbol{a}(\omega) = -\sqrt{\kappa}\,\mathbf{a}_{\text{in}}(\omega). \tag{13.62}$$

We can eliminate the internal fields to find a relation between in and out fields

$$\boldsymbol{a}_{\text{out}}(\omega) = -[A + (i\omega + \kappa/2)I]\left[A + \left(i\omega - \frac{\kappa}{2}\right)I\right]^{-1}\boldsymbol{a}_{\text{in}}(\omega). \tag{13.63}$$

Let us examine an empty one sided cavity as our first application. We have the system Hamiltonian

$$H_{\text{Sys}} = \hbar\omega_0 a^{\dagger}a. \tag{13.64}$$

We then have

$$\boldsymbol{A} = \begin{pmatrix} -i\omega_0 & 0 \\ 0 & i\omega_0 \end{pmatrix}. \tag{13.65}$$

Equation (13.63) then gives

$$\boldsymbol{a}_{\text{out}}(\omega) = \frac{\kappa/2 + i(\omega - \omega_0)}{\kappa/2 - i(\omega - \omega_0)}\boldsymbol{a}_{\text{in}}(\omega), \tag{13.66}$$

in which we see that the in and out fields differ only by a phase. The relationship between the input and the internal field is

$$\boldsymbol{a}(\omega) = \frac{\sqrt{\gamma}}{\kappa/2 - i(\omega - \omega_0)}\boldsymbol{a}_{\text{in}}(\omega), \tag{13.67}$$

which leads to a Lorentzian of width $\gamma/2$ for the intensity transmission function.

A two-sided cavity has two partially transparent mirrors with associated loss coefficients κ_1 and κ_2, as shown in figure 13.3. In this case, there are two input ports and two output ports.

The equation of motion for the internal field is then given by an obvious generalization as

$$\frac{da(t)}{dt} = -i\omega_0 a(t) - \frac{1}{2}(\kappa_1 + \kappa_2)a(t) + \sqrt{\kappa_1}\,a_{\text{IN}}(t) + \sqrt{\kappa_2}\,b_{\text{IN}}(t). \tag{13.68}$$

The relationship between the internal and input field frequency components for an empty cavity is then

$$\boldsymbol{a}(\omega) = \frac{\sqrt{\kappa_1}\,a_{\text{in}}(\omega) + \sqrt{\kappa_2}\,b_{\text{in}}(\omega)}{\left(\frac{\kappa_1 + \kappa_2}{2}\right) - i(\omega - \omega_0)}. \tag{13.69}$$

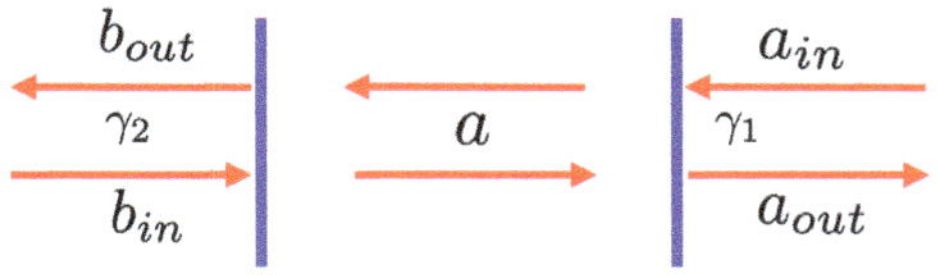

Figure 13.3. Fields involved in the input–output theory for a double-sided cavity.

Next we apply the boundary conditions at each mirror 13.54,

$$a_{\text{out}}(t) + a_{\text{in}}(t) = \sqrt{\kappa_1}\, a(t), \tag{13.70}$$

$$b_{\text{out}}(t) + b_{\text{in}}(t) = \sqrt{\kappa_2}\, a(t). \tag{13.71}$$

We find

$$\boldsymbol{a}_{\text{out}}(\omega) = \frac{\left[\frac{\kappa_1 - \gamma_2}{2} + i(\omega - \omega_0)\right]\boldsymbol{a}_{\text{in}}(\omega) + \sqrt{\kappa_1\kappa_2}\, b_{\text{in}}(\omega)}{\frac{\kappa_1 + \gamma_2}{2} - i(\omega - \omega_0)}. \tag{13.72}$$

Before continuing considering the interactions within the cavity, we derive some general relations connecting the two-time correlation functions inside and outside the cavity.

13.5 Two time correlation functions

We integrate equation (13.53) over ω, and find

$$a_{\text{in}}(t) = \frac{\sqrt{i}}{2} a(t) - \frac{1}{\sqrt{2\pi}} \int_{-x}^{x} d\omega b(\omega, t). \tag{13.73}$$

If $c(t)$ is any system operator, we have

$$[c(t), \sqrt{\kappa}\, a(t)] = \frac{\kappa}{2}[c(t), a(t)]. \tag{13.74}$$

Because $c(t)$ only depends on $a_{\text{in}}(t')$ for earlier times $t' < t$ and the input field operators must commute at different times, we have

$$[c(t), \sqrt{\kappa}\, a_{\text{in}}(t')] = 0, \quad t' > t. \tag{13.75}$$

Also

$$[c(t), \sqrt{\kappa}\, a_{\text{out}}(t')] = 0, \quad t' < t. \tag{13.76}$$

We can then show

$$[c(t), \sqrt{\gamma}\, a_{\text{in}}(t')] = \gamma[c(t), a(t)], \quad t' < t \tag{13.77}$$

and putting it all together we have

$$[c(t), \sqrt{;\kappa}\, a_{\mathbf{a}}(t')] = \kappa)(t - t')[c(t), a(t')], \tag{13.78}$$

where $\theta(t)$ is the Heavyside step function

$$\theta(t) = \begin{cases} 1 & t > 0 \\ \frac{1}{2} & t = 0. \\ 0 & t < 0 \end{cases} \tag{13.79}$$

The output field commutator is

$$\left[a_{\text{out}}(t),\, a_{\text{out}}^{\dagger}(t')\right] = \left[a_{\text{in}}(t),\, a_{\text{in}}^{\dagger}(t')\right] \tag{13.80}$$

as required. For simple inputs such as a vacuum or a coherent state, everything is determined by the properties of the intracavity field of those of the internal system. All moments of the involving products of the input and intracavity fields will factorize. Using our boundary condition equation (13.54) we find

$$\left\langle a_{\text{out}}^{+}(t),\, a_{\text{out}}(t')\right\rangle = \imath\langle a^{\dagger}(t),\, a(t')\rangle, \tag{13.81}$$

where we define

$$\langle A,\, B\rangle \equiv \langle AB\rangle - \langle A\rangle\langle B\rangle, \tag{13.82}$$

just looking at fluctuations about the mean. In this case, there is a direct relationship between the two-time correlation between the output field and the internal field. Consider now the phase-dependent two-time correlation function

$$\begin{aligned} \langle a_{\text{out}}(t),\, a_{\text{out}}(t')\rangle &= \left\langle a_{\text{in}}(t) - \sqrt{\gamma}a(t),\, a_{\text{in}}(t') - \sqrt{\gamma}a(t')\right\rangle \\ &= \gamma\langle a(t),\, a(t')\rangle - \sqrt{\gamma}\langle [a_{\text{in}}(t'),\, a(t)]\rangle \\ &= \gamma\langle a(t),\, a(t')\rangle + \gamma\theta(t' - t)\langle [a(t'),\, a(t)]\rangle. \end{aligned} \tag{13.83}$$

These results indicate that the usual spectrum of the output field will be identical to the spectrum of the cavity field. The photon statistics of the output field will also be the same as those of the intracavity field. Where a difference will arise, is in the phase-sensitive spectrum such as in squeezing experiments.

13.6 Spectrum of squeezing for an OPO

A degenerate parametric oscillator (DPO) involves a nonlinear medium within an optical cavity where a strong pump field (at frequency ω_p) drives the generation of signal photons at half the pump frequency ($\omega_s = \omega_p/2$). Below the oscillation threshold, the system remains in the quantum amplification regime.

The input–output relations for the degenerate parametric oscillator below the threshold (so there are stable steady states) are derived as follows. The cavity contains a signal mode (a_s) and a pump mode (a_p). For a nondegenerate twin beam set-up, we would have a signal and an idler field,

$$H_{\text{cav}} = \hbar\omega_p a_p^{\dagger} a_p + \hbar\omega_s a_s^{\dagger} a_s. \tag{13.84}$$

The pump field generates pairs of degenerate photons in the signal mode. The sum of the energy of the two photons is the energy of a pump photon, and they are generated at the same time. The interaction Hamiltonian is

$$H_{\text{int}} = i\hbar\chi\left(a_p a_s^{\dagger 2} - a_p^\dagger a_s^2\right), \tag{13.85}$$

where χ is the nonlinear coupling constant. The coupling of the cavity modes to their external baths is

$$H_{\text{bath}} = \sum_{j=s,p} \hbar \int d\omega \left[g_j\left(b_j^\dagger(\omega) a_j + a_j^\dagger b_j(\omega)\right)\right], \tag{13.86}$$

where $b_j(\omega)$ are external field operators for the pump ($j = p$) and signal ($j = s$) fields. The Heisenberg equations for the intracavity modes are derived using $\dot{O} = \frac{i}{\hbar}[H, O]$. We find for the signal field

$$\dot{a}_s = -\left(i\omega_s + \frac{\kappa_s}{2}\right)a_s - 2\chi a_p a_s^\dagger + \sqrt{\kappa_s} b_{\text{in},s}. \tag{13.87}$$

We now treat the pump field as a strong classical field, $a_p \approx \alpha_p e^{-i\omega_p t}$, where α_p is the amplitude. Its dynamics are typically assumed to be steady-state:

$$a_p \approx \text{constant}. \tag{13.88}$$

Below the threshold, the nonlinear interaction is weak, and we linearize the equations for small perturbations around the steady-state solution. Let $a_s = \bar{a}_s + \delta a_s$, where $\bar{a}_s \approx 0$ is the steady-state value (zero below threshold) and δa_s are fluctuations. The linearized equation for the signal field becomes

$$\delta\dot{a}_s = -\left(i\omega_s + \frac{\kappa_s}{2}\right)\delta a_s - 2\chi\alpha_p \delta a_s^\dagger + \sqrt{\kappa_s} b_{\text{in},s}. \tag{13.89}$$

The cavity–output relation for the signal mode is

$$b_{\text{out},s} = b_{\text{in},s} + \sqrt{\kappa_s}\delta a_s. \tag{13.90}$$

To determine the output field properties, we substitute the linearized equation of motion for δa_s. We use a Fourier transform as

$$\delta a_s(t) = \frac{1}{\sqrt{2\pi}} \int_{-\infty}^{\infty} \delta a_s(\omega) e^{-i\omega t} d\omega \tag{13.91}$$

and similarly for $b_{\text{in},s}(t)$, we find

$$\delta a_s(\omega) = \frac{\sqrt{\kappa_s} b_{\text{in},s}(\omega) - 2\chi\alpha_p \delta a_s^\dagger(-\omega)}{i(\omega - \omega_s) + \kappa_s/2}. \tag{13.92}$$

The coupled equations for $\delta a_s(\omega)$ and $\delta a_s^\dagger(-\omega)$ can be solved to give

$$\delta a_s(\omega) = \frac{\sqrt{\kappa_s}\left[b_{\text{in},s}(\omega) + G b_{\text{in},s}^\dagger(-\omega)\right]}{\Delta_s}, \tag{13.93}$$

where $G = 2\chi\alpha_p/(i(\omega - \omega_s) + \kappa_s/2)$ is the gain, and $\Delta_s = (i(\omega - \omega_s) + \frac{\kappa_s}{2})^2 - |G|^2$. The output field is given by

$$b_{\text{out},s}(\omega) = b_{\text{in},s}(\omega) + \sqrt{\kappa_s}\,\delta a_s(\omega). \tag{13.94}$$

The input–output relation for the degenerate parametric oscillator below threshold is

$$b_{\text{out},s}(\omega) = b_{\text{in},s}(\omega) + \frac{\kappa_s\left[b_{\text{in},s}(\omega) + Gb^{\dagger}_{\text{in},\,s}(-\omega)\right]}{\Delta_s}. \tag{13.95}$$

This relation shows the interplay between the input signal, the parametric gain G, and the loss.

To calculate the spectrum of squeezing for the output field look at the quadratures of the output field. These are

$$\begin{aligned} X_s^{\text{out}}(\omega) &= \frac{1}{\sqrt{2}}\left(b_{\text{out},s}(\omega) + b^{\dagger}_{\text{out},\,s}(-\omega)\right) \\ P_s^{\text{out}}(\omega) &= \frac{1}{\sqrt{2}i}\left(b_{\text{out},s}(\omega) - b^{\dagger}_{\text{out},\,s}(-\omega)\right). \end{aligned} \tag{13.96}$$

The spectrum of squeezing is determined by calculating the variance of these quadratures. From the input–output relation

$$b_{\text{out},s}(\omega) = b_{\text{in},s}(\omega) + \frac{\kappa_s\left[b_{\text{in},s}(\omega) + Gb^{\dagger}_{\text{in},\,s}(-\omega)\right]}{\Delta_s}, \tag{13.97}$$

where

$$G = \frac{2\chi\alpha_p}{i(\omega - \omega_s) + \kappa_s/2} \tag{13.98}$$

is the parametric gain, and $\Delta_s = (i(\omega - \omega_s) + \frac{\kappa_s}{2})^2 - |G|^2$ is an effective detuning. The variance of the amplitude quadrature is

$$\left\langle \Delta X_s^{\text{out}}(\omega)^2\right\rangle = \frac{1}{2}\left(\left\langle b_{\text{out},s}(\omega)b^{\dagger}_{\text{out},\,s}(\omega)\right\rangle + \left\langle b^{\dagger}_{\text{out},\,s}(-\omega)b_{\text{out},s}(-\omega)\right\rangle\right). \tag{13.99}$$

Similarly, the variance of the phase quadrature is

$$\left\langle \Delta P_s^{\text{out}}(\omega)^2\right\rangle = \frac{1}{2}\left(\left\langle b_{\text{out},s}(\omega)b^{\dagger}_{\text{out},\,s}(\omega)\right\rangle - \left\langle b^{\dagger}_{\text{out},\,s}(-\omega)b_{\text{out},s}(-\omega)\right\rangle\right). \tag{13.100}$$

For simplicity, assume that the input noise operators obey, as for an ordinary quantum vacuum at $T = 0$,

$$\begin{aligned} \left\langle b_{\text{in},s}(\omega)b^{\dagger}_{\text{in},\,s}(\omega')\right\rangle &= \delta(\omega - \omega'), \\ \left\langle b^{\dagger}_{\text{in},\,s}(\omega)b_{\text{in},s}(\omega')\right\rangle &= 0. \end{aligned} \tag{13.101}$$

Substituting the input–output relation for $b_{\text{out},s}(\omega)$ into these expressions we find the variances. The spectrum of squeezing is defined as the normalized variance of a given quadrature. For the amplitude quadrature $X_s^{\text{out}}(\omega)$, the squeezing spectrum is

$$S_X(\omega) = 1 - \frac{4\kappa_s\,|G|^2}{[(\kappa_s/2)^2 - |G|^2]^2 + (\omega - \omega_s)^2}. \tag{13.102}$$

Below the threshold, $|G|^2 < (\kappa_s/2)^2$, ensuring stability. The squeezing is maximum at $\omega = \omega_s$ and decreases the off-resonance. For the phase quadrature $P_s^{\text{out}}(\omega)$, the squeezing spectrum becomes

$$S_P(\omega) = 1 + \frac{4\kappa_s\,|G|^2}{[(\kappa_s/2)^2 - |G|^2]^2 + (\omega - \omega_s)^2}. \tag{13.103}$$

The squeezing in the total field may be found by integrating over all ω. At the threshold, we find

$$S_2^{\text{tot}} = \int\left(1 - \frac{\gamma_1^2}{\gamma_1^2 + \omega^2}\right)d\omega = \frac{\gamma_1}{2}. \tag{13.104}$$

Hence, the squeezing in the internal field is

$$V(X_2) = \frac{1}{2}. \tag{13.105}$$

Thus, the internal field mode is 50% squeezed, in agreement with the calculations of Milburn and Walls [10]. This can be surpassed in the individual frequency components of the output field, which have 100% squeezing for $\omega = 0$. It is the squeezing in the individual frequency components of the output field that may be measured by a spectrum analyzer following a homodyne detection scheme.

In the double-ended cavity, we have the spectrum of squeezing from one output port to be

$$S_X^{\text{double, one port}}(\omega) = 1 - \frac{2\gamma\eta\epsilon^2}{\gamma^2\omega^2 + (4\epsilon^2 + \frac{\gamma^2}{4} - \omega^2)^2}, \tag{13.106}$$

which is half the result for the single-ended system. If we look at just one port, the total squeezing is reduced by two as well.

13.7 Cascaded systems

Deriving the master equation for an open quantum system A, driven by the output of an open quantum system B, involves several steps. The gist of this is once the light leaves system A and passes down the optical path, the interaction with B has no effect on A. This can be assured by the use of an optical diode that prevents light from coming back to A. This process is rooted in the framework of quantum optics and open quantum systems, specifically using the input–output formalism and the concept of cascaded quantum systems. Carmichael's theory of cascaded open quantum systems [11] describes a theoretical framework for understanding the dynamics of quantum systems that are sequentially coupled, where the output of one quantum system serves as the input to another. This approach is particularly useful in describing quantum optics and quantum networks, where quantum information or energy propagates unidirectionally between different subsystems (figure 13.4).

The complete Hamiltonian for $A \oplus B \oplus R$ is

$$H = H_A + H_B + H_R + H_{AR} + H_{BR}, \tag{13.107}$$

with

$$H_{AR} = i\hbar(2\kappa_A)^{1/2}[aE^{\dagger}(0) - \text{h.c.}] \tag{13.108}$$

$$H_{BR} = i\hbar(2\kappa_B)^{1/2}[bE^{\dagger}(l) - \text{h.c.}], \tag{13.109}$$

where κ_A and κ_B are the cavity linewidths, and a and b are annihilation operators for the cavity modes. H describes two systems interacting with the same reservoir. We can eliminate the spatial separation between the evaluation of the bath terms, as we are assuming that the two are not statistically independent, but are indeed the same bath. We relate $E(l)$ to $E(0)$ via

$$U_A(\tau)E(l)U_A^{-1}(\tau) = E(0) + \frac{1}{2}(2\kappa_A)^{1/2}a, \tag{13.110}$$

where

$$U_A(\tau) = \exp[(i/\hbar)(H_A + H_R + H_{AR})\tau]. \tag{13.111}$$

Therefore, the field at l is the time-retarded field at 0 plus a piece radiated by A. We may then define

$$\chi'(t) = U_A(\tau)\chi(t)U_A^{-1}(\tau). \tag{13.112}$$

We can show that $\chi'(t)$ satisfies the Liouville equation with the Hamiltonian

$$H' = H_S + H_R + H_{SR}, \tag{13.113}$$

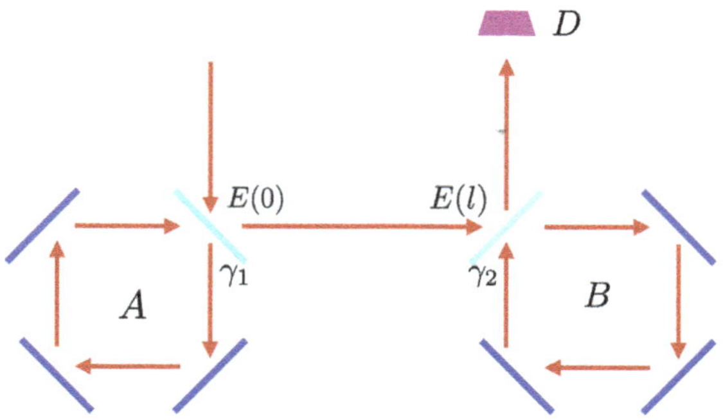

Figure 13.4. Schematic of cascaded optical systems.

where

$$H_S = H_A + H_B + i\hbar(\kappa_A \kappa_B)^{1/2}(a^\dagger b - \text{h.c.}) \tag{13.114}$$

$$H_{\text{SR}} = i\hbar\{[(2\kappa_A)^{1/2}a + (2\kappa_B)^{1/2}b] - \text{h.c.}\}. \tag{13.115}$$

One finds that an effective Hamiltonian term appears:

$$-\frac{i\hbar}{2}\sqrt{\kappa_A\kappa_B}\left(c_A^\dagger c_B - c_B^\dagger c_A\right). \tag{13.116}$$

Where does this term come from? In a cascaded system, one quantum system's output mode acts as the input to another system without any back-propagation. Concretely, if system A is upstream and system B is downstream, the field leaving A then drives B, but B does not send signals back to A. This unidirectionality is the core assumption in the Gardiner–Carmichael theory of cascaded systems. This term is purely imaginary (notice the factor of $-i\hbar/2$) and describes how quanta from A can be coherently funneled into B. If there were bidirectional coupling, you would get real coupling terms (such as $c_A^\dagger c_B + c_B^\dagger c_A$), but the unidirectional flow modifies the sign and introduces the i-factor. Notice that A and B now couple to the bath at the same location and they also couple directly, with the coupling constant $(\kappa_A \kappa_B)^{1/2}$. Each mode obeys a Langevin-like equation. For concreteness, assume that each has a free Hamiltonian $\hbar\omega_j 2c_j^\dagger c_j$, plus damping

$$\dot{c}_A(t) = -\left(i\omega_A + \frac{\kappa_A}{2}\right)c_A(t) + \sqrt{\kappa_A}\, c_{A,\text{in}}(t), \tag{13.117}$$

$$\dot{c}_B(t) = -\left(i\omega_B + \frac{\kappa_B}{2}\right)c_B(t) + \sqrt{\kappa_B}\, c_{B,\text{in}}(t). \tag{13.118}$$

But now we impose cascading

$$c_{B,\text{in}}(t) = c_{A,\text{out}}(t) = c_{A,\text{in}}(t) + \sqrt{\kappa_A}\, c_A(t). \tag{13.119}$$

Then

$$\dot{c}_B(t) = -\left(i\omega_B + \frac{\kappa_B}{2}\right)c_B(t) + \sqrt{\kappa_B}\, c_{A,\text{in}}(t) + \underbrace{\sqrt{\kappa_B}\sqrt{\kappa_A}\, c_A(t)}_{\text{extra driving from } A}. \tag{13.120}$$

To see where the effective Hamiltonian piece $-(i\hbar/2)\sqrt{\kappa_A\kappa_B}\,(c_A^\dagger c_B - c_B^\dagger c_A)$ comes from, note that in the Heisenberg equation,

$$\dot{c}_B = (i/\hbar)[H_{\text{eff}}, c_B] - \frac{\kappa_B}{2}c_B + (\text{noise terms}). \tag{13.121}$$

For the effective Hamiltonian to be correct, we want the commutator $[H_{\text{eff}}, c_B]$ to yield a term $\propto\sqrt{\kappa_A\kappa_B}\, c_A$. One can verify that choosing

$$H_{\text{eff}} = H_A + H_B - \frac{i\hbar}{2}\sqrt{\kappa_A\kappa_B}\left(c_A^\dagger c_B - c_B^\dagger c_A\right) \tag{13.122}$$

does exactly that. We focus on the cross-coupling piece,

$$H_{\text{int}} = -\frac{i\hbar}{2}\sqrt{\kappa_A\kappa_B}\left(c_A^\dagger c_B - c_B^\dagger c_A\right). \tag{13.123}$$

Then

$$\frac{i}{\hbar}[H_{\text{int}}, c_B] = \frac{i}{\hbar}\left[-\frac{i\hbar}{2}\sqrt{\kappa_A\kappa_B}\left(c_A^\dagger c_B - c_B^\dagger c_A\right), c_B\right]. \tag{13.124}$$

The constants factor out, and the bosonic commutators among c_A, c_B give precisely This matches the 'coherent' part of the driving that appears in the Langevin equation for c_B. Crucially, there is no reverse term in $\dot{c}_A$ from c_B. The imaginary sign in H_{int} ensures the coupling is effectively 'one-way' when you look at both equations of motion. In a fully bidirectional coupling, you would have a real coupling $\propto(c_A^\dagger c_B + c_B^\dagger c_A)$ instead. If we assume $c_{A,\text{in}}(t)$ is just vacuum noise (or otherwise uncorrelated), then the only 'coherent' driving of B comes from $\sqrt{\kappa_A}\,c_A$. Each mode obeys a Langevin-like equation. The last term is the key. It says 'system B is driven by c_A' in a one-way fashion. The master equation for the cascaded system is

$$\frac{d\rho}{dt} = -\frac{i}{\hbar}[H_{\text{eff}}, \rho] + \mathcal{D}[c_A]\rho + \mathcal{D}[c_B]\rho - \sqrt{\kappa_A\kappa_B}\left(\left[c_B^\dagger, c_A\rho\right] + \left[\rho c_A^\dagger, c_B\right]\right). \tag{13.125}$$

Those last two terms come from eliminating the traveling field that links A to B.

Let us apply this to an atom in free space driven by the output of an OPO as well as a coherent field. When the OPO output is a squeezed vacuum (with non-zero N and squeezing parameter M) and also sees a coherent field, the atom's master equation (in a frame rotating at the atomic frequency) becomes

$$\dot{\rho} = -\frac{i}{\hbar}\left[\frac{\hbar\Omega}{2}(\sigma_+ + \sigma_-), \rho\right] \tag{13.126}$$

$$+\frac{\gamma}{2}(N+1)(2\sigma_-\rho\sigma_+ - \sigma_+\sigma_-\rho - \rho\sigma_+\sigma_-) \tag{13.127}$$

$$+\frac{\gamma}{2}N(2\sigma_+\rho\sigma_- - \sigma_-\sigma_+\rho - \rho\sigma_-\sigma_+) \tag{13.128}$$

$$-\gamma M(\sigma_+\rho\sigma_+ + \sigma_-\rho\sigma_-), \tag{13.129}$$

where Ω is the effective coherent drive, γ is the spontaneous decay rate of the atom, $N = \sinh^2 r$ is the effective thermal photon number in the squeezed vacuum (with squeezing parameter r), and $M = \cosh r \sinh r$ (which we assume is real by choosing the squeezing phase appropriately, e.g. $\theta = 0$). From this master equation, one can derive the Bloch equations,

$$u = \langle\sigma_x\rangle = \langle\sigma_- + \sigma_+\rangle, \tag{13.130}$$

$$v = \langle \sigma_y \rangle = i\langle \sigma_- - \sigma_+ \rangle, \tag{13.131}$$

$$w = \langle \sigma_z \rangle. \tag{13.132}$$

A straightforward but longish calculation yields the following modified Bloch equations

$$\dot{u} = -\frac{\gamma}{2}[(2N+1) - 2M]u, \tag{13.133}$$

$$\dot{v} = -\frac{\gamma}{2}[(2N+1) + 2M]v + \Omega w, \tag{13.134}$$

$$\dot{w} = -\gamma(2N+1)(w+1) - 2\Omega v. \tag{13.135}$$

In the absence of squeezing ($N = M = 0$), these are reduced to the standard optical Bloch equations for a coherently driven two-level atom:

$$\dot{u} = -\frac{\gamma}{2}u, \tag{13.136}$$

$$\dot{v} = -\frac{\gamma}{2}v + \Omega w \tag{13.137}$$

$$\dot{w} = -\gamma(w+1) - 2\Omega v. \tag{13.138}$$

The squeezed vacuum modifies the decay rates: the u (i.e. σ_x) component decays at a rate $\frac{\gamma}{2}[(2N+1) - 2M]$ while the v (i.e. σ_y) component decays at $\frac{\gamma}{2}[(2N+1) + 2M]$. This is the signature that the noise in the squeezed reservoir is not isotropic. The inversion w relaxes with an effective rate $\gamma(2N+1)$, and the steady state (in the absence of drive) is modified by the squeezed environment. The steady-state solutions are

$$u_{ss} = 0 \tag{13.139}$$

$$v_{ss} = -\frac{2\Omega\gamma(2N+1)}{\gamma^2(2N+1)[(2N+1)+2M] + 4\Omega^2} \tag{13.140}$$

$$w_{ss} = -\frac{\gamma^2(2N+1)[(2N+1)+2M]}{\gamma^2(2N+1)[(2N+1)+2M] + 4\Omega^2}. \tag{13.141}$$

13.8 Modes of the Universe

Here, we present another path to the input–output theory [12, 13]. It is more in the spirit of a typical electromagnetism problem. For a sphere with a charge inside, for example, one writes down the field using the appropriate special functions. The expansion coefficients are different inside the sphere than outside. The coefficients

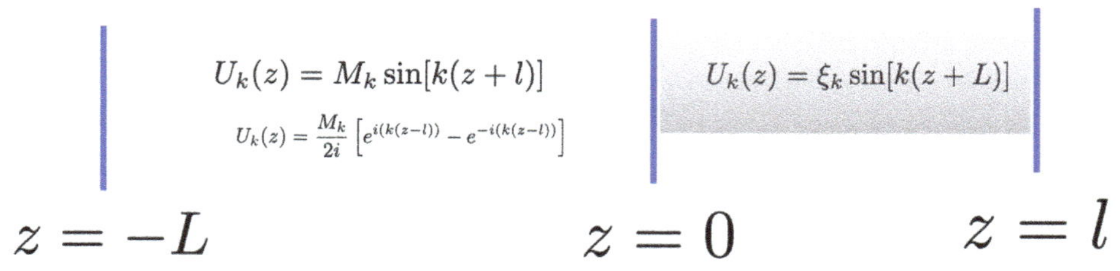

Figure 13.5. Fields involved in the modes of the Universe.

are then found using appropriate boundary conditions. For instance, the electric field and potential must be continuous at the surface. There is a boundary term if there are free charges on the surface of the sphere. You have done many of these types of problems, or you will (figure 13.5).

The electric field is expressed as

$$E(z, t) = \sum_k \mathcal{E}_k(t) U_k(z) + \text{c.c.}, \tag{13.142}$$

where $\mathcal{E}_k(t)$ are the mode amplitudes, and the $U_k(z)$ are the spatial mode functions. The mode functions $U_k(z)$ are defined piecewise, depending on the position z

$$U_k(z) = \begin{cases} \xi_k \sin[k(z+L)] & \text{for } z < 0 (\text{inside the cavity}) \\ M_k \sin[k(z-l)] & \text{for } z > 0 (\text{outside the cavity}) \end{cases}, \tag{13.143}$$

where L is the length of the cavity on the left side, l is the length on the right side, and ξ_k and M_k are coefficients ensuring the continuity of the field across the boundaries. The boundary conditions at the mirrors require the continuity of the electric field at $z = 0$, and continuity of the derivative of the electric field. At $z = 0$, the fields inside and outside the cavity are matched. These are expressed as

$$E_{\text{cav}}(0, t) = E_{\text{out}}(0, t) \tag{13.144}$$

and

$$\left.\frac{\partial}{\partial z} E_{\text{cav}}(z, t)\right|_{z=0} = \left.\frac{\partial}{\partial z} E_{\text{out}}(z, t)\right|_{z=0}. \tag{13.145}$$

Applying these to our mode expansions, we find from the continuity of E

$$\sum_k E_k(t) \xi_k \sin(kL) = -\sum_k E_k(t) M_k \sin(kl). \tag{13.146}$$

Equating the coefficients for each mode k, we obtain

$$\xi_k \sin(kL) = -M_k \sin(kl). \tag{13.147}$$

Continuity of the derivative yields

$$\left.\frac{\partial E_{\text{cav}}(z, t)}{\partial z}\right|_{z=0} = \left.\frac{\partial E_{\text{out}}(z, t)}{\partial z}\right|_{z=0}. \tag{13.148}$$

Equating the coefficients for each mode k, we find

$$\xi_k \cos(kL) = M_k \cos(kl). \tag{13.149}$$

Dividing these equations eliminates ξ_k and gives the ratio

$$\frac{\sin(kL)}{\cos(kL)} = -\frac{\sin(kl)}{\cos(kl)} \tag{13.150}$$

or

$$\tan(kL) = -\tan(kl). \tag{13.151}$$

We can rewrite our pair of equations as

$$M_k = -\xi_k \frac{\sin(kL)}{\sin(kl)} \tag{13.152}$$

$$M_k = \xi_k \frac{\cos(kL)}{\cos(kl)}. \tag{13.153}$$

We define the input field and output field as

$$\begin{aligned} b_{\text{in}}(t) &= \sum_k M_k b_k e^{-i\omega_k t}, \\ b_{\text{out}}(t) &= \sum_k \xi_k b_k e^{-i\omega_k t}. \end{aligned} \tag{13.154}$$

Using the boundary conditions derived above:

$$\xi_k = M_k \frac{\cos(kl)}{\cos(kL)}. \tag{13.155}$$

The intracavity mode operator is

$$2a(t) = \sum_k \xi_k 2b_k e^{-i\omega_k t}, \tag{13.156}$$

so we obtain the input–output relation

$$2b_{\text{out}}(t) = 2b_{\text{in}}(t) + \sqrt{2\kappa}\, 2a(t), \tag{13.157}$$

where the cavity decay rate is

$$\kappa = \frac{\pi\, |M_k|^2 D(\omega_c)}{2}. \tag{13.158}$$

This is the form we found in our derivation of a damped cavity in the dissipative terms of our Lindblad equation. A coupling constant times a density of states, the latter evaluated at the line center and assumed to be flat. The Born–Markov approximations are made here as well. Let us connect this to the more standard form of κ from optics. The cavity photon decay rate κ is related to the energy loss per round trip. A photon inside the cavity undergoes multiple reflections before leaking out through the partially transmitting mirrors. The cavity has length L and refractive index n, so the optical round-trip path is $2Ln$, and the speed of light in the medium is $v = \frac{c}{n}$, so the round-trip time is

$$t_{\mathrm{RT}} = \frac{2Ln}{c}. \tag{13.159}$$

The reflectivity of the cavity mirrors is R, meaning that in each round trip, a fraction $(1 - R)$ of the intensity is lost due to transmission through the mirror. The fraction of energy lost per round trip is proportional to $(1 - R)$. Because the decay time τ_{cav} is the time scale over which the cavity field amplitude decays, it is related to the round-trip loss:

$$\kappa = \frac{(1 - R)}{t_{\mathrm{RT}}}. \tag{13.160}$$

Substituting $t_{\mathrm{RT}} = 2Ln/c$

$$\kappa = \frac{c(1 - R)}{2Ln}. \tag{13.161}$$

13.9 Correlation functions

Consider two coupled damped oscillators, with one driven:

$$\begin{aligned} \dot{a} &= -\gamma_a a - igb + F + \eta_a(t), \\ \dot{b} &= -\gamma_b b - iga + \eta_b(t), \end{aligned} \tag{13.162}$$

$\eta_a(t)$ and $\eta_b(t)$ are noise terms (assumed Gaussian and delta-correlated) with

$$\left\langle \eta_i(t)\eta_j^{\dagger}(t') \right\rangle = D_i \delta_{ij} \delta(t - t'), \tag{13.163}$$

where $i, j \in \{a, b\}$. The drive F usually establishes a non-zero steady state for the driven oscillator. We denote the steady states by a_s and b_s (with b_s possibly zero if only a is driven). We then write

$$a(t) = a_s + \delta a(t), \tag{13.164}$$

$$b(t) = b_s + \delta b(t). \tag{13.165}$$

Inserting these into the equations and assuming that the fluctuations $\delta a(t)$ and $\delta b(t)$ are small, we obtain the linearized equations:

$$\begin{aligned} \frac{d}{dt}\delta a &= -\gamma_a \delta a - ig\delta b + \eta_a(t), \\ \frac{d}{dt}\delta b &= -\gamma_b \delta b - ig\delta a + \eta_b(t). \end{aligned} \tag{13.166}$$

In the matrix form, defining the fluctuation vector

$$\delta\mathbf{x}(t) = \begin{pmatrix} \delta a(t) \\ \delta b(t) \end{pmatrix}. \tag{13.167}$$

We have

$$\frac{d}{dt}\delta\mathbf{x}(t) = \mathbf{M}\delta\mathbf{x}(t) + \eta(t) \tag{13.168}$$

with the drift (stability) matrix

$$\mathbf{M} = \begin{pmatrix} -\gamma_a & -ig \\ ig & -\gamma_b \end{pmatrix} \tag{13.169}$$

and the noise vector

$$\eta(t) = \begin{pmatrix} \eta_a(t) \\ \eta_b(t) \end{pmatrix}. \tag{13.170}$$

The formal solution for the fluctuations is given by

$$\delta\mathbf{x}(t+\tau) = e^{\mathbf{M}\tau}\delta\mathbf{x}(t) + \int_t^{t+\tau} e^{\mathbf{M}(t+\tau-t')}\eta(t')dt'. \tag{13.171}$$

In the stationary state, the correlation functions depend only on the time difference τ. We focus on the first term (since the noise is delta-correlated and contributes to the steady-state covariance). The two-time correlation function between the fluctuations in the driven oscillator is given by

$$\langle\delta a(t)\delta a^{\dagger}(t+\tau)\rangle = [e^{\mathbf{M}\tau}\mathbf{V}]_{11}, \tag{13.172}$$

where $\mathbf{V}$ is the stationary covariance matrix,

$$\mathbf{V} = \langle\delta\mathbf{x}(t)\delta\mathbf{x}^{\dagger}(t)\rangle, \tag{13.173}$$

which is determined from the Lyapunov equation

$$\mathbf{M}\mathbf{V} + \mathbf{V}\mathbf{M}^{\dagger} = -\mathbf{D} \tag{13.174}$$

with

$$\mathbf{D} = \begin{pmatrix} D_a & 0 \\ 0 & D_b \end{pmatrix}. \tag{13.175}$$

The 1-1th element of the product $e^{\mathbf{M}\tau}\mathbf{V}$ then provides the correlation of interest.

13.10 Input–output (slight return)

When we looked at a driven two-level atom in a simple dipole approximation (and ignoring geometry),

$$E^{(+)}(t) \propto \sigma_{-}(t) = \sigma_{-}(t). \tag{13.176}$$

Plus the dipole spatial pattern if needed. But that is not the full story. To obtain the actual photon operator in the far field, you typically perform an input–output boundary condition

$$a_{\text{out}}(t) = a_{\text{in}}(t) + \sqrt{\Gamma}\sigma_{-}(t). \tag{13.177}$$

Generally, you have a continuum of modes a_k, and $\sigma_{-}(t)$ sources them. Only after you specify how the atomic dipole couples to the modes do you get the full field operators $E_{\text{out}}^{(+)}(t)$. However, this simple approach is appropriate in some cases. When there is essentially one emitter in the problem and the detector is measuring only the light coming from that emitter (no competing scattering or multi-mode interference). Also, if you are typically looking at the radiation in the far field, the retarded time factor is well defined, and the geometry is such that the emitter's dipole pattern predominantly sets the field amplitude.

For a two-level system in a broadband squeezed bath (on resonance with the atomic transition), the leading modifications to the master equation can be summarized as follows,

$$\begin{aligned}\mathcal{L}_{\text{squeezed}}\rho = \Gamma\Bigg((N+1)\mathcal{D}[\sigma_{-}]\rho + N\mathcal{D}[\sigma_{+}]\rho - M\left[\sigma_{-}\rho\sigma_{-} - \frac{1}{2}\left\{\sigma_{-}^{2}, \rho\right\}\right]\Bigg) \\ - M^{*}[\sigma_{+}\rho\sigma_{+}-].\end{aligned} \tag{13.178}$$

The terms $(N+1)\mathcal{D}[\sigma_{-}]$ and $N\mathcal{D}[\sigma_{+}]$ can be seen as emission and absorption, respectively, similar to a thermal-bath interpretation but with possibly very large 'effective temperature' if N is large. The additional 'two-photon' terms $\propto M$, M^{*} couple the $\sigma_{-}\rho\sigma_{-}$ and $\sigma_{+}\rho\sigma_{+}$ channels, reflecting the phase-sensitive nature of squeezing. If the dissipation of the two-level atom comes from a squeezed vacuum (with zero displacement amplitude), that bath is *not* trivial. The environment has an enhanced 'effective photon number' $N = \sinh^{2}(r)$ and two-photon correlations $\langle aa\rangle \neq 0$ (the essence of squeezing)

Hence, the output field (or the field at a detector) is a *superposition* of the atom's emission (σ_{-}), and the squeezed reservoir itself, which is *not* vacuum and has non-zero fluctuations/correlations in certain quadratures. You can write the single-mode input–output relation as

$$a_{\text{out}}(t) = a_{\text{in}}(t) + \sqrt{\Gamma}\sigma_{-}(t). \tag{13.179}$$

But now $a_{\text{in}}(t)$ is a squeezed state (not vacuum). Thus,

$$E_{\text{out}}^{(+)}(t) \propto \underbrace{a_{\text{in}}(t)}_{\text{squeezed}} + \underbrace{\sqrt{\Gamma}\sigma_{-}(t)}_{\text{atomic emission}}. \tag{13.180}$$

If you compute correlation functions such as $\langle E_{\text{out}}^{(-)}E_{\text{out}}^{(+)}\rangle$ or higher-order noise spectra, *both* terms matter. You cannot simply identify the total field with σ_{-}. Part of your detected signal may be photons 'already present' in the squeezed reservoir (due to its phase-sensitive fluctuations) and part re-emitted by the atom. A squeezed bath has terms such as $\langle a_{\text{in}}a_{\text{in}}\rangle \neq 0$. These can mix with the atomic operator σ_{-} in the correlation functions. Furthermore, you cannot ignore a_{in}. Even though the squeezed vacuum has 'no coherent amplitude' (i.e. $\backslash\langle a_{\mathrm{in}}\rangle = 0$), it *does* have non-zero second-moment and two-photon correlations. These drastically affect the total field's fluctuations. One must keep them in any detailed calculation of $g^{(2)}(\tau)$, quadrature noise, etc.

References

[1] van Kampen N G 1981 Itô versus Stratonovich *J. Stat. Phys.* **24** 175–87

[2] Gardiner C and Zoller P 2004 *Quantum Noise: A Handbook of Markovian and non-Markovian Quantum Stochastic Methods with Applications to Quantum Optics* (Springer Series in Synergetics) (New York: Springer)

[3] Carmichael H J 1993 *An Open Systems Approach to Quantum Optics* (Berlin: Springer)

[4] Carmichael H J 1999 *Statistical Methods in Quantum Optics 1: Master Equations and Fokker–Planck Equations* (Berlin: Springer)

[5] Carmichael H J 2007 *Statistical Methods in Quantum Optics 2: Non-Classical Fields* (Theoretical and Mathematical Physics) (Berlin: Springer)

[6] Breuer H-P and Petruccione F 2002 *The Theory of Open Quantum Systems* (Oxford: Oxford University Press)

[7] Steck D 2007 *Quantum and atom optics notes The Quantum Well* https://publish.obsidian.md/myquantumwell/Literature/Steck%2C+D.+A.%2CQuantum+and+Atom+Optics%2C+2007+(last+revised+September+2021)

[8] van Kampen N G 2007 *Stochastic Processes in Physics and Chemistry* 3rd edn (North Holland: Elsevier)

[9] Walls D F and Milburn G J 2008 *Quantum Optics* 2nd edn (Berlin: Springer)

[10] Milburn G and Walls D F 1981 Production of squeezed states in a degenerate parametric amplifier *Opt. Commun.* **39** 401–4

[11] Carmichael H J 1993 Quantum trajectory theory for cascaded open systems *Phys. Rev. Lett.* **70** 2273–6

[12] Oche J G-B, Lu N, Pedrotti L M, Prasad S, Scully M O and Wódkiewicz K 1990 Treatment of the spectrum of squeezing based on the modes of the Universe. I. Theory and a physical picture *Phys. Rev.* A **41** 369–80

[13] Gea-Banacloche J, Lu N, Pedrotti L M, Prasad S, Scully M O and Wódkiewicz K 1990 Treatment of the spectrum of squeezing based on the modes of the Universe. II Applications *Phys. Rev.* A **41** 381–7

IOP Publishing

An Introduction to Quantum Optics (Second Edition)

An open systems approach

Perry Rice

Chapter 14

The Schwinger–Keldysh formalism

14.1 Preliminaries

In this chapter, we make connections to more traditional methods in quantum statistical mechanics for nonequilibrium systems. We will

- Find a path integral representation for the density operator.
- Include sources and identify Green's functions.
- Determine ways to extract averages and correlation functions.

Along the way, we will contrast these nonequilibrium methods with the approach for a system in thermal equilibrium, which perhaps you are somewhat familiar with. For the relatively simple quantum optics systems, this may seem like using a bulldozer to dig a small hole. But the power of this method comes in handy when looking at complex systems with many degrees of freedom, as in many-body physics. This includes the study of Bose–Einstein and Fermi gases, semiconductors, and large networks of devices.

Let us start with a collection of harmonic oscillators and define the partition function

$$Z = \sum_n e^{-\beta E_n}. \tag{14.1}$$

The relative probability of a level occupation is $p_n = \exp(-\beta E_n)$ in thermal equilibrium, so Z is the sum of these. The actual probability of a level occupation is $P_n = p_n / Z$ so that

$$\sum_n P_n = 1. \tag{14.2}$$

Our collection of harmonic oscillators has energy levels $E_n = n\hbar\omega$ (ignoring zero point energy), which becomes

doi:10.1088/978-0-7503-6014-2ch14

$$
\begin{aligned}
Z &= \sum_n e^{-\beta n\hbar\omega} \\
&= \sum_n x^n \\
&= \frac{1}{1-x}
\end{aligned}
\tag{14.3}
$$

with $x \equiv e^{\beta\hbar\omega}$. So

$$
Z = \frac{1}{e^{-\beta\hbar\omega} - 1}. \tag{14.4}
$$

We can calculate the average energy

$$
\begin{aligned}
\langle E \rangle &= \frac{1}{Z}\sum_n E_n e^{-\beta E_n} \\
&= \frac{-1}{Z}\frac{\partial Z}{\partial\beta} \\
&= \frac{\hbar\omega}{e^{\beta\hbar\omega} - 1}.
\end{aligned}
\tag{14.5}
$$

Often, one defines

$$
W = ln(Z) \tag{14.6}
$$

from which we can find

$$
\langle E \rangle = \frac{\partial W}{\partial\beta}. \tag{14.7}
$$

In the classical limit, $\beta\hbar\omega \to 0$ ($kT \gg \hbar\omega$), the energy level splitting is small compared to kT, and we find

$$
\langle E \rangle = kT. \tag{14.8}
$$

This is the result if we used an integral over continuous energies ($\hbar \to 0$). Half of this average is due to the kinetic energy ($\langle mv^2/2\rangle$), and half is due to the potential energy ($\langle k_{\rm spr}x^2/2\rangle$). This is the basis of the equipartition theorem, where each degree of freedom has an average energy of $kT/2$, or for one mole $N_A kT/2 = RT/2$. In our example, we have two degrees of freedom, each contributing $kT/2$ to the average energy. In three dimensions, we would have $3R$, as in the case of a noble gas. The specific heat at a constant volume is

$$
\begin{aligned}
C_V &= \frac{\partial\langle E \rangle}{\partial T} \\
&= k\left(\frac{\hbar\omega}{kT}\right)^2 \frac{e^{\hbar\omega/kT}}{(e^{\hbar\omega/kT} - 1)^2},
\end{aligned}
\tag{14.9}
$$

which goes to zero for low temperature, due to $kT < \hbar\omega$, and the higher energy levels freeze out. In the high-temperature limit, or as $\hbar \to 0$, we find $C_v = k$, or for a mole of three dimensional oscillators with six degrees of freedom, $C_V = 3R$, the classical result of Dulong and Petit. The low-temperature behavior of the specific heat of a diatomic gas provided early evidence of the need to have quantized energy levels. The main lesson here is that the partition function will contain all knowable information for a system, and by taking functional derivatives, we can extract expectation values; also two-time correlation functions as we will see below.

We define the partition function

$$Z(t) = \mathrm{tr}(\rho(t)). \tag{14.10}$$

This is the essential connection between the density matrix and the partition function. This is guided by the choice in equilibrium statistical mechanics

$$Z(t) = \mathrm{tr}(\rho_{\mathrm{thermal}}) = \mathrm{tr}(e^{-\beta H}). \tag{14.11}$$

This is derived from the definition of a density matrix and the use of the Boltzmann distribution

$$\rho = \sum_{mn} R_{nm} |E_n\rangle\langle E_m| \tag{14.12}$$

and

$$R_{nm} = e^{-\beta E_n}\delta_{mn}. \tag{14.13}$$

These lead to

$$\begin{aligned}\rho &= \sum_n e^{-\beta E_n} |E_n\rangle\,\langle E_m| \\ &= \sum_n e^{-\beta H} |E_n\rangle\,\langle E_m| \\ &= e^{-\beta H} \sum_n |E_n\rangle\,\langle E_m| \\ &= e^{-\beta H}.\end{aligned} \tag{14.14}$$

For time-dependent systems, we have

$$\begin{aligned}\langle O(t)\rangle &= \eta\,\mathrm{Tr}(O\rho(t)) \\ &= \frac{\mathrm{Tr}(OU_{-\infty,\,t}\rho(-\infty)U^{\dagger}{}_{-\infty,\,t})}{\mathrm{Tr}(\rho_{-\infty})},\end{aligned} \tag{14.15}$$

where we have the propagator in time for a density operator

$$U(t - t') = e^{-i(t-t')\mathcal{L}} \tag{14.16}$$

$$U^{\dagger}(t - t') = e^{i(t-t')\mathcal{L}}, \tag{14.17}$$

where we have used the time evolution for the density operator, as demanded by

$$\dot{\rho} = \mathcal{L}\rho. \tag{14.18}$$

Notice that the time evolution operator acts on the density matrix from the right *and* the left, two copies. We keep the normalization of the density matrix fluid ($\mathrm{Tr}(\rho)$) for convenience. If the evolution is non-unitary, we still require it to preserve the trace of the density operator under the Liouvillian evolution, but it will be convenient to make this choice. If the Hamiltonian (and hence Liouvillian) is time dependent, we must use

$$U_{t,t_0} = \begin{cases} \mathcal{T}\exp\left[-\frac{i}{\hbar}\int_{t_0}^{t} \mathcal{L}(t_1)dt_1\right] & : t \geqslant t_0 \\ \tilde{\mathcal{T}}\exp\left[-\frac{i}{\hbar}\int_{t_0}^{t} \mathcal{L}(t_1)dt_1\right] & : t < t_0 \end{cases}, \tag{14.19}$$

where $\mathcal{T}$ and $\tilde{\mathcal{T}}$ are the time ordering and anti-time ordering operators, respectively. We notice that in equation (14.19) that the Liouvillian operator acts both to the right and the left of the density matrix, as usual.

14.2 Green's functions

Formally, a Green's function is a solution of a linear differential equation with a Dirac delta inhomogeneous source with homogeneous boundary conditions [1, 2]. The differential equation we will be concerned with has a special inhomogeneity function, the Dirac delta $\delta(x - x')$. In other words, this object is defined to be zero when $x \neq x'$, and infinite at $x = x'$:

$$\delta(x - x') = \begin{cases} \infty, & \mathrm{at}\, x = x' \\ 0, & \mathrm{otherwise} \end{cases}. \tag{14.20}$$

Rigorously, the Dirac delta is not a function, since it would require a definite value for each point in its domain, but is instead classified as a distribution. Its most important property is

$$f(x) = \int_{-\infty}^{\infty} \delta(x - x')f(x')dx', \tag{14.21}$$

where $f(x)$ is any continuous function of x. Now shall we return to the classical Green's functions. To put the mathematical problem in perspective, imagine one would like to solve a partial linear inhomogeneous differential equation, say,

$$Df(x) = g(x), \tag{14.22}$$

where D a linear differential operator, $f(x)$ is the desired solution, and $h(x)$ is the inhomogeneity source. The particular solution $f(x)$ can be formally found with the aid of a function $G(x, x')$

$$f(x) = \int G(x, x')h(x')dx', \tag{14.23}$$

where the Green's function $G(x, x')$ is defined as the solution of a differential equation with a delta inhomogeneity:

$$DG(x, x') = \delta(x - x'). \tag{14.24}$$

To verify this, act with D on both sides of equation and use the Dirac delta fundamental property, equation (14.20). Note that D acts on the x coordinate, keeping x' fixed.

We consider the Green's functions as a 'building block' for finding the particular solution $f(x)$, since they are solutions to the delta-impulse equations. In signal processing fields, the Green's function is often referred to as a response function, connecting a perturbation or 'input signal' $h(x)$ to the 'output' $f(x)$.

Let us consider an example from classical electromagnetics. Poisson's equation is given by

$$\nabla^2\Phi(\mathbf{r}) = -\rho(\mathbf{r}), \tag{14.25}$$

where $\Phi(\mathbf{r})$ is the scalar potential, $\rho(\mathbf{r})$ is the source term (e.g. charge density in electrostatics), and ∇^2 is the Laplacian operator. The Green's function $G(\mathbf{r}, \mathbf{r}')$ satisfies

$$\nabla^2 G(\mathbf{r}, \mathbf{r}') = -\delta(\mathbf{r} - \mathbf{r}'). \tag{14.26}$$

So that the solution to Poisson's equation is formally given by

$$\Phi(\mathbf{r}) = \int G(\mathbf{r}, \mathbf{r}')\rho(\mathbf{r}')d^d\mathbf{r}', \tag{14.27}$$

which represents the convolution of the Green's function with the source distribution. We know from our experience that by associating the Dirac delta $\delta(\mathbf{r} - \mathbf{r}')$ with a point charge at $\mathbf{r}'$, we can find a corresponding potential from Coulomb's law; the electric potential is simply

$$\varphi(\mathbf{r}) = \frac{1}{4\pi\lambda_0}\frac{1}{|\mathbf{r} - \mathbf{r}'|}. \tag{14.28}$$

We also have by the principal of superposition

$$\varphi(\mathbf{r}) = \frac{1}{4\pi\lambda_0}\int\frac{\rho(\mathbf{r}')}{|\mathbf{r} - \mathbf{r}'|}d\mathbf{r}'. \tag{14.29}$$

Therefore, the appropriate Green's function to the localized charge problem of is

$$G(\mathbf{r}, \mathbf{r}') = -\frac{1}{4\pi}\frac{1}{|\mathbf{r} - \mathbf{r}'|}. \tag{14.30}$$

What about the delta function? The equation to solve in spherical coordinates is

$$\frac{1}{r^2}\frac{d}{dr}\left(r^2\frac{dG}{dr}\right) = -\delta(\mathbf{r} - \mathbf{r}'). \tag{14.31}$$

For $r \neq 0$, the right-hand side is zero, so we solve

$$\frac{1}{r^2}\frac{d}{dr}\left(r^2\frac{dG}{dr}\right) = 0. \tag{14.32}$$

Multiplying by r^2 and integrating once we find

$$\frac{d}{dr}\left(r^2\frac{dG}{dr}\right) = C_1 \tag{14.33}$$

and integrating again,

$$r^2\frac{dG}{dr} = C_1 r + C_2. \tag{14.34}$$

Dividing by r^2:

$$\frac{dG}{dr} = \frac{C_1}{r} + \frac{C_2}{r^2}. \tag{14.35}$$

Integrating once more:

$$G(r) = C_1 \ln r - \frac{C_2}{r} + C_3. \tag{14.36}$$

Because we expect $G(r)$ to be finite at infinity and decay at large distances, we set $C_1 = 0$, so we obtain

$$G(r) = -\frac{C}{r}. \tag{14.37}$$

To determine C, we integrate both sides of the equation over a small sphere of radius λ around $r = 0$ and use Gauss's theorem:

$$\int_{\text{sphere}} \nabla^2 G dV = -\int_{\text{sphere}} \delta(\mathbf{r} - \mathbf{r}')dV = -1. \tag{14.38}$$

Using Gauss's theorem

$$\oint_{\text{sphere}} \nabla G \cdot d\mathbf{S} = \oint_{\text{sphere}} \frac{dG}{dr} r^2 d\Omega = -1 \tag{14.39}$$

and substituting $G(r) = -\frac{C}{r}$:

$$\frac{dG}{dr} = \frac{C}{r^2}. \tag{14.40}$$

So,

$$\oint_{\text{sphere}} \frac{C}{r^2} r^2 d\Omega = C \oint_{\text{sphere}} d\Omega. \tag{14.41}$$

Because the integral of $d\Omega$ over a full sphere is 4π,

$$C(4\pi) = 1. \tag{14.42}$$

Thus,

$$C = \frac{1}{4\pi}. \tag{14.43}$$

Thus, the Green's function is

$$G(\mathbf{r}, \mathbf{r}') = \frac{1}{4\pi|\mathbf{r} - \mathbf{r}'|} \tag{14.44}$$

as we anticipated.

We can also use a formal operator relation. Using the inverse operator of the Laplacian, the Green's function is formally expressed as

$$G = \frac{1}{\nabla^2} = [\ \nabla^2\ ]^{-1}. \tag{14.45}$$

This means that the potential is given by

$$\Phi = [\ \nabla^2\ ]^{-1}\rho. \tag{14.46}$$

This inverse operator notation is formal and must be interpreted carefully.

For the time-independent Schrödinger equation:

$$(H - E)\psi(\mathbf{r}) = \delta(\mathbf{r} - \mathbf{r}'). \tag{14.47}$$

The Green's function $G(\mathbf{r}, \mathbf{r}'; E)$ satisfies

$$(H - E)G(\mathbf{r}, \mathbf{r}'; E) = \delta(\mathbf{r} - \mathbf{r}'), \tag{14.48}$$

where H is the Hamiltonian. If the Schrödinger equation can be solved exactly for eigenfunctions $\psi_n(x)$ with eigenvalues E_n, we can express the Green's function as

$$G(x, x'; E) = \sum_n \frac{\psi_n(x)\psi_n^*(x')}{E_n - E}. \tag{14.49}$$

A common approach is to split the Hamiltonian into a free part and a potential,

$$H = H_0 + V, \tag{14.50}$$

where $H_0 = -\frac{\hbar^2}{2m}\frac{d^2}{dx^2}$ is the kinetic energy operator. Then, Green's function can be rewritten as

$$G = \frac{1}{H_0 + V - E}. \tag{14.51}$$

Using the Dyson series expansion [3], the normal Taylor expansion but keeping track of the order of noncommuting operators, we can express G in terms of the free Green's function G_0,

$$G = \frac{1}{H_0 - E} + \frac{1}{H_0 - E}V\frac{1}{H_0 - E} + \frac{1}{H_0 - E}V\frac{1}{H_0 - E}V\frac{1}{H_0 - E} + \ldots \tag{14.52}$$

or more compactly as the resolvent expansion

$$G = G_0 + G_0VG_0 + G_0VG_0VG_0 + \ldots, \tag{14.53}$$

where G_0 is the free-particle Green's function:

$$G_0 = \frac{1}{H_0 - E}. \tag{14.54}$$

Equation (14.52) can often be translated into a series of Feynman diagrams [4, 5] by associating the correct terms with propagation, vertices, and the like. Instead of summing the series explicitly, we can write a formal integral equation:

$$G = G_0 + G_0 VG, \tag{14.55}$$

which can be solved iteratively. In position space, this becomes

$$G(x, x'; E) = G_0(x, x'; E) + \int G_0(x, x''; E)V(x'')G(x'', x'; E)dx''. \tag{14.56}$$

This is the Lippmann–Schwinger equation for the Green's function.

The propagator $K(\mathbf{r}, t; \mathbf{r}', t')$ describes how a quantum state evolves from one point in spacetime to another:

$$\psi(\mathbf{r}, t) = \int d^3r' K(\mathbf{r}, t; \mathbf{r}', t')\psi(\mathbf{r}', t'). \tag{14.57}$$

It is given by

$$K(\mathbf{r}, t; \mathbf{r}', t') = \langle \mathbf{r}| e^{-iH(t-t')/\hbar} |\mathbf{r}'\rangle. \tag{14.58}$$

This means that the propagator is the fundamental solution to the time-dependent Schrödinger equation. The time-dependent propagator is related to the time-independent Green's function through a Fourier transform:

$$G(\mathbf{r}, \mathbf{r}'; E) = \int_0^\infty dt e^{iEt/\hbar} K(\mathbf{r}, t; \mathbf{r}', 0). \tag{14.59}$$

Conversely, the propagator can be obtained from Green's function:

$$K(\mathbf{r}, t; \mathbf{r}', 0) = \int_{-\infty}^\infty \frac{dE}{2\pi i} e^{-iEt/\hbar} G(\mathbf{r}, \mathbf{r}'; E). \tag{14.60}$$

Thus, the propagator and Green's function are Fourier transforms of each other, or the propagator is the Green's function in the time domain. In the path integral formalism, the propagator (two-point function) is

$$G^{(2)}(x, y) = \frac{\int \mathcal{D}\phi e^{iS[\phi]} \phi(x)\phi(y)}{\int \mathcal{D}\phi e^{iS[\phi]}}, \tag{14.61}$$

where S is the action, the integral of the Lagrangian. In terms of the quantum field operators, the propagator (two-point correlation function) is

$$D_F(x - y) = \langle 0| T\{\phi(x)\phi(y)\} |0\rangle, \tag{14.62}$$

with $\phi(y)$ creates a particle at spacetime point y, $\phi(x)$ annihilates a particle at spacetime point x. This is an intuitive and physically meaningful way to think about the propagator in quantum field theory (QFT). The propagator describes the amplitude of a particle to be created at some spacetime point, travel through spacetime, and be annihilated at another point The propagator gives the probability amplitude for the particle to travel from y to x. This is analogous to the position-space representation of the evolution of a quantum mechanical particle. The Green's functions are related to the effective action via the partition function (or generating functional)

$$Z[J] = \int \mathcal{D}\phi e^{iS[\phi]+i\int J\phi}. \tag{14.63}$$

The two-point function (which is still the propagator in an interacting theory) is

$$G_c^{(2)}(x, y) = \left. \frac{\delta^2 W[J]}{\delta J(x)\delta J(y)} \right|_{J=0}, \tag{14.64}$$

where $W[J] = -i \ln Z[J]$ In the interacting theory, the propagator differs from the free case because the interactions modify it via self-energy corrections.

14.3 Path integral representation of the density matrix

In this section, we will find a path integral representation of the density matrix in terms of an action, the Schwinger–Keldysh (SK) action (figure 14.1). The path integral will be of the form [6–8, 9–11]

$$\rho\left(\phi_+, \phi'_-\right) = \int d\phi_+ d\phi'_- \rho\left(\phi_+, \phi'_-\right) \int D\phi_+ D\phi_- \exp\{iS[\phi, \phi/]\}. \tag{14.65}$$

Here $S[\phi_+, \phi_-]$ is the Schwinger–Keldysh action, which involves two c-number fields ϕ_+ and ϕ_-. We integrate the complex exponential of this action (to be determined) over all possible 'paths' in (ϕ_+, ϕ_-) space. Operators seem to have

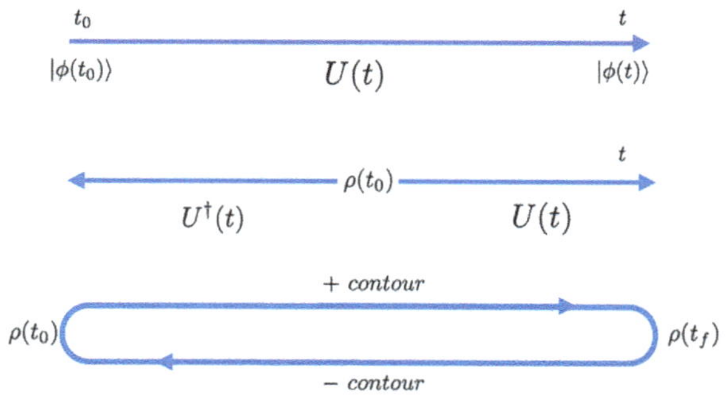

Figure 14.1. Time evolution of the wave function, density matrix, and the Schwinger–Keldysh contour.

vanished; we will find that this path integral is over all possible coherent states, and we use $a|\alpha\rangle = \alpha|\alpha\rangle$. For unitary evolution, we have

$$S_{\mathrm{SK}} = S\left[\phi_+\right] - S[\phi_-] \tag{14.66}$$

and only one branch of time evolution is needed.

Let us recall how we obtain information from the density matrix,

$$\langle \mathcal{O}(t)\rangle = \frac{1}{\mathrm{Tr}\{\rho(t)\}}\mathrm{Tr}\left\{\underbrace{\mathcal{U}_{-\infty,t}}_{\text{forward}}\ \mathcal{O}\ \underbrace{\mathcal{U}_{t,-\infty}}_{\text{backward}}\ \rho(-\infty)\right\}. \tag{14.67}$$

We often seek the nonequilibrium steady state. Let us recall the identity

$$I = U_{t,\infty}U_{\infty,t}. \tag{14.68}$$

Inserting this into our expression for $\langle O(t)\rangle$:

$$\langle O(t)\rangle = \frac{\mathrm{Tr}\left\{U_{-\infty,\infty}U_{\infty,t}OU_{t,-\infty}\rho(-\infty)\right\}}{\mathrm{Tr}\{\rho(-\infty)\}}. \tag{14.69}$$

The general plan is to have the propagator on the right side of ρ evolve from $-\infty \to t$ as usual (figure 14.2). The propagator on the left side starts at $-\infty$, goes to ∞, and then returns to t, avoiding the specification of a final state. This means that we should use the integration contour in figure 14.3. As we are interested in the

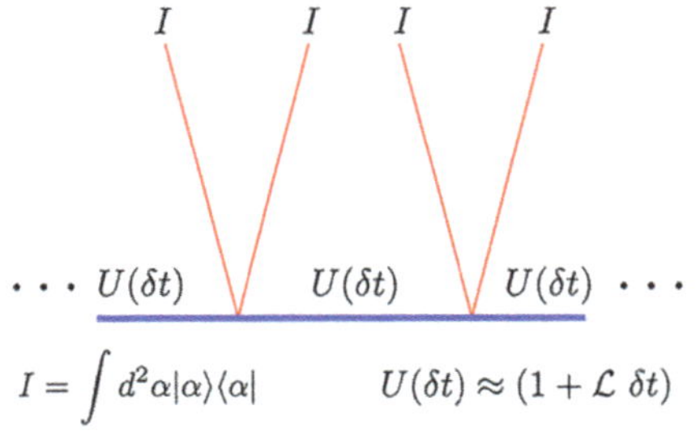

Figure 14.2. Closeup of identities on the Schwinger–Keldysh contour.

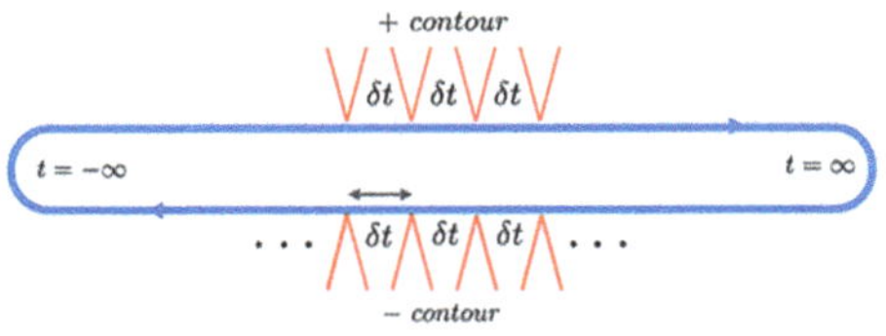

Figure 14.3. Insertion along the whole contour of identities on the Schwinger–Keldysh contour.

steady state, we extend the time integration back to $-\infty$ and ignore the initial conditions, assuming that the transient effects have died out. We have done the same in the quantum regression theorem, assuming the system has reached a steady state, and solve for the time dependence of fluctuations about the steady state. The interpretation of such an expression is as follows: we come from $t = -\infty$, at t we get O inserted, evolve to $t = +\infty$ and then we close the contour going from ∞ to $-\infty$.

14.4 Lindblad to Keldysh

Substituting the coherent state resolution of identity into the Lindblad equation, the evolution of $\rho(\phi_+^*, \phi_-, t)$ becomes [12]

$$\begin{aligned}\frac{d\rho\left(\phi_+^*, \phi_-, t\right)}{dt} =& - i(H(\phi_+) - H(\phi_-))\rho\left(\phi_+^*, \phi_-, t\right)\\ &+ \sum_i \mathcal{D}_i[\phi_+, \phi_-]\rho\left(\phi_+^*, \phi_-, t\right),\end{aligned} \tag{14.70}$$

where $H(\phi)$ is the Hamiltonian in the coherent state basis, and $\mathcal{D}_i[\phi_+, \phi_-]$ encapsulates the dissipative Lindblad terms,

$$\mathcal{D}_i[\phi_+, \phi_-] = L_i(\phi_+)\rho L_i^\dagger(\phi_+) - \frac{1}{2}\left(L_i^\dagger L_i(\phi_+) + L_i^\dagger L_i(\phi_-)\right)\rho. \tag{14.71}$$

We accomplish this via a Trotter decomposition of the evolution into finite time steps, and place a coherent state identity before and after the short time evolution

$$e^{\Delta t\mathcal{L}} \approx e^{\Delta t\mathcal{L}_H}e^{\Delta t\mathcal{L}_D}, \tag{14.72}$$

where $\mathcal{L}_H(\rho) = -i[H, \rho]$ represents the Hamiltonian evolution, and $\mathcal{L}_D(\rho)$ contains the dissipative terms. At each time step, we insert the coherent state resolutions of the identity into the time-evolution operator. The time evolution can now be expressed as a path integral over the coherent state variables. This is the same as the Feynman path integral [4] where one inserts identities of the form

$$\mathbb{I} = \int dx|x\rangle\langle x| \tag{14.73}$$

and ends up with a path integral over all possible trajectories in space. For a single time step, the density matrix evolves as

$$\rho\left(\phi_+^*, \phi_-, t + \Delta t\right) = \int\frac{d^2\phi_0}{\pi}\langle\phi_+|e^{\Delta t\mathcal{L}}|\phi_0\rangle\rho\left(\phi_0^*, \phi_0, t\right)\langle\phi_0|e^{\Delta t\mathcal{L}}\,|\phi_-\rangle^*. \tag{14.74}$$

After N time steps ($t = N\Delta t$), the density matrix evolves as

$$\rho(t) = (e^{\mathcal{L}\Delta t})^N \rho(0). \tag{14.75}$$

The forward evolution corresponds to $|\phi_+\rangle$, and for a single time slice

$$\left\langle \phi_+^{(j+1)} \middle| e^{-iH\Delta t} \middle| \phi_+^{(j)} \right\rangle \approx exp\left(\phi_+^* + \frac{\phi_+^{(j+1)} - \phi_+^{(j)}}{\Delta t} - H[\phi_+] \right)\Delta t. \tag{14.76}$$

where j indexes the time steps. At each time step, we insert the coherent state resolution of the identity

$$\mathbb{I} = \int \frac{d^2\phi_+^{(j)} d^2\phi_-^{(j)}}{\pi^2} \left| \phi_+^{(j)} \right\rangle \left\langle \phi_-^{(j)} \right|. \tag{14.77}$$

In the continuum limit ($\Delta t \to 0$), the forward branch action becomes

$$S_+ = \int dt \left[\phi_+^* \dot{\phi}_+ - H[\phi_+] \right]. \tag{14.78}$$

The backward evolution corresponds to $|\phi_-\rangle$, and a similar derivation yields

$$S_- = \int dt \left[\phi_-^* \dot{\phi}_- - H[\phi_-] \right]. \tag{14.79}$$

Taking the trace of the density matrix, the partition function can then be expressed as an integral over all possible trajectories of the forward and backward coherent states:

$$Z = \int \prod_{j=1}^N \frac{d^2\phi_+^{(j)} d^2\phi_-^{(j)}}{\pi^2} e^{i(S_+ - S_-)}, \tag{14.80}$$

where S_+ and S_- are the actions for the forward and backward branches, respectively, and we define the measure

$$D\phi_+ D\phi_- = \prod_{j=1}^N \frac{d^2\phi_+^{(j)} d^2\phi_-^{(j)}}{\pi^2}. \tag{14.81}$$

The Lindblad terms contribute to both damping and noise. In the path integral, these appear as a linear term coupling ϕ_+ and ϕ_-, and a quadratic term in $\phi_q = \phi_+ - \phi_-$, representing noise. Notice that these two terms couple the forward and backward evolution, resulting in a non-unitary evolution.

We now transform into the Keldysh basis (called a Keldysh rotation) and define the classical and quantum components of the fields:

$$\phi_c = \frac{\phi_+ + \phi_-}{2}, \quad \phi_q = \phi_+ - \phi_-. \tag{14.82}$$

The quantum component has a zero mean and thus involves fluctuations, the classical component is related to expectation values. We rewrite the partition function in terms of ϕ_c and ϕ_q,

$$\mathcal{Z} = \int \mathcal{D}[\phi_c, \phi_q] e^{iS_{\text{SK}}[\phi_c, \phi_q]}. \tag{14.83}$$

For the Hamiltonian contribution to the action, we find

$$S_H = \int dt \phi_q \left(i\frac{\partial \phi_c}{\partial t} - H(\phi_c) \right). \tag{14.84}$$

The dissipative Lindbladian terms yield

$$S_{\text{diss}} = \int dt \phi_q \mathcal{L}[\phi_c] - \frac{i}{2} \phi_q^2 \mathcal{D}[\phi_c]. \tag{14.85}$$

The quadratic term in ϕ_q corresponds to the environmental noise, and it imposes a penalty on the action for fluctuating away from the steady state. Combining all terms, the SK action becomes:

$$S_{\text{SK}} = \int dt \left[\phi_q (\dot{\phi}_c - H(\phi_c) - \mathcal{L}[\phi_c]) - \frac{i}{2} \phi_q^2 \mathcal{D}[\phi_c] \right]. \tag{14.86}$$

14.5 Feynman–Vernon influence functional

Equivalently, we can obtain a path integral description as

$$\int D\phi_R D\phi_L e^{i(S(\phi) - S(\phi'))} e^{iS_{\text{FV}}}, \tag{14.87}$$

where S_{FV} is the Feynman–Vernon influence function. The Feynman–Vernon formalism is widely used to study open quantum systems [13], particularly systems coupled to a bath (or reservoir) of harmonic oscillators. Here, we focus on the quantum oscillator coupled to a reservoir of oscillators at zero temperature and derive the Feynman–Vernon path integral representation for the reduced density matrix of the system. We consider a single quantum harmonic oscillator coupled to a bath of harmonic oscillators. The system is initially decoupled from the bath (Born approximation, factorization of the density matrix, but the system + bath evolves as a closed system. At some point, we trace out the bath degrees of freedom to focus on the system's dynamics. The coupling between the system and the bath occurs through the usual linear coupling model. The total Hamiltonian is

$$H_{\text{total}} = H_S + H_B + H_{\text{int}}, \tag{14.88}$$

where the system Hamiltonian H_S is

$$H_S = \hbar\omega_0\left(a^\dagger a + \frac{1}{2}\right), \tag{14.89}$$

where a and $a^\dagger$ are the annihilation and creation operators for the system oscillator, satisfying

$$[a, a^\dagger] = 1. \tag{14.90}$$

The bath Hamiltonian H_B is

$$H_B = \sum_k \hbar\omega_k\left(b_k^\dagger b_k + \frac{1}{2}\right), \tag{14.91}$$

where b_k and $b_k^\dagger$ are the annihilation and creation operators for the kth bath mode, satisfying

$$\left[b_k, b_k^\dagger\right] = 1 \tag{14.92}$$

with interaction Hamiltonian H_{int}. Under the rotating wave approximation, we retain only the energy-conserving terms $ab_k^\dagger$ and $a^\dagger b_k$, discarding counter-rotating terms ab_k and $a^\dagger b_k^\dagger$. The interaction Hamiltonian becomes

$$H_{\text{int}} = \sum_k g_k\left(ab_k^\dagger + a^\dagger b_k\right), \tag{14.93}$$

where g_k are the coupling constants. The total density matrix evolves according to

$$\rho_{\text{total}}\left(x_f, x_f', t\right) = \int dx_i dx_i' \rho_{\text{total}}\left(x_i, x_i', 0\right)\mathcal{K}\left(x_f, x_f', t; x_i, x_i', 0\right). \tag{14.94}$$

In the path integral formalism, the total propagator can be written as

$$\mathcal{K} = \int DxDq \exp\left(\frac{i}{\hbar}S_{\text{total}}\right), \tag{14.95}$$

where $S_{\text{total}} = S_S[x] + S_B[q] + S_{\text{int}}[x, q]$ is the action of the total system and x and q are the coordinates for the a and b oscillators. The bath degrees of freedom $\{q_j\}$ appear quadratically and can be integrated out exactly, leading to an effective influence functional for the system dynamics.

The bath degrees of freedom $\{q_j\}$ appear quadratically and can be integrated out exactly, leading to an effective influence functional for the system dynamics. To derive the reduced density matrix of the system, we trace out the bath degrees of freedom

$$\rho_S\left(x_f, x_f', t\right) = \int dx_i dx_i' \rho_S(x_i, x_i', 0)\mathcal{F}[x, x'] \int_{x_i}^{x_f} \mathcal{D}x \int_{x_i'}^{x_f'} \mathcal{D}x' e^{\frac{i}{\hbar}S_S[x] - \frac{i}{\hbar}S_S[x']}. \tag{14.96}$$

The influence functional $\mathcal{F}[x, x']$ encapsulates the effect of the bath on the system and takes the form

$$\mathcal{F}[x, x'] = \exp\left(-\frac{1}{\hbar}\int_0^t ds \int_0^t ds'[L(x(s), x'(s'))]\right). \tag{14.97}$$

For a harmonic bath at zero temperature, the influence functional introduces a non-local effect in time (arising from the memory kernel of the bath). Specifically, it can be expressed as

$$L(x(s), x'(s')) = \eta(s - s')[x(s) - x'(s)][x(s') - x'(s')], \tag{14.98}$$

where $\eta(s - s')$ is the bath correlation function, which is determined by the spectral density of the bath. In the Markovian limit

$$\eta(s - s') \approx 2\hbar\kappa\delta(s - s'). \tag{14.99}$$

This reduces the influence functional to

$$\mathcal{F}[x, x'] = \exp\left(-\frac{2\kappa}{\hbar}\int_0^t ds[x(s) - x'(s)]^2\right). \tag{14.100}$$

We see that it imposes a penalty on the action for the differences in the forward and backward evolution. The total Feynman–Vernon path integral for the reduced density matrix of the system becomes

$$\rho_S\left(x_f, x_f', t\right) = \int dx_i dx_i' \rho_S(x_i, x_i', 0) \int_{x_i}^{x_f} \mathcal{D}x \int_{x_i'}^{x_f'} \mathcal{D}x' \mathcal{F}[x, x'] e^{\frac{i}{\hbar}S_S[x] - \frac{i}{\hbar}S_S[x']}, \tag{14.101}$$

which agrees with the Schwinger–Keldysh approach with

$$S_{\rm SK} = S_S(x) - S_S(x') + i\left(\frac{2\kappa}{\hbar}\int_0^t ds[x(s) - x'(s)]^2\right). \tag{14.102}$$

14.6 Correlation functions

Now, we are in general interested in the two-time correlation functions

$$\langle O_1(t)O_2(t')\rangle . \tag{14.103}$$

To do this, we must in general find the two-time correlation functions:

Lesser Green's function: $G^{<}(\tau) = i\langle a^\dagger(\tau)a(0)\rangle$.

Greater Green's function: $G^{>}(\tau) = -i\langle a(0)a^\dagger(\tau)\rangle$.

Retarded Green's function: $G^R(\tau) = -i\theta(\tau)\langle\{a(\tau), a^\dagger(0)\}\rangle$.

Advanced Green's function: $G^A(\tau) = i\theta(-\tau)\langle\{a(\tau), a^\dagger(0)\}\rangle$.

Let us outline the procedure:

- Pass from a Lindblad master equation to an SK action, as we have done.
- Read off the inverse of the Green's function matrix from the SK action.
- Invert that matrix and identify the Green's functions.

Lots of fussing, but it is not very bad. When we are interested in a stationary state but would like to obtain information on temporal correlation functions at arbitrarily long time differences, it is useful to perform the limit $t_0 \to -\infty$, $t_f \to +\infty$. As stated earlier, we are interested in fluctuations in the steady state, which we extract from

$$Z = \int \mathcal{D}\left[\phi_+, \phi_+^*, \phi_-, \phi_-^*\right] e^{iS} \tag{14.104}$$

$$S = \int_{-\infty}^{\infty} dt\left(\phi_+^* i\partial_t\phi_+ - \phi_-^* i\partial_t\phi_- - i\mathcal{L}\left(\phi_+^*, \phi_+, \phi_-^*, \phi_-\right)\right). \tag{14.105}$$

As anticipated above, correlation functions can be obtained by introducing the source terms $J_\pm = (j_\pm, j_\pm^*)$ that couple to the fields $\phi_\pm = (\phi_\pm, \phi_\pm^*)$. We start with the path integral definition

$$Z[J] = \int \mathcal{D}\psi^\dagger \mathcal{D}\psi e^{iS[\psi,\, \psi^\dagger] + i\int dt(J^\dagger\psi + \psi^\dagger J)}. \tag{14.106}$$

By defining the connected generating function as

$$W[J] = -i \ln Z[J]. \tag{14.107}$$

Taking the functional derivatives of $W[J]$ with respect to the source J yields connected correlation functions, defined as

$$\frac{\delta W[J]}{\delta J(t)} = \frac{1}{Z[J]} \int \mathcal{D}\phi\phi(t) e^{iS[\phi] + i\int dtJ\phi} = \langle \phi(t) \rangle_J, \tag{14.108}$$

which gives the expectation value of the field in the presence of the source J. A second derivative

$$\frac{\delta^2 W[J]}{\delta J(t)\delta J(t')} = \langle \phi(t)\phi(t')\rangle_J - \langle \phi(t) \rangle_J \langle \phi(t')\rangle_J \tag{14.109}$$

gives the connected two-point correlation function

$$G_c(t, t') = \langle T_c\phi(t)\phi(t')\rangle - \langle \phi(t)\rangle \langle \phi(t')\rangle, \tag{14.110}$$

where T_c is contour ordering in the Keldysh contour. This makes contact with the method we used for the classical partition functions. For an interacting theory, the action has additional terms

$$S = S_{\text{free}} + S_{\text{int}}, \tag{14.111}$$

where S_{int} could be, for example, a quartic interaction

$$S_{\text{int}} = -\frac{g}{2} \int dt\psi^\dagger(t)\psi^\dagger(t)\psi(t)\psi(t). \tag{14.112}$$

Coming from a Lindblad master equation, our dissipative terms are already in the action, and possibly a driving term if we include it in H. Now, if the partition function is no longer Gaussian, we can expand it perturbatively

$$Z[J] = \int \mathcal{D}\psi^\dagger \mathcal{D}\psi e^{iS_{\text{free}}} e^{iS_{\text{int}} + i\int dtJ^\dagger\psi + i\int dt\psi^\dagger J}. \tag{14.113}$$

Expanding e^{iS} as a power series, we obtain an expression for $W[J]$ that includes the interaction corrections. The full Green's function now satisfies the Dyson equation

$$G(t, t') = G_0(t, t') + \int dt_1 dt_2 G_0(t, t_1)\Sigma(t_1, t_2)G(t_2, t') + \cdots, \tag{14.114}$$

where $\Sigma(t_1, t_2)$ is the self-energy, computed from the functional derivatives of $W[J]$. In quantum field theory, the Green's function (or propagator) describes the response of a system to an external perturbation. It satisfies an equation of the form The Green's function satisfies the equation

$$(i\partial_t - H)G(t, t') = \delta(t - t'). \tag{14.115}$$

This means that the operator $(i\partial_t - H)$ acts as the inverse Green's function

$$G^{-1}(t, t') = (i\partial_t - H)\delta(t - t'). \tag{14.116}$$

A general quadratic action for a noninteracting system is written in the form

$$S = \int dt dt' \psi^\dagger(t)K(t, t')\psi(t'), \tag{14.117}$$

where $K(t, t')$ is the kernel that determines the dynamics of $\psi(t)$. We can identify $K(t, t')$ as

$$K(t, t') = G^{-1}(t, t') = (i\partial_t - H)\delta(t - t'). \tag{14.118}$$

Thus, the action takes the form

$$S = \int dt dt' \psi^\dagger(t)(i\partial_t - H)\delta(t - t')\psi(t'). \tag{14.119}$$

Because the delta function enforces $t' = t$, we simplify the integral

$$S = \int dt \psi^\dagger(t)(i\partial_t - H)\psi(t). \tag{14.120}$$

This is the standard form of the quadratic action for free particles. We will deal with those types of actions until we consider fermions. To verify that this action correctly describes the dynamics, we computed the functional variation

$$\frac{\delta S}{\delta \psi^\dagger(t)} = \int dt' (i\partial_t - H)\delta(t - t')\psi(t'). \tag{14.121}$$

Using the property of the delta function, this simplifies to

$$(i\partial_t - H)\psi(t) = 0. \tag{14.122}$$

Because we identified G^{-1} as the kernel

$$G^{-1}(t, t') = (i\partial_t - H)\delta(t - t') \tag{14.123}$$

we see that the full propagator $G(t, t')$ is obtained by inverting this operator

$$G(t, t') = (G^{-1})^{-1} = (i\partial_t - H)^{-1}. \tag{14.124}$$

To fully explain why the propagator naturally follows as the inverse of the quadratic term in the action, let us break it down in the context of Gaussian path integrals. In quantum field theory (QFT), the partition function is given by the functional integral

$$Z = \int \mathcal{D}[\bar{\psi}, \psi] e^{iS}, \tag{14.125}$$

where S is the action. This integral defines the probability amplitude for different field configurations. The term e^{iS} acts as a weight function configurations with larger action contribute less due to rapid oscillations in the exponent. For a free (non-interacting) theory, the action is quadratic in the fields. In matrix notation

$$S = \int dt dt' \sum_{i,j} \bar{\psi}_i(t) G_{ij}^{-1}(t, t') \psi_j(t'). \tag{14.126}$$

This is analogous to a Gaussian integral in ordinary statistical mechanics

$$Z = \int d\vec{x} e^{-\frac{1}{2}\vec{x}^T A \vec{x}}. \tag{14.127}$$

In this case, the result of the Gaussian integration is

$$Z \sim \frac{1}{\sqrt{\det A}}. \tag{14.128}$$

For the SK action, the equivalent formula involves a functional determinant

$$Z = \det(G^{-1}). \tag{14.129}$$

Since the quadratic term in the exponent is G^{-1}, it determines the fluctuations of the system. The key result from Gaussian integration is

$$\langle x_i x_j \rangle = A_{ij}^{-1}. \tag{14.130}$$

To confirm this form of the action, we perform a functional variation. Varying the action with respect to $\psi^\dagger$

$$\frac{\delta S}{\delta \psi^\dagger(t)} = \int dt' G^{-1}(t, t') \psi(t'). \tag{14.131}$$

Setting $\delta S = 0$ gives the equation of motion

$$G^{-1}\psi = 0. \tag{14.132}$$

We now extend the previous derivation to include interactions with external fields and coupling to an environment at zero temperature. This means that the system is not purely free, and the Green's function is modified by both self-energy corrections and external field interactions. For an interacting system in an external potential $V(x, t)$, the action takes the form:

$$S = \int dt dt' \psi^\dagger(t) G^{-1}(t, t') \psi(t') + S_{\text{int}} + S_{\text{env}}. \tag{14.133}$$

Here

$$G^{-1}(t,\, t') = (i\partial_t - H - V(x,\, t))\delta(t - t') \tag{14.134}$$

is the inverse Green's function in the presence of an external field $V(x, t)$, S_{int} describes interactions, and S_{env} describes coupling to an external environment.

A common way to model an environment (e.g. phonons or a bath of harmonic oscillators) is via a coupling term

$$S_{\text{env}} = \int dtdt'dx\psi^{\dagger}(x,\, t)\Sigma_{\text{env}}(t,\, t')\psi(x,\, t'), \tag{14.135}$$

where $\Sigma_{\text{env}}(t, t')$ is the self-energy correction due to the bath. These are defined via

$$\Sigma_{++}(t,\, t') = -i\,\langle T_C H_{\text{int}}(t) H_{\text{int}}(t')\rangle \tag{14.136}$$

$$\Sigma_{--}(t,\, t') = -i\,\langle \tilde{T}_C H_{\text{int}}(t) H_{\text{int}}(t')\rangle \tag{14.137}$$

$$\Sigma_{+-}(t,\, t') - -i\,\langle H_{\text{int}}(t) H_{\text{int}}(t')\rangle \tag{14.138}$$

$$\Sigma_{-+}(t,\, t') = -i\,\langle H_{\text{int}}(t') H_{\text{int}}(t)\rangle\,. \tag{14.139}$$

For coupling to a bath, these are essentially the correlation functions for the reservoir. Again, if we are coming from a master equation that term is already present in the action via the terms that resemble Lindblad dissipators. It is also easy to evaluate these for a driving field. We now compute

$$\frac{\delta^2 S}{\delta\psi^{\dagger}(t)\delta\psi(t')}. \tag{14.140}$$

From the noninteracting part, we obtain

$$\frac{\delta^2 S_0}{\delta\psi^{\dagger}(t)\delta\psi(t')} = G^{-1}(t,\, t') = (i\partial_t - H - V(x,\, t))\delta(t - t'). \tag{14.141}$$

From the interaction term, differentiating gives an additional self-energy contribution

$$\frac{\delta^2 S_{\text{int}}}{\delta\psi^{\dagger}(t)\delta\psi(t')} = \Sigma_{\text{int}}(t,\, t'), \tag{14.142}$$

where Σ_{int} is the self-energy correction. A particular term can be in H in the noninteracting action, or separated out and put in the self-energies, as in the case of a constant drive, as we will see. From the environmental coupling, differentiating gives

$$\frac{\delta^2 S_{\text{env}}}{\delta\psi^{\dagger}(t)\delta\psi(t')} = \Sigma_{\text{env}}(t,\, t'). \tag{14.143}$$

Thus, the full second derivative of the action is

$$\frac{\delta^2 S_{\text{eff}}}{\delta\psi^{\dagger}(t)\delta\psi(t')} = G^{-1}(t,\, t') - \Sigma_{\text{int}}(t,\, t') - \Sigma_{\text{env}}(t,\, t'), \tag{14.144}$$

where $S_{\rm eff}$ is the effective action, including particle interactions, dissipation, and driving. Now, inverting this equation gives the full interacting Green's function. In our passing from a master equation, the environment terms are already included, as well as a drive if such is applied. If not, we start with

$$G(t, t') = (G^{-1}(t, t') - \Sigma_{\rm int}(t, t') - \Sigma_{\rm env}(t, t'))^{-1}. \tag{14.145}$$

This is the Dyson equation, which relates the full Green's function to the noninteracting propagator, the interaction self-energy, and the environmental contribution. This is a way of adding interactions to the bare Green's functions. We will say more about the interaction and environment terms in the examples. The key thing to note is that the noninteracting formalism captures the heart of the argument. For a system where we map a known Lindblad master equation to an action, interactions and the environment are included, so we will read off the effective action. We can form the full Greens function matrix from the 'bare' Greens's functions and self-energies, the latter encoding interactions, such as external fields and bath interactions

$$\mathcal{G}_0 = \begin{bmatrix} G^0_{++} & G^0_{+-} \\ G^0_{-+} & G^0_{--} \end{bmatrix} \tag{14.146}$$

and

$$\Sigma_{\pm} = \begin{bmatrix} \Sigma_{++} & \Sigma_{+-} \\ \Sigma_{-+} & \Sigma_{--} \end{bmatrix}. \tag{14.147}$$

The self-energies include interactions, including external driving and bath interactions. For Markovian dissipation, we have

$$\Sigma_{+-} = -i\gamma \begin{bmatrix} & 0 \\ 0 & 0 \end{bmatrix} \tag{14.148}$$

and for a constant drive of the form $Y(a + a^\dagger)$ in the Hamiltonian, we have

$$\Sigma_{+-}(\omega) = \begin{bmatrix} 0 & -iY \\ iY & 0 \end{bmatrix}. \tag{14.149}$$

The 2×2 matrix notation originated with Schwinger in 1960. Going back to the matrix of Green's functions, also known as two-point functions, we have

$$G_{\rm contour}(t, t') = \begin{pmatrix} G_{++}(t, t') & G_{+-}(t, t') \\ G_{-+}(t, t') & G_{--}(t, t') \end{pmatrix}, \tag{14.150}$$

where $G_{\rm contour}$ is the matrix of Green's functions in terms of the fields on the upper and lower contours, and

$$G_{++}(t, t') = -i\left\langle \mathcal{T}_c a_+(t) a_+^\dagger(t') \right\rangle \tag{14.151}$$

$$G_{--}(t, t') = -i\left\langle \tilde{\mathcal{T}}_c a_-(t) a_-^\dagger(t') \right\rangle \tag{14.152}$$

$$G_{+-}(t,\, t') = -i\langle a_{-}(t')a_{+}(t)\rangle \tag{14.153}$$

$$G_{-+}(t,\, t') = -i\langle a_{-}(t)a_{+}(')\rangle. \tag{14.154}$$

Notice that

$$G_{++}(t,\, t') + G_{--}(t,\, t') = G_{+-}(t,\, t') + G_{-+}(t,\, t'). \tag{14.155}$$

So there is some redundancy.

The contour Green's functions can be transformed into the Keldysh basis (classical and quantum fields). The Green's function matrix in the Keldysh basis is related to the contour Green's function matrix by a unitary transformation

$$\mathcal{G}_{\text{Keldysh}} = U\mathcal{G}_{\text{contour}}U^{\dagger}, \tag{14.156}$$

where

$$U = \frac{1}{\sqrt{2}}\begin{pmatrix} & 1 \\ & -1 \end{pmatrix}. \tag{14.157}$$

This is the transformation that takes us from $\pm$ fields to quantum and classical fields,

$$\Phi = \begin{pmatrix} \phi_{+} \\ \phi_{-} \end{pmatrix} = U\Phi = \begin{pmatrix} \phi_{c} \\ \phi_{q} \end{pmatrix}. \tag{14.158}$$

After this transformation, the Green's function matrix in the Keldysh basis becomes

$$\mathcal{G}_{\text{Keldysh}}(t,\, t') = \begin{pmatrix} G^{K}(t,\, t') & G^{R}(t,\, t') \\ G^{A}(t,\, t') & 0 \end{pmatrix}. \tag{14.159}$$

Note in particular the coupling of the classical field $\Phi_c = (\phi_c,\, \phi_c^*)$ to the quantum source $J_q = (j_q,\, j_q^*)$, and vice versa. Also, there is no more redundancy. We also define the self-energy

$$\Sigma_{\alpha\beta}(t,\, t') = i\frac{\delta^2\Gamma}{\delta G_{\alpha}(t)\delta G_{\beta}(t')} \tag{14.160}$$

with

$$\Gamma[G] = W[J] - \int dt \sum_{\alpha} J_{\alpha}, \tag{14.161}$$

which are equivalent to our earlier definitions; these are in the rotated basis, however. Then, we need to perform matrix inversion to find the Green's functions

$$\mathcal{G}^{-1} = \mathcal{G}_0 - \Sigma. \tag{14.162}$$

In the rotated basis, the inverse Green's function follows the block structure

$$\mathcal{G}^{-1}(\omega) = \begin{pmatrix} 0 & (G^{A})^{-1} \\ (G^{R})^{-1} & (G^{K})^{-1} \end{pmatrix}. \tag{14.163}$$

To solve for $\mathcal{G}$, we must invert $\mathcal{G}^{-1}$, which is uniquely defined as long as G^R and G^A are invertible. Thus, the Green's function matrix is

$$\mathbf{G} = \begin{bmatrix} G^R & G_K \\ 0 & G^A \end{bmatrix}. \tag{14.164}$$

So we take the action we have and read off the matrix of Greens's functions, invert it, and we have our correlation functions. We also have the relation

$$G^R - G^A = G^{>} - G^{<}. \tag{14.165}$$

The first correlation function we are interested in is

$$\begin{aligned} g^{(1)}(\tau) &= \frac{\langle a^{\dagger}(0)a(\tau)\rangle}{\langle a^{\dagger}(0)a(0)\rangle} \\ &= \frac{G^{<}(0, \tau)}{G^{<}(0, 0)}. \end{aligned} \tag{14.166}$$

In terms of Green's functions,

$$\langle a^{\dagger}(0)a(t)\rangle = iG^{<}(0, t). \tag{14.167}$$

What about four point correlators, as in

$$g^{(2)}(\tau) = \frac{\langle a^{\dagger}(0)a^{\dagger}(\tau)a(\tau)a(0)\rangle}{\langle a^{\dagger}(0)a(0)\rangle^{2}}, \tag{14.168}$$

where $t = 0$ is after the system is in the steady state?

Wicks theorem [14] states that for a system in a Gaussian (or quasi-free) state described by a nonequilibrium density matrix, the expectation value of a time-ordered product of field operators can be expressed as the sum of the products of the pairwise contractions. It is essentially an expression of the Gaussian moment theorem, that for a Gaussian probability distribution, odd correlations are zero and even correlations can be expressed as the products of two operator correlations. If we choose variables such that the center is at zero,

$$\langle x^{2n+1}\rangle = \int_{-\infty}^{\infty} x^{2n+1}e^{x^2/2\sigma^2} = 0. \tag{14.169}$$

Even correlations can be found via

$$\langle x^{2n}\rangle = \int_{0}^{\infty} x^{2n}e^{-ax^2}dx = \frac{1\cdot 3\cdot 5\cdots(2n-1)}{2^{n+1}a^{n}}\left(\frac{\pi}{a}\right)^{1/2} \tag{14.170}$$

or in a form that can be used recursively

$$\langle X^{2n}\rangle (2n-1)\sigma^{2}\langle X^{2n-2}\rangle. \tag{14.171}$$

If the average is zero, the only parameter left to describe things is the standard deviation.

In the Keldysh formalism, we define the contour-ordered expectation value of the four operators ψ_1, ψ_2, ψ_3, ψ_4 as

$$\langle T_c\psi_1\psi_2\psi_3\psi_4\rangle, \tag{14.172}$$

where T_c denotes contour ordering along the Keldysh time contour, which includes both the forward and backward branches. By Wick's theorem, this expectation value can be decomposed into a sum of the products of the two-point functions. For a system with a Gaussian (noninteracting) action, Wick's theorem states that any n-point function can be decomposed into a sum of the products of two-point functions

$$\langle T_c\psi_1\psi_2\psi_3\psi_4\rangle = \langle T_c\psi_1\psi_2\rangle\langle T_c\psi_3\psi_4\rangle + \langle T_c\psi_1\psi_3\rangle\langle T_c\psi_2\psi_4\rangle + \langle T_c\psi_1\psi_4\rangle\langle T_c\psi_2\psi_3\rangle. \quad (14.173)$$

This allows us to contract the correlation functions, shown schematically in figure 14.4. For fermions, the second term is negative. In the derivation, you pick up a sign due to commutators for each field permutation. It is always one for bosons, and -1 for fermions, so an odd number of perturbations (one in this case) yields a minus sign. This will show up when we discuss resonance fluorescence. In the Keldysh basis, the contour-ordered Green's functions are decomposed into different real-time components.

For any bosonic operators a and $a^\dagger$, Wick's theorem expands the four-operator correlation function into a sum of two-point functions

$$\langle a^\dagger(0)a^\dagger(\tau)a(\tau)a(0)\rangle, \quad (14.174)$$

which expands as

$$\langle a^\dagger(0)a(\tau)\rangle\langle a^\dagger(\tau)a(0)\rangle + \langle a^\dagger(0)a(0)\rangle\langle a^\dagger(\tau)a(\tau)\rangle + \langle a^\dagger(0)a^\dagger(\tau)\rangle\left\langle a(\tau)a(0)\right\rangle. \quad (14.175)$$

The last term contains the anomalous correlations

$$\langle a^\dagger(0)a^\dagger(\tau)\rangle, \qquad \left\langle a(0)a(\tau)\right\rangle. \quad (14.176)$$

In a system where the particle number is conserved, the expectation value of the two creation operators or two annihilation operators is zero:

$$\langle a^\dagger(0)a^\dagger(\tau)\rangle = 0, \qquad \left\langle a(0)a(\tau)\right\rangle = 0. \quad (14.177)$$

The anomalous Green's function is related to the normal Green's function via Bogoliubov transformations in many cases (e.g. in squeezing or Bose–Einstein condensates). In a driven-damped oscillator, as we consider next, we should find this to be zero as photons do not interact directly.

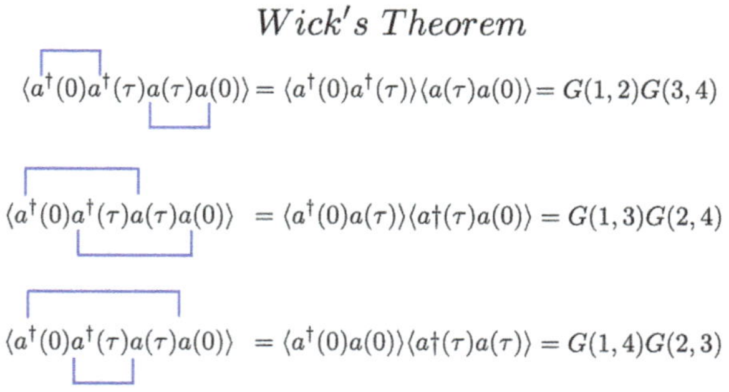

Figure 14.4. Application of Wick contractions.

14.7 The Keldysh zoo

Before we continue, it is prudent to list the Greens functions and other functions that appear in the story. Recall that in the unrotated basis we have

$$G_{\text{contour}}(t, t') = \begin{pmatrix} G_{++}(t, t') & G_{+-}(t, t') \\ G_{-+}(t, t') & G_{--}(t, t') \end{pmatrix} \tag{14.178}$$

and the self-energy terms

$$\Sigma_{\pm} = \begin{bmatrix} \Sigma_{++} & \Sigma_{+-} \\ \Sigma_{-+} & \Sigma_{--} \end{bmatrix}. \tag{14.179}$$

Defined in terms of the correlation functions of the $\pm$ fields and interactions, respectively. We are primarily interested in

$$G^R(t, t') = -i\theta(t - t')\langle\{\sigma_-(t), \sigma_+(t')\}\rangle \tag{14.180}$$

$$G^A(t, t') = i\theta(t' - t)\langle\{\sigma_-(t), \sigma_+(t')\}\rangle. \tag{14.181}$$

G^R and G^A describe response functions, they tell you how the system reacts to a perturbation applied before. They contain spectral information via the spectral function

$$A(\omega) = -\frac{1}{\pi}\text{Im}G^R(\omega). \tag{14.182}$$

We also have

$$G^<(t, t') = i\langle\sigma_+(t')\sigma_-(t)\rangle. \tag{14.183}$$

This is a normally ordered correlation function. It appears in the photon emission spectra and population dynamics:

$$G^>(t, t') = -i\langle\sigma_-(t)\sigma_+(t')\rangle. \tag{14.184}$$

This also involves normal ordering but in reverse time order, anti-time ordering. We have the relation

$$G^> - G^< = G^R - G^A, \tag{14.185}$$

which indicates that normally ordered correlations (encoded in $G^<$, $G^>$) are not directly contained in G^R, G^A, but are related to them through the Keldysh equation. In the original Schwinger–Keldysh contour formalism, the Green's functions are defined as

$$G_{ab}(t, t') = -i\left\langle T_C\sigma_a(t)\sigma_b^\dagger(t')\right\rangle, \quad a, b \in \{+, -\}. \tag{14.186}$$

These are related to the $\pm$ ones via

$$G^R = G_{++} - G_{+-} \tag{14.187}$$

$$G^A = G_{++} - G_{-+} \tag{14.188}$$

$$G^K = G_{++} + G_{--}. \tag{14.189}$$

We also have the fluctuation–dissipation theorem (true in nonequilibrium) that defines the Keldysh Green's function

$$G^K = G^R \Sigma^K G^A, \tag{14.190}$$

where the self-energy Σ^K determines environmental influence on fluctuations (squeezing, noise, and emission spectra). The Dyson equation naturally relates system fluctuations (G^K) to the external influences encoded in Σ^K). This relation comes from inverting the inverse Greens function matrix in the block diagonal form in the rotated basis. For a system interacting with a thermal bath or vacuum, the fluctuation–dissipation theorem (FDT) relates the Keldysh self-energy to the retarded self-energy

$$\Sigma^K(\omega) = 2i\Gamma(\omega)(1 + 2n_B(\omega)), \tag{14.191}$$

where $\Gamma(\omega)$ is the decay rate and $n_B(\omega)$ is the occupation function of the bath. At zero temperature ($T = 0$), and in the Markov approximation, these simplifies to

$$\Sigma^K(\omega) = -2i\gamma. \tag{14.192}$$

14.8 Damped driven cavity

From the Lindblad master equation

$$\frac{d\rho}{dt} = -i\left[\omega\left(a^\dagger a + \frac{1}{2}\right) - F(a + a^\dagger), \rho\right] + \kappa\left(a\rho a^\dagger - \frac{1}{2}\{a^\dagger a, \rho\}\right) \tag{14.193}$$

we find the SK action

$$S = \int dt \begin{bmatrix} a_c^* & a_q^* \end{bmatrix} \begin{bmatrix} 0 & (i\partial_t + \kappa/2) \\ (i\partial_t + \kappa/2) & i\kappa \end{bmatrix} \begin{bmatrix} a_c \\ a_q \end{bmatrix} + \int dt \left(F^*(t) a_q + F(t) a_q^*\right). \tag{14.194}$$

From the Keldysh action, only looking at the quadratic terms

$$S = \int dt \begin{bmatrix} a_c^* & a_q^* \end{bmatrix} \begin{bmatrix} 0 & \left(i\partial_t + \frac{\kappa}{2}\right) \\ \left(i\partial_t + \frac{\kappa}{2}\right) & i\kappa \end{bmatrix} \begin{bmatrix} a_c \\ a_q \end{bmatrix} \tag{14.195}$$

the inverse Green's function matrix is

$$G^{-1}(\omega) = \begin{bmatrix} 0 & (\omega + i\kappa/2) \\ (\omega + i\kappa/2) & i\kappa \end{bmatrix}. \tag{14.196}$$

To obtain the Green's functions, we invert this matrix. The inverse matrix is

$$G(\omega) = \begin{bmatrix} G^K(\omega) & G^R(\omega) \\ G^A(\omega) & 0 \end{bmatrix} \tag{14.197}$$

and explicitly

$$\begin{aligned} G^R(\omega) &= \frac{1}{\omega + i\kappa/2} \\ G^A(\omega) &= \frac{1}{\omega - i\kappa/2} \\ G^K(\omega) &= \frac{i\kappa}{(\omega + i\kappa/2)(\omega - i\kappa/2)}. \end{aligned} \tag{14.198}$$

Since

$$S_{\text{out}}(\omega) = -\frac{1}{\pi}\text{Im}G^K(\omega) \tag{14.199}$$

we obtain

$$S_{\text{out}}(\omega) = \frac{\kappa}{\pi}\frac{1}{\omega^2 + (\kappa/2)^2}. \tag{14.200}$$

What about the effect of the driving? The total field in the cavity can be decomposed into

$$a(t) = \alpha + \Delta a(t), \tag{14.201}$$

where α is the coherent part due to the classical drive, and $\Delta a(t)$ represents quantum fluctuations around the mean field. Since the system is driven resonantly by a classical field, the steady-state solution gives a coherent amplitude,

$$\alpha = \frac{F}{i\kappa/2}. \tag{14.202}$$

The output spectrum is given by

$$S_{\text{out}}(\omega) = \int dte^{i\omega t}\langle a^\dagger(t)a(0)\rangle. \tag{14.203}$$

Expanding this,

$$\langle a^\dagger(t)a(0)\rangle = |\alpha|^2 + \langle \Delta a^\dagger(t)\Delta a(0)\rangle. \tag{14.204}$$

Taking the Fourier transform, the first term contributes a delta function at zero frequency

$$S_{\text{coh}}(\omega) = |\alpha|^2\delta(\omega), \tag{14.205}$$

while the second term gives the incoherent Lorentzian component

$$S_{\text{incoh}}(\omega) = \frac{\kappa}{\pi}\frac{1}{\omega^2 + (\kappa/2)^2}. \tag{14.206}$$

The normalized second-order intensity correlation function $g^{(2)}(t)$ is defined as follows:

$$g^{(2)}(t) = \frac{\langle a^\dagger(0)a^\dagger(t)a(t)a(0)\rangle}{\langle a^\dagger(0)a(0)\rangle^2}. \tag{14.207}$$

To express this in terms of Keldysh Green's functions, we use Wick's theorem and the two-particle Green's function formalism using the definition of lesser and greater Green's functions. Since the action is Gaussian, we use Wick's theorem to decompose the four-point function

$$\langle a^\dagger(0)a^\dagger(t)a(t)a(0)\rangle \approx \langle a^\dagger(0)a(0)\rangle\langle a^\dagger(t)a(t)\rangle + |\langle a^\dagger(0)a(t)\rangle|^2. \tag{14.208}$$

In terms of Green's functions

$$\begin{aligned}\langle a^{\dagger}(0)a(0)\rangle &= -\, iG^{<}(0),\\ \langle a^{\dagger}(t)a(t)\rangle &= -\, iG^{<}(0),\\ |\langle a^{\dagger}(0)a(t)\rangle|^{2} &= |G^{<}(t)|^{2}.\end{aligned} \tag{14.209}$$

Thus, the second-order correlation function is

$$g^{(2)}(t) = 1 + \frac{|G^{<}(t)|^{2}}{|G^{<}(0)^{2}|} \tag{14.210}$$

and explicitly

$$g^{(2)}(t) = 1 + e^{-\kappa|t|}. \tag{14.211}$$

This can be written as

$$g^{(2)}(t) = 1 + |g^{(1)}(t)|^{2} \tag{14.212}$$

a general result if no particles are created in the cavity, and no lasing or parametric processes are present.

14.9 Optical parametric oscillator

We now consider a nonlinear system. For a degenerate parametric oscillator below the threshold of a phase transition, the relevant degrees of freedom are described by the annihilation operator a, and the primary effect at zero temperature is damping due to losses. The master equation for the density matrix ρ is

$$\frac{d\rho}{dt} = -\frac{i}{\hbar}[H, \rho] + \kappa\mathcal{D}[a]\rho, \tag{14.213}$$

where in the interaction picture

$$H = \frac{\hbar\lambda}{2}(a^{\dagger 2} + a^{2}), \tag{14.214}$$

where the cavity loss rate is κ, and the parametric driving is λ. For below-threshold conditions ($\lambda < \kappa/2$), the system remains in a dissipative regime. Because there *is* particle production, correlated photon pairs, we have to adapt our methods. We will need to solve for anomalous Greens functions such as $\langle a(0)a(\tau)\rangle$ which we have ignored before. We would need to have terms in the action that go as a^2 and $a^{\dagger 2}$, which means adding source terms proportional to these quantities in the action. The most convenient way to do this is to go to the Keldysh–Nambu space, a doubling of the degrees of freedom, essentially treating $\phi_{\pm}$ as complex fields. We define a two-component field, each of which will have a component for the + and $^-$ contours:

$$\Psi = \begin{bmatrix} a \\ a^{\dagger} \end{bmatrix}, \quad \Psi^{\dagger} = \begin{bmatrix} a^{\dagger} & a \end{bmatrix}. \tag{14.215}$$

This leads to a 4×4 structure for the action to account for the two branches of the Keldysh contour. We generally do not perform a Keldysh rotation in the Nambu basis:

$$S = \frac{1}{2}\int \frac{d\omega}{2\pi}\begin{pmatrix}\Phi_{cl}^{\dagger}(\omega) & \Phi_{q}^{\dagger}(\omega)\end{pmatrix}\begin{pmatrix}0 & [G^{A}(\omega)]^{-1}\\ [G^{R}(\omega)]^{-1} & [G^{K}(\omega)]^{-1}\end{pmatrix}\begin{pmatrix}\Phi_{cl}(\omega)\\ \Phi_{q}(\omega)\end{pmatrix}, \qquad (14.216)$$

where each component is itself a 2×2 Nambu matrix:

$$G^{R}(\omega) = \begin{bmatrix} G^{R}(\omega) & F^{R}(\omega)\\ \bar{F}^{R}(\omega) & \bar{G}^{R}(\omega)\end{bmatrix} \qquad (14.217)$$

$$G^{A}(\omega) = \begin{bmatrix} G^{A}(\omega) & F^{A}(\omega)\\ \bar{F}^{A}(\omega) & \bar{G}^{A}(\omega)\end{bmatrix} \qquad (14.218)$$

$$G^{K}(\omega) = \begin{bmatrix} G^{K}(\omega) & F^{K}(\omega)\\ \bar{F}^{K}(\omega) & \bar{G}^{K}(\omega)\end{bmatrix}, \qquad (14.219)$$

where G^R, G^A, G^K are the normal Green's functions and F^R, F^A, F^K are the anomalous (pairing) Green's functions. These can be read right off the action. Recall that the Keldysh action in the Nambu basis can be written as

$$S = \int d\omega \Psi^{\dagger}(\omega)G^{-1}(\omega)\Psi(\omega), \qquad (14.220)$$

where the inverse Green's function takes the block form

$$G^{-1}(\omega) = \begin{bmatrix} 0 & (G^{R})^{-1}(\omega)\\ (G^{A})^{-1}(\omega) & G^{K}(\omega)\end{bmatrix}, \qquad (14.221)$$

where each entry is a 2×2 matrix in the Nambu space. In terms of the correlation functions, we have

$$G^{R}(\omega) = \begin{bmatrix} G^{R}_{aa^{\dagger}}(\omega) & G^{R}_{aa}(\omega)\\ G^{R}_{a^{\dagger}a^{\dagger}}(\omega) & G^{R}_{a^{\dagger}a}(\omega)\end{bmatrix} \qquad (14.222)$$

$$G^{A}(\omega) = \begin{bmatrix} G^{A}_{aa^{\dagger}}(\omega) & G^{A}_{aa}(\omega)\\ G^{A}_{a^{\dagger}a^{\dagger}}(\omega) & G^{A}_{aa^{\dagger}a}(\omega)\end{bmatrix} \qquad (14.223)$$

$$G^{K}(\omega) = \begin{bmatrix} G^{K}_{aa^{\dagger}}(\omega) & G^{K}_{aa}(\omega)\\ G^{K}_{a^{\dagger}a^{\dagger}}(\omega) & G^{K}_{a^{\dagger}a}(\omega)\end{bmatrix}, \qquad (14.224)$$

where

$$G_{ij}^{R}(t) = -i\Theta(t)\Big\langle\Big\{\Psi_i(t),\, \Psi_j^{\dagger}(0)\Big\}\Big\rangle. \tag{14.225}$$

$$G_{ij}^{A}(t) = i\Theta(-t)\Big\langle\Big\{\Psi_i(t),\, \Psi_j^{\dagger}(0)\Big\}\Big\rangle. \tag{14.226}$$

$$G_{ij}^{K}(t) = -i\Big\langle\Big[\Psi_i(t),\, \Psi_j^{\dagger}(0)\Big]\Big\rangle. \tag{14.227}$$

In finding the inverse of the blocks, we will have coupled equations for the normal and anomalous Green's functions.

Let us invert the matrix for G_R^{-1} to find

$$G_R(\omega) = \frac{1}{\left(\omega + i\frac{\kappa}{2}\right)^2 - \lambda^2}\begin{bmatrix} \omega + i\frac{\kappa}{2} & \lambda \\ \lambda & \omega + i\frac{\kappa}{2} \end{bmatrix}. \tag{14.228}$$

Expanding the denominator

$$\left(\omega + i\frac{\kappa}{2}\right)^2 - \lambda^2 = \omega^2 - \lambda^2 + i\kappa\omega. \tag{14.229}$$

Thus, the retarded Green's function is

$$G_R(\omega) = \frac{1}{\omega^2 - \lambda^2 + i\kappa\omega}\begin{bmatrix} \omega + i\frac{\kappa}{2} & \lambda \\ \lambda & \omega + i\frac{\kappa}{2} \end{bmatrix}. \tag{14.230}$$

The advanced block $G_0^{-1,A}(\omega)$ is the Hermitian conjugate $[G_0^{-1,R}(\omega)]^{\dagger}$. Finally, we have the Keldysh Green's function

$$G^{K}(\omega) = \frac{-i\kappa}{(\omega^2 - \lambda^2)^2 + \kappa^2\omega^2}\begin{bmatrix} \omega^2 - \lambda^2 + \frac{\kappa^2}{4} & 2\lambda\omega \\ \lambda\omega & \omega^2 - \lambda^2 + \frac{\kappa^2}{4} \end{bmatrix}. \tag{14.231}$$

A crucial property of this degenerate OPO is that when $|\lambda|$ grows large enough, the denominator $\Delta(\omega)$ can vanish at some real frequency ω. Physically, this indicates an oscillation threshold (the parametric instability) and is a second-order quantum phase transition. The quantum part is that the 'knob' that controls the phase is not the temperature, but the size of the ratio of λ/κ. The threshold condition in the simpler form

$$\lambda = \left(\frac{\kappa}{2}\right). \tag{14.232}$$

Above the threshold, the system develops a macroscopic oscillation amplitude (a coherent state in the mode), and the simple quadratic theory must be expanded to include nonlinear saturation.

Using Green's functions, we can express this in terms of the lesser Green's function $G^{<}(\omega)$, which is related to the Keldysh Green's function $G^{K}(\omega)$ via

$$G^{<}(\omega) = \frac{1}{2}G^{K}(\omega). \tag{14.233}$$

Thus, the spectrum is given by

$$S(\omega) = \mathrm{Im}G^{<}(\omega) = \frac{1}{2}\mathrm{Im}G^{K}(\omega), \tag{14.234}$$

then the incoherent spectrum is given by

$$S(\omega) = \frac{1}{2}\mathrm{Im}G_{11}^{K}(\omega). \tag{14.235}$$

We extract the imaginary part:

$$S(\omega) = \frac{\kappa}{2}\frac{\omega^2 - \lambda^2 + \dfrac{\kappa^2}{4}}{(\omega^2 - \lambda^2)^2 + \kappa^2\omega^2}. \tag{14.236}$$

For negligible λ this is essentially a Lorentzian because the parametric process has little impact on the bare cavity result. As one approaches the threshold, however, the spectrum gets narrower and narrower, indicative of critical slowing down near a phase transition.

The spectrum of squeezing is defined in terms of the fluctuations of the quadratures of the field. The quadrature operators are

$$X_\theta = ae^{-i\theta} + a^\dagger e^{i\theta}. \tag{14.237}$$

The squeezing spectrum is given by the Fourier transform of the symmetrized correlation function

$$S_\theta(\omega) = \int_{-\infty}^{\infty} d\tau e^{i\omega\tau}\langle\{X_\theta(\tau), X_\theta(0)\}\rangle. \tag{14.238}$$

Using Keldysh Green's functions, we can express this in terms of the normal and anomalous Green's functions. Expanding the quadrature correlation,

$$\begin{aligned}\langle\{X_\theta(\tau), X_\theta(0)\}\rangle = e^{-i\theta}\langle\{a(\tau), a(0)\}\rangle + e^{i\theta}\langle\{a^\dagger(\tau), a^\dagger(0)\}\rangle \\ + 2\cos\theta\langle\{a(\tau), a^\dagger(0)\}\rangle.\end{aligned} \tag{14.239}$$

In Fourier space, the squeezing spectrum is

$$S_\theta(\omega) = 2\mathrm{Re}\left[e^{-i\theta}G_{12}^{<}(\omega) + e^{i\theta}G_{21}^{<}(\omega) + 2\cos\theta G_{11}^{<}(\omega)\right]. \tag{14.240}$$

For $\theta = 0$ we have

$$S_0(\omega) = 2\mathrm{Re}\left[G_{11}^{<}(\omega) + G_{12}^{<}(\omega) \right]. \tag{14.241}$$

For $\theta = \frac{\pi}{2}$ the result is

$$S_{\pi/2}(\omega) = 2\mathrm{Re}\left[G_{11}^{<}(\omega) - G_{12}^{<}(\omega) \right]. \tag{14.242}$$

With the < Green's functions, we have

$$\begin{aligned} G_{11}^{<}(\omega) &= \frac{-i\kappa}{2} \frac{\omega^2 - \lambda^2 + \frac{\kappa^2}{4}}{(\omega^2 - \lambda^2)^2 + \kappa^2\omega^2} \\ G_{12}^{<}(\omega) &= \frac{-i\kappa}{2} \frac{2\lambda\omega}{(\omega^2 - \lambda^2)^2 + \kappa^2\omega^2}. \end{aligned} \tag{14.243}$$

Thus

$$\begin{aligned} S_0(\omega) &= 2\,\mathrm{Re}\left[\frac{-i\kappa}{2} \frac{\omega^2 - \lambda^2 + \frac{\kappa^2}{4} + 2\lambda\omega}{(\omega^2 - \lambda^2)^2 + \kappa^2\omega^2} \right] \\ S_{\pi/2}(\omega) &= 2\,\mathrm{Re}\left[\frac{-i\kappa}{2} \frac{\omega^2 - \lambda^2 + \frac{\kappa^2}{4} - 2\lambda\omega}{(\omega^2 - \lambda^2)^2 + \kappa^2\omega^2} \right]. \end{aligned} \tag{14.244}$$

Since the denominator is real, the imaginary part comes from the numerator. Extracting the real part, we obtain

$$S_0(\omega) = \frac{\kappa\left(\omega^2 - \lambda^2 + \dfrac{\kappa^2}{4}\right)}{(\omega^2 - \lambda^2)^2 + \kappa^2\omega^2} + \frac{\kappa(2\lambda\omega)}{(\omega^2 - \lambda^2)^2 + \kappa^2\omega^2} \tag{14.245}$$

$$S_{\pi/2}(\omega) = \frac{\kappa\left(\omega^2 - \lambda^2 + \dfrac{\kappa^2}{4}\right)}{(\omega^2 - \lambda^2)^2 + \kappa^2\omega^2} - \frac{\kappa(2\lambda\omega)}{(\omega^2 - \lambda^2)^2 + \kappa^2\omega^2}. \tag{14.246}$$

With some algebraic fiddling, this is the same result we found in chapter 13 using the input–output theory. The denominator

$$(\omega^2 - \lambda^2)^2 + \kappa^2\omega^2 \tag{14.247}$$

can be rewritten as

$$\left[\left(\omega^2 + \frac{\kappa^2}{4} - \lambda^2 \right)^2 + \kappa^2\omega^2 \right]. \tag{14.248}$$

Factoring the numerator terms

$$\omega^2 - \lambda^2 + \frac{\kappa^2}{4} + 2\lambda\omega \tag{14.249}$$

we separate out the dependence on ω^2, λ, and κ.

$$S_1^{\text{out}}(\omega) = 1 + \frac{2\gamma_1|\varepsilon|}{\left(\frac{\gamma_1}{2} - |\varepsilon|\right)^2 + \omega^2} \tag{14.250}$$

$$S_2^{\text{out}}(\omega) = 1 - \frac{2\gamma_1|\varepsilon|}{\left(\frac{\gamma_1}{2} + |\varepsilon|\right)^2 + \omega^2}, \tag{14.251}$$

which was our original result.

What about the photon statistics via the second-order intensity correlation function? We use

$$g^{(2)}(t) = 1 + \frac{|G^{<}(t)|^2}{|G^{<}(0)^2|}. \tag{14.252}$$

We know those Green's functions and find

$$\begin{aligned} g^{(2)}(\tau) = {} & 1 + [\exp(-\kappa|\tau|)\lambda^2] \\ & \times [(\kappa + \lambda)^2 \exp(2\epsilon|\tau|) + (\kappa - \lambda)^2 \exp(-2\lambda|\tau|)]. \end{aligned} \tag{14.253}$$

At $\tau = 0$ we have

$$g^{(2)}(0) = \left(3 \sinh^2 \frac{1}{2}s + 1\right)/\sinh^2(s/2), \tag{14.254}$$

where we use the squeezing parameter

$$\exp(s) = \frac{\kappa + \lambda}{\kappa - \lambda}. \tag{14.255}$$

We can also express

$$g^{(2)}(\tau) = 1 + [g^{(2)}(0) - 1]e^{-2(\kappa-\lambda)\tau}. \tag{14.256}$$

The main physical result is that the field is always antibunched, sometimes extremely so. As the DOPO makes photons in pairs, this is expected. Furthermore, the fluctuations continue to grow as the parametric drive is increased toward threshold, diverging at the phase transition.

14.10 Interlude on fermions and Grassman variables

For resonance fluorescence, we need to deal with the fermionic nature of the pseudospin. This can be problematic. Unlike a harmonic oscillator, a two-level system is saturable $\sigma_-^2|\psi\rangle = 0$. Strictly speaking, an algebra cannot be made fully equivalent to a free bosonic field in Nambu space. Mapping a spin-1/2 operator to a Schwinger-boson doublet is route to 'bosonize' the two-level system while keeping track of the Hilbert space constraint. In the Schwinger-boson representation, we introduce two bosonic operators $b_\uparrow$ and $b_\downarrow$ that obey the usual bosonic commutation relations, but they are subject to the single-occupancy constraint

$$b_\uparrow^\dagger b_\uparrow + b_\downarrow^\dagger b_\downarrow = 1. \tag{14.257}$$

We write the spin operators as

$$\sigma_+ = b_\uparrow^\dagger b_\downarrow, \quad \sigma_- = b_\downarrow^\dagger b_\uparrow, \quad \sigma_z = b_\uparrow^\dagger b_\uparrow - b_\downarrow^\dagger b_\downarrow. \tag{14.258}$$

In this language, the two-level system is represented by a bosonic doublet

$$\Psi_b = \begin{pmatrix} b_\uparrow \\ b_\downarrow \end{pmatrix}. \tag{14.259}$$

With the physical Hilbert space being the one-boson sector. Essentially, one boson operator kills an atom in the ground state and adds one in the excited state (or vice versa). This is similar to the electron—hole pair formulation. Once the spin is represented in this way, one can develop a Nambu space (or doubled) Green's function formalism for the Schwinger-boson doublet. For instance, one might define a Nambu spinor as

$$\Psi(\omega) = \begin{pmatrix} b_\uparrow(\omega) \\ b_\downarrow^\dagger(-\omega) \end{pmatrix}, \tag{14.260}$$

or use a four-component formalism if you wish to retain both $b_\uparrow$ and $b_\downarrow$ as independent degrees of freedom along with their conjugates. The precise choice depends on which operators (or linear combinations thereof) you want to study. In this representation, the effective action (or the inverse Green's functions) will be written in terms of these bosonic fields and will include an effective self-energy term that arises from spontaneous emission as well as from the cavity coupling if the atom is embedded in a driven cavity. The key point is that if you try to 'bosonize' a two-level system exactly, you run into the problem that the spin operators do not obey bosonic commutation relations. For example, if we attempt to represent the atomic lowering operator as a bosonic field b, we must deal with the nonlinear relation

$$\sigma_- = b^\dagger \sqrt{1 - b^\dagger b}, \tag{14.261}$$

so that (in particular) $\sigma_-^2 = 0$ is built in. This nonlinearity would lead to trilinear (or higher-order) terms when you couple the atom to the cavity, for instance, in the Jaynes–Cummings Hamiltonian

$$H_{\text{int}} = g(a^\dagger \sigma_- + \sigma_+ a). \tag{14.262}$$

If you substitute $\sigma_+ \sim \sqrt{1 - b^\dagger b}\; b$ you obtain terms that are not bilinear in the bosonic fields. A common way to avoid this complication is to use a Schwinger-boson or Holstein–Primakoff representation and then linearize the theory in the low-excitation limit. In the Holstein–Primakoff representation (for example), one writes

$$\sigma_+ = b^\dagger\sqrt{1 - b^\dagger b}, \quad \sigma_- = \sqrt{1 - b^\dagger b}\, b \tag{14.263}$$

with the constraint

$$b^\dagger b \leqslant 1. \tag{14.264}$$

In the low-excitation (or weak-driving) limit, we have $\langle b^\dagger b\rangle \ll 1$, so we can approximate

$$\sqrt{1 - b^\dagger b} \approx 1, \tag{14.265}$$

which leads to the simple linearized relations

$$\sigma_+ \approx b^\dagger, \quad \sigma_- \approx b. \tag{14.266}$$

Then, the Jaynes–Cummings coupling simplifies to

$$H_{\text{int}} \approx g(a^\dagger b + b^\dagger a), \tag{14.267}$$

which is bilinear in the cavity field a and the atomic field b. Once you have a bilinear Hamiltonian (and similarly a bilinear dissipative part, with cavity decay at rate κ and atomic spontaneous emission at rate γ), you can write down a quadratic Keldysh action. To capture both normal and anomalous correlations (important for computing, for example, the squeezing spectrum), you double the degrees of freedom in Nambu space. For example, you define for the cavity field

$$\Psi_a(\omega) = \begin{pmatrix} a(\omega) \\ a^\dagger(-\omega) \end{pmatrix}. \tag{14.268}$$

However, we have thrown out a lot of physics, mainly the saturable absorber nature of the atom so we have coupled linear oscillators. If we have N atoms, for weak excitation only a few get excited and one can use a system size expansion in $1/N$ as in chapter 12. For the atomic field (now bosonized via Holstein–Primakoff or Schwinger bosons in the low excitation limit)

$$\Psi_b(\omega) = \begin{pmatrix} b(\omega) \\ b^\dagger(-\omega) \end{pmatrix}. \tag{14.269}$$

Then, you write down the Dyson equation for the two-point functions. Schematically,

$$G = G_0 + G_0\Sigma G = G_0 + G_0\Sigma G_0 + G_0\Sigma G_0\Sigma G_0 + \cdots, \tag{14.270}$$

where G_0 is the bare Green's function (a 4×4 matrix in combined Keldysh–Nambu space) and Σ is the self-energy arising from the nonlinear interactions. This notation is commonly used and 'hides' the integrals

$$\begin{aligned} G(k, \omega) = G_0(k, \omega) &+ \int G_0(k, \omega)\Sigma(k, \omega)G(k, \omega)d\omega \\ &+ \int\int G_0(k, \omega)\Sigma(k, \omega)G_0(k, \omega')\Sigma(k, \omega')G_0(k, \omega')d\omega d\omega' + \cdots. \end{aligned} \tag{14.271}$$

Grassmann variables are a special kind of 'number' used primarily in the context of quantum field theory and supersymmetry. They do not have a representation in terms of real or complex numbers; they are an entirely different beast. They have unusual algebraic properties compared to ordinary (commuting) numbers, and their integration is defined in a way that parallels differentiation. If θ_i and θ_j are two Grassmann variables, they satisfy

$$\theta_i\theta_j = -\theta_j\theta_i. \tag{14.272}$$

This means that swapping the order introduces a minus sign reflecting the fact that the fermion operators anticommute. Because of the anticommutation property, any Grassmann variable squares to zero

$$\theta_i^2 = 0. \tag{14.273}$$

More generally, any product that contains the same variable more than once vanishes. This is also a reflection of the Pauli exclusion principle for fermions, where there can be only one excitation. For a set of N independent Grassmann variables, any polynomial function can be at most of the order N (i.e. it will have a finite expansion). This is because higher-order products necessarily involve repeated variables that vanish. The integration over the Grassmann variables is defined in such a way that it mimics differentiation, for a Grassmann variable θ and numbers (or Grassmann-even objects) a and b:

$$\int d\theta(a\theta + b) = a\int d\theta\theta + b\int d\theta 1. \tag{14.274}$$

The integration rules for a single Grassmann variable are chosen as

$$\int d\theta 1 = 0, \quad \int d\theta\theta = 1. \tag{14.275}$$

These rules ensure consistency with the idea that integration over Grassmann variables behaves like differentiation (note that differentiation of θ gives 1). For multiple variables, the integration is defined as an iterated integral. For example, for two Grassmann variables θ_1 and θ_2, one typically defines

$$\int d\theta_2 d\theta_1\theta_1\theta_2 = 1. \tag{14.276}$$

The order of integration (and the order of the variables in the integrand) is important because of the anticommutation relations. A key application in field theory is the evaluation of Gaussian integrals over Grassmann variables. If θ_i are Grassmann variables and M is a matrix, one finds that

$$\int\prod_i \ d\theta_i \exp\left(-\frac{1}{2}\sum_{ij}\theta_i M_{ij}\theta_j\right) \propto \sqrt{\det M}. \tag{14.277}$$

Note that in contrast to the bosonic (commuting) Gaussian integrals, the result involves a determinant (or, in some cases, the inverse of a determinant), reflecting the anticommuting nature of the variables. In the Keldysh theory for bosons, we used the bosonic coherent state, so what do we use in the fermionic case? A fermionic coherent state for a two-level (or single fermionic mode) system is typically defined as an eigenstate of the annihilation operator with a Grassmann eigenvalue. For a fermionic creation operator $c^\dagger$ and its corresponding annihilation operator c, with the vacuum state $|0\rangle$ (which is annihilated by c), one defines the Grassmann coherent state $|\xi\rangle$ by

$$c|\xi\rangle = \xi|\xi\rangle, \tag{14.278}$$

where ξ is a Grassmann number. A common way to construct this state is by exponentiating the creation operator

$$|\xi\rangle = e^{-c^\dagger\xi}|0\rangle. \tag{14.279}$$

Just as for bosons, we can find a way to generate any state by the appropriate operator acting on the vacuum. Because the Grassmann numbers (and the fermionic operators) are nilpotent, the exponential truncates at first order

$$e^{-c^\dagger\xi} = 1 - c^\dagger\xi, \tag{14.280}$$

so that

$$|\xi\rangle = (1 - c^\dagger\xi)|0\rangle. \tag{14.281}$$

In this representation, the state expands in the two available states of the two-level system: The vacuum $|0\rangle$ (no particle), and the one-particle state $|1\rangle = c^\dagger|0\rangle$. Thus, the coherent state is a linear combination

$$|\xi\rangle = |0\rangle - \xi|1\rangle. \tag{14.282}$$

This state indeed satisfies

$$c|\xi\rangle = c(|0\rangle - \xi|1\rangle) = 0 - \xi c|1\rangle. \tag{14.283}$$

Since $c|1\rangle = cc^\dagger|0\rangle$ and using the anticommutation relation $\{c, c^\dagger\} = 1$, we find that

$$c|1\rangle = |0\rangle, \tag{14.284}$$

so that

$$c|\xi\rangle = -\xi|0\rangle. \tag{14.285}$$

The Grassmann coherent state is used in path integrals. As with the bosonic coherent states, the fermionic states are overcomplete

$$\langle\xi|\xi'\rangle = e^{\bar{\xi}\xi' - \bar{\xi}\xi - \bar{\xi}'\xi'}. \tag{14.286}$$

But we can still resolve the identity, as in the bosonic case,

$$I = \int d\bar{\xi}d\xi e^{-\bar{\xi}\xi}|\xi\rangle\langle\xi|. \tag{14.287}$$

This is what is inserted (well a pair separated by evolution over δt) at each time step in the contour integral, to turn it into a path integral over the fermionic coherent states.

14.11 Resonance fluorescence

Now we are ready to proceed with a single damped driven atom. The mechanics of the derivation for the Fermionic actions mimic the Bosonic case, only using Grassman fields, so we shall not belabor it here. We define the Nambu spinor

$$\Psi = \begin{bmatrix} \sigma_- \\ \sigma_+ \end{bmatrix} \tag{14.288}$$

and its conjugate

$$\bar{\Psi} = \begin{bmatrix} \sigma_+ & \sigma_- \end{bmatrix}. \tag{14.289}$$

Now, in the Keldysh formalism, we must double the degrees of freedom to include both the forward (+) and backward (−) time branches, leading to the extended Nambu–Keldysh spinor

$$\Psi_{\text{Keldysh}} = \begin{bmatrix} \Psi^c \\ \Psi^q \end{bmatrix} = \begin{bmatrix} \Psi_+ \\ \Psi_- \end{bmatrix} = \begin{bmatrix} \sigma_-^+ \\ \sigma_+^+ \\ \sigma_-^- \\ \sigma_+^- \end{bmatrix}. \tag{14.290}$$

Thus, the Green's function in the Nambu–Keldysh space is now a 4 × 4 matrix

$$G(\omega) = \begin{bmatrix} G^R & G^K \\ 0 & G^A \end{bmatrix}, \tag{14.291}$$

where each entry is a 2 × 2 matrix in Nambu space:

$$Z_{\text{SK}} = \int (d\bar{\eta}_+ d\eta_+)(d\bar{\eta}_- d\eta_-) \exp\{iS_{\text{SK}}[\eta_\pm, \bar{\eta}_\pm]\}. \tag{14.292}$$

Because the Hamiltonian is linear in c, $c^\dagger$, namely $H = Y(c + c^\dagger)$, the fermionic action is quadratic. Dissipation from the spontaneous emission enters through the usual SK self-energy (coming from integrating out the 'bath' of vacuum electromagnetic modes). At zero temperature, the bath contributes a retarded decay $i(\gamma/2)$ to the atom's Green's function plus a Keldysh part that fixes the noise correlations. We duplicate fields on two branches of the time contour, we perform the usual Keldysh rotation and have $c_+(t)$, $\bar{c}_+(t)$ on the forward branch ('+'), and $c_-(t)$, $\bar{c}_-(t)$ on the backward branch ('−'). They are Grassmann-valued fields (anticommuting). The free (closed-system) action for each branch looks like

$$S_\pm[c_\pm, \bar{c}_\pm] = \int_{t_i}^{t_f} dt[\bar{c}_\pm i\partial_t c_\pm - H_{\text{TLS}}(c_\pm, \bar{c}_\pm)], \tag{14.293}$$

where H_{TLS} is written in terms of $\bar{c}_{\pm}$ (the analog of $c^{\dagger}$) and $c_{\pm}$. Concretely, inserting $H_{\text{TLS}} = \Delta c^{\dagger}c + \Omega(c + c^{\dagger})$, you obtain

$$S_{\pm}[c_{\pm}, \bar{c}_{\pm}] = \int_{t_i}^{t_f} dt[\bar{c}_{\pm}i\partial_t c_{\pm} - \Delta\bar{c}_{\pm}c_{\pm} - \Omega(\bar{c}_{\pm} + c_{\pm})]. \tag{14.294}$$

The full Keldysh weight is then

$$\exp\{i[S_{+}[c_{+}, \bar{c}_{+}] - S_{-}[c_{-}, \bar{c}_{-}]]\}, \tag{14.295}$$

plus any environmental/bath terms. To include spontaneous emission (damping) into the free-space modes $\{a_k\}$, you add a photon-bath action $S_{\text{bath}}[a_{k,\pm}]$, and the interaction $S_{\text{int}} = \int dt \sum_k [g_k \bar{c}_{\pm} a_{k,\pm} + \text{h.c.}]$. The total path integral is over $\{c_{\pm}, \bar{c}_{\pm}\}$ and $\{a_{k,\pm}\}$. Formally,

$$\mathcal{Z} = \int \mathcal{D}[c_{\pm}, \bar{c}_{\pm}, a_{k,\pm}] e^{i[S_{+}[c_{+}, \bar{c}_{+}] - S_{-}[c_{-}, \bar{c}_{-}] + S_{\text{bath}} + S_{\text{int}}]}. \tag{14.296}$$

We can include the driving term $Y(\sigma_{-} + \sigma_{+})$ with

$$Y(\bar{c}_{+}(t) + c_{+}(t)) + Y(\bar{c}_{-}(t) + c_{(t)}), \tag{14.297}$$

reflecting a contribution on both the forward and backward parts of the contour. Under standard Markov approximations (broad reservoir, Born–Markov, rotating-wave), one can integrate the photonic fields $\{a_{k,\pm}\}$ to obtain an effective action for the single fermion

$$S_{\text{eff}}[c_{\pm}, \bar{c}_{\pm}] = S_{\text{TLS}}[c_{\pm}, \bar{c}_{\pm}] + \Delta S_{\text{diss}}[c_{\pm}, \bar{c}_{\pm}], \tag{14.298}$$

where ΔS_{diss} is $\Sigma^R(\omega) = -i\Gamma/2$, giving the lifetime of the excited fermion state, and a Keldysh component that sets the noise/occupation relations (though for a zero-temperature bath at optical frequencies, the emission is mostly 'one-way'). Thus, after integrating the environment, the single mode c remains Grassmann-valued but now has a dissipative propagator. This sets up the usual Keldysh Green's functions for the driven-damped fermion. The effective action is

$$S_{\text{SK}} = \int \frac{d\omega}{2\pi} \bar{\Psi}(\omega) \begin{pmatrix} 0 & G_A^{-1}(\omega) \\ G_R^{-1}(\omega) & \Sigma^K(\omega) \end{pmatrix} \Psi(\omega), \tag{14.299}$$

where

$$\Psi(\omega) = \begin{pmatrix} c(\omega) \\ c^{\dagger}(-\omega) \end{pmatrix}, \tag{14.300}$$

where here we *have* used a Keldysh rotation, which is why we have Σ_K and not G^K. From the Dyson equation, resulting from undoing the Keldysh rotation, we have

$$G = (G_0^{-1} - \Sigma)^{-1}, \tag{14.301}$$

$$G_R^{-1}(\omega) = \begin{pmatrix} \omega + i\dfrac{\gamma}{2} & -Y \\ Y & \omega + i\dfrac{\gamma}{2} \end{pmatrix} \tag{14.302}$$

and

$$\Sigma^K(\omega) \approx i\gamma\mathbf{I}. \tag{14.303}$$

Recall G_A^{-1} is the complex conjugate of G_R^{-1}. We have

$$S(\omega) = -\frac{1}{\pi}\mathrm{Im}G^<(\omega) \tag{14.304}$$

and

$$G^<(\omega) = \frac{-i\gamma}{(\omega - Y)^2 + \left(\dfrac{\gamma}{2}\right)^2} + \frac{-i\gamma}{(\omega + Y)^2 + \left(\dfrac{\gamma}{2}\right)^2}. \tag{14.305}$$

This is the sum of two terms, how does a Mollow triplet arise? Adding the two terms, we find

$$G^<(\omega) = -i\gamma\left[\frac{(\omega + Y)^2 + \left(\dfrac{\gamma}{2}\right)^2 + (\omega - Y)^2 + \left(\dfrac{\gamma}{2}\right)^2}{\left[(\omega - Y)^2 + \left(\dfrac{\gamma}{2}\right)^2\right]\left[(\omega + Y)^2 + \left(\dfrac{\gamma}{2}\right)^2\right]}\right]. \tag{14.306}$$

Expanding the numerator, we find

$$2\omega^2 + 2Y^2 + \gamma^2/2. \tag{14.307}$$

For the denominator, we do the same thing and find

$$G^<(\omega) = -i\gamma\frac{2\omega^2 + 2Y^2 + \gamma^2/2}{(\omega^2 - Y^2)^2 + (\omega^2\gamma^2) + 2\omega^2Y^2}. \tag{14.308}$$

This can be put in the form

$$G^<(\omega) = -i\gamma\left(\frac{A_0}{\omega^2 + (\gamma/2)^2} + \frac{A_+}{(\omega - Y)^2 + (\gamma/2)^2} + \frac{A_-}{(\omega + Y)^2 + (\gamma/2)^2}\right), \tag{14.309}$$

where the weight factors A_0, A_+, and A_- depend on the driving strength and give the usual results:

$$A_0 = \frac{2}{1 + \dfrac{3\gamma^2}{4Y^2}} \tag{14.310}$$

$$A_+ = A_- = \frac{1}{1 + \dfrac{3\gamma^2}{4Y^2}}. \tag{14.311}$$

A similar thing occurs in the spectrum of squeezing, giving the expected result.

It is worth commenting on $g^{(2)}(\tau)$. For the bosons, we found

$$g^{(2)}(t) = 1 + \frac{|G^{<}(t)|^2}{|G^{<}(0)|^2}. \tag{14.312}$$

This is unity plus a positive definite quantity, and thus there can be no antibunching. What about for the fermions? The answer lies in the application of Wick's theorem. For a Gaussian fermionic theory, the four-operator expectation value factorizes into the sums of the products of the two-point correlators. Symbolically, for normal-ordered operators (and ignoring c-number shifts), we have

$$\langle c^{\dagger}(1)c^{\dagger}(2)c(3)c(4)\rangle = \langle c^{\dagger}(1)c(3)\rangle\langle c^{\dagger}(2)c(4)\rangle - \langle c^{\dagger}(1)c(4)\rangle\langle c^{\dagger}(2)c(3)\rangle. \tag{14.313}$$

Plus terms that vanish if some operators coincide (as in $\sigma_{-}(0)\sigma_{-}(0)$, or if we are imposing a particular ordering. The minus sign arises from the exchange of fermionic fields. With that minus sign comes the possibility of antibunching, as we find

$$g^{(2)}(t) = 1 - \frac{|G^{<}(t)|^2}{|G^{<}(0)|^2}, \tag{14.314}$$

which is always zero at zero delay. A single atom cannot emit two photons at once. If we want information about σ_z, in terms of Nambu fields, $\sigma_z = \Psi^{\dagger}\tau_z\Psi$ with

$$\tau_z = \begin{bmatrix} & 0 \\ 0 & -1 \end{bmatrix} \tag{14.315}$$

and

$$\langle \sigma_z \rangle = \mathrm{Tr}[\tau_z G^{<}(t, t)]. \tag{14.316}$$

What about correlation functions such as

$$C_z(t, t') = \langle \sigma_z(t)\sigma_z(t') \rangle . \tag{14.317}$$

For nonequilibrium cases, the correlation function is

$$C_z(t, t') = \langle \sigma_z(t)\sigma_z(t') \rangle = \mathrm{Tr}[\tau_z G^{<}(t, t')\tau_z G^{>}(t', t)], \tag{14.318}$$

which depends on the full Keldysh Green's functions, which satisfy

$$\begin{aligned} G^{<} &= G^{R}\Sigma^{<}G^{A}. \\ G^{>} &= G^{R}\Sigma^{>}G^{A}. \end{aligned} \tag{14.319}$$

Thus, the correlation function in nonequilibrium is

$$C_z(t, t') = \mathrm{Tr}[\tau_z G^{R}(t, t')\Sigma^{<}G^{A}(t', t)\tau_z G^{R}(t', t)\Sigma^{>}G^{A}(t, t')]. \tag{14.320}$$

Another example is

$$C_{z-}(t, t') = \langle \sigma_z(t)\sigma_{-}(t') \rangle. \tag{14.321}$$

Using $\sigma_- = \psi_\downarrow^\dagger \psi_\uparrow$, we obtain

$$C_{z-}(t, t') = \mathrm{Tr}[\tau_z G^<(t, t') G^>(t', t)]. \tag{14.322}$$

The nonequilibrium expression in terms of $G^<$ and $G^>$ is

$$C_{z-}(t, t') = \mathrm{Tr}[\tau_z G^R(t, t') \Sigma^< G^A(t', t) G^>(t', t)]. \tag{14.323}$$

14.12 Single-atom cavity electrodynamics

This is a two-level atom in a cavity, the Jaynes–Cummings model, with the addition of driving and dissipation by both the atom and the cavity. Here we cannot progress very far, but remember that there is no exact solution of the master equation for all parameters either. The central difficulty is that we have a complex field for the field and another for the atom. This results in an 8×8 matrix of the inverse Green's functions. The general structure is

$$\begin{aligned} S = \int \frac{d\omega}{2\pi} \Bigg\{ & \underbrace{\left(\bar{\Psi}_a^c(\omega)\ \bar{\Psi}_a^q(\omega)\right) \begin{pmatrix} 0 & \left[G_a^{-1,A}(\omega)\right] \\ \left[G_a^{-1,R}(\omega)\right] & \Sigma_a^K(\omega) \end{pmatrix} \begin{pmatrix} \Psi_a^c(\omega) \\ \Psi_a^q(\omega) \end{pmatrix}}_{\text{Cavity sector}} \\ & + \underbrace{\left(\bar{\psi}^c(\omega)\ \bar{\psi}^q(\omega)\right) \begin{pmatrix} 0 & \left[G_\psi^{-1,A}(\omega)\right] \\ \left[G_\psi^{-1,R}(\omega)\right] & \Sigma_\psi^K(\omega) \end{pmatrix} \begin{pmatrix} \psi^c(\omega) \\ \psi^q(\omega) \end{pmatrix}}_{\text{Atomic (Grassmann) sector}} \\ & - \underbrace{g\ [\bar{\Psi}_a(\omega)\ \tau_x\ \Psi_\psi(\omega) + \bar{\Psi}_\psi(\omega)\ \tau_x\ \Psi_a(\omega)]}_{\text{Coherent drive}} \\ & + \underbrace{(F^*(\omega) a(\omega) + F(\omega) a^\dagger(-\omega))}_{\text{Driving term}} \Bigg\}, \end{aligned} \tag{14.324}$$

where each matrix element is a 2×2 matrix. We will just make a few comments here, mainly on the applications of Wick's theorem. As we have seen, as we permute operators, we need to take care of a sign, $-$ for every permutation of a fermionic field and $+$ for every bosonic field.

In an ordinary Wick expansion, you sum all the ways of pairing the four operators into two-point contractions. In a graded (boson + fermion) system, the only subtlety is that swapping two fermionic operators can introduce a minus sign; this procedure is sometimes known as 'super Wick'. In Keldysh nonequilibrium Green's) theory, mixed correlation functions such as

$$\langle A^\dagger(0) B^\dagger(\tau) B(\tau) A(0) \rangle \tag{14.325}$$

can be computed using the contour-ordered Green's functions on the Keldysh contour. Since the Keldysh formalism is designed for nonequilibrium problems, time arguments must be placed on the Keldysh contour $\mathcal{C}$. The correlation function can be rewritten in terms of the contour-ordered expectation values

$$\langle A^\dagger(0)B^\dagger(\tau)B(\tau)A(0)\rangle = \mathrm{Tr}[\rho_0 T_{\mathcal{C}} A^\dagger(0)B^\dagger(\tau)B(\tau)A(0)] \tag{14.326}$$

where $T_{\mathcal{C}}$ denotes contour ordering and ρ_0 is the initial density matrix. To compute this function, one typically expresses it in terms of the contour-ordered Green's functionation function to be factorized into a product of Green's functions:

$$\langle A^\dagger(0)B^\dagger(\tau)B(\tau)A(0)\rangle = G_{A^\dagger A}(0,\, 0)G_{B^\dagger B}(\tau,\, \tau) - G_{A^\dagger B}(0,\, \tau)G_{B^\dagger A}(\tau,\, 0). \tag{14.327}$$

For interacting systems, one must use the full diagrammatic expansion. We might be interested in the cross-correlation

$$\langle a^\dagger(0)\sigma_+(\tau)\sigma_-(\tau)a(0)\rangle \approx G_{a^\dagger a}(0,\, 0)G_{\sigma_+\sigma_-}(\tau,\, \tau) - G_{a^\dagger\sigma_-}(0,\, \tau)G_{\sigma_+ a}(\tau,\, 0), \tag{14.328}$$

where $G_{a^\dagger a}(t,\, t')$ is the photon Green's function, $G_{\sigma_+\sigma_-}(t,\, t')$ is the qubit excitation correlation function, and $G_{a^\dagger\sigma_-}(t,\, t')$ and $G_{\sigma_+ a}(t,\, t')$ are the cross Green's functions between the qubit and the cavity. This gives the probability of detecting a fluorescent photon at time τ conditioned on a detection in transmission at $\tau = 0$. The Keldysh Green's function is a 2×2 matrix:

$$G = \begin{bmatrix} G^R & G^K \\ 0 & G^A \end{bmatrix}. \tag{14.329}$$

Because the photon and atom(s) interact, their Green's functions must satisfy a coupled Dyson equation

$$G^R(\omega) = G_0^R(\omega) + G_0^R(\omega)\Sigma^R(\omega)G^R(\omega), \tag{14.330}$$

where $G_0^R(\omega)$ is the bare Green's function (without interactions), and $\Sigma^R(\omega)$ is the self-energy, which accounts for the interactions and dissipation. For the Jaynes–Cummings model, the self-energies are

$$\begin{aligned} \Sigma_a^R(\omega) &= \frac{g^2}{\omega - \omega_q + i\gamma/2} \\ \Sigma_\sigma^R(\omega) &= \frac{g^2}{\omega - \omega_c + i\kappa/2}. \end{aligned} \tag{14.331}$$

These are it not constants of the form $i\kappa$. This form arises in systems where the decay process is Markovian, meaning the environment has no memory, and the damping is independent of the system's internal structure. The atom is non-Markovian with respect to the field, and vice versa. If one of the decay rates is the largest rate in the problem, that variable can be adiabatically eliminated as we did back in chapter 10 for the bad cavity limit of the cavity QED. However, in the Jaynes–Cummings model, the self-energy does not come from an external bath alone, but instead from the coherent interaction between the cavity and the qubit. Since this interaction is energy-conserving (not just a simple loss process), it introduces a frequency-dependent self-energy. To find the photon's retarded self-energy $\Sigma_a^R(\omega)$, we integrate the qubit's degrees of freedom. Because the atom has a bare propagator

$$G_\sigma^{0,R}(\omega) = \frac{1}{\omega - \omega_q + i\gamma/2}, \tag{14.332}$$

the interaction-induced self-energy for the photon becomes

$$\Sigma_a^R(\omega) = g^2 G_\sigma^{0,R}(\omega) = \frac{g^2}{\omega - \omega_q + i\gamma/2}. \tag{14.333}$$

Similarly, for the atom, the self-energy due to its interaction with the cavity is

$$\Sigma_\sigma^R(\omega) = \frac{g^2}{\omega - \omega_c + i\kappa/2}. \tag{14.334}$$

These self-energies are not purely imaginary constants because the atom does not see a flat Markovian bath, nor does the cavity. They depend on the frequency ω, reflecting the fact that the interaction is dispersive (not just a simple decay).

The dressed retarded Green's functions are then

$$G_a^R(\omega) = \frac{1}{\omega - \omega_c - \Sigma_a^R(\omega) + i\kappa/2} \tag{14.335}$$

$$G_\sigma^R(\omega) = \frac{1}{\omega - \omega_q - \Sigma_\sigma^R(\omega) + i\gamma/2}. \tag{14.336}$$

We now rewrite the correlation function

$$\langle a^\dagger(0)\sigma_+(\tau)\sigma_-(\tau)a(0)\rangle \tag{14.337}$$

in terms of Keldysh Green's functions.

The first term involves the photon–photon and atom–atom correlations

$$\begin{aligned} \langle a^\dagger(0)a(0)\rangle &= -iG_a^<(0, 0). \\ \langle \sigma_+(\tau)\sigma_-(\tau)\rangle &= iG_\sigma^<(\tau, \tau). \end{aligned} \tag{14.338}$$

Thus, the first term is

$$\left(-iG_a^<(0, 0)\right)\left(iG_\sigma^<(\tau, \tau)\right). \tag{14.339}$$

Next up is the photon–atom cross correlations

$$\begin{aligned} \langle a^\dagger(0)\sigma_-(\tau)\rangle &= -iG_{a^\dagger\sigma_-}^<(0, \tau), \\ \langle \sigma_+(\tau)a(0)\rangle &= iG_{\sigma_+a}^<(\tau, 0). \end{aligned} \tag{14.340}$$

This results in

$$\left(-iG_{a^\dagger\sigma_-}^<(0, \tau)\right)\left(iG_{\sigma_+a}^<(\tau, 0)\right). \tag{14.341}$$

The third term also represents the atom–photon cross correlations

$$\begin{aligned} \langle a^\dagger(0)\sigma_+(\tau)\rangle &= -iG^<_{a^\dagger\sigma_+}(0, \tau), \\ \langle \sigma_-(\tau)a(0)\rangle &= iG^<_{\sigma_- a}(\tau, 0). \end{aligned} \tag{14.342}$$

The result is

$$\left(-iG^<_{a^\dagger\sigma_+}(0, \tau)\right)\left(iG^<_{\sigma_- a}(\tau, 0)\right). \tag{14.343}$$

The final expression for the correlation function is

$$\begin{aligned} \langle a^\dagger(0)\sigma_+(\tau)\sigma_-(\tau)a(0)\rangle = G^<_a(0, 0)G^<_\sigma(\tau, \tau) &+ G^<_{a^\dagger\sigma_-}(0, \tau)G^<_{\sigma_+ a}(\tau, 0) \\ &+ G^<_{a^\dagger\sigma_+}(0, \tau)G^<_{\sigma_- a}(\tau, 0), \end{aligned} \tag{14.344}$$

where each Keldysh Green's function is obtained from

$$G^< = G^R\Sigma^<G^A. \tag{14.345}$$

The self-energies $\Sigma^<$ capture damping and driving effects.

14.13 Three-point correlation functions

So far, we have only seen two or four-operator correlation functions. When looking at the field–intensity correlation $h_\theta(\tau)$ we need to find things to evaluate a correlation such as

$$\langle a^\dagger(0)[a(t) + a^\dagger(t)]a(0)\rangle. \tag{14.346}$$

In the Keldysh method, calculating three-point correlators such as

$$\langle A^\dagger(0)B(\tau)A(0)\rangle \tag{14.347}$$

requires careful handling of operators on the Keldysh contour, which involves forward and backward time branches. The contour-ordered expectation value is given by

$$\langle T_C A^\dagger(0)B(\tau)A(0)\rangle, \tag{14.348}$$

where T_C is the contour ordering operator. The key idea in the Keldysh formalism is that each operator can be defined on either the forward (+) or backward (–) time branches. Thus, to compute this correlator, we need to consider the placement of $A^\dagger(0)$, $B(\tau)$, and $A(0)$ on the contour. The three-point function decomposes into various real-time correlation functions, which means that depending on whether each operator is on the forward or backward branch, the correlator can be expressed in terms of real-time Green's functions

$$G^{abc}(t_1, t_2, t_3) = \langle A^a(t_1)B^b(t_2)A^c(t_3)\rangle, \tag{14.349}$$

where $a, b, c \in \{+, -\}$ specify whether an operator is on the forward or backward branch. Each of these components has its own physical meaning, such as retarded/advanced components or lesser/greater components. If the system is noninteracting or can be treated using perturbation theory, you can apply Wick's theorem to express the three-point function in terms of two-point Green's functions,

$$\langle A^{\dagger}(0)B(\tau)A(0)\rangle = G_{A^{\dagger}A}(\tau)G_B(\tau) + \text{vertex corrections}, \tag{14.350}$$

where the second term arises in the interacting case. For an interacting system, the three-point function involves a vertex correction term

$$\Gamma^{(3)}(t_1, t_2, t_3) = \int dt_4 dt_5 dt_6 G(t_1, t_4)G(t_2, t_5)G(t_3, t_6)V(t_4, t_5, t_6), \tag{14.351}$$

where V represents the interaction corrections. In steady-state or time-translation-invariant systems, it is often useful to transform into frequency space:

$$G^{(3)}(\omega_1, \omega_2) = \int dt_1 dt_2 e^{i\omega_1 t_1} e^{i\omega_2 t_2} \langle A^{\dagger}(0)B(t_1)A(t_2)\rangle. \tag{14.352}$$

In this representation, the three-point function can be related to the susceptibilities and interaction vertices. The general formula is

$$G^{(3)}(t_1, t_2, t_3) = \left.\frac{\delta^3 Z[J]}{\delta J_A^{\dagger}(t_1)\delta J_B(t_2)\delta J_A(t_3)}\right|_{J=0}. \tag{14.353}$$

Let us consider

$$\sigma_- \to a, \quad \sigma_+ \to a^{\dagger}. \tag{14.354}$$

We rewrite the expression explicitly as

$$\langle a^{\dagger}(0)a(\tau)a(0)\rangle + \langle a^{\dagger}(0)a^{\dagger}(\tau)a(0)\rangle. \tag{14.355}$$

Therefore, we need to compute the two separate terms, $\langle a^{\dagger}(0)a(\tau)a(0)\rangle$ (annihilation operator at τ) and $\langle a^{\dagger}(0)a^{\dagger}(\tau)a(0)\rangle$ (creation operator at τ). Using the Keldysh contour formalism, we express these correlators using real-time Green's functions.

Consider the first term. This corresponds to a three-point function in terms of the lesser/greater Green's functions:

$$G_{a^{\dagger}aa}(t_1, t_2, t_3) = \langle T_C a^{\dagger}(t_1)a(t_2)a(t_3)\rangle. \tag{14.356}$$

For a noninteracting system, we can use Wick's theorem to factorize this

$$\langle a^{\dagger}(0)a(\tau)a(0)\rangle = \langle a^{\dagger}(0)a(0)\rangle\langle a(\tau)\rangle + \langle a^{\dagger}(0)a(\tau)\rangle\langle a(0)\rangle. \tag{14.357}$$

If the expectation value $\langle a(\tau)\rangle = 0$, the first term vanishes, leaving

$$\langle a^{\dagger}(0)a(\tau)a(0)\rangle = \langle a^{\dagger}(0)a(\tau)\rangle\langle a(0)\rangle. \tag{14.358}$$

Again, if $\langle a^{\dagger}(\tau)\rangle = 0$ then the first term vanishes. The second term also vanishes since $\langle a(0)\rangle = 0$. For interacting systems, Wick's theorem does not hold, and we must include a vertex correction $\Gamma^{(3)}$,

$$G^{(3)}(t_1, t_2, t_3) = \int dt_4 dt_5 dt_6 G(t_1, t_4)G(t_2, t_5)G(t_3, t_6)\Gamma^{(3)}(t_4, t_5, t_6). \tag{14.359}$$

This depends on the interaction strength and the system's dynamics. We want to compute the Dyson series for

$$G_{a\sigma_-}(\omega) = \langle a(0)\sigma_-(\tau)\rangle, \tag{14.360}$$

using Feynman diagrams, and we introduce a brief sketch here. Before introducing the interactions, we define the bare (noninteracting) Green's functions. For the cavity mode, we have

$$G^0_{aa}(\omega) = \frac{1}{\omega - i\kappa} \tag{14.361}$$

and for the atom we have

$$G^0_{\sigma_-\sigma_-}(\omega) = \frac{1}{\omega - i\gamma}. \tag{14.362}$$

The interaction term in the Jaynes–Cummings Hamiltonian is

$$H_{\text{int}} = g(\sigma_+ a + \sigma_- a^\dagger). \tag{14.363}$$

This contributes a vertex in Feynman diagrams where the black dot represents the coupling strength g. The lowest-order term in the Dyson expansion corresponds to a single interaction vertex

$$G^{(1)}_{a\sigma_-}(\omega) = G^0_{aa}(\omega) g G^0_{\sigma_-\sigma_-}(\omega). \tag{14.364}$$

This corresponds to the equation

$$G^{(1)}_{a\sigma_-}(\omega) = \frac{g}{(\omega - i\kappa)(\omega - i\gamma)}. \tag{14.365}$$

The next-order correction includes an intermediate photon propagation before re-interacting

$$G^{(2)}_{a\sigma_-}(\omega) = G^0_{aa}(\omega) g G^0_{\sigma_-\sigma_-}(\omega) g G^0_{aa}(\omega). \tag{14.366}$$

This extends to all orders, forming a geometric Dyson series

$$G_{a\sigma_-}(\omega) = G^0_{aa}(\omega) g G^0_{\sigma_-\sigma_-}(\omega) + G^0_{aa}(\omega) g G^0_{\sigma_-\sigma_-}(\omega) g G^0_{aa}(\omega) g G^0_{\sigma_-\sigma_-}(\omega) + \dots, \tag{14.367}$$

which sums up to the Dyson equation

$$G_{a\sigma_-}(\omega) = \frac{g}{(\omega - i\kappa)(\omega - i\gamma) - g^2}. \tag{14.368}$$

Taking the inverse Fourier transform, we obtain the real-time response

$$G^R_{a\sigma_-}(t) = -i\Theta(t) g e^{-i\Omega_R t} e^{-(\kappa+\gamma)t/2}, \tag{14.369}$$

where the Rabi frequency is

$$\Omega_R = \sqrt{g^2 - \left(\frac{\kappa - \gamma}{2}\right)^2}. \tag{14.370}$$

14.14 Keldysh methods for out-of-time order correlations

We talked about out-of-time correlators in chapter 10. They are *not* in normal ordering, so our usual methods do not work.

Let us consider

$$H = \hbar\omega a^{\dagger}a + g(a^{\dagger}a)^2. \tag{14.371}$$

The standard OTOC we need to compute is

$$C(t) = \langle [a(t), a^{\dagger}(0)][a(t), a^{\dagger}(0)]^{\dagger} \rangle, \tag{14.372}$$

which is typically expressed as

$$C(t) = \langle a^{\dagger}(0)a(t)a^{\dagger}(0)a(t) \rangle - \langle a(t)a^{\dagger}(0)a(t)a^{\dagger}(0) \rangle. \tag{14.373}$$

In the Keldysh formalism, the four-point function is computed using the contour-ordered Green's functions. We can use a folded contour, as in figure 14.5 and do the path integral ordered along the contour.

Because there are four branches, the full Green's function is a 4 × 4 matrix

$$\mathbf{G}(t, t') = \begin{pmatrix} G_{11}(t, t') & G_{12}(t, t') & G_{13}(t, t') & G_{14}(t, t') \\ G_{21}(t, t') & G_{22}(t, t') & G_{23}(t, t') & G_{24}(t, t') \\ G_{31}(t, t') & G_{32}(t, t') & G_{33}(t, t') & G_{34}(t, t') \\ G_{41}(t, t') & G_{42}(t, t') & G_{43}(t, t') & G_{44}(t, t') \end{pmatrix}. \tag{14.374}$$

Each matrix element $G_{ij}(t, t')$ encodes correlations between operators on branch i at time t and on branch j at time t'. Then, the OTOC $C(t)$ is constructed from the appropriate products of these matrix elements that connect the branches on which V and W are inserted. For example, one might schematically write

$$C(t) = G_{12}(t, 0)G_{34}(t, 0) + \cdots, \tag{14.375}$$

where the full answer would involve a sum of all the ways to contract the operators according to the folded contour ordering. The out-of-time ordering is reflected in the 'cross' terms—that is, in the Green's function matrix elements that connect operators from what you might call the 'first pair' of contour segments (say, branches 1 and 2) with those from the 'second pair' (branches 3 and 4). In a folded

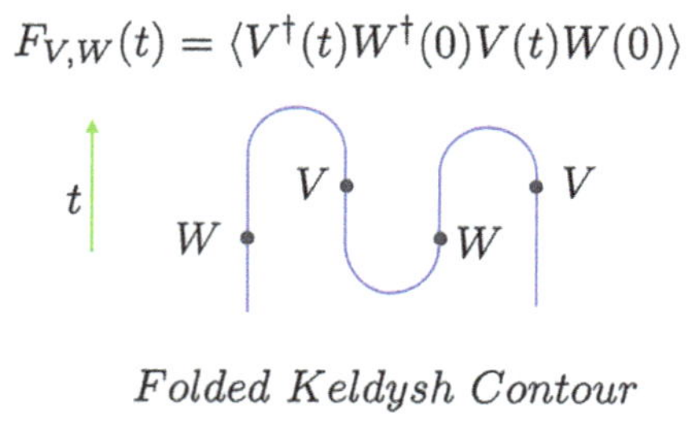

Figure 14.5. Folded Keldysh contour to calculate OTOCs.

Keldysh contour for OTOCs, the nontrivial time ordering (which is neither purely time-ordered nor anti-time-ordered) arises because operators inserted on different branches are not simply ordered by real time; instead, their ordering on the closed contour can lead to contributions where an operator on an early branch is later 'contracted' with one on a later branch, even if their real-time labels are interleaved. The 4×4 Green's function matrix generally has off-diagonal elements that connect, for example, branch 1 to branch 3 or branch 2 to branch 4, and these cross terms are essential to capture the full out-of-time-ordered structure of the correlator. These terms reflect the interference between the two 'halves' of the folded contour and are what make OTOCs sensitive to scrambling and chaos. Here we look at the calculation of an OTOC for the anharmonic oscillator

$$H = \hbar\omega_0 a^\dagger a + g(a^\dagger a)^2. \tag{14.376}$$

Using perturbation theory (to leading order in g) to obtain the self-energy correction. In this treatment, we assume that the oscillator is coupled to a bath via a collapse operator

$$L = \sqrt{\kappa}a, \tag{14.377}$$

so that the master equation is

$$\dot{\rho} = -\frac{i}{\hbar}[H, \rho] + \kappa\left(a\rho a^\dagger - \frac{1}{2}\{a^\dagger a, \rho\}\right). \tag{14.378}$$

We then wish to compute the OTOC:

$$C(t) = \langle a^\dagger(0)a(t)a^\dagger(0)a(t)\rangle - \langle a(t)a^\dagger(0)a(t)a^\dagger(0)\rangle. \tag{14.379}$$

For a quadratic (Gaussian) system, one could factorize the four-point function using Wick's theorem. (In fact, in a Gaussian theory the OTOC can be written in terms of two-point functions.) However, when $g \neq 0$ the Hamiltonian is nonlinear and corrections must be included via the self-energy. In the absence of interactions ($g = 0$), the retarded Green's function at zero temperature (with Lindblad decay) is

$$G_0^R(\omega) = \frac{1}{\omega - \omega_0 + i\kappa/2}. \tag{14.380}$$

Then, in the time domain,

$$G^R(t) = -i\Theta(t)e^{-i(\omega_0 + \delta\omega)t}e^{-\kappa t/2}. \tag{14.381}$$

At higher-order (order g^2) additional diagrams will contribute corrections that are nontrivial functions of time; these corrections will eventually modify the four-point function. One obtains for the four-point function an approximate factorization

$$\langle a^\dagger(0)a(t)a^\dagger(0)a(t)\rangle \approx \langle a^\dagger(0)a(t)\rangle^2, \tag{14.382}$$

and similarly for the other ordering. The difference between the two orderings encodes the out-of-time ordering. To the leading order, the OTOC is given by

$$C(t) \approx |G^R(t)|^2. \tag{14.383}$$

Thus, with the above $G^R(t)$ we have

$$|G^R(t)|^2 = \Theta(t)e^{-\kappa t}. \tag{14.384}$$

This is the result for $g = 0$ just to illustrate the start of the procedure.

References

[1] Jackson J D 1999 *Classical Electrodynamics* 3rd edn (New York: Wiley)
[2] Weber H, Harris F and Arfken G 2012 *Mathematical Methods for Physicists* (Amsterdam: Elsevier)
[3] Dyson F J 1949 The *S* matrix in quantum electrodynamics *Phys. Rev.* **75** 1736–55
[4] Feynman R P 1949 The theory of positrons *Phys. Rev.* **76** 749–59
[5] Feynman R P and Hibbs A R 1965 *Quantum Mechanics and Path Integrals* International Series in Pure and Applied Physics (New York: McGraw-Hill)
[6] Kamenev A 2023 *Field Theory of Non-Equilibrium Systems* (Cambridge: Cambridge University Press)
[7] Prigogine I 2017 *Non-Equilibrium Statistical Mechanics* (New York: Dover)
[8] Rammer J 2007 *Quantum Field Theory of Non-equilibrium States* (Cambridge: Cambridge University Press)
[9] Calzetta E and Hu B L 1999 Stochastic dynamics of correlations in quantum field theory: from the Schwinger–Dyson to Boltzmann–Langevin equation *Phys. Rev.* D **61** 025012
[10] Stefanucci G and van Leeuwen R 2013 *Nonequilibrium Many-Body Theory of Quantum Systems* (Cambridge: Cambridge University Press)
[11] Martin P C and Schwinger J 1959 Theory of many-particle systems. I *Phys. Rev.* **115** 1342–73
[12] Sieberer L M, Buchhold M and Diehl S 2016 Keldysh field theory for driven open quantum systems *Rep. Progr. Phys.* **79** 096001
[13] Feynman R P and Vernon F L 1963 The theory of a general quantum system interacting with a linear dissipative system *Ann. Phys.* **24** 118–73
[14] Wick G C 1950 The evaluation of the collision matrix *Phys. Rev.* **80** 268–72

IOP Publishing

An Introduction to Quantum Optics (Second Edition)
An open systems approach
Perry Rice

Chapter 15

Double-sided Feynman diagrams

15.1 Density matrix perturbation theory

We will introduce a perturbative treatment that is very useful for multidimensional nonlinear spectroscopy [1–5]. For a pure state that satisfies the Schrödinger equation

$$\frac{d\Psi}{dt} = \frac{-i}{\hbar}H\Psi \tag{15.1}$$

we have the following equation for the density operator $\rho = |\Psi\rangle\langle\Psi|$:

$$\dot{\rho} = \frac{-i}{\hbar}[H, \rho]. \tag{15.2}$$

We use this in anticipation of including dephasing and decay processes, where a wave function does not exist. This can be written in integral form as

$$\rho(t) - \rho(0) = \frac{-i}{\hbar}\int_0^t dt'[H(t'), \rho(t')]. \tag{15.3}$$

We then iterate this solution and find

$$\rho(t) = \rho^{(0)} + \rho^{(1)} + \rho^{(2)} + \rho^{(3)} \cdots, \tag{15.4}$$

with

$$\rho^{(0)} = \rho(t = 0) = \rho_0 \tag{15.5}$$

$$\rho^{(1)} = \left(\frac{-i}{\hbar}\right)\int_0^t dt'[H(t'), \rho(0)] \tag{15.6}$$

$$\rho^{(2)} = \left(\frac{-i}{\hbar}\right)^2\int_0^t dt'\int_0^{t'} dt''[H(t''), [H(t')\rho(0)]] \tag{15.7}$$

doi:10.1088/978-0-7503-6014-2ch15

$$\rho^{(3)} = \left(\frac{-i}{\hbar}\right)^3 \int_0^t dt' \int_0^{t'} dt'' \int_0^{t''} dt''[H(t'''), [H(t''), [H(t')\rho(0)]]] \tag{15.8}$$

$$\rho^{(4)} = \left(\frac{-i}{\hbar}\right)^4 \int_0^t dt' \int_0^{t'} dt'' \int_0^{t''} \int_0^{t'''} dt''''[H(t''''), [H(t'''), [H(t''), [H(t')\rho(0)]]]] \tag{15.9}$$

and one can continue on. We stop here as we need go no further than fourth order for two-photon absorption.

Let us examine $\rho^{(1)}$ in the interaction picture as a place to start and also to retrieve the linear response. The required commutator is

$$[H(t'), \rho_0] = ((V(t') + V^{\dagger}(t'))\rho_0 - \rho_0(V(t') + V^{\dagger}(t'))). \tag{15.10}$$

The requirement that the first interaction needs to be absorptive if we start in the ground state at $t = 0$ and recalling that V comprises raising operators, we have

$$[H(t'), \rho_0] = E(V(t')\rho_0 - \rho_0 V^{\dagger}(t')). \tag{15.11}$$

We then write down the first-order contribution to the density matrix

$$\rho^{(1)} = \frac{-i}{\hbar} \int_0^t dt' \cos(\omega t')\big(V(t')\rho_0 - \rho_0 V^{\dagger}(t')\big). \tag{15.12}$$

Realizing that $V \approx e^{+i\omega_{ij}t}$ for a given raising operator, we can use the rotating wave approximation to arrive at

$$\rho^{(1)} = \frac{-iE_0}{\hbar} \int_0^t dt'(V(t')e^{-i\omega t}\rho_0 - \rho_0 V^{\dagger}(t')e^{i\omega_t}). \tag{15.13}$$

Since we are starting from the ground state, the only terms that remain are V_{eg} and $V_{eg}^{\dagger}$ and corresponding operators for $e'g$, we have

$$\rho^{(1)} = \rho_{eg}^{(1)} \tag{15.14}$$

and

$$\begin{aligned} \rho_{eg}^{(1)} &= \frac{-i}{\hbar} \int_0^t dt'(V(t')e^{i\omega t'}\rho_0 - \rho_0 V^{\dagger}(t')e^{-i\omega t'}) \\ &= \frac{-i\mu E_0}{\hbar} \int_0^t (dt'\big|e^{i(\omega-\omega_{eg})t'}\rho_0 - \rho_0\big|e^{-i(\omega-\omega_{eg})t'}), \end{aligned} \tag{15.15}$$

where we have specified

$$V(t) = -\vec{\mu}\cdot\vec{E}(t), \tag{15.16}$$

where we have suppressed any orientational or polarization effects arising from the dot product. Using $\rho_0 = |g\rangle\langle g|$ we have

$$\rho_{eg}^{(1)} = \frac{-iE_0}{\hbar} \int_0^t dt'(|e\rangle\langle g|e^{i(\omega-\omega_{eg})t'} - |g\rangle\langle e'|e^{-i(\omega-\omega_{e'g})t'}). \tag{15.17}$$

To ensure that the time integrals converge, and more importantly to model a real system, we introduce phenomenological decay/dephasing rates; we have seen how these arise from bath couplings in previous chapters. Typically, one introduces an ϵ to move the pole off the real axis, and then takes the limit $\epsilon \rightarrow 0$. The decay/dephasing will keep the poles off the real axis, so once they are introduced, we have no need for ϵ. This is exactly what we would find if we used Liouvillian time evolution, as the Green's function (propagator) is

$$G(T) = e^{-\gamma_{eg}T}e^{i\omega_{eg}T}. \tag{15.18}$$

This leaves us with

$$\rho_{eg}^{(1)} = \frac{-i}{\hbar}\int_0^t dt' e^{-\gamma_{eg}t'}(e^{i(\omega-\omega_{eg})t'} - e^{-i(\omega-\omega_{e'g})t'}). \tag{15.19}$$

This is the proper Green's function for a homogeneously broadened system, where all the atoms see the same environment. In the case of Doppler broadening, where the speeds of the molecules obey a Maxwell–Boltzmann distribution, we would have

$$f(v) = \frac{1}{\sqrt{2\pi\sigma_v^2}}e^{-v^2/(2\sigma_v^2)}. \tag{15.20}$$

This leads to a Gaussian distribution of frequencies

$$f(\omega) = \frac{1}{\sqrt{2\pi\sigma^2}}e^{-(\omega-\omega_0)^2/(2\sigma^2)}. \tag{15.21}$$

The Greens function can be found via

$$G(T) = \int_{-\infty}^{\infty} f(\omega)e^{-i\omega t}d\omega, \tag{15.22}$$

which results in

$$\begin{aligned} G_{\text{dop}}(T) &= e^{-\gamma_{eg}T}e^{i\omega_{eg}T} \\ &= \frac{1}{\sqrt{2\pi\sigma^2}}\int_{-\infty}^{\infty} e^{-(\omega-\omega_0)^2/(2\sigma^2)}e^{-i\omega t}d\omega, \end{aligned} \tag{15.23}$$

which results in

$$G(t) = e^{-i\omega_0 t}e^{-\frac{1}{2}\sigma^2 t^2}. \tag{15.24}$$

Sometimes we have a convolution of the Gaussian and Lorentzian distributions,

$$V(\omega; \sigma, \gamma) = \int_{-\infty}^{\infty} G(\omega'; \sigma)\, L(\omega - \omega'; \gamma)\, d\omega'. \tag{15.25}$$

There is no simple closed form for this distribution. It is appropriate for vapors at room temperature where we have Doppler (Gaussian) + collisional (Lorentzian) broadening; or for solid-state systems where we have Inhomogeneous crystal fields + homogeneous dephasing.

If we excite with a $E = E_0\,\delta(0)$, the time integral just time stamps the interaction terms, here at $t = 0$, and we find for the time evolution of the density matrix

$$\rho_{eg}^{(1)}(T) = \frac{-i\mu E_0}{\hbar} e^{-\gamma_{eg}T}(|e\rangle\,\langle g|e^{i(\omega-\omega_{eg})T} - |g\rangle\,\langle e'|e^{-i\omega_{e'g}T}). \tag{15.26}$$

If I want to find the average dipole moment due to this interaction, we need to calculate

$$\begin{aligned}\langle|e\rangle\langle g|\rangle &= \mathrm{Tr}(\rho^{(1)}(t)|e\rangle\langle g|)\\ &= \mathrm{Tr}((|e\rangle\langle e| + |g\rangle\langle g|))e^{-\gamma_{eg}T}e^{i\omega_{eg}T}\\ &= e^{-\gamma_{eg}T}e^{i\omega_{eg}T},\end{aligned} \tag{15.27}$$

where we have suppressed the size of the dipole moment μ and magnitude of applied pulse E_0. Adding the complex conjugate from the second diagram, we have

$$\left\langle\mu_{eg}\right\rangle = e^{-\gamma_{eg}t}\cos\left(\omega_{eg}t\right). \tag{15.28}$$

We introduce the notation V_i for a dipole operator acting at $t = t_i$. It is always a raising operator acting to the right. If the current state is $|e\rangle$, the raising operator will promote you to $|f\rangle$, the state to be raised is obvious from the diagram in question. In this notation

$$\rho^{(1)} = V_1\rho_0 - \rho_0 V_1^{\dagger}, \tag{15.29}$$

where V_1 has associated with it a phase, a response function, and a raising operator

$$V_i = \mu E_0 e^{-\gamma_i t}e^{i(k_i z - \omega_i t)}\sigma_+^i. \tag{15.30}$$

For ultrashort pulses, the effect of the interaction an impulse response, and the delta functions kill the integrals. Hence, we can write down the contribution from a pulse as

$$R(t) = \mu E_0 e^{-\gamma_i t}e^{\pm i\omega_i t}, \tag{15.31}$$

where t above is the time since the pulse arrived, and i denotes the transition being raised or lowered. In frequency space, which we obtain from a Fourier transform of the above, we have

$$R(\omega) = \frac{\mu E_0}{\pi}\frac{\gamma_i}{\gamma_i \mp i(\omega - \omega_i)}. \tag{15.32}$$

15.2 Double-sided Feynman diagrams

At this point we discuss rules for diagrammatic representations of the needed integrals. The perturbative expansion of the density matrix is

$$\begin{aligned}\rho_I(t) = \rho_I(t_0) + \sum_{n=1}^{\infty}\left(-\frac{i}{\hbar}\right)^n \int_{t_0}^{t} d\tau_n \int_{t_0}^{\tau_n} d\tau_{n-1} \ldots \int_{t_0}^{\tau_2} d\tau_1\\ \left[H_I'(\tau_n), \left[H_I'(\tau_{n-1}), \ldots\left[H_I'(\tau_1), \rho_I(t_0)\right]\ldots\right]\right],\end{aligned} \tag{15.33}$$

where we have interactions time ordered $\tau_n > \tau_{n-1}$.... With the dipole interactions, we have

$$\rho^{(n)}(t) = \left(-\frac{i}{\hbar}\right)^n \int_{-\infty}^{t} d\tau_n \int_{-\infty}^{\tau_n} d\tau_{n-1} \ldots \int_{-\infty}^{\tau_2} d\tau_1 E(\tau_n)E(\tau_{n-1}) \cdot \ldots \cdot E(\tau_1) \cdot \\ U_0(t, t_0) \cdot [\mu_I(\tau_n), [\mu_I(\tau_{n-1}), \ldots[\mu_I(\tau_1), \rho(-\infty)]\ldots]] \cdot U_0^\dagger(t, t_0). \quad (15.34)$$

Here, we envision a sequence of pulses incident on the system. Let us go back to the wave functions for a second. To the lowest order we have

$$C_g = C_g^{(0)} = 1 \quad (15.35)$$

$$C_e = C_e^{(0)} = 0. \quad (15.36)$$

To the first order we find

$$C_g = C_g^{(0)} + C_g^{(1)} = 1 \quad (15.37)$$

$$C_e = C_e^{(0)} + C_e^{(1)} \propto E \quad (15.38)$$

so at first we excite a dipole moment. A nonzero population in the excited state shows up at the second order

$$P_e = |C_e|^2 \propto E^2. \quad (15.39)$$

We can represent this with a diagram with the evolution of a bra on the left side and the evolution of a ket on the right side, as shown in figure 15.1. The second event is radiation from the dipole, and we should properly think of the entire process as producing a signal at t_1. This can be a spontaneous process; here we have a second pulse generating a stimulated emission (figure 15.2)

$$\begin{aligned} S^{(1)}(t_1) &= -\frac{i}{\hbar}\langle\mu(t_1)[\mu(0), \rho(-\infty)]\rangle \\ &= -\frac{i}{\hbar}(\langle\mu(t_1)\mu(0)\rho(-\infty)\rangle - \langle\mu(t_1)\rho(-\infty)\mu(0)\rangle) \\ &= -\frac{i}{\hbar}(\langle\mu(t_1)\mu(0)\rho(-\infty)\rangle - \langle\rho(-\infty)\mu(0)\mu(t_1)\rangle) \\ &= -\frac{i}{\hbar}(\langle\mu(t_1)\mu(0)\rho(-\infty)\rangle - \langle\mu(t_1)\mu(0)\rho(-\infty)\rangle^*). \end{aligned} \quad (15.40)$$

From our discussion so far, we have enumerated the rules for ladder diagrams (double-sided Feynman diagrams):

1. Vertical lines represent the time evolution of the ket and bra of the density matrix. Time is running from the bottom to the top.
2. Interactions are represented by arrows. The last interaction, which originates from the trace $P^{(n)}(t) = \langle\mu\rho^{(n)}(t)\rangle$, yields the signal.
3. Each diagram has a sign $(-1)^n$, where n is the number of interactions from the right. This is due to the use of the commutation relation $[a, a^\dagger] = 1$. The last interaction is excluded.

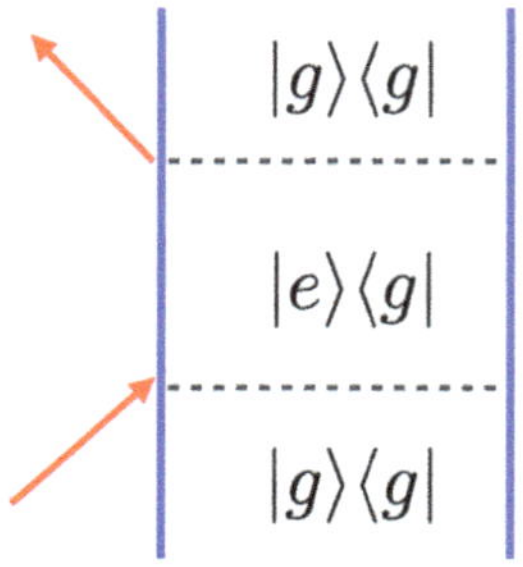

Figure 15.1. Linear interaction.

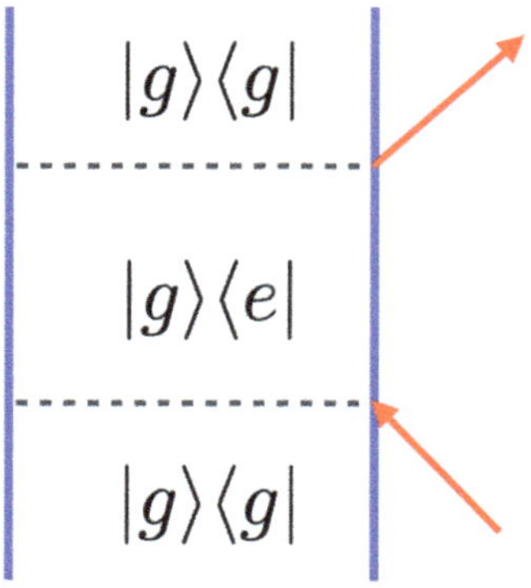

Figure 15.2. Conjugate diagram.

4. Arrows pointing to the right represent the interaction factor $e^{-i\omega t+ikr}$, while an arrow pointing to the left represents the interaction factor $e^{+i\omega t-ikr}$. The last line conserves energy and momentum.
5. Arrows pointing toward the system represent an excitation, and arrows pointing away from the line represents de-excitation. These stem from the use of the rotating wave approximation and the time evolution of the raising and lowering atoms. The last interaction always points away from the system.

Corresponding to the double-sided Feynman diagrams, we also have Lee–Albrecht ladder diagrams (figure 15.3). These are much like the usual cartoon description of an optical process you are familiar with, but we use a dashed line to keep track of the interactions on the bra side of evolution (figure 15.4).

We must keep track of the electric field as well, for each pulse we have two terms

$$(e^{-i\omega t} + e^{+i\omega t}). \tag{15.41}$$

If we have three pulses at times t_1, t_2, t_3 (yielding six terms each), and four interactions, we have 864 terms in

$$\int dt_1 \int dt_2 \int dt_3 E(t - t_3)E(t - t_3 - t_2)E(t - t_3 - t_2 - t_1) \cdot S(t_3, t_2, t_1). \tag{15.42}$$

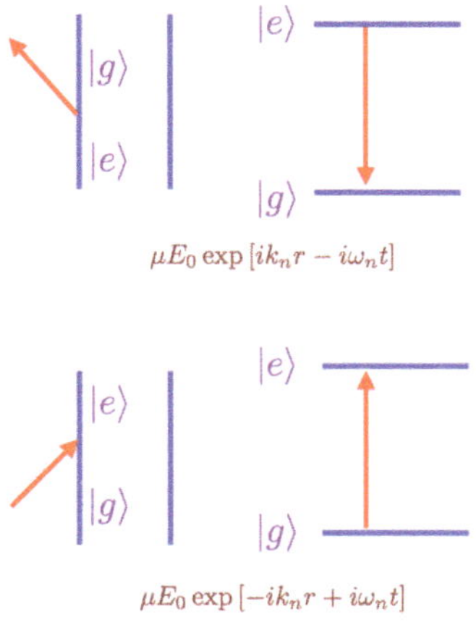

Figure 15.3. Ket side Lee–Albrecht diagrams.

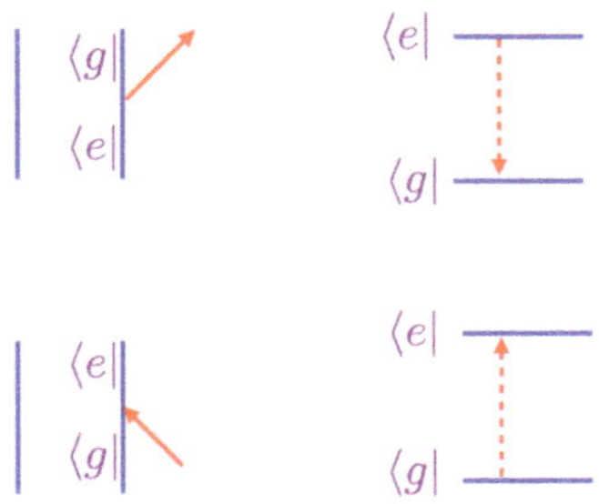

Figure 15.4. Bra-side Lee–Albrecht diagrams.

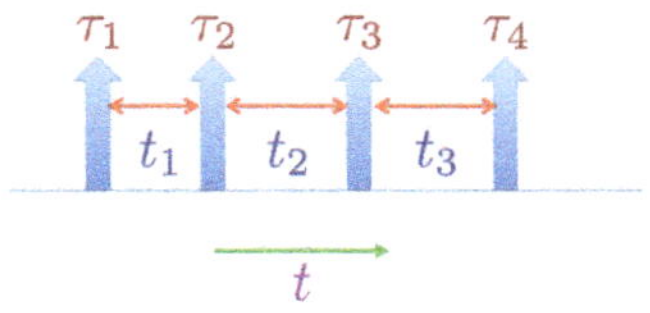

Figure 15.5. Pulse train.

Yikes. OK, we assume the pulses do not overlap, and further that they are delta function pulses (figure 15.5). Really that requires pulses short compared to the time scales you are investigating. We are now at 32 terms, as there are only two terms per pulse ($2^3 = 8$) and four interactions including the final emission. The rotating wave reduces this to on term per pulse, and we are now at four terms. The delta function kills the integrations, leaving us with a relatively simple problem.

For our linear example above, at $t = 0$ we generate a dipole moment on the transition, which evolves via oscillations and decay until emission at t_1. With multiple pulses and systems with more than two levels, we can interrogate the transition energies, lifetimes, and line strength, making this a powerful technique for examining molecular and condensed matter systems. For each interaction, we will pick up a phase, leading to a total phase of

$$k = \pm k_1 \pm k_2 \pm k_3. \tag{15.43}$$

But due to the rotating wave approximation, only on term survives when looking at a particular emission direction, allowing us to further separate signals.

15.3 Two-dimensional Fourier spectroscopy

So far, this is all standard spectroscopy. Here we discuss a technique that yields more information than just the standard output of a spectrometer [6]. Here, we send a train of two to four pulses at a sample. Each pulse interacts with the system (we assume a delta function form), which then evolves, and the next pulse interacts with the evolved signal. The simplest for this is pump–probe spectroscopy, where you use two pulses and monitor fluorescence or use a fancier detection scheme. These pulses can be colinear, or they may come in with different k vectors.

The latter will pick out various processes spatially. When we have three to four pulses, we sweep two of the delay times and take a 2D Fourier transform (2DFS) with respect to those delay variables. If one sweeps three pulses, you would have a 3D spectrum; the 2D case is easier to deal with. One can always choose different pairs of pulse delays to sweep, and thus build up a fuller picture if desired. At short t_2 (middle delay), one usually finds an elliptical line shape, with the major axis proportional to the inhomogeneous lifetime, and the minor axis is proportional to the homogeneous linewidth. At larger t_2, we end up with a circular shape due to spectral diffusion (figure 15.6).

Along the diagonal are the 'usual' features one expects, off the diagonal we get cross peaks which contain information on the couplings between transitions. These peaks appear when the system emits a signal that oscillates at a different frequency with respect to the excitation signal. A cross peak means that two states of the system are coupled because when the pulses pump one state, the system responds with emission from a different state. This coupling can be related to an energy transfer or charge transfer process between molecules. Depending on the detection protocol, one has access to the real and imaginary parts of the response, as well as the magnitude.

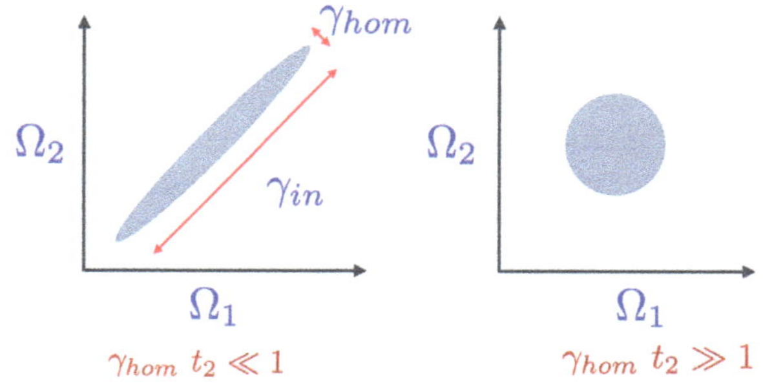

Figure 15.6. Spectral diffusion.

Various Liouville pathways have standard names, some of which refer back to nuclear magnetic resonance spectroscopy. We have

1. *Ground state bleaching*. This refers to the bleaching (reduction) of the ground state's absorption due to the population being excited to an upper state. After excitation by the following pulse, the ground state is depleted, reducing its ability to absorb light. It is characterized by one excitation to a higher state and subsequent evolution as if the system remains in a superposition or returns to the ground state.
2. *Stimulated emission.* The same process we have already discussed. After excitation, the system returns to the ground state by emitting a photon, stimulated by the probe pulse.
3. *Excited state absorption.* This is further absorption after the system has been excited, typically from a singly excited state to a higher-lying excited state.
4. *Double quantum coherence.* This occurs when two interactions promote the system coherently from the ground to a doubly excited state, exciting a dipole between the ground state and a double excited state, and the system evolves in that coherence.
5. *Rephasing.* This pathway will reverse the inhomogeneous dephasing as in a photon echo. The signal rephrases during the detection period, producing an echo that helps isolate the homogeneous broadening.
6. *Non-rephasing.* No compensation for inhomogeneous dephasing; retains information about non-echo dynamics.

The difference between dephasing and non-rephasing processes (NRP) is whether the phase of the signal gets reversed or not after τ_1, as shown in figure 15.7.

A common signal to measure is the overall absorption,

$$A = \mathrm{Re}(\mathrm{RP} + \mathrm{NRP}). \tag{15.44}$$

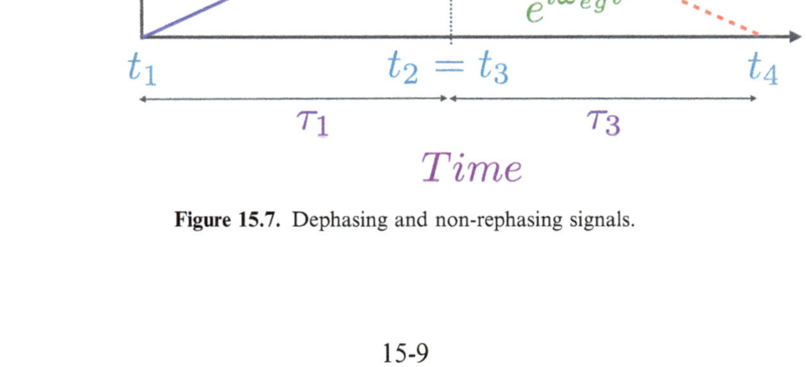

Figure 15.7. Dephasing and non-rephasing signals.

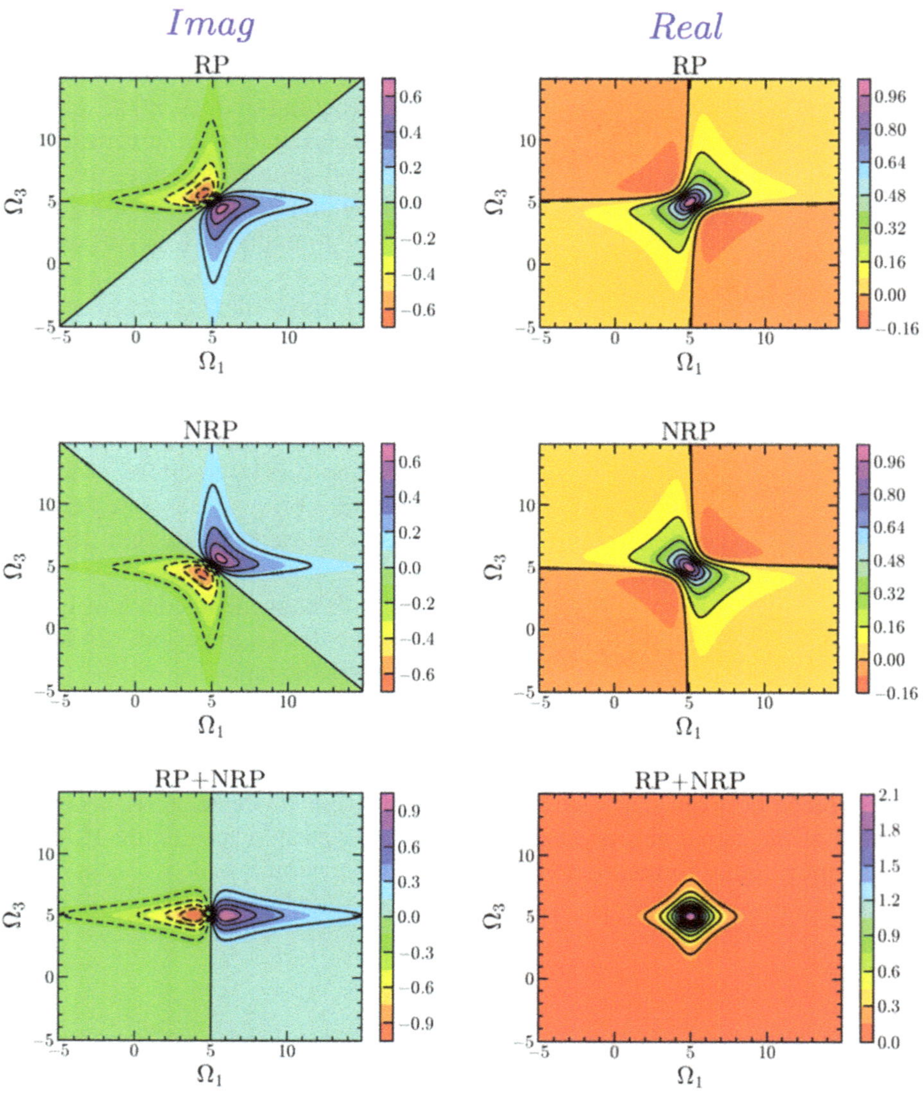

Figure 15.8. Signals associated with one transition.

We now look schematically at the 2DFS signals for a pair of two-level systems with the upper levels coupled. Figure 15.8 shows typical plots for the rephasing process (RP) and non-rephasing process (NRP) signals, plus their sum. The imaginary part resembles a butterfly, and there is a $\pi/2$ rotation between RP and NRP. The real part resembles a dog bone and has the same rotation. The sum plots are symmetric about both axes.

In figure 15.9 we show a model system with two-level systems with upper-level coupling.

In figure 15.10 we plot the sum of the RP and NRP signals for the coupled system, the real and imaginary parts as well as the magnitude. We see strong diagonal peaks and weaker cross peaks resulting from the upper-level coupling α.

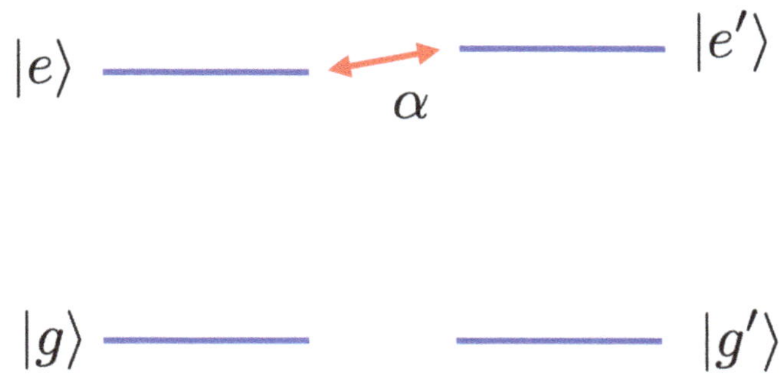

Figure 15.9. Coupled two-level systems.

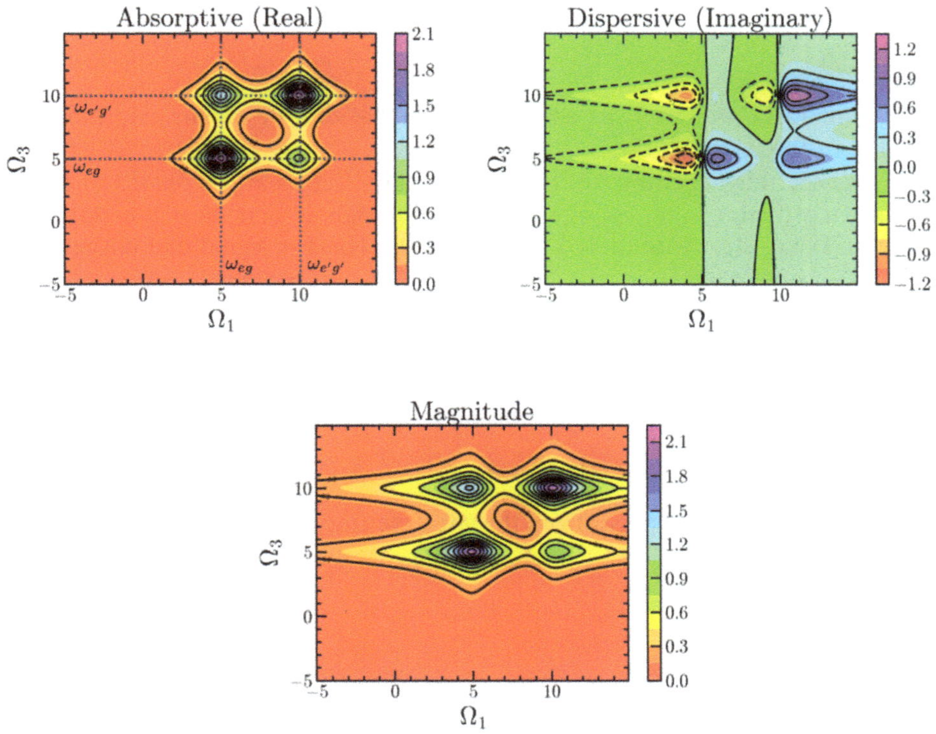

Figure 15.10. Total signal for coupled two-level systems.

15.4 Nonlinear interactions

Going beyond the linear response [7], we begin with an evaluation of the commutator

$$\begin{aligned}[H(t''), [H(t')\rho(0)]] &= (V_2 + V_2^\dagger)(V_1\rho_0 - \rho_0 V_1^\dagger) - ((V_1\rho_0 - \rho_0 V_1^\dagger)(V_2 + V_2^\dagger)) \\ &= V_2 V_1 \rho_0 + V_2^\dagger V_1 \rho_0 - V_2 \rho_0 V_1^\dagger \\ &\quad - V_1 \rho_0 V_2^\dagger - \rho_0 V_1^\dagger V_2 + \rho_0 V_1^\dagger V_2^\dagger. \end{aligned} \tag{15.45}$$

Here, we have introduced the notation V_i for a dipole operator acting at $t = t_i$. It is a raising operator acting to the right. If the current state is $|e\rangle$, the raising operator will promote you to $|f\rangle$, the state to be raised is obvious from the diagram in question. Next, we examine

$$\begin{aligned}[H(t'''), [H(t''), [H(t')\rho(0)]]] = &(V_3 + V_3^\dagger)[H(t''), [H(t')\rho(0)]] \\ &- [H(t''), [H(t')\rho(0)]](V_3 + V_3^\dagger).\end{aligned} \tag{15.46}$$

After some machination, we arrive at 16 terms:

$$\begin{aligned}[H(t'''), [H(t''), [H(t')\rho(0)]]] = &V_3^\dagger V_2 V_1 \rho_0 + V_3 V_2^\dagger V_1 \rho_0 - V_3 V_2 \rho_0 V_1^\dagger - V_3^\dagger V_2 \rho_0 V_1^\dagger \\ &- V_3 V_1 \rho_0 V_2^\dagger - V_3^\dagger V_1 \rho_0 V_2^\dagger + V_3 \rho_0 V_1^\dagger V_2 + V_3 \rho_0 V_1^\dagger V_2^\dagger \\ &- V_2 V_1 \rho_0 V_3^\dagger - V_3^\dagger V_1 \rho_0 V_2^\dagger + V_3 \rho_0 V_1^\dagger V_2 + V_3 \rho_0 V_1^\dagger V_2^\dagger \\ &- V_2 V_1 \rho_0 V_3^\dagger - V_2^\dagger V_1 \rho_0 V_3^\dagger + \rho_0 V_1^\dagger V_2 V_3 + \rho_0 V_1^\dagger V_2^\dagger V_3 .\end{aligned} \tag{15.47}$$

In figure 15.11 we show the associated double-sided Feynman diagrams corresponding to the terms above.

OK, we see this is getting out of hand quickly. Let us apply phases to the four pulses to tag them (ϕ_1, ϕ_2, ϕ_3, ϕ_4) and look for signals that have a certain phase condition. Let us start with those known as NRP. This is a signal that has an overall phase

$$\Phi = \phi_1 - \phi_2 + \phi_3 + \phi_4. \tag{15.48}$$

From the signs of these phases, we can identify which interactions will lead to signals with this signature. For the above condition, the interactions that occur are

$$V_1,\ V_2^\dagger,\ V_3,\ V_4^\dagger. \tag{15.49}$$

The first interaction must occur from the left if we start from the ground state (recall V_is are raising operators)

$$\rho_1 = V_1 \rho_0. \tag{15.50}$$

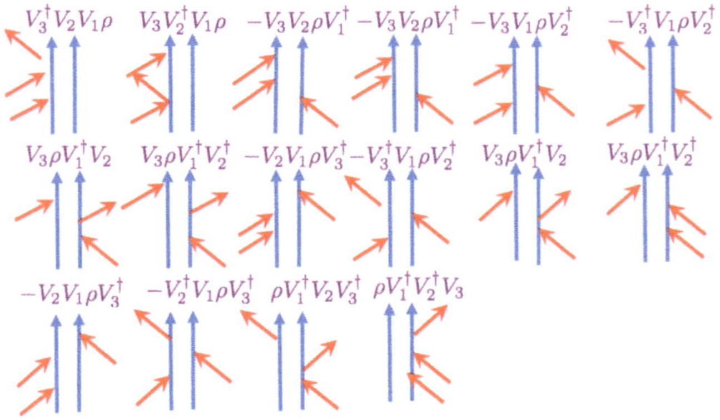

Figure 15.11. Liouville pathways for three pulses.

Then the next interaction can occur from either side, taking away a photon from the left side or adding one to the right

$$\rho_2 = V_2^\dagger V_1 \rho_0 + V_1 \rho_0 V_2^\dagger. \tag{15.51}$$

For the third interaction, we find that the result is

$$\rho_3 = V_3 V_2^\dagger V_1 \rho_0 + V_3 V_1 \rho_0 V_2^\dagger + V_1 \rho_0 V_2^\dagger V_3. \tag{15.52}$$

After the fourth interaction, we find

$$\begin{aligned}\rho_4 = \; & V_4^\dagger V_3 V_2^\dagger V_1 \rho_0 + V_4^\dagger V_3 V_1 \rho_0 V_2^\dagger + V_4^\dagger V_1 \rho_0 V_2^\dagger V_3 \\ & + V_3 V_2^\dagger V_1 \rho_0 V_4^\dagger + V_3 V_1 \rho_0 V_2^\dagger V_4^\dagger + V_1 \rho_0 V_2^\dagger V_3 V_4^\dagger.\end{aligned} \tag{15.53}$$

The first and fifth terms leave you in the ground state, leaving four nonzero contributions

$$\mathrm{GSB}_2 = -V_3 \rho_0 V_1^\dagger V_2 V_4^\dagger \tag{15.54}$$

$$\mathrm{SE}_2 = -V_2 \rho_0 V_1^\dagger V_3 V_4^\dagger \tag{15.55}$$

$$\mathrm{ESA}_{21} = -V_4^\dagger V_3 V_2 \rho_0 V_1^\dagger \tag{15.56}$$

$$\mathrm{ESA}_{22} = -V_3 V_2 \rho_0 V_1^\dagger V_4^\dagger. \tag{15.57}$$

The names are ground state bleaching (GSB), stimulated emission (SE), and excited state absorption (ESA). The signs are determined thusly. We know that interactions $V_4 V_3 V_2 V_1 \rho$ are positive. Every interaction on the left comes from the negative terms in the commutator. We can also look at signals with

$$\Phi = \phi_1 - \phi_2 + \phi_3 - \phi_4 \tag{15.58}$$

called NRP. The interactions are $(V_1,\ V_2^\dagger,\ V_3,\ V_4^\dagger)$. This yields

$$\mathrm{GSB}_1 = V_3 V_2^\dagger V_1 \rho_0 V_4^\dagger \tag{15.59}$$

$$\mathrm{SE}_1 = V_1 \rho_0 V_2^\dagger V_3 V_4^\dagger \tag{15.60}$$

$$\mathrm{ESA}_{11} = V_4^\dagger V_3 V_1 \rho_0 V_2^\dagger \tag{15.61}$$

$$\mathrm{ESA}_{12} = V_3 V_1 \rho_0 V_2^\dagger V_4^\dagger. \tag{15.62}$$

Another phase signature of interest is

$$\Phi = \phi_1 + \phi_2 - \phi_3 - \phi_4, \tag{15.63}$$

known as a double quantum coherence (DQC) signal because a dipole developed between the biexciton state and the ground state. The interactions are $(V_1,\ V_2,\ V_3^\dagger,\ V_4^\dagger)$, leading to

$$\mathrm{DQC}_1 = V_3^\dagger V_2 V_1 \rho_0 V_4^\dagger \tag{15.64}$$

$$\mathrm{DQC}_{21} = V_4^\dagger V_2 V_1 \rho V_3^\dagger \tag{15.65}$$

$$\mathrm{DQC}_{22} = V_2 V_1 \rho_0 V_3^\dagger V_4^\dagger. \tag{15.66}$$

One often looks at the sum of these four signals, with a factor for the quantum yield Γ of the $|f\rangle$ state,$S = (\mathrm{RP} + \mathrm{NRP} + (1 - \Gamma)\mathrm{ESA}_{11})$. We define

$$G_i^\pm = e^{-\gamma_i t} e^{\mp i\omega_i t}. \tag{15.67}$$

We had 16 double-sided Feynman diagrams, added an interaction, and ended up with 11 terms? These are shown in figures 15.12–15.15.

This reduction is due to the restrictions of these phases, for example $V_4 V_3^\dagger V_2 V_1 \rho_0$ which has a phase $\phi_1 + \phi_2 - \phi_3 + \phi_4$. We calculate the population of the biexciton state as

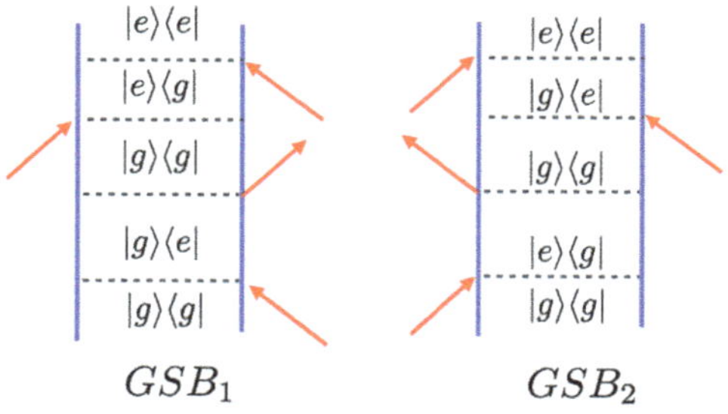

Figure 15.12. Liouville pathways for ground state bleaching.

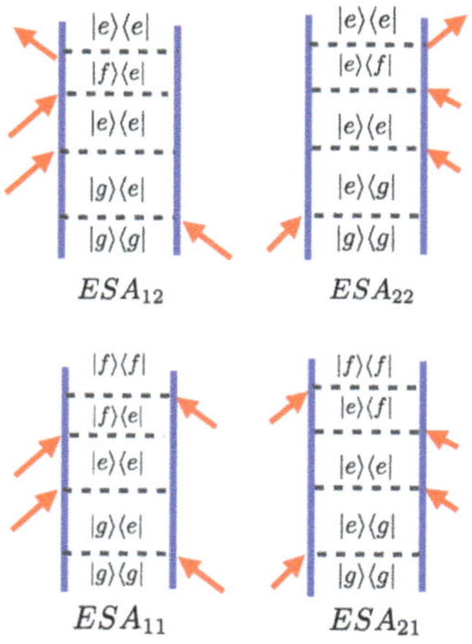

Figure 15.13. Liouville pathways for excited state absorption.

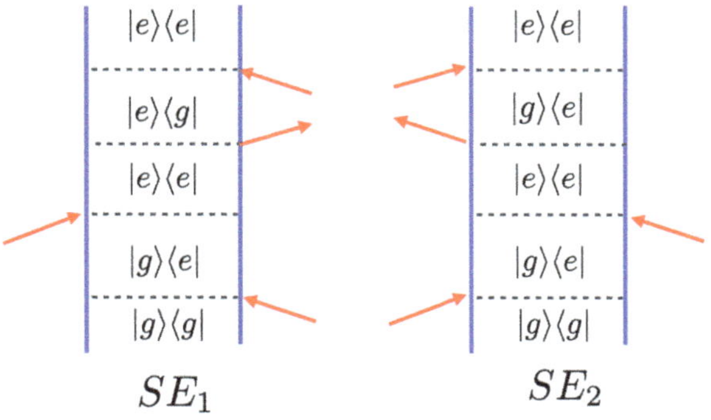

Figure 15.14. Liouville pathways for stimulated absorption.

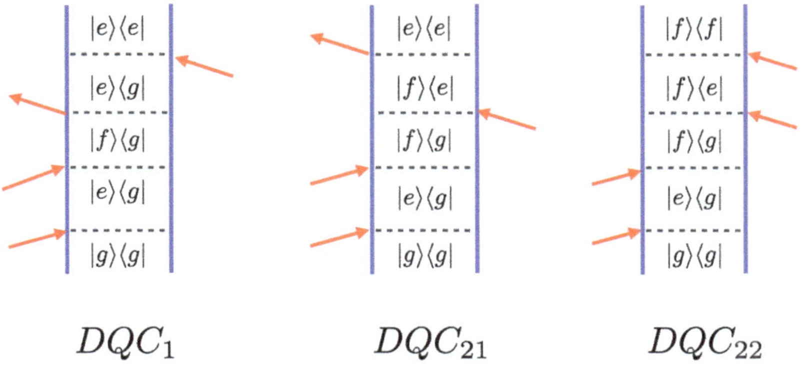

Figure 15.15. Liouville pathways for double quantum coherence.

$$\begin{aligned} P_f &= \mathrm{Tr}(\rho(t)|f\rangle\langle f|) \\ &= \mathrm{Tr}\,(\rho^{(4)}(t)|f\rangle\langle f|) \\ &= \mathrm{Tr}\,(\rho^{(3)}(t)V|e\rangle\langle f|) \end{aligned} \tag{15.68}$$

and the fluorescence from this state is detected. To do this, we restrict ourselves to processes that leave us with a nonzero ρ_{ff}. There is one signal each from NRP, RP, and DQC. We denote these as

$$\mathrm{NRP}_{ff} = \mathrm{ESA}_{12} \tag{15.69}$$

$$\mathrm{RP}_{ff} = \mathrm{ESA}_{22} \tag{15.70}$$

$$\mathrm{DQC}_{ff} = \mathrm{DQC}_{22}. \tag{15.71}$$

For now, we focus on diagrams that leave us in state $|fe\rangle$ or $|fe'\rangle$, a fourth pulse can create a biexciton (state $|f\rangle$) and then the fluorescent signal is monitored. We could have written down these diagrams; all pathways that lead to $\rho = |f\rangle\langle f|$ involve two different interactions each on the bra and ket side, the interactions are of the form $V_i V_j \rho_0 V_k^\dagger V_l^\dagger$ which has six unique pathways. Three are shown in figures 15.16–15.18, and there are the corresponding conjugate diagrams.

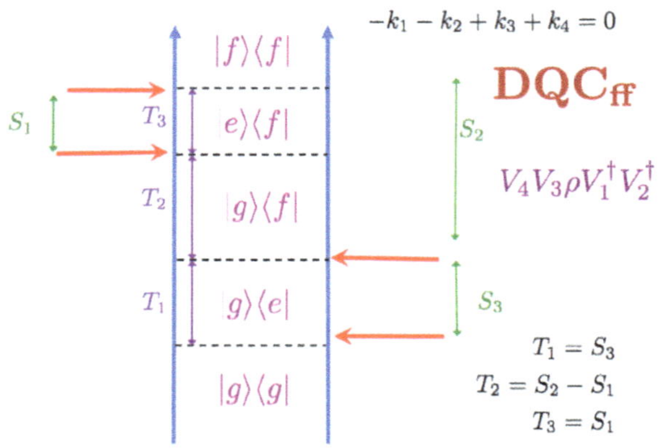

Figure 15.16. DQC process for TPA, with interactions and phase information. The S time labels are for future use.

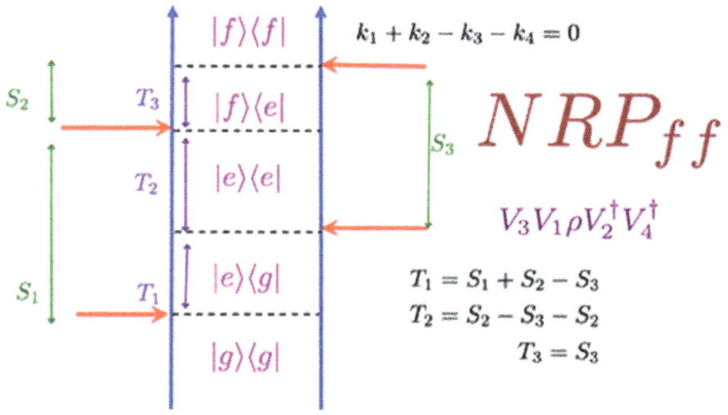

Figure 15.17. NRP process for TPA, with interactions and phase information. The S time labels are for future use.

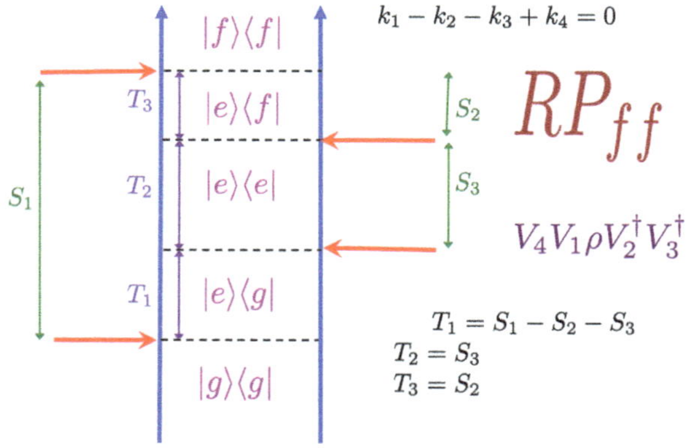

Figure 15.18. RP process for TPA, with interactions and phase information. The S time labels are for future use.

We calculate the population of the biexciton state as

$$\begin{aligned} P_f &= \mathrm{Tr}(\rho(t)|f\rangle\langle f|) \\ &= \mathrm{Tr}(\rho^{(3)}(t)|f\rangle\langle f|) \\ &= \langle f|\rho^{(3)}(t)|f\rangle \end{aligned} \tag{15.72}$$

and the fluorescence from this state is detected.

15.5 Liouville pathways

We now address the sum over the intermediate states. These are shown in figure 15.19.

In the case of DQC_{ff}, we go from $|gg\rangle \to |ge\rangle \to |gf\rangle \to |ef\rangle \to |ee\rangle$, which results in the two pathways in figure 15.20. The RP/NRP pathways are shown in figure 15.21.

Essentially, the intermediate state can be $|eg\rangle$ or $|e'g\rangle$. As they are indistinguishable, we sum the amplitudes. This results in

$$\mathrm{DQC}_{ff}(t_i) = \sum_{e,\, e'}(e^{-\gamma_{eg}t_1}e^{-i\omega_{eg}t_1} + e^{-\gamma_{e'g}t_1}e^{i\omega_{e'g}t_1}). \tag{15.73}$$

RP(t) and NRP(t) excite the ee states, $|ee\rangle$, $|ee'\rangle$, $|e'e\rangle$, and $|e'e'\rangle$ as shown in figure 15.21.

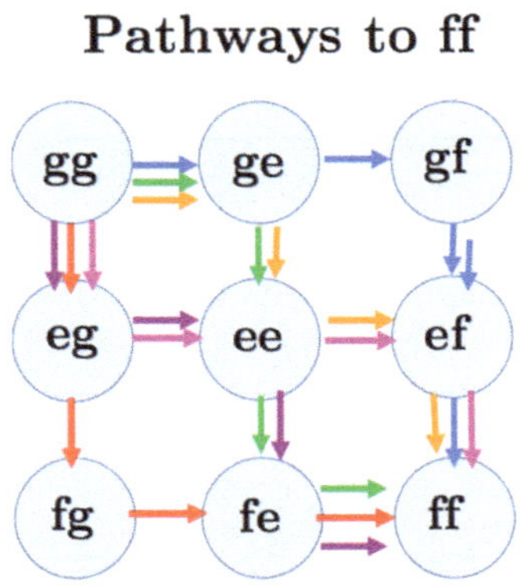

Figure 15.19. TPA pathways.

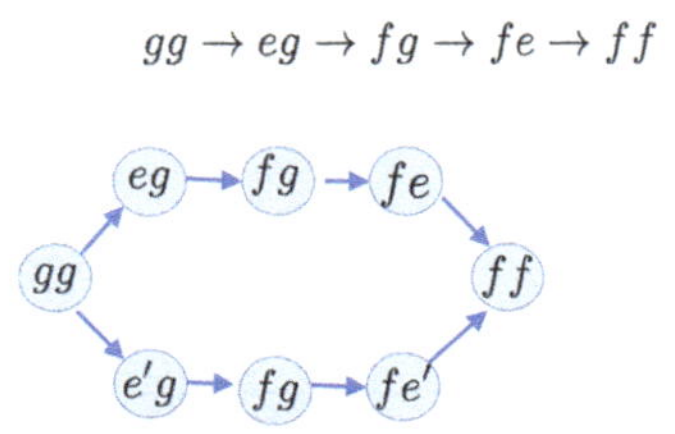

Figure 15.20. TPA pathways for DQC.

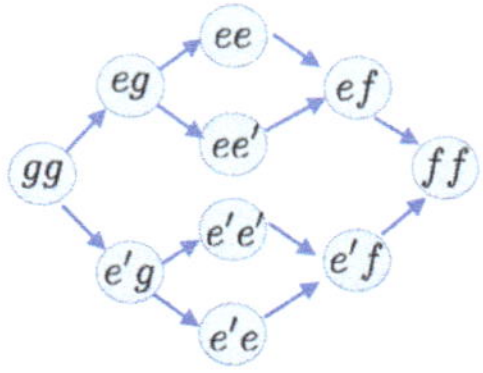

Figure 15.21. TPA pathways for RP/NRP.

In the frequency domain, we have

$$\begin{aligned}\mathrm{DQC}_{ff}(\omega_i) = &\left(\frac{1}{i\gamma_{eg} - (\omega_{eg} - \omega_1)} + \frac{1}{i\gamma_{e'g} - (\omega_{e'g} - \omega_1)}\right)\\ &\times\left(\frac{1}{i\gamma_{ef} + (\omega_{ef} - \omega_3)} + \frac{1}{i\gamma_{e'f} + (\omega_{e'f} - \omega_3)}\right)\\ &\times\left(\frac{1}{i\gamma_{fg} - (\omega_{fg} - \omega_2)}\right)\end{aligned} \tag{15.74}$$

and

$$\begin{aligned}\mathrm{RP}_{ff}(\omega_i) = &\sum\nolimits_{e,\,e'}\left(\frac{1}{i\gamma_{eg} - (\omega_{eg} - \omega_1)}\right)\left(\frac{1}{i\gamma_{ee} + (\omega_{ee} - \omega_2)}\right)\left(\frac{1}{i\gamma_{ef} + (\omega_{ef} - \omega_3)}\right)\\ = &\left(\frac{1}{i\gamma_{eg} - (\omega_{eg} - \omega_1)} + \frac{1}{i\gamma_{eg} - (\omega_{eg} - \omega_1)}\right)\\ &\times\left(\frac{1}{i\gamma_{ee} + (\omega_{ee} - \omega_2)} + \frac{1}{i\gamma_{e'e} + (\omega_{ee'} - \omega_2)}\right.\\ &\left.+\frac{1}{i\gamma_{e'e} + (\omega_{e'e} - \omega_2)} + \frac{1}{i\gamma_{e'e'} + (\omega_{ee'} - \omega_2)}\right)\\ &\times\left(\frac{1}{i\gamma_{ef} - (\omega_{ef} - \omega_3)} + \frac{1}{i\gamma_{e'f} - (\omega_{e'f} - \omega_3)}\right)\end{aligned} \tag{15.75}$$

$$\begin{aligned}\mathrm{NRP}_{ff}(\omega_i) = &\sum\nolimits_{e,\,e'}\left(\frac{1}{i\gamma_{eg} - (\omega_{eg} - \omega_1)}\right)\left(\frac{1}{i\gamma_{ee} + (\omega_{ee} - \omega_2)}\right)\left(\frac{1}{i\gamma_{ef} + (\omega_{ef} - \omega_3)}\right)\\ = &\left(\frac{1}{i\gamma_{eg} - (\omega_{eg} - \omega_1)} + \frac{1}{i\gamma_{eg} - (-\omega_{eg} - \omega_1)}\right)\\ &\times\left(\frac{1}{i\gamma_{ee} + (\omega_{ee} - \omega_2)} + \frac{1}{i\gamma_{e'e} + (\omega_{ee'} - \omega_2)}\right.\\ &\left.+\frac{1}{i\gamma_{e'e} + (\omega_{e'e} - \omega_2)} + \frac{1}{i\gamma_{e'e'} + (\omega_{ee'} - \omega_2)}\right)\\ &\times\left(\frac{1}{i\gamma_{ef} + (\omega_{ef} - \omega_3)} + \frac{1}{i\gamma_{e'f} + (\omega_{e'f} - \omega_3)}\right).\end{aligned} \tag{15.76}$$

15.6 Classical protocols $T_2 = 0$

If the two middle pulses are simultaneous, $T_2 = 0$, we have in the time domain

$$\begin{aligned}\mathrm{DQC}_{ff}(t_i) = & (e^{-\gamma_{eg}t_1}e^{i\omega_{e'g}t_1} + e^{-\gamma_{e'g}t_1}e^{i\omega_{e'g}t_1}) \\ & \times (e^{-\gamma_{ef}t_3}e^{i\omega_{ef}t_3} + e^{-\gamma_{e'f}t_3}e^{i\omega_{e'f}t_3}) \\ & \times \left(\frac{1}{i\gamma_{fg} - (\omega_{fg} - \omega_2)}\right)\end{aligned} \tag{15.77}$$

and

$$\begin{aligned}\mathrm{RP}_{ff}(t_i) = & (e^{-\gamma_{eg}t_1}e^{-(i-\omega_{eg})t_1} + e^{-\gamma_{e'g}t_1}e^{-i\omega_{e'g}t_1}) \\ & \times (e^{-\gamma_{ef}t_3}e^{i\omega_{ef}t_3} + e^{-\gamma_{e'f}t_3}e^{i\omega_{e'f}t_3})\end{aligned} \tag{15.78}$$

$$\begin{aligned}\mathrm{NRP}_{ff}(t_i) = & (e^{-\gamma_{eg}t_1}e^{-(i-\omega_{eg})t_1} + e^{-\gamma_{e'g}t_1}e^{-i\omega_{e'g}t_1}) \\ & \times (e^{-\gamma_{ef}t_3}e^{-i\omega_{ef}t_3} + e^{-\gamma_{e'f}t_3}e^{-i\omega_{e'f}t_3}).\end{aligned} \tag{15.79}$$

In the frequency domain,

$$\begin{aligned}\mathrm{DQC}_{ff}(\omega_i) = & \left(\frac{1}{i\gamma_{eg} - (\omega_{eg} - \omega_1)} + \frac{1}{i\gamma_{e'g} - (\omega_{e'g} - \omega_1)}\right) \\ & \times \left(\frac{1}{i\gamma_{ef} + (\omega_{ef} - \omega_3)} + \frac{1}{i\gamma_{e'f} + (\omega_{e'f} - \omega_3)}\right)\end{aligned} \tag{15.80}$$

and transitions eg and ef are excited. The middle pulse takes you from $|gf\rangle \to |ef\rangle$.

We also have

$$\begin{aligned}\mathrm{RP}_{ff}(\omega_i) = & \sum\nolimits_{e,\,e'}\left(\frac{1}{i\gamma_{eg} - (\omega_{eg} - \omega_1)}\right)\left(\frac{1}{i\gamma_{ee} + (\omega_{ee} - \omega_2)}\right)\left(\frac{1}{i\gamma_{ef} + (\omega_{ef} - \omega_3)}\right) \\ = & \left(\frac{1}{i\gamma_{eg} - (\omega_{eg} - \omega_1)} + \frac{1}{i\gamma_{eg} - (\omega_{eg} - \omega_1)}\right) \\ & \times \left(\frac{1}{i\gamma_{ee} + (\omega_{ee} - \omega_2)} + \frac{1}{i\gamma_{e'e} + (\omega_{ee'} - \omega_2)}\right. \\ & \left. + \frac{1}{i\gamma_{e'e} + (\omega_{e'e} - \omega_2)} + \frac{1}{i\gamma_{e'e'} + (\omega_{e'e} - \omega_2)}\right).\end{aligned} \tag{15.81}$$

In figure 15.22 we show the NRP signal as γ_{eg} is varied, at smaller values we can resolve all of the excitations. The sum of NRP and RP are shown under the same circumstances in figure 15.23.

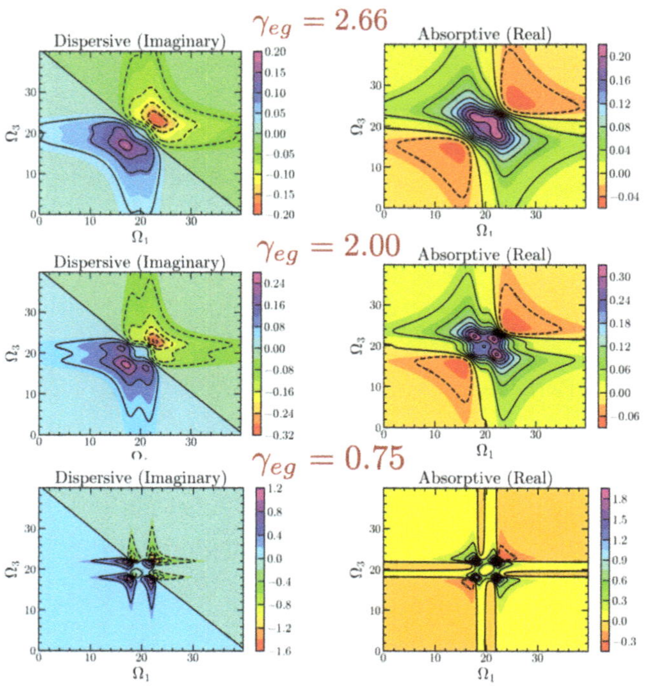

Figure 15.22. NRP for $t_{23} = 0$, varying γ_{eg}.

15.7 Keldysh contour versus real-time contour

In the double-sided Feynman (or Liouville space) approach, or the ladder approach, you explicitly follow the evolution of the density matrix by 'attaching' the interactions (in our case, four δ-pulses) to either the ket (right side) or the bra (left side) of the density operator. Each pulse is represented as an arrow, and the diagrams are drawn with time running vertically. The diagrammatic rules tell you how to order the interactions on each 'side' and how to combine the contributions. In this picture, the time ordering is built in by the way the arrows are drawn along the two branches. For example, for four δ-pulses at times t_1, t_2, t_3, and t_4, you would have to consider all the different ways in which these pulses can act on the bra and the ket (subject to the overall causal structure) and then sum over these 'pathways'. This is particularly useful in nonlinear optical spectroscopy, where the different diagrams can be associated with different physical processes (such as rephasing and non-rephasing contributions). The contour-ordered (or Keldysh) formalism, on the other hand, as shown in figure 15.24, uses a closed-time contour to automatically incorporate the proper time ordering. All operators (whether they eventually act on the ket or the bra) are placed on a single time contour that goes forward in time and then backward. The time ordering is then imposed along this contour (i.e. 'contour ordering'). When the perturbation is made up of δ pulses, the time integrals become trivial (the δ functions 'pick out' the interaction times), but you still have to sum over the different possibilities of placing the interactions on the forward and backward

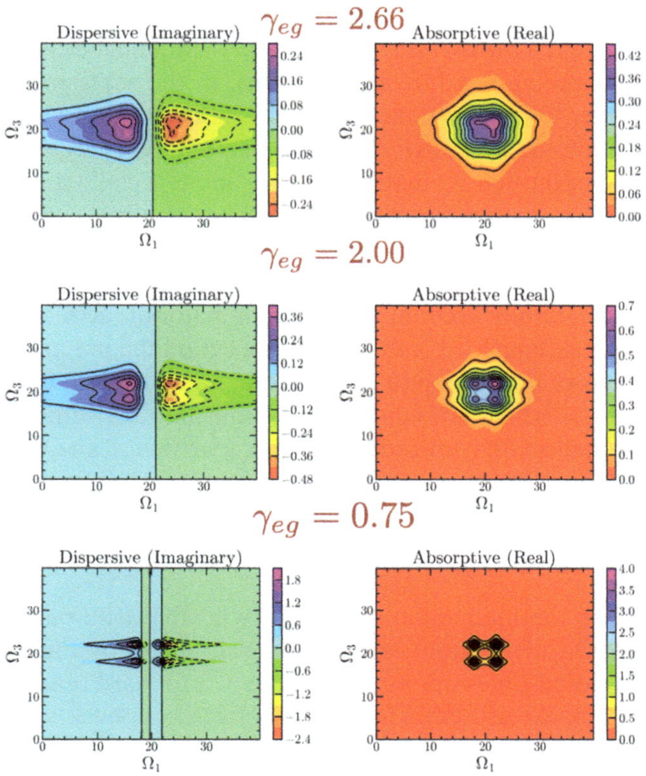

Figure 15.23. Total signal for $t_{23} = 0$, varying γ_{eg}.

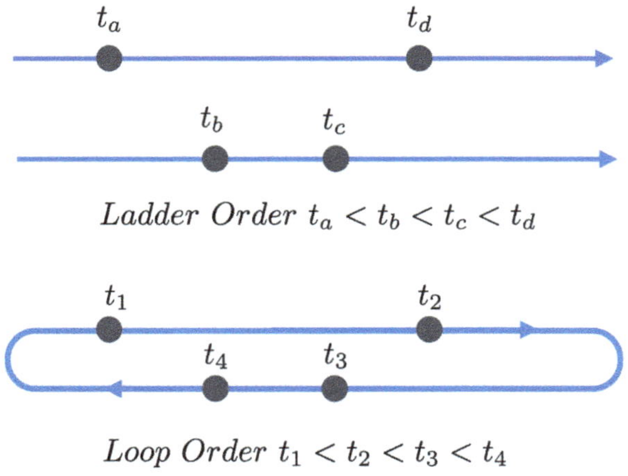

Figure 15.24. Different event ordering in LAP and LOP.

parts of the contour. The contour-ordered approach is particularly powerful when dealing with nonequilibrium problems or when one needs to include effects such as initial state correlations.

In the ladder diagram approach (which we shall now refer to as LAP), the time ordering is split between the two sides (bra and ket) and must be handled 'manually' by drawing and summing the different diagrams. In the contour-ordered approach (which we now refer to as LOP), the time ordering is enforced automatically along the closed-time contour. The LAP diagrams make it very clear which interaction acts on which part of the density matrix, and they are especially useful in interpreting the physical processes in spectroscopy. The LOP formalism, while more abstract, provides a compact notation and is often more convenient for systematic perturbative calculations and generalizations to interacting many-body systems. In the LAP method, one must carefully account for the combinatorial factors and relative signs that come from the different ways the pulses can act on the bra and ket. In the LOP method, these factors are built into the contour time-ordering operator; however, one must be cautious when 'unfolding' the contour into real-time contributions.

In the LOP formalism, the nth-order response is written as an expectation value of the time-ordered operators along a closed-time contour C. For example, for a four-pulse experiment, you have something like

$$R(t_4, t_3, t_2, t_1) = \langle T_C V(t_4)V(t_3)V(t_2)V(t_1)\rangle, \tag{15.82}$$

where the $V(t_j)$ are the interaction operators (with the delta-pulses 'selecting' the times t_1, t_2, t_3, t_4). On the contour, all times are 'ordered' by the contour-ordering operator T_C. The closed-time contour is made up of a forward branch (from $t = -\infty$ to $t = +\infty$) and a backward branch (returning from $+\infty$ back to $-\infty$). Unfolding means that you rewrite the contour integral as a sum of the integrals on the two real-time branches. In practice, this involves assigning a label each time t_j with a branch index p_j (say $+1$ for the forward branch and -1 for the backward branch). Replace the contour-ordering operator with ordinary real-time ordering together with theta functions (e.g. $\theta(t_i - t_j)$) that enforce the proper time ordering on each branch. The full expression becomes a sum of contributions where each t_j is integrated along either the forward or the backward branch with the appropriate sign factors and ordering constraints. Once the expression is written in terms of real-time integrals, the next step is to 'pick out' the DQC contribution (for example). In a typical DQC experiment, it is expected that two of the pulses create a coherence between the ground state and a doubly excited state, and that a later pulse converts that double quantum coherence back into a single quantum coherence. Each term in the unfolded expression carries phase factors that come from the evolution between pulses (for instance, factors such as $e^{\pm i\omega t}$). The DQC contribution will have a characteristic phase signature, often something like a frequency given by the sum of two excitation energies. The ordering of the pulses on the forward and backward branches will determine which terms correspond to having two interactions 'on the same side' (which create the double quantum coherence) followed by the interactions that convert it. In your algebraic expression, the theta functions (or step functions) will enforce these orderings. Collect all the terms that share the same phase-matching condition (or frequency combination) that defines the DQC signal. In many practical cases, the DQC signal is isolated by selecting the terms that, after

Fourier transformation, contribute at the 'double quantum' frequency (typically the sum of the frequencies of the two pulses that created the coherence). For example, if in the double-sided picture the DQC term is generated when pulses 1 and 2 create the double quantum coherence and pulse 3 converts it, then in the unfolded contour expression you would identify those contributions where the branch indices and time-ordering enforce that pattern.

When you measure a multidimensional spectrum, you typically scan certain time delays and then perform Fourier transform with respect to those delays. In the LAP protocol, the Fourier conjugate variables are directly related to the waiting times between pulses (e.g. coherence times and population times), so the resulting spectrum shows resonances that correspond directly to the energy differences that govern those time intervals. In contrast, in the LOP protocol, the delays are defined along the closed-time contour. The Fourier transform is now taken with respect to a different set of time variables. These variables can mix the contributions from interactions that, in the LAP picture, would be separated. As a consequence, the frequency axes (and hence the resonances that appear) reflect different combinations of the system's transition frequencies. Because of these different Fourier 'conjugate pairs', the two display protocols will emphasize different resonant features. For example, one protocol might make it easier to see resonances associated with a DQC pathway, while the other might highlight resonances corresponding to simple ground state bleaching or stimulated emission. In other words, although the overall signal is the same, the way in which the resonances (peaks) appear in the multidimensional spectrum depends on which set of delay times you choose to scan and transform.

The spectra (LOP with $t_2 = 0$) would not be the same as those from a ladder (LAP) experiment where $t_2 = 0$ because the delay time scanning procedure differs. This means that choosing between the LOP and LAP display protocols fundamentally changes how the experimental results appear, even though they contain the same underlying physics. What would the lap delay times be if the lop ones were t_1, $t_2 = t_3$, and t_4?

If the LOP (loop) protocol defines the time delays as t_1, $t_2 = t_3$, and t_4, then the corresponding LAP protocol delays (t_1', t_2', t_3') can be determined by understanding how the LOP time variables map onto the physical evolution times used in the ladder (double-sided) approach.

In the LOP protocol:

1. t_1 corresponds to the initial delay between the first two interactions.
2. $t_2 = t_3$ implies that the second and third delays are equal, meaning that whatever time evolution happens after the second interaction is identical to that after the third interaction.
3. t_4 is the final evolution time.

Because the LOP protocol mixes time evolution intervals from both the forward and backward branches, its time variables do not directly correspond to the delays used in the LAP protocol. In the LAP protocol, the natural time variables are:

1. t_1': The coherence time (delay between the first and second pulse).
2. t_2': The population time (evolution between the second and third pulses).
3. t_3': The dephasing time (delay before the fourth pulse interaction).

The approximate mapping is

$$t_1' = t_1 \tag{15.83}$$

$$t_2' = t_2 + t_3 = 2t_2 \tag{15.84}$$

$$t_3' = t_4. \tag{15.85}$$

Thus, the LAP delay times corresponding to the LOP time variables (t_1, $t_2 = t_3$, t_4) are

$$t_1' = t_1 \tag{15.86}$$

$$t_2' = 2t_2 \tag{15.87}$$

$$t_3' = t_4. \tag{15.88}$$

If the middle two times are equal in LAP and in LOP, they explore two different sets of dynamics, as shown in figure 15.25.

The population time t_2' in the LAP protocol is twice the value of t_2 in the LOP protocol. This is because, in LOP, $t_2 = t_3$ means that the contour variable is effectively 'split' over the forward and backward branches, whereas in LAP, t_2' is a real waiting time in the density matrix evolution. If an experiment were performed using LAP delay times, scanning t_2' would involve twice the waiting time compared to an equivalent LOP scan. Since frequency is the inverse of time, in the LAP protocol, the conjugate frequency axis Ω_2 will differ from the equivalent Fourier conjugate variable in the LOP protocol, shifting how peaks appear. This means that a

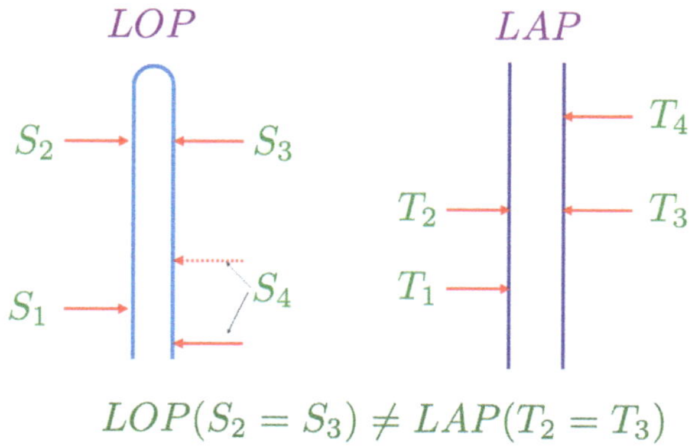

Figure 15.25. Equal middle event times yield different physics in LAP and LOP. Note the ambiguity of the time of the fourth pulse in LOP as opposed to LAP.

measurement using these two protocols would not correspond to the same time delays, which further confirms that the two methods may highlight different resonances when transformed into the frequency space. The difference in time variables between the LOP and LAP protocols skews the LAP plot, making it look different from the LOP plot when plotted in frequency space. In 2D spectroscopy, each time delay has a Fourier conjugate frequency (e.g. $t_1' \to \Omega_1$, $t_2' \to \Omega_2$, $t_3' \to \Omega_3$). Since $t_2' = 2t_2$ in LAP, the Fourier transform of t_2' results in a frequency axis Ω_2 that is scaled by a factor of 1/2 compared to the equivalent frequency in LOP. This means that the spectral features in the Ω_2 direction in the LAP protocol will be compressed (lower in frequency) compared to the LOP spectrum. Because the LAP protocol stretches time delays differently, when we perform the Fourier transform, some peaks appear at shifted positions in the frequency space compared to the LOP. The relative spacing between features along the frequency axes changes, making some peaks appear closer together or further apart than in the LOP case. Some resonances that may be clearly separated in LOP might overlap in LAP, or vice versa, and this can make interpretation non-trivial if one is trying to compare the two spectra directly, as shown in figure 15.26.

The core difficulty when working on the LOP contour is that the atomic response function depends on the relative ordering of the interactions along the contour. In the Keldysh formalism, the time variable runs along a closed contour with distinct forward and backward branches. As a result, the position of an interaction on the contour determines whether it is on the 'ket' (forward) or 'bra' (backward) side of the density matrix. This affects the phase factors and decay factors associated with the interaction. In other words, to perform the single integral over the entire Keldysh contour, you must account for the fact that different segments of the contour correspond to different 'normal time orderings' of the interactions. These differences promote the six distinct Liouville pathways (RP+, RP–, NRP+, NRP–, DQC+, DQC–). Each pathway carries its own phase evolution and dephasing characteristics, so the atomic response is not uniform over the whole contour—it depends

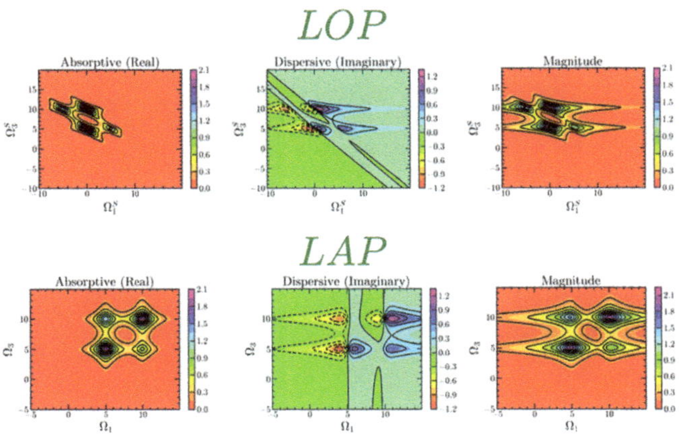

Figure 15.26. Same data plotted with LAP and LOP protocols.

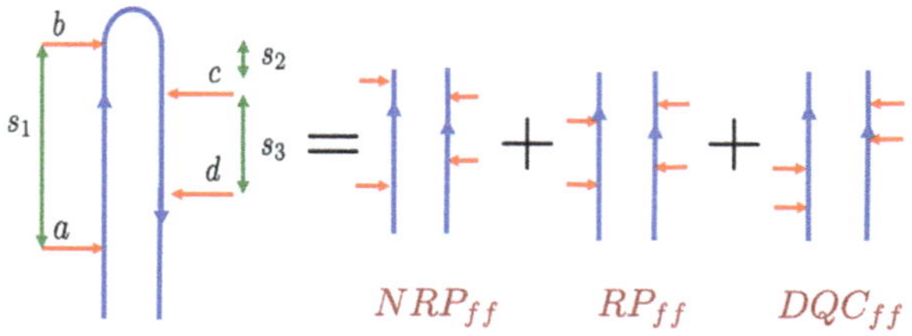

Figure 15.27. LOP is the sum of LAP diagrams.

critically on where the interaction falls with respect to the forward and backward branches. This single integral is formally equivalent to summing over all six time orderings and thus recovers the total 2D spectrum for the entangled-photon experiment. Therefore, the advantage of LOP is the conceptual unification and systematic inclusion of all pathways, but when it comes to the actual evaluation of the response, breaking the integral into parts (as in ladder diagrams) is generally required. This is shown in figure 15.27.

In other words, *they contain the same physics, but differ in path selectivity and display protocols.*

References

[1] Mukamel S, Tanimura Y and Hamm P 2009 Coherent multidimensional optical spectroscopy *Acc. Chem. Res.* **42** 1207–9

[2] Mukamel S 2000 *Principles of Nonlinear Optics and Spectroscopy* (Oxford: Oxford University Press)

[3] Hamm P and Zanni M 2011 *Concepts and Methods of 2D Infrared Spectroscopy* (Cambridge: Cambridge University Press)

[4] Cho M 2008 Coherent two-dimensional optical spectroscopy *Chem. Rev.* **108** 1331–418

[5] Tokmakoff A 2011 *Nonlinear and Two-dimensional Spectroscopy* (Chicago, IL: University of Chicago)

[6] Jonas D M 2003 Two-dimensional femtosecond spectroscopy *Annu. Rev. Phys. Chem.* **54** 425–63

[7] Boyd R W 2020 *Nonlinear Optics* 4th edn (New York: Academic)

IOP Publishing

An Introduction to Quantum Optics (Second Edition)

An open systems approach

Perry Rice

Chapter 16

Entanglement

16.1 No cloning

You have probably heard of quantum computing, cryptography, communication, etc. Likely a reason you are reading this book. For a deeper presentation than presented here, I recommend books by Jaeger [1], Wilde [2], Vedral [3], and the classic book by Nielson and Chaung [4] commonly referred to as 'Mike and Ike' (MI) using a play on the author's names and a classic candy. There are scads more, and new ones appear daily. Classically, if you want to send a message, say an email, you send multiple copies and do error correction on the receiving end, like a majority vote of the info in the copies. It is quite amazing really that one can send a string of 1s and 0s across the world through lossy channels, and reconstruct a video on the receiving end. Modern error correction is quite advanced, and the power of processors allows it to be quite involved. One of the main reasons quantum communications is difficult is the no-cloning theorem, it is impossible to clone an arbitrary unknown state. So one cannot just send multiple copies to ensure the message gets through. Of course, this entails some security, perhaps, if someone mucks with your quantum system in transit, it should be possible to tell. Let us assume that we have a machine that makes copies of states, a Qurox cloner

$$\mathrm{QC}|\psi\rangle \otimes |0\rangle = |\psi\rangle \otimes |\psi\rangle, \tag{16.1}$$

where $|0\rangle$ is some fiducial starting state. How would this work for a superposition state? We suppress normalization for clarity

$$\begin{aligned}\mathrm{QC}(|1\rangle + |0\rangle) \otimes |0\rangle &= |1\rangle \otimes |1\rangle + |0\rangle \otimes |0\rangle \\ &\neq (|1\rangle + |0\rangle) \otimes (|0\rangle + |1\rangle),\end{aligned} \tag{16.2}$$

where all we have used is the linearity of quantum mechanics. It is not possible to clone a quantum state. If you have a *particular* quantum state such as an excited atom, you can prepare another atom in the excited state,b ut you cannot take a state unknown to yourself and create a copy.

16.2 Bell inequalities

Fortunately for us, quantum mechanics allows correlations between systems that are stronger than those allowed by classical mechanics. Quantum taketh, but giveth back in a different, richer, but more difficult to implement form. These results underlie many cooler aspects of quantum information science, including quantum cryptography and quantum computing. Bell showed that if quantum mechanics were replaced by a theory that is both local and realistic, then certain correlations between measurements on entangled particles must satisfy specific inequalities, which are known as Bell inequalities [5]. A local hidden-variable theory will not reproduce correctly the results of quantum mechanics. A Franson interferometer [6] is described in chapter 7. Because of the large path difference, single-photon interference is washed out (the paths exceed the photon coherence length). However, two-photon interference can still occur in the joint detection (coincidence counts), leading to observable interference fringes in the coincidence rate as a function of $\phi_A + \phi_B$. Consider

$$S = E(A,\, B) + E(A,\, B') + E(A',\, B) - E(A',\, B'), \tag{16.3}$$

which must satisfy $|S| \leqslant 2$ for any local hidden-variable model. However, quantum mechanics predicts that $|S|$ can reach up to $2\sqrt{2} \approx 2.828$, which is called the Tsirel'son bound [7]. In a typical Bell test, each party (traditionally called Alice and Bob) must be able to choose between two different measurement settings-often labeled A and A' for Alice, and B and B' for Bob. It is *crucial* that the operators for the two measurements (A and A') do *not* commute. This means that there is an uncertainty principle for the two operators, and we cannot know their value simultaneously. For example, if your qubit is in an eigenstate of σ_x and you measure σ_y, it is no longer in an eigenstate of σ_x, and a measure of that will actually yield equal probabilities of up and down for σ_x. Measurement affects the system. This is the only aspect of quantum mechanics needed here. Classically, the outcomes A and A' have definite values no matter what measurements may be performed, and they may be known simultaneously. This is not allowed for non-commuting operators. A commonly tested form of Bell's inequality is the Clauser–Horne–Shimony–Holt (CHSH) inequality [8]. If one designates the measurement settings (e.g. different polarization angles, or phases in a Franson set-up) as $\{A,\, A'\}$ for one side and $\{B,\, B'\}$ for the other, then the CHSH expression. We will look at the Franson interferometer discussed in chapter 7. Here a signal and idler are generated in a nonlinear crystal, and each is sent to an interferometer. Consider the initial entangled state of two photons, say in a maximally entangled state (also called a Bell state):

$$|\Psi\rangle = \frac{1}{\sqrt{2}}(|H\rangle_1|V\rangle_2 + |V\rangle_1|H\rangle_2), \tag{16.4}$$

where $|H\rangle$ and $|V\rangle$. We find that we get detection coincidences when the signal and idler both take the long or short legs of their interferometers. If one blocks the long leg of one and the short length of the other, there are no coincidence counts. The addition of the path information wipes out the fringe. After their generation, the two photons do not interact in any way. Because of the large path difference, single-photon interference is washed out (the paths exceed the photon coherence length). However, two-photon interference can still occur in the joint detection (coincidence counts), leading to observable interference fringes in the coincidence rate as a function of $\phi_A + \phi_B$.

Recall that the output state at the detectors is a superposition of the possible polarization states. Using the fact that the probability of detecting coincidence events at both detectors D_1 and D_2 is related to the interference pattern, we can write the probability $P(\phi_1, \phi_2)$ of detecting coincidences as a function of the phase shifts ϕ_1 and ϕ_2. This probability can be expressed as

$$P(\phi_1, \phi_2) = \frac{1}{2}[1 + \cos(\phi_1 - \phi_2)]. \tag{16.5}$$

The key thing is that the fringe pattern for the joint system can be affected by a phase adjustment on one of the arms, no matter how long they are. Because each photon can travel either the short or the long path, the only indistinguishable two-photon events contributing to interference in coincidence measurements are when both photons take the short paths or both photons take the long paths, leading to an overall relative phase in the coincidence probability of the form $\phi_A + \phi_B$. In a CHSH test, by toggling between ϕ_A versus ϕ'_A and ϕ_B versus ϕ'_B, one can sample the coincidence rates at four different values of the phase sum. This procedure, if the system is indeed entangled and well-isolated from loopholes, will typically show correlations that exceed the local-realist bound of 2. In a standard CHSH scenario, each party must be able to randomly (or at least independently) select between two settings. Concretely, Alice's interferometer has a controllable phase ϕ_A, and Alice can choose between two possible phase values: $\phi_A \equiv A$ or $\phi'_A \equiv A'$. Bob's interferometer also has a controllable phase ϕ_B and he can choose between two possible phase values: $\phi_B \equiv B$ or $\phi'_B \equiv B'$. There are then four possible pairings of settings: (A, B), (A, B'), (A', B), and (A', B'). These correspond to the four terms needed in the CHSH correlation function. Each photon detection at Alice's station is assigned either $+$ or $-$ based on which detector fires, and similarly at Bob's station, binary measurements. Each choice of settings (A, B) gives a distribution of joint outcomes $(+, +)$, $(+, -)$, $(-, +)$, $(-, -)$. From the measured coincidence counts N_{++}, N_{+-}, N_{-+}, N_{--} we define the expectation value of the product of the outcomes:

$$S(A, B) = \frac{N_{++} + N_{--} - N_{+-} - N_{-+}}{N_{++} + N_{+-} + N_{-+} + N_{--}}h \tag{16.6}$$

- N_{++} = number of coincidences with Alice outcome + and Bob outcome +,
- N_{+-} = number of coincidences with Alice outcome + and Bob outcome –,
- N_{-+} = number of coincidences with Alice outcome – and Bob outcome +,
- N_{--} = number of coincidences with Alice outcome – and Bob outcome –.

To test the Bell–CHSH inequality, we combine the expectation values at the four setting pairs as:

- $A = 0$,
- $A' = \pi/2$,
- $B = \pi/4$,
- $B' = -\pi/4$.

One finds that $S(A, B) = 2\sqrt{2}$, which maximally violates the CHSH inequality and saturates Tsirel'son's bound. OK, what is $S(A, B)$ in simple terms, and why the minus sign in the one term? Consider the combination

$$\langle (A + A')B + (A - A')B' \rangle, \tag{16.7}$$

where the brackets $\langle \cdot \rangle$ denote a classical average over many iterations of the experiment:

$$\langle (A + A')B + (A - A')B' \rangle_{av} = \frac{1}{n}\sum_{j=1}^{n}\left(\left(a_j + a_j'\right)b_j + \left(a_j - a_j'\right)b_j'\right) \tag{16.8}$$

with

$$a_j^2 = a_j'^2 = b_j^2 = b_j'^2 = 1. \tag{16.9}$$

If $a_j = a_j' = \pm 1$ then $|\left(a_j + a_j'\right)| = 2$ and $|\left(a_j - a_j'\right) = 0|$. If $a_j = a_j'$ then the first term is 0 and the second term is 2. The assumption we have made is that the a_j, etc, all have a definite value at all times, which is true in the classical world, and $S(A, B)$ is cleverly chosen to exploit this with terms that go as 1 ± 1, and leads to a Bell violation. In quantum physics, we use the expectation values of non-commuting operators calculated using wave functions or density operators. In the case of the Franson interferometer, the four terms are given by terms such as

$$P(\phi_1, \phi_2) = \frac{1}{2}[1 + \cos(\phi_1 - \phi_2)]. \tag{16.10}$$

If we substitute the angles mentioned in 16.2, we find $2\sqrt{2}$, a clear violation of the CHSH inequality.

David Mermin [9–11] proved a more astounding result, that one experiment should give 1 in a classical world and -1 in a quantum world. It involves an entangled state of three qubits (two-level system) invented by Greenberger, Horne, and Zeilinger (GHZ) [12]:

$$|\text{GHZ}\rangle = \frac{1}{\sqrt{2}}(|000\rangle + |111\rangle). \tag{16.11}$$

Each of Alice, Bob, and the new person Charlie holds one qubit of this state. Each party can choose to measure one of two observables, typically taken to be Pauli matrices σ_x or σ_y. Denote Alice's choice by A_x or A_y, Bob's by B_x or B_y, and Charlie's by C_x or C_y. These are defined by

$$\sigma_x = \begin{pmatrix} 0 & 1 \\ & 0 \end{pmatrix}, \quad \sigma_y = \begin{pmatrix} 0 & -i \\ i & 0 \end{pmatrix}. \tag{16.12}$$

We will look at four different joint measurement settings:

- $\sigma_x \otimes \sigma_y \otimes \sigma_y$,
- $\sigma_y \otimes \sigma_x \otimes \sigma_y$,
- $\sigma_y \otimes \sigma_y \otimes \sigma_x$,
- $\sigma_y \otimes \sigma_y \otimes \sigma_y$.

We will compare what local realism (classical hidden variables) says versus what quantum mechanics predicts for the product of the three measurement outcomes in each setting. A local (hidden-variable) model would assign definite outcomes to each one-qubit measurement. For instance, each qubit has predetermined answers for:

- 'What will I output if measured in σ_x?' (either $+1$ or -1),
- 'What will I output if measured in σ_y?' (either $+1$ or -1),

and these answers cannot depend on which observables the other two qubits choose. Denote each qubit's hidden-variable-determined outcomes as follows:

- For Alice's qubit $A_x(\lambda)$, $A_y(\lambda) \in \{\pm 1\}$,
- For Bob's qubit: $B_x(\lambda)$, $B_y(\lambda) \in \{\pm 1\}$,
- For Charlie's qubit: $C_x(\lambda)$, $C_y(\lambda) \in \{\pm 1\}$,

where λ is the hidden variable (or set of variables). Under local realism, the product for each of the above settings is simply:

- $A_x(\lambda) \times B_y(\lambda) \times C_y(\lambda)$,
- $A_y(\lambda) \times B_x(\lambda) \times C_y(\lambda)$,
- $A_y(\lambda) \times B_y(\lambda) \times C_x(\lambda)$,
- $A_y(\lambda) \times B_y(\lambda) \times C_y(\lambda)$.

A key point in the GHZ argument is that from the first three measurements, if they happen to yield products of $+1$ (classically or quantumly)-local realism forces a prediction about the fourth product. Specifically, if you multiply together the first three products, you effectively obtain

$$[A_x(\lambda)B_y(\lambda)C_y(\lambda)] \times [A_y(\lambda)B_x(\lambda)C_y(\lambda)] \times [A_y(\lambda)B_y(\lambda)C_x(\lambda)]. \tag{16.13}$$

Notice that each qubit's local choice appears exactly twice in that triple-product (e.g. $A_x(\lambda)$ appears once, $A_y(\lambda)$ appears twice overall, etc). Because $A_x(\lambda)^2 = +1$ (the outcome is ± 1, so squaring it gives $+1$), all the squares drop out. Eventually, you find that the product of the first three settings must be equal to the fourth setting's product $A_y(\lambda)B_y(\lambda)C_y(\lambda)$. Concretely,

$$[A_x(\lambda)B_y(\lambda)C_y(\lambda)] \times [A_y(\lambda)B_x(\lambda)C_y(\lambda)] \times [A_y(\lambda)B_y(\lambda)C_x(\lambda)] = A_y(\lambda)B_y(\lambda)C_y(\lambda). \tag{16.14}$$

Thus, if you observe (or assume) that the first three products are +1, classical reasoning says that the fourth product must also be +1.

What does quantum mechanics predict for these four measurements on the state

$$|\mathrm{GHZ}\rangle = \frac{1}{\sqrt{2}}(|000\rangle + |111\rangle). \tag{16.15}$$

A standard calculation shows:

$$(\sigma_x \otimes \sigma_y \otimes \sigma_y)|\mathrm{GHZ}\rangle = +|\mathrm{GHZ}\rangle \tag{16.16}$$

$$(\sigma_y \otimes \sigma_x \otimes \sigma_y)|\mathrm{GHZ}\rangle = +|\mathrm{GHZ}\rangle \tag{16.17}$$

$$(\sigma_y \otimes \sigma_y \otimes \sigma_x)|\mathrm{GHZ}\rangle = +|\mathrm{GHZ}\rangle. \tag{16.18}$$

In each case, the GHZ state is an eigenstate of these operators with eigenvalue +1. Hence, the measured product of outcomes is +1 with certainty (since an eigenvalue of +1 means the possible measurement outcomes multiply to +1). However, if we measure σ_y on each qubit,

$$(\sigma_y \otimes \sigma_y \otimes \sigma_y)|\mathrm{GHZ}\rangle = -|\mathrm{GHZ}\rangle. \tag{16.19}$$

Thus $|\mathrm{GHZ}\rangle$ is an eigenstate of $\sigma_y \otimes \sigma_y \otimes \sigma_y$ with eigenvalue −1. This means that if all three parties measure σ_y, the product of their outcomes is guaranteed to be −1. No averaging or inequality is needed to determine this mismatch. This is an 'all-or-nothing' demonstration of the impossibility of local hidden variables reproducing the GHZ correlations.

16.3 Entangled states

While we will not be examining quantum computation, or simulation, we shall be examine *entanglement*, which is a resource for the aforementioned tasks [5, 13, 14]. Entanglement is a property of two quantum systems that are interacting or have interacted. If the two systems do not interact, their wave function is a *product* state. Consider two two-level systems, or *qubits*. Each system lives in a two-dimensional Hilbert space, and the state of both systems lives in a four-dimensional space. The product space is

$$\mathcal{H}_4 = \mathcal{H}_2^1 \otimes \mathcal{H}_2^2, \tag{16.20}$$

where $\mathcal{H}_\varepsilon^\rangle$ is a two-dimensional Hilbert space appropriate for the individual systems. A state vector in the larger space for independent qubits would be

$$|\Psi_4\rangle = |\psi_2\rangle \otimes |\phi_2\rangle. \tag{16.21}$$

Over the last two decades, studies into the foundations of quantum mechanics have shown us that entanglement is an inherently quantum mechanical property, which may be used as a resource for tasks such as factoring a number that is the product of two large prime numbers, search protocols, and quantum teleportation.

The measure of entanglement is well defined only in the case of two interacting qubits. Consider a two-qubit state

$$|\psi\rangle = C_{00}|11\rangle + C_{01}|01\rangle + C_{10}|10\rangle + C_{11}|00\rangle. \tag{16.22}$$

If the system is in the product state

$$|\Psi\rangle = (A_0|0\rangle + A_1|1\rangle)(B_0|0\rangle + B_1|1\rangle). \tag{16.23}$$

Then the coefficients will satisfy

$$\mathcal{D} = C_{00}C_{11} - C_{01}C_{10} = 0. \tag{16.24}$$

This can be written as det $M = 0$ with

$$\mathcal{D} = \det\begin{pmatrix} C_{00} & C_{01} \\ C_{10} & C_{11} \end{pmatrix}. \tag{16.25}$$

If this relation is *not* satisfied then the state is entangled. Let us consider some examples. Consider two atoms in an equal superposition of the ground and excited states,

$$|\psi_i\rangle = \left(\frac{1}{\sqrt{2}}\right)(|g\rangle + |e\rangle), \tag{16.26}$$

then the product state of these would be

$$|\Phi\rangle = \left(\frac{1}{2}\right)(|g\rangle + |e\rangle)\otimes(|g\rangle + |e\rangle) \tag{16.27}$$

$$= \frac{1}{2}(|gg\rangle + |ge\rangle + |eg\rangle + |ee\rangle), \tag{16.28}$$

where

$$|eg\rangle = |e\rangle \otimes |g\rangle. \tag{16.29}$$

We see that $\mathcal{D} = 0$, as it should for a product state a state where atom 1 is in the excited state and atom 2 is in the ground state.

Now consider a state

$$|+\rangle = \left(\frac{1}{2}\right)(|eg\rangle + |ge\rangle). \tag{16.30}$$

Here $\mathcal{D} = 1/4$ and the state is entangled. This is a state where the atoms are anticorrelated, in the sense that if one is in the ground state, it is known that the other atom is in the excited state and vice versa. Therefore, if I measure the state of atom 1, I have an equal chance of obtaining ground or excited, random. But I know the state of the other atom instantly. If the two atoms are quite far apart, this seems like *spooky action at a distance*, but it can not be used to send information faster than the speed of light as the result of the first measurement is random. It is like a classical particle blowing up into two chunks, the momentum is conserved so if I know where one particle is, I know where the other chunk is. There is no magic.

What about the operators? If we want an operator that raises atom 1 and lowers atom 2, that would be

$$\begin{aligned} O_{-+} &= (\sigma_- \otimes I)(I \otimes \sigma_+) \\ &= \left[\begin{pmatrix} 0 & 0 \\ 1 & 0 \end{pmatrix} \otimes \begin{pmatrix} 1 & 0 \\ 0 & 1 \end{pmatrix}\right]\left[\begin{pmatrix} 1 & 0 \\ 0 & 1 \end{pmatrix} \otimes \begin{pmatrix} 0 & 1 \\ 0 & 0 \end{pmatrix}\right] \\ &= \begin{pmatrix} 0 & 0 & 0 & 0 \\ 0 & 1 & 0 & 0 \\ 0 & 0 & 1 & 0 \\ 0 & 0 & 0 & 0 \end{pmatrix}. \end{aligned} \tag{16.31}$$

What kind of operator could create the entangled state $|+\rangle$ from $|gg\rangle$? That operator would be

$$O_E = \left(\frac{1}{\sqrt{2}}\right)\begin{pmatrix} 0 & 0 & 0 & 0 \\ 0 & 0 & 0 & 1 \\ 0 & 0 & 0 & 1 \\ 0 & 0 & 0 & 0 \end{pmatrix}. \tag{16.32}$$

It is *not* possible to write this as the outer product of the two operators $O_E = O_1 \bigotimes O_2$. This is a conditional operator that will put atom 1 in the excited state if the second atom is in the ground state, and vice versa. The concurrence is a mathematical measure used to quantify the degree of entanglement between two quantum systems, particularly in a two-qubit system. It is derived from the density matrix of a quantum state and is used to determine how strongly the qubits are entangled.

The concurrence C is computed using the following formula:

$$C = \max\left(0, \sqrt{\lambda_1} - \sqrt{\lambda_2} - \sqrt{\lambda_3} - \sqrt{\lambda_4}\right), \tag{16.33}$$

where λ_1, λ_2, λ_3, λ_4 are the eigenvalues (in descending order) of the matrix $\rho\tilde{\rho}$, ρ is the density matrix of the quantum state, and $\tilde{\rho} = (\sigma_y \otimes \sigma_y)\rho^*(\sigma_y \otimes \sigma_y)$, where ρ^* is the complex conjugate of ρ, and σ_y is the Pauli Y matrix. For the two-qubit state $|\psi\rangle = A|00\rangle + \sqrt{1 - A^2}\,|11\rangle$, the concurrence is calculated as follows: the density matrix for $|\psi\rangle$ is

$$\rho = |\psi\rangle\langle\psi| = \begin{bmatrix} A^2 & 0 & 0 & A\sqrt{1 - A^2} \\ 0 & 0 & 0 & 0 \\ 0 & 0 & 0 & 0 \\ A\sqrt{1 - A^2} & 0 & 0 & 1 - A^2 \end{bmatrix}. \tag{16.34}$$

We define $\tilde{\rho} = (\sigma_y \otimes \sigma_y)\rho^*(\sigma_y \otimes \sigma_y)$, where σ_y is the Pauli Y matrix. Now calculate the eigenvalues of $\rho\tilde{\rho}$ and use the following formula for concurrence:

$$C = \max(0, \sqrt{\lambda_1} - \sqrt{\lambda_2} - \sqrt{\lambda_3} - \sqrt{\lambda_4}), \tag{16.35}$$

where λ_1, λ_2, λ_3, λ_4 are the eigenvalues of $\rho\tilde{\rho}$ in descending order. For the state $|\psi\rangle = A|00\rangle + \sqrt{1 - A^2}\,|11\rangle$, it can be shown that

$$C = 2|A\sqrt{1 - A^2}\,|. \tag{16.36}$$

It is maximized (at $C = 1$) for $A = 1/\sqrt{2}$, and minimized (at $C = 0$) for $A = 0$ or 1.

The von Neumann entropy is a measure of the uncertainty or disorder associated with a quantum state. The von Neumann entropy is defined for a quantum state described by the density matrix ρ as

$$S(\rho) = -\mathrm{Tr}(\rho \ln \rho). \tag{16.37}$$

In terms of the eigenvalues of ρ, $\{p_i\}$, the entropy can be expressed as

$$S(\rho) = -\sum_i p_i \ln p_i. \tag{16.38}$$

This is directly analogous to the Shannon entropy for a classical probability distribution $\{p_i\}$. For a pure state, $\rho^2 = \rho$ and the eigenvalues of ρ are 1 (with one eigenvalue being 1 and the rest 0). In this case, the von Neumann entropy is zero

$$S(\rho) = 0. \tag{16.39}$$

For a composite quantum system divided into subsystems A and B, if the total state $|\psi_{AB}\rangle$ is pure, then the von Neumann entropy of the reduced density matrix of either subsystem is a direct measure of the entanglement between them. The state of the subsystem A is described by the reduced density matrix

$$\rho_A = \mathrm{Tr}_B(|\psi_{AB}\rangle\langle\psi_{AB}|) \tag{16.40}$$

and similarly, for the subsystem B,

$$\rho_B = \mathrm{Tr}_A(|\psi_{AB}\rangle\langle\psi_{AB}|). \tag{16.41}$$

The von Neumann entropy for the subsystem A is

$$S(\rho_A) = -\mathrm{Tr}(\rho_A \ln \rho_A). \tag{16.42}$$

A key result is that for a pure state,

$$S(\rho_A) = S(\rho_B). \tag{16.43}$$

This is known as the entanglement entropy, and it quantifies how much the two subsystems are entangled. If $S(\rho_A) = 0$, the reduced state is pure, which means that the total state $|\psi_{AB}\rangle$ is a product state, and separable. If $S(\rho_A) > 0$ the reduced state is mixed, which indicates that there is entanglement between A and B. The larger the entropy, the greater the entanglement. For maximally entangled states, the entropy reaches its maximum possible value given the dimensions of the subsystems. It tells you that you need state A to fully characterize B, and vice versa. For the state

$$|\psi\rangle = A|00\rangle + \sqrt{1 - A^2}\,|11\rangle. \tag{16.44}$$

Its density matrix is

$$\begin{aligned}\rho = |\psi\rangle\langle\psi| = A^2|00\rangle\langle 00| + A\sqrt{1 - A^2}\,|00\rangle\langle 11| + A\sqrt{1 - A^2}\,|11\rangle\langle 00| \\ + (1 - A^2)|11\rangle\langle 11|.\end{aligned} \tag{16.45}$$

Tracing out the second qubit means computing

$$\rho_A = \mathrm{Tr}_B(\rho). \tag{16.46}$$

Recall that

$$\mathrm{Tr}_B(|ij\rangle\langle kl|) = |i\rangle\langle k|\delta_{j,l}. \tag{16.47}$$

The term $A^2|00\rangle\langle 00|$ gives

$$\mathrm{Tr}_B(A^2|00\rangle\langle 00|) = A^2|0\rangle\langle 0|. \tag{16.48}$$

The cross terms $A\sqrt{1 - A^2}\,|00\rangle\langle 11|$ and its conjugate vanish upon tracing out the second qubit because $\delta_{0,1} = 0$. The term $(1 - A^2)|11\rangle\langle 11|$ gives

$$\mathrm{Tr}_B((1 - A^2)|11\rangle\langle 11|) = (1 - A^2)|1\rangle\langle 1|. \tag{16.49}$$

Thus, the reduced density matrix is

$$\rho_A = A^2|0\rangle\langle 0| + (1 - A^2)|1\rangle\langle 1|. \tag{16.50}$$

Since ρ_A is diagonal with eigenvalues

$$\lambda_1 = A^2 \quad \text{and} \quad \lambda_2 = 1 - A^2, \tag{16.51}$$

the entropy is

$$S(\rho_A) = -\lambda_1 \log_2 \lambda_1 - \lambda_2 \log_2 \lambda_2. \tag{16.52}$$

Substituting the eigenvalues, we obtain

$$S(\rho_A) = -A^2 \log_2 A^2 - (1 - A^2)\log_2(1 - A^2). \tag{16.53}$$

For a mixed state, the eigenvalues p_i are such that $0 < p_i < 1$ and more than one p_i is non-zero. The entropy $S(\rho)$ is greater than zero, indicating a lack of complete knowledge about the state. For the mixed states of two qubits, there are several well-established measures of entanglement. These measures help quantify the degree of entanglement in a bipartite quantum system described by a density matrix ρ. We will define them and work out some useful examples. The most commonly used measures include the concurrence, but we also have others. The entanglement of formation $E_F(\rho)$ quantifies the amount of entanglement required to prepare a given state ρ. For two qubits, it is computed from the concurrence C as

$$E_F(\rho) = h\left(\frac{1 + \sqrt{1 - C^2}}{2}\right), \tag{16.54}$$

where $h(x)$ is the binary entropy function:

$$h(x) = -x \log_2(x) - (1 - x)\log_2(1 - x). \tag{16.55}$$

For pure states, the entanglement of formation equals the von Neumann entropy. Negativity is a measure based on the Peres–Horodecki criterion (partial transpose test) [15, 16]. It quantifies the extent to which the partial transpose ρ^{T_A} of the density matrix has negative eigenvalues. It is defined as

$$\mathcal{N}(\rho) = \frac{\|\rho^{T_A}\|_1 - 1}{2}, \tag{16.56}$$

where $\|\rho^{T_A}\|_1$ is the trace norm of the partial transpose ρ^{T_A}. The logarithmic negativity is a related measure, given by

$$E_N(\rho) = \log_2 \|\rho^{T_A}\|_1. \tag{16.57}$$

Let us compute the negativity for the state we used The density matrix corresponding to $|\psi\rangle$ is

$$\begin{aligned}\rho = |\psi\rangle\langle\psi| &= A^2|00\rangle\langle 00| + A\sqrt{1 - A^2}\,|00\rangle\langle 11| \\ &+ A\sqrt{1 - A^2}\,|11\rangle\langle 00| + (1 - A^2)|11\rangle\langle 11|.\end{aligned} \tag{16.58}$$

We work in the computational basis ordered as $\{|00\rangle, |01\rangle, |10\rangle, |11\rangle\}$. The partial transpose with respect to the second subsystem, denoted by ρ^{T_B}, is defined by transposing only the indices corresponding to the second qubit. In the matrix elements, if we write

$$\rho = \sum_{i,j,k,l} \rho_{ij,kl}|ij\rangle\langle kl|, \tag{16.59}$$

then the partial transpose is

$$\rho^{T_B} = \sum_{i,j,k,l} \rho_{ij,kl}|il\rangle\langle kj|. \tag{16.60}$$

For the term $\rho_{1,1} = A^2$, $|00\rangle\langle 00|$ remains unchanged under the partial transpose because both the ket and bra for the second subsystem are 0. Thus, the element at position $(|00\rangle, |00\rangle)$ is A^2. The term $|00\rangle\langle 11|$ under partial transpose becomes $|01\rangle\langle 10|$ (the second qubit indices are swapped). In our ordered basis, $|01\rangle$ is the second basis vector and $|10\rangle$ is the third. Therefore, we obtain

$$(\rho^{T_B})_{2,3} = A\sqrt{1 - A^2} \tag{16.61}$$

and $\rho_{4,1} = A\sqrt{1 - A^2}$.

The term $|11\rangle\langle 00|$ becomes $|10\rangle\langle 01|$ under the partial transpose, so

$$(\rho^{T_B})_{3,2} = A\sqrt{1 - A^2}. \tag{16.62}$$

Term $\rho_{4,4} = 1 - A^2$, and the term $|11\rangle\langle 11|$ remains unchanged, yielding the element at $(|11\rangle, |11\rangle)$ as $1 - A^2$. Thus, for $\{|00\rangle, |01\rangle, |10\rangle, |11\rangle\}$, the partial transpose ρ^{T_B} is given by

$$\rho^{T_B} = \begin{pmatrix} A^2 & 0 & 0 & 0 \\ 0 & 0 & A\sqrt{1 - A^2} & 0 \\ 0 & A\sqrt{1 - A^2} & 0 & 0 \\ 0 & 0 & 0 & 1 - A^2 \end{pmatrix}. \tag{16.63}$$

The matrix ρ^{T_B} is block-diagonal. Its blocks are, a 1×1 block with the element A^2, a 1×1 block with the element $1 - A^2$, and a 2×2 block:

$$M = \begin{pmatrix} 0 & A\sqrt{1 - A^2} \\ A\sqrt{1 - A^2} & 0 \end{pmatrix}. \tag{16.64}$$

The eigenvalues of the matrix M are found by solving

$$\det(M - \lambda I) = \lambda^2 - (A\sqrt{1 - A^2})^2 = 0, \tag{16.65}$$

which gives

$$\lambda = \pm A\sqrt{1 - A^2}. \tag{16.66}$$

Thus, the four eigenvalues of ρ^{T_B} are

- $\lambda_1 = A^2$,
- $\lambda_2 = 1 - A^2$,
- $\lambda_3 = A\sqrt{1 - A^2}$,
- $\lambda_4 = -A\sqrt{1 - A^2}$.

The negativity $\mathcal{N}$ is defined as the sum of the absolute values of the negative eigenvalues of ρ^{T_B}. Here, the only negative eigenvalue is

$$\lambda_4 = -A\sqrt{1 - A^2}. \tag{16.67}$$

Thus, the negativity is

$$\mathcal{N} = |-A\sqrt{1 - A^2}| = A\sqrt{1 - A^2}. \tag{16.68}$$

Alternatively, if you use the formula

$$\mathcal{N} = \frac{\|\rho^{T_B}\|_1 - 1}{2} \tag{16.69}$$

you would first compute the trace norm (sum of the absolute values of all eigenvalues)

$$\|\rho^{T_B}\|_1 = A^2 + (1 - A^2) + A\sqrt{1 - A^2} + A\sqrt{1 - A^2} = 1 + 2A\sqrt{1 - A^2}. \quad (16.70)$$

Then,

$$\mathcal{N} = \frac{(1 + 2A\sqrt{1 - A^2}) - 1}{2} = A\sqrt{1 - A^2}. \quad (16.71)$$

While this seems a tad torturous, it is easier to compute (indeed it is calculable) than the relative entropy of entanglement or other measures that must be maximized or minimized with respect to a set of states.

The relative entropy of entanglement measures the distance (defined by an inner product $\mathrm{Tr}[\langle\psi_{\mathrm{sep}}|\rho|\psi_{\mathrm{sep}}])$ between the mixed state ρ and the closest separable state σ. This is difficult to calculate as you have to check all separable states, to find the minimum

$$E_R(\rho) = \min_{\sigma \in \text{separable}} \mathrm{Tr}(\rho \log \rho - \rho \log \sigma). \quad (16.72)$$

For a pure state, this is also equal to the von Neumann entropy

16.4 Schmidt basis

The Schmidt decomposition is a way to express any bipartite pure state in a particularly simple form that directly reveals the entanglement between the two subsystems. The Schmidt decomposition follows directly from the singular value decomposition (SVD) of the coefficient matrix that describes the state in some product basis. This can be seen as a generalization of eigenvalues and eigenvectors for a non-square matrix. Consider a pure state $|\psi\rangle$ in a bipartite Hilbert space $\mathcal{H}_A \otimes \mathcal{H}_B$. Choose the orthonormal bases $\{|i\rangle_A\}$ for $\mathcal{H}_A$ and $\{|j\rangle_B\}$ for $\mathcal{H}_B$. Then, the state can be written as

$$|\psi\rangle = \sum_{i,j} C_{ij} |i\rangle_A \otimes |j\rangle_B, \quad (16.73)$$

where C_{ij} are complex coefficients. You can think of the coefficients C_{ij} as the entries of a matrix C. The SVD theorem tells us that any complex matrix C can be decomposed as

$$C = UDV^\dagger, \quad (16.74)$$

where U is a unitary matrix (i.e. $U^\dagger U = I$) acting on the indices corresponding to the subsystem A, V is a unitary matrix acting on the indices corresponding to the subsystem B, and D is a diagonal matrix with non-negative real entries d_k (the singular values). In the component form, this decomposition reads

$$C_{ij} = \sum_k U_{ik} d_k (V^\dagger)_{kj} = \sum_k U_{ik} d_k V^*_{jk}. \quad (16.75)$$

This is a generalization of diagonalizing a square matrix.

Substituting the SVD expression for C_{ij} back into the state $|\psi\rangle$

$$|\psi\rangle = \sum_{i,j}\left(\sum_k U_{ik} d_k V^*_{jk}\right)|i\rangle_A \otimes |j\rangle_B \tag{16.76}$$

and interchanging the order of the summation

$$|\psi\rangle = \sum_k d_k\left(\sum_i U_{ik}|i\rangle_A\right) \otimes \left(\sum_j V^*_{jk}|j\rangle_B\right). \tag{16.77}$$

We then define the new vectors in $\mathcal{H}_A$ and $\mathcal{H}_B$

$$|u_k\rangle_A = \sum_i U_{ik}|i\rangle_A, \quad |v_k\rangle_B = \sum_j V^*_{jk}|j\rangle_B. \tag{16.78}$$

Because U and V are unitary, the sets $\{|u_k\rangle_A\}$ and $\{|v_k\rangle_B\}$ are orthonormal bases for their respective spaces. Using this new basis, the state $|\psi\rangle$ can now be written as

$$|\psi\rangle = \sum_k d_k\ |u_k>_A \otimes |v_k>_B. \tag{16.79}$$

This is the Schmidt decomposition of the state. $d_k \geqslant 0$ are the Schmidt coefficients (which are the singular values from the SVD). The number of non-zero d_k is known as the Schmidt rank of the state, and the vectors $|u_k\rangle_A$ and $|v_k\rangle_B$ form orthonormal sets. For a normalized state, the coefficients satisfy

$$\sum_k d_k^2 = 1. \tag{16.80}$$

If the system is a product state, there is only one Schmidt coefficient equal to unity. If the Schmidt coefficients are equal, the state is maximally entangled. For example, for a Bell state, let us consider one of the common Bell states, for example, the state

$$|\Phi^+\rangle = \frac{1}{\sqrt{2}}(|00\rangle + |11\rangle). \tag{16.81}$$

For $|\Phi^+\rangle$, we can identify $|u_1\rangle = |0\rangle$ and $|v_1\rangle = |0\rangle$, $|u_2\rangle = |1\rangle$, and $|v_2\rangle = |1\rangle$, with the corresponding coefficients $\lambda_1 = \frac{1}{\sqrt{2}}$ and $\lambda_2 = \frac{1}{\sqrt{2}}$. Thus, the Schmidt decomposition is already given by

$$|\Phi^+\rangle = \frac{1}{\sqrt{2}}|0\rangle \otimes |0\rangle + \frac{1}{\sqrt{2}}|1\rangle \otimes |1\rangle. \tag{16.82}$$

Since both coefficients are equal, the state is maximally entangled. The entanglement entropy (von Neumann entropy of the reduced state) is

$$S = -\left(\lambda_1^2 \ln \lambda_1^2 + \lambda_2^2 \ln \lambda_2^2\right) = -2\left(\frac{1}{2}\ln\frac{1}{2}\right) = \ln 2 \tag{16.83}$$

$$|\psi_{AB}\rangle = \sum_i \sqrt{p_i}\ |u_i\rangle_A \otimes |v_i\rangle_B, \tag{16.84}$$

where $\{p_i\}$ are the Schmidt coefficients (with $\sum_i p_i = 1$). The reduced density matrices become

$$\rho_A = \sum_i p_i |u_i\rangle\langle u_i|, \quad \rho_B = \sum_i p_i |v_i\rangle\langle v_i|. \tag{16.85}$$

Then, the von Neumann entropy is

$$S(\rho_A) = S(\rho_B) = -\sum_i p_i \ln p_i. \tag{16.86}$$

This entropy directly quantifies the amount of entanglement in the state.

16.5 Bell-state basis

It is useful to define the basis of the maximally entangled states for the two qubits. This is one basis, one can add phase shifts to find variations of this:

$$|\Psi_\pm\rangle = \left(\sqrt{\frac{1}{2}}\right)(|eg\rangle \pm e^{i\theta_\Psi}|ge\rangle) \tag{16.87}$$

$$|\Phi_\pm\rangle = e^{i\theta_{\Phi\Psi}}\left(\sqrt{\frac{1}{2}}\right)(|ee\rangle \pm e^{i\theta_\Phi}|gg\rangle). \tag{16.88}$$

With the choice $\theta_\Psi = \theta_\Phi = \theta_{\Phi\Psi} = 0$, we have the often used basis

$$|\Psi_\pm\rangle = \left(\sqrt{\frac{1}{2}}\right)(|eg\rangle \pm |ge\rangle) \tag{16.89}$$

$$|\Phi_\pm\rangle = \left(\sqrt{\frac{1}{2}}\right)(|ee\rangle \pm |gg\rangle). \tag{16.90}$$

16.6 Teleportation

The quantum teleportation protocol allows the transfer of a qubit's quantum state $|\psi\rangle = \alpha|0\rangle + \beta|1\rangle$ This process does not physically transfer the qubit itself but rather transfers its quantum state to another qubit. You do not 'beam Kirk up'. You beam the quantum state of Kirk to something, a vat of organic material, a corpse, it is not clear. Also the original Kirk is destroyed. There is no 'storing' Kirk in the transporter buffer. Say party A has the qubit in state $|\psi\rangle = \alpha|0\rangle + \beta|1\rangle$. A and B share an entangled Bell pair,

$$|\Phi^+\rangle_{AB} = \frac{1}{\sqrt{2}}(|00\rangle + |11\rangle), \tag{16.91}$$

where A holds qubit A_2, and B holds qubit B_1. The combined initial state of the system is

$$|\psi\rangle_{A_1} \otimes |\Phi^+\rangle_{A_2B_1} \tag{16.92}$$

The total state of the system (qubits A_1, A_2, and B_1) is

$$|\Psi\rangle = (\alpha|0\rangle + \beta|1\rangle) \otimes \frac{1}{\sqrt{2}}(|00\rangle + |11\rangle). \tag{16.93}$$

Expanding this, we find

$$|\Psi\rangle = \frac{1}{\sqrt{2}}(\alpha|0\rangle|00\rangle + \alpha|0\rangle|11\rangle + \beta|1\rangle|00\rangle + \beta|1\rangle|11\rangle). \tag{16.94}$$

We rewrite $|\Psi\rangle$ in the Bell basis for qubits A_1 and A_2,

$$|\Psi\rangle = \frac{1}{2}\sum_{i=0}^{3}|\Phi_i\rangle_{A_1A_2} \otimes |\psi_i\rangle_{B_1}, \tag{16.95}$$

where $|\Phi_i\rangle_{A_1A_2}$ are the four Bell states, and $|\psi_i\rangle_{B_1}$ are the corresponding states of qubit B_1 (dependent on the measurement outcome). Then, party A performs a Bell-state measurement on qubits A_1 and A_2, collapsing them into one of the four Bell states

$$|\Phi^+\rangle = \frac{1}{\sqrt{2}}(|00\rangle + |11\rangle) \tag{16.96}$$

$$|\Phi^-\rangle = \frac{1}{\sqrt{2}}(|00\rangle - |11\rangle) \tag{16.97}$$

$$|\Psi^+\rangle = \frac{1}{\sqrt{2}}(|01\rangle + |10\rangle) \tag{16.98}$$

$$|\Psi^-\rangle = \frac{1}{\sqrt{2}}(|01\rangle - |10\rangle). \tag{16.99}$$

The result of the Bell measurement determines which correction operation B must apply to qubit B_1 to recover the original state $|\psi\rangle$. Then A sends the two classical bits corresponding to the Bell measurement outcome to B. This communication must be transmitted at a speed less than c, so there is nothing superluminal involved here. These two classical bits specify which correction B needs to be applied to their qubit. Based on the measurement outcome, B applies one of the following Pauli operations to qubit B_1:

Outcome $|\Phi^+\rangle$: apply I (no operation).
Outcome $|\Phi^-\rangle$: apply Z.
Outcome $|\Psi^+\rangle$: apply X.
Outcome $|\Psi^-\rangle$: apply XZ (or ZX).

After applying the correction, the state of B_1 becomes $|\psi\rangle = \alpha|0\rangle + \beta|1\rangle$. Consider the initial state

$$|\psi\rangle = \frac{1}{\sqrt{2}}(|0\rangle + |1\rangle), \tag{16.100}$$

shared with a Bell pair $|\Phi^{+}\rangle = \frac{1}{\sqrt{2}}(|00\rangle + |11\rangle)$. Assume a measurement outcome: $|\Psi^{+}\rangle$. then B applies X to B_1, recovering $|\psi\rangle = \frac{1}{\sqrt{2}}(|0\rangle + |1\rangle)$. And the qubit whose state has been teleported is undefined, as it must be via the no-cloning theorem.

Fidelity is a key metric used to certify the performance of a quantum teleportation protocol. It quantifies how well the teleported quantum state matches the original state. Fidelity is the overlap between the original state $|\psi\rangle$ and the teleported state $|\psi_{\text{teleported}}\rangle$. For a given quantum state $|\psi\rangle$, the fidelity is

$$F = |\langle\psi|\psi_{\text{teleported}}\rangle|^2. \tag{16.101}$$

For mixed states (where the teleported state is represented by a density matrix $\rho_{\text{teleported}}$)

$$F = \langle\psi|\rho_{\text{teleported}}|\psi\rangle. \tag{16.102}$$

If $F = 1$, the teleportation is perfect, meaning that the teleported state exactly matches the original state. Lower fidelity indicate imperfections or errors. The fidelity is used to determine whether the teleportation process surpasses what could be achieved by classical means alone (e.g. using measurement and reconstruction). The classical fidelity limit serves as a benchmark; for qubits, the classical fidelity limit is $F_{\text{classical}} = 2/3$. If an eavesdropper learns the two classical bits sent by Alice, it does them no good without access to the entangled qubit Bob possesses. This is the maximum fidelity achievable using classical communication and measurements without exploiting entanglement.

16.7 Quantum dense coding

Quantum dense coding (or simply dense coding) is a quantum communication protocol that enables the transmission of two classical bits of information using only one quantum bit (qubit). This protocol uses quantum entanglement to achieve this efficiency and illustrates the power of quantum communication compared to that of classical communication.

As in teleportation, two parties, typically named Alice and Bob, share a pair of entangled qubits. For example, they might share a Bell state

$$|\Phi^{+}\rangle = \frac{1}{\sqrt{2}}(|00\rangle + |11\rangle). \tag{16.103}$$

Alice has one qubit, and Bob has the others. Alice wants to send two classical bits of information (00, 01, 10, or 11) to Bob. To encode the information, Alice applies one of the four unitary operations to her qubit, corresponding to the two bits she wants to transmit:

I (identity) for 00,

X (Pauli-X) for 01,

Z (Pauli-Z) for 10,

XZ (Pauli-X followed by Z) for 11.

After encoding the information, Alice sends her qubit to Bob. Now, Bob has two qubits, the classically transmitted one and his member of the shared Bell pair. Bob then performs a joint measurement on the two qubits in the Bell basis. The outcome of this measurement allows Bob to determine which operation Alice applied, thereby decoding the two classical bits of information. It is essentially a teleportation protocol run backward. Again, this does not allow for faster than light communication. The protocol leverages the shared entanglement between Alice and Bob, which allows them to encode and decode more information than is classically possible with a single-qubit transmission.

16.8 Quantum key distribution

There are many versions of this, but we concentrate on the BB84 protocol [17]. To prepare for the quantum key distribution (QKD) protocol Alice randomly prepares qubits in one of the four states using two nonorthogonal bases.

In the Z-basis ($\{|0\rangle, |1\rangle\}$) we have $|0\rangle$ (representing binary 0) and $|1\rangle$ (representing binary 1). In the X-basis ($\{|+\rangle, |-\rangle\}$) we have $|+\rangle = \frac{1}{\sqrt{2}}(|0\rangle + |1\rangle)$, $|-\rangle = \frac{1}{\sqrt{2}}(|0\rangle - |1\rangle)$.

Alice chooses a random sequence of these basis states and sends the corresponding qubits to Bob over a quantum channel. Then, Bob randomly chooses a measurement basis (Z or X) for each received qubit. If his basis matches Alice's preparation basis, the measurement result will correctly reflect Alice's bit. If his basis differs, the result will be random. After Bob measures all the qubits, Alice and Bob communicate over a public classical channel (which can be monitored but not tampered with) to announce the bases they used for each qubit. They then discard all measurement results where their bases do not match, and the remaining results form a raw key shared by Alice and Bob.

If an eavesdropper, Eve, tries to intercept and measure the qubits, the quantum states collapse due to her measurement. This introduces errors in Bob's measurements, which Alice and Bob can detect during error checking. To detect and correct errors (caused by noise or potential eavesdropping), Alice and Bob compare a small subset of their raw key bits publicly. If too many discrepancies exist, an error rate above 25%, they abort the protocol. Quantum mechanics prevents Eve from copying qubits perfectly, ensuring the security of the transmitted qubits. The final secret key can be used in a one-time pad encryption system, which is probably secure. As an example, consider Alice decides on a bit string

$$[0, 1, 1, 0, 0, 1]. \tag{16.104}$$

She makes a choice of bases

$$[Z, X, Z, X, X, Z]. \tag{16.105}$$

She then sends qubits

$$[|0\rangle, |-\rangle, |1\rangle, |+\rangle, |+\rangle, |1\rangle]. \tag{16.106}$$

Bob makes a choice for his bases,

$$[X, X, Z, Z, X, Z]. \tag{16.107}$$

Suppose he obtains measurement results

$$[|+\rangle, |-\rangle, |1\rangle, |+\rangle, |+\rangle, |1\rangle]. \tag{16.108}$$

Alice and Bob then discover the results where their bases differ (e.g. the first and fourth qubits). The remaining key bits (where bases match) are [1, 0, 1]. Then, Alice and Bob test a small portion of their key for errors and perform privacy amplification to produce the final key. QKD is secure against any computational or technological advances due to the laws of quantum mechanics. Eve's presence introduces detectable anomalies in the protocol. There are of course many ways in practice of defeating this, the simplest being to cut the fiber optic between Alice and Bob, but that leaves a mark! More sneaky, one can spy on Alice and learn (either in real time or after) what sequence she chose. In some of the early experiments, an observer in the hallway, with the door closed, could discern the sequence by listening to the click and clack of the polarizers being rotated.

16.9 Positive operator-valued measures

A positive operator-valued measure (POVM) is a general framework for describing quantum measurements. Unlike the standard projective measurements, which are restricted to orthogonal projectors, POVMs allow for a broader range of measurements that are not necessarily orthogonal, enabling greater flexibility in describing physical systems. The formal definition of a POVM is a set of positive semi-definite operators $\{E_i\}$ on a Hilbert space $\mathcal{H}$ satisfying positivity $E_i \geqslant 0$ (each operator E_i is positive semi-definite), and completeness, the operators sum to the identity operator:

$$\sum_i E_i = I. \tag{16.109}$$

When a POVM measurement is applied to a quantum state described by the density matrix ρ, the probability of obtaining outcome i is given by

$$P(i) = \mathrm{Tr}(\rho E_i). \tag{16.110}$$

Consider a qubit system with the following POVM elements

$$E_1 = \frac{1}{2}|0\rangle\langle 0|, \tag{16.111}$$

$$E_2 = \frac{1}{2}|1\rangle\langle 1|, \tag{16.112}$$

$$E_3 = \frac{1}{2}(|+\rangle\langle +|), \tag{16.113}$$

where $|+\rangle = \frac{1}{\sqrt{2}}(|0\rangle + |1\rangle)$. Each E_i is positive semi-definite:

$$E_1 \geqslant 0,\ E_2 \geqslant 0,\ E_3 \geqslant 0 \tag{16.114}$$

and

$$E_1 + E_2 + E_3 = \frac{1}{2}(|0\rangle\langle 0| + |1\rangle\langle 1| + |+\rangle\langle +|). \tag{16.115}$$

After verifying the algebra, this sums to the identity operator for a qubit system. This POVM allows for outcomes associated with $|0\rangle$, $|1\rangle$, and $|+\rangle$, extending the traditional orthogonal projective measurement basis. The probability of each outcome is determined by the state of the qubit.

16.10 Mixed states Schmidt basis

We consider a quantum system consisting of two subsystems A and B with dimensions $\dim \mathcal{H}_A = d$ and $\dim \mathcal{H}_B = d'$, so that the Hilbert space of the composite system is $\mathcal{H} \equiv \mathcal{H}_A \otimes \mathcal{H}_B$. We will also call $\mathcal{H}$ a $d \times d'$-dimensional system. Henceforth we assume $d \leqslant d'$. With orthonormal bases $\{|a\rangle\}$, $\{|b\rangle\}$ a pure state $\psi \in \mathcal{H}_A \otimes \mathcal{H}_B$ can be written as

$$|\psi\rangle = \sum_{a=1}^{d}\sum_{b=1}^{d'} \psi_{ab}|a\rangle \otimes |b\rangle \equiv \sum_{a,b} \psi_{ab}|ab\rangle. \tag{16.116}$$

Pure bipartite states have the important property that there are always orthonormal bases $\{|j_A\rangle\}$, $\{|j_B\rangle\}$ such that [16, 17]

$$|\psi\rangle = \sum_{j=1}^{r(\psi)} \sqrt{\lambda_j}|jj\rangle \tag{16.117}$$

with real positive numbers λ_j, the so-called Schmidt coefficients. The Schmidt rank $r(\psi) \leqslant d$ corresponds to the rank of the reduced density matrices of the subsystems $r(\psi) = \mathrm{rank}(\mathrm{Tr}_B|\psi\rangle\langle\psi|) = \mathrm{rank}(\mathrm{Tr}_A|\psi\rangle\langle\psi|)$. With this, we note that a pure bipartite state is separable if its Schmidt rank equals 1, otherwise it is entangled. In particular, we may define the maximally entangled state in d dimensions

$$|\Psi_d\rangle = \sum_{j=1}^{d} \frac{1}{\sqrt{d}}|jj\rangle. \tag{16.118}$$

16.11 Quantum relative and conditional entropies

The quantum relative entropy is a measure of the 'distance' or distinguishability between two quantum states ρ and σ. It is defined as

$$S(\rho\|\sigma) = \mathrm{Tr}(\rho \log \rho) - \mathrm{Tr}(\rho \log \sigma). \tag{16.119}$$

where ρ and σ are density matrices and log refers to the matrix logarithm. It has the property of non-negativity: $S(\rho\|\sigma) \geqslant 0$, with equality if and only if $\rho = \sigma$. It is asymmetric, $S(\rho\|\sigma) \neq S(\sigma\|\rho)$ and it has no direct interpretation as a distance. It is not a metric because it does not satisfy symmetry or the triangle inequality.

Consider two-qubit states

$$\rho = \begin{bmatrix} 0.7 & 0 \\ 0 & 0.3 \end{bmatrix}, \quad \sigma = \begin{bmatrix} 0.5 & 0 \\ 0 & 0.5 \end{bmatrix}. \tag{16.120}$$

$S(\rho\|\sigma)$ is found to be

$$S(\rho\|\sigma) = \mathrm{Tr}(\rho \log \rho) - \mathrm{Tr}(\rho \log \sigma). \tag{16.121}$$

Using eigenvalues, the result is

$$S(\rho\|\sigma) = 0.7 \log \frac{0.7}{0.5} + 0.3 \log \frac{0.3}{0.5} \approx 0.088. \tag{16.122}$$

The quantum conditional entropy describes the uncertainty of one quantum system A, given the knowledge of another system B, for a bipartite quantum state ρ_{AB}. It is defined as

$$S(A|B) = S(\rho_{AB}) - S(\rho_B), \tag{16.123}$$

where $S(\rho) = -\mathrm{Tr}(\rho \log \rho)$ is the von Neumann entropy, and $\rho_B = \mathrm{Tr}_A(\rho_{AB})$ is the reduced density matrix of B. Unlike the classical conditional entropy, $S(A|B)$ can take negative values. Negativity implies entanglement—if $S(A|B) < 0$, ρ_{AB} is entangled.

Consider the pure entangled state $|\psi\rangle_{AB} = \frac{1}{\sqrt{2}}(|00\rangle + |11\rangle)$. The joint state $\rho_{AB} = |\psi\rangle\langle\psi|$, and the reduced states $\rho_A = \rho_B \frac{1}{2} I$. Next, we compute $S(A|B)$ and we find $S(\rho_{AB}) = 0$ (pure state), and $S(\rho_B) = 1$ (maximal entropy for a two-level system):

$$S(A|B) = 0 - 1 = -1. \tag{16.124}$$

So it is entangled, as it should be as a Bell state.

16.12 Quantum mutual information

Quantum mutual information quantifies the total correlation (both classical and quantum) between the two subsystems A and B of a composite quantum state ρ_{AB}. It is defined as

$$I(A{:}\, B) = S(\rho_A) + S(\rho_B) - S(\rho_{AB}), \tag{16.125}$$

where $S(\rho) = -\mathrm{Tr}(\rho \log \rho)$ is the von Neumann entropy, $\rho_A = \mathrm{Tr}_B(\rho_{AB})$ and $\rho_B = \mathrm{Tr}_A(\rho_{AB})$ are the reduced density matrices of subsystems A and B. ρ_{AB} is the density matrix of the combined system.

Let us examine a product state (no correlation)

$$\rho_{AB} = \rho_A \otimes \rho_B, \tag{16.126}$$

where

$$\rho_A = \frac{1}{2}\begin{bmatrix} 1 & 0 \\ 0 & 1 \end{bmatrix}, \quad \rho_B = \frac{1}{2}\begin{bmatrix} 1 & 0 \\ 0 & 1 \end{bmatrix} \tag{16.127}$$

$S(\rho_A) = 1$, $S(\rho_B) = 1$, $S(\rho_{AB}) = S(\rho_A) + S(\rho_B) = 2$. The mutual information is

$$I(A{:}\ B) = S(\rho_A) + S(\rho_B) - S(\rho_{AB}) = 1 + 1 - 2 = 0, \tag{16.128}$$

so no correlation exists between A and B.

How about a maximally entangled state

$$|\psi_{AB}\rangle = \frac{1}{\sqrt{2}}(|00\rangle + |11\rangle), \tag{16.129}$$

with

$$\rho_{AB} = |\psi_{AB}\rangle\langle\psi_{AB}|. \tag{16.130}$$

Here, we have $S(\rho_A) = S(\rho_B) = 1$ (reduced states are maximally mixed), and $S(\rho_{AB}) = 0$ as it is a pure state. The mutual information is

$$I(A{:}\ B) = S(\rho_A) + S(\rho_B) - S(\rho_{AB}) = 1 + 1 - 0 = 2. \tag{16.131}$$

Thus, A and B are maximally correlated (both quantum and classical).

What about a classically correlated state, both in 1 or both in 0. The density matrix is

$$\rho_{AB} = \frac{1}{2}(|00\rangle\langle 00| + |11\rangle\langle 11|), \tag{16.132}$$

with no off diagonal density matrix elements. We find $S(\rho_A) = S(\rho_B) = 1$, and $S(\rho_{AB}) = 1$ (entropy of the mixed state). The mutual information is

$$I(A{:}\ B) = S(\rho_A) + S(\rho_B) - S(\rho_{AB}) = 1 + 1 - 1 = 1. \tag{16.133}$$

Therefore, A and B share classical correlations, but no quantum entanglement.

16.13 Three-qubit states

For three-qubit pure states, the entanglement measures are more nuanced than those for two-qubit systems because multipartite entanglement can involve more complex relationships between the qubits. There are two types of entangled states of interest. First, the GHZ state

$$|\mathrm{GHZ}\rangle = \frac{1}{\sqrt{2}}(|000\rangle + |111\rangle). \tag{16.134}$$

If we make a measurement on any of the qubits, we obtain 0 or 1 with equal probability, leaving the remaining two qubits in a product state. If we measure qubit 1 to be 0, the remains state is a product state

$$|\Psi\rangle = |00\rangle . \tag{16.135}$$

Several measures have been proposed, depending on the type of entanglement being quantified. The three-tangle (τ_3) quantifies the genuine tripartite entanglement in three-qubit pure states. For a three-qubit state $|\psi\rangle$, the three-tangle is computed using the coefficients of the state. In the computational basis we have

$$|\psi\rangle = \sum_{i,j,k=0}^{1} c_{ijk}|i\rangle|j\rangle|k\rangle . \tag{16.136}$$

The three-tangle is defined as

$$\tau_3 = 4\left| \det\begin{pmatrix} c_{000} & c_{011} \\ c_{101} & c_{110} \end{pmatrix} \right|^2 . \tag{16.137}$$

The range of $\tau_3 \in [0, 1]$, where $\tau_3 = 1$ means maximum tripartite entanglement (e.g. GHZ state). If $\tau_3 = 0$, there is no tripartite entanglement (e.g. W state or separable states). For the GHZ state $|\mathrm{GHZ}\rangle = \frac{1}{\sqrt{2}}(|000\rangle + |111\rangle)$ the three-tangle

$$\tau_3 = 1, \tag{16.138}$$

indicating maximal genuine tripartite entanglement. For the W state, $|W\rangle = \frac{1}{\sqrt{3}}(|001\rangle + |010\rangle + |100\rangle)$ we have

$$\tau_3 = 0, \tag{16.139}$$

because $|\mathrm{W}\rangle$ has no genuine tripartite entanglement although it exhibits strong bipartite correlations. For a mixture of $|\mathrm{GHZ}\rangle$ and $|\mathrm{W}\rangle$,

$$\rho = p|\mathrm{GHZ}\rangle\langle\mathrm{GHZ}| + (1-p)|\mathrm{W}\rangle\langle\mathrm{W}|, \tag{16.140}$$

the three-tangle $\tau_3(\rho)$ depends on p, of course If $p = 1$, $\tau_3(\rho) = 1$, and if $p = 0$, $\tau_3(\rho) = 0$ as we might expect.

Bipartite measures quantify the entanglement between one qubit and the other pair. Compute the reduced density matrix for a single qubit by tracing out the other two qubits, e.g. $\rho_A = \mathrm{Tr}_{BC}(|\psi\rangle\langle\psi|)$, then use measures such as the concurrence or von Neumann entropy of the reduced density matrix:

$$S(\rho_A) = -\mathrm{Tr}(\rho_A \log_2 \rho_A). \tag{16.141}$$

Svetlichny [18] introduced an inequality to determine that detects true tripartite entanglement, as opposed to a split between a bipartite state and a single qubit.

Using generalized tripartite states

$$|\psi_g\rangle = \cos\theta_1|000\rangle + \sin\theta_1|111\rangle \tag{16.142}$$

$$|\psi_s\rangle = \frac{1}{\sqrt{2}}\{|000\rangle + |11\rangle(\cos\theta_3|0\rangle + \sin\theta_3|1\rangle)\}. \tag{16.143}$$

ψ_s is known as the maximal slice state. As in the Mermin system, we use spin projections onto unit vectors $A = \overrightarrow{\sigma_1}\cdot\overrightarrow{a}$ or $A' = \overrightarrow{\sigma_1}\cdot\overrightarrow{a'}$ on qubit 1, $B = \overrightarrow{\sigma_2}\cdot\overrightarrow{b}$ or $B' = \overrightarrow{\sigma_2}\cdot\overrightarrow{b'}$ on qubit 2, and $C = \overrightarrow{\sigma_3}\cdot\overrightarrow{c}$ or $C' = \overrightarrow{\sigma_3}\cdot\overrightarrow{c'}$ on the third qubit. If a theory is consistent with the hybrid nonlocal-local realism, then the quantum prediction for any three-qubit state $|\Psi\rangle$ is bounded by Svetlichny's inequality

$$|\langle\Psi|S|\Psi\rangle| \equiv S(\Psi) \leqslant 4, \tag{16.144}$$

where the Svetlichny's operator S is defined as

$$S = A(DC + D'C') + A'(D'C - DC') = M + M', \tag{16.145}$$

where $D = B + B'$ and $D' = B - B'$, and $\langle M\rangle \leqslant 2$ and $\langle M'\rangle \leqslant 2$ are Mermin's inequalities. The latter can be expressed as

$$|M| = |E(ABC') + E(AB'C) + E(A'BC) - E(A'B'C')| \leqslant 2. \tag{16.146}$$

For a generalized GHZ state

$$E_{\mathrm{GHZ}}(ABC) = \langle \mathrm{GHZ}|\sigma(\mathbf{n}_1) \otimes \sigma(\mathbf{n}_2) \otimes \sigma(\mathbf{n}_3)|\mathrm{GHZ}\rangle \tag{16.147}$$

$$= \cos(\phi_1 + \phi_2 + \phi_3). \tag{16.148}$$

For the choice of angles $\phi_1 + \phi_2 + \phi_3 = [n + (1/2)]\pi$ and $\phi_i' = \phi_i + (\pi/2)$, we find $|M| = 4$, $|M'| = 0$, and $|S_{\mathrm{V}}| = 4$. So states with a maximal Mermin violation do not violate Svetlichny's inequality. This means we must go beyond expectation values of the form $E(ABC)$ to investigate the Svetlichny inequality. For the other class of tripartite entangled states, the generalized W state $\sigma(\theta_1) \otimes \sigma(\theta_2) \otimes \sigma(\theta_3)$ we have

$$E_{\mathrm{W}}(\theta_1, \theta_2, \theta_3) = -\frac{2}{3}\cos(\theta_1 + \theta_2 + \theta_3) - \frac{1}{3}\cos\theta_1\cos\theta_2\cos\theta_3.$$

The maximum violation of the Svetlichny inequality for a generalized GHZ state is

$$S(\psi_{g\mathrm{GHZ}}) \leqslant \begin{cases} 4\sqrt{1 - \tau(\psi_{g\mathrm{GHZ}})}, & \tau(\psi_g) \leqslant 1/3 \\ 4\sqrt{2\tau(\psi_{g\mathrm{GHZ}})}, & \tau(\psi_g) \geqslant 1/3 \end{cases}. \tag{16.149}$$

When $\tau \leqslant 1/2$, GGHZ states do not violate Svetlichny's inequality. For the maximal slice states we have

$$S(\psi_s) \leqslant 4\sqrt{1+\tau(\psi_s)}. \tag{16.150}$$

So we see that the Svetlichny inequality accesses a type of nonclassicality not accessible by the three-tangle τ. A reminder that once we get past two qubits the terrain is not flat and certain.

16.14 Tripartite basis sets

For a two-qubit system, we had the product basis, the angular momentum basis (eigenstates of J_z and J^2), or the Bell-state basis. There are several interesting choices for a three-qubit system. There is a boring product state basis,

$$|1\rangle == |000\rangle \tag{16.151}$$

$$|2\rangle == |001\rangle \tag{16.152}$$

$$|3\rangle == |010\rangle \tag{16.153}$$

$$|4\rangle == |100\rangle \tag{16.154}$$

$$|5\rangle == |110\rangle \tag{16.155}$$

$$|6\rangle == |101\rangle \tag{16.156}$$

$$|7\rangle == |011\rangle \tag{16.157}$$

$$|8\rangle == |111\rangle, \tag{16.158}$$

an angular momentum basis reflecting $1/2 \otimes 1/2 \otimes 1/2 = 3/2 \oplus 1/2 \oplus 1/2$, there are two sets of states with $J_z \pm 1/2$. A mathematician may write this in terms of the dimensions of spaces, $2 \otimes 2 \otimes 2 = 4 \oplus 2 \oplus 2$

$$\left|\frac{3}{2}\frac{3}{2}\right\rangle = |\uparrow\uparrow\uparrow\rangle \tag{16.159}$$

$$\left|\frac{3}{2}\frac{1}{2}\right\rangle = \frac{1}{\sqrt{3}}(|\uparrow\uparrow\downarrow\rangle + |\uparrow\downarrow\uparrow\rangle + |\downarrow\uparrow\uparrow\rangle) \tag{16.160}$$

$$\left|\frac{3}{2} - \frac{1}{2}\right\rangle = \frac{1}{\sqrt{3}}(|\downarrow\downarrow\uparrow\rangle + |\downarrow\uparrow\downarrow\rangle + |\uparrow\downarrow\downarrow\rangle) \tag{16.161}$$

$$\left|\frac{3}{2} - \frac{3}{2}\right\rangle = |\downarrow\downarrow\downarrow\rangle \tag{16.162}$$

$$\left|\frac{1}{2}\frac{1}{2}\right\rangle_{12} = \frac{1}{\sqrt{2}}(|\uparrow\downarrow\rangle - |\downarrow\uparrow\rangle)|\uparrow\rangle \tag{16.163}$$

$$\left|\frac{1}{2} - \frac{1}{2}\right\rangle_{12} = \frac{1}{\sqrt{2}}(|\uparrow\downarrow\rangle - |\downarrow\uparrow\rangle)|\downarrow\rangle \tag{16.164}$$

$$\left|\frac{1}{2}\frac{1}{2}\right\rangle_{23} = \frac{1}{\sqrt{2}}|\uparrow\rangle(|\uparrow\downarrow\rangle - |\downarrow\uparrow\rangle) \tag{16.165}$$

$$\left|\frac{1}{2} - \frac{1}{2}\right\rangle_{23} = \frac{1}{\sqrt{2}}|\downarrow(|\uparrow\downarrow\rangle - |\downarrow\uparrow\rangle). \tag{16.166}$$

We can also have a basis with the three types of entanglement, GHZ, W, and bipartite splits,

$$|\mathrm{GHZ}_+\rangle = \frac{1}{\sqrt{2}}(|\uparrow\uparrow\uparrow\rangle + |\downarrow\downarrow\downarrow\rangle) \tag{16.167}$$

$$|\mathrm{GHZ}_-\rangle = \frac{1}{\sqrt{2}}(|\uparrow\uparrow\uparrow\rangle - |\downarrow\downarrow\downarrow\rangle) \tag{16.168}$$

$$|W_1\rangle = \frac{1}{\sqrt{3}}(|\uparrow\uparrow\downarrow\rangle + |\uparrow\downarrow\uparrow\rangle + |\downarrow\uparrow\uparrow\rangle) \tag{16.169}$$

$$|W_2\rangle = \frac{1}{\sqrt{3}}(|\downarrow\downarrow\uparrow\rangle + |\downarrow\uparrow\downarrow\rangle + |\uparrow\downarrow\downarrow\rangle) \tag{16.170}$$

$$|\mathrm{Bi}_{u12}\rangle = \frac{1}{\sqrt{2}}(|\uparrow\downarrow\rangle - |\downarrow\uparrow\rangle)|\uparrow\rangle \tag{16.171}$$

$$|\mathrm{Bi}_{d12}\rangle = \frac{1}{\sqrt{2}}(|\uparrow\downarrow\rangle - |\downarrow\uparrow\rangle)|\downarrow\rangle \tag{16.172}$$

$$|\mathrm{Bi}_{u23}\rangle = \frac{1}{\sqrt{2}}|\uparrow\rangle(|\uparrow\downarrow\rangle - |\downarrow\uparrow\rangle) \tag{16.173}$$

$$|\mathrm{Bi}_{d23}\rangle = \frac{1}{\sqrt{2}}|\downarrow(|\uparrow\downarrow\rangle - |\downarrow\uparrow\rangle), \tag{16.174}$$

or entirely in GHZ type states,

$$|\psi_1\rangle = (|000\rangle + |111\rangle)/\sqrt{2} \tag{16.175}$$

$$|\psi_2\rangle = (|000\rangle - |111\rangle)/\sqrt{2} \tag{16.176}$$

$$|\psi_3\rangle = (|011\rangle + |100\rangle)/\sqrt{2} \quad (16.177)$$

$$|\psi_4\rangle = (|011\rangle - |100\rangle)/\sqrt{2} \quad (16.178)$$

$$|\psi_5\rangle = (|010\rangle + |101\rangle)/\sqrt{2} \quad (16.179)$$

$$|\psi_6\rangle = (|010\rangle - |101\rangle)/\sqrt{2} \quad (16.180)$$

$$|\psi_7\rangle = (|001\rangle + |110\rangle)/\sqrt{2} \quad (16.181)$$

$$|\psi_8\rangle = (|001\rangle - |110\rangle)/\sqrt{2}. \quad (16.182)$$

16.15 Qudits

Rungta *et al* [19] have extended the definition of concurrence, by generalizing the spin-flip superoperator of Wooters to a 'universal inverter', which acts on quantum systems of arbitrary dimension, which is often denoted 'qudits'. They introduced the I-concurrence for joint pure states of $D_1 \times D_2$ bipartite quantum systems. We call this generalized concurrence the I-concurrence to emphasize its relation to the universal inverter. Wooters utilized

$$\mathcal{S}_2(\rho) = \sigma_y \rho^* \sigma_y = \frac{1}{2}(I - \vec{P} \cdot \vec{\sigma}).$$

As we have seen the concurrence of a pure state $|\Psi_{AB}\rangle$ of two qubits is

$$C_2(\Psi_{AB}) \equiv \sqrt{\langle \Psi_{AB}|\mathcal{S}_2 \otimes \mathcal{S}_2(|\Psi_{AB}\rangle\langle\Psi_{AB}|)|\Psi_{AB}\rangle} = \left|\langle \Psi_{AB}|\sigma_y \otimes \sigma_y|\Psi^*_{AB}\rangle\right|.$$

Rungta *et al* [20] showed that if the concurrence is to be generated by a product superoperator, the only suitable superoperator to go into the tensor product is what they defined 'universal inverter'. The action of the universal inverter on a qudit state ρ is given by

$$\mathcal{S}_D(\rho) = \nu_D(I - \rho),$$

where ν_D is a positive constant. The corresponding generalized concurrence for a joint pure state $|\Psi_{AB}\rangle$ of a $D_1 \times D_2$ system is given by

$$C(\Psi_{AB}) \equiv \sqrt{\langle \Psi_{AB}|\mathcal{S}_{D_1} \otimes \mathcal{S}_{D_2}(|\Psi_{AB}\rangle\langle\Psi_{AB}|)|\Psi_{AB}\rangle} = 2\nu_{D_1}\nu_{D_2}\left[1 - \mathrm{tr}\left(\rho_A^2\right)\right].$$

16.16 Quantum trajectory entanglement

This section has been reproduced with permission from [23].

Here we consider a method for quantifying entanglement for an open system, suggested by Nha and Carmichael [21, 22]. This attempts to quantify what one can about entanglement given a measurement record. As to the viability of this approach we quote them.In this Letter, we consider mixed state entanglement from a new

perspective. For pure states, the accepted measure of entanglement is settled. ... We consider entanglement in bipartite open systems, specifically, where we resolve the arbitrariness by limiting attention to physically relevant decompositions—those for which pi is the probability of a classical record comprising information read from the environment. This information may be read in many ways. For each, there is a different decomposition of, and hence a different degree of entanglement [21].

We will first examine an example from cavity quantum electrodynamics, a two-level atom in a driven cavity [23]. The quantum trajectory wave function and non-Hermitian Hamiltonian are

$$|\Psi_c(t)\rangle = \sum_n^{\infty}(C_{g,n}(t)e^{-iE_{g,n}t}|g, n\rangle + C_{e,n}(t)e^{-iE_{e,n}t}|e, n\rangle) \tag{16.183}$$

$$H = \hbar g(a^\dagger\sigma_- + a\sigma_+) - iKa^\dagger a - i\frac{Y}{2}\sigma_+\sigma_- + i\hbar\varepsilon(a^\dagger - a), \tag{16.184}$$

with the collapse operators

$$C_{\text{cav}} = \sqrt{\kappa}\,a \tag{16.185}$$

$$C_{\text{at}} = \sqrt{\frac{Y}{2}}\sigma_-. \tag{16.186}$$

The details of the math are back in chapter 11. Because the probability of an excited state is small, so is the probability of a jump downward, so the system persists in this state for some time. The quantities p and q would be one for the coupled harmonic oscillators. The entanglement of the formation for this system is calculated from the density matrix after tracing over the field variables

$$\begin{aligned}\rho_{\text{atom}} &= \text{Tr}_{\text{field}}|\Psi\rangle\langle\Psi| \\ &= \begin{pmatrix} 1 + A_{1,g}^2 + A_{2,g}^2 & A_{1,e}A_{1,g} + A_{0,e} \\ A_{1,e}A_{1,g} + A_{0,e} & A_{1,e}^2 + A_{0,e}^2 \end{pmatrix}.\end{aligned} \tag{16.187}$$

The eigenvalues of this matrix are, to the lowest nonvanishing order [24],

$$\begin{aligned}\lambda_1 &= (A_{1,g}A_{0,e} - A_{1,e})^2 \\ &= |A_{1,g}|^2|A_{0,e}|^2(q - 1)^2 \\ &= \left(\frac{\varepsilon}{\kappa}\right)^4\xi^2\end{aligned} \tag{16.188}$$

$$\begin{aligned}\lambda_2 &= 1 - (A_{1,g}A_{0,e} - A_{1,e})^2 \\ &= 1 - \left(\frac{\varepsilon}{\kappa}\right)^4 \xi^2,\end{aligned} \tag{16.189}$$

where ε is the driving field and q is defined as

$$q = \frac{1 + 2C}{1 + 2C - 2C'} \tag{16.190}$$

which depends on decay rates (γ for the atom and κ for the cavity), atom–field coupling g, and number of atoms N, as detailed in chapter 11. We have defined

$$\xi = \frac{2g}{\gamma(1 + 2C_1)^2}(q - 1). \tag{16.191}$$

The entropy $E = -\lambda_1 \log_2 \lambda_1 - \lambda_2 \log_2 \lambda_2$ is then (again to lowest leading order)

$$\begin{aligned}E &= -\left(\frac{\varepsilon}{K}\right)^4 \xi^2 \log_2\left[\left(\frac{\varepsilon}{K}\right)^4 \xi^2\right] - \left(1 - \left(\frac{\varepsilon}{K}\right)^4 \xi^2\right) \log_2\left[1 - \left(\frac{\varepsilon}{K}\right)^4 \xi^2\right] \\ &\approx -\left(\frac{\varepsilon}{K}\right)^4 \xi^2 \left(\log_2\left[\left(\frac{\varepsilon}{K}\right)^4\right] + \log_2[\xi^2] - 1\right) \\ &\approx -\left(\frac{\varepsilon}{K}\right)^4 \log_2\left[\left(\frac{\varepsilon}{K}\right)^4\right] \xi^2,\end{aligned} \tag{16.192}$$

where we have taken the weak field limit, ε being the smallest rate in the problem, so $\varepsilon/\kappa \ll 1$. The approximation will hold provided $(\varepsilon/\kappa)^2 \ll |\xi|$. This entropy is the same as that obtained using the density matrix for the field alone, traced over the atomic degrees of freedom.

The concurrence, can also be used to characterize the entanglement between two quantum systems of arbitrary dimension. The concurrence for our system is

$$\begin{aligned}C &= \sqrt{2\left(1 - \mathrm{Tr}\,\rho_{\mathrm{atom}}^2\right)} \\ &= \sqrt{4(A_{1,\,g}A_{0,\,e} - A_{1,\,e})^2} \\ &= 2\left(\frac{\varepsilon}{\kappa}\right)^2 |\xi| \quad .\end{aligned} \tag{16.193}$$

In the presence of a non-zero vacuum contribution (as any real quantum state will have), one can learn nothing about entanglement simply by measuring one-excitation amplitudes or probabilities. For example, the state $|0, g\rangle + \alpha(|1, g\rangle + |0, e\rangle)$ is entangled, but only if it is certain that the probability amplitudes for higher excitation are truly zero. A state of the form $|0, g\rangle + \alpha(|1, g\rangle + |0, e\rangle) + O(\varepsilon^2)$ cannot be said to be entangled without information on the relative size of the probability amplitude $A_{1,e}$. In terms of the system parameters, the concurrence is

$$C = |2\alpha\beta(q - 1)| = \frac{16g^3\varepsilon^2\kappa}{(2g^2 + \gamma\kappa)^2(2g^2 + \kappa(\gamma + 2\kappa))}. \tag{16.194}$$

Although the rates of decay could be the same through the two reservoirs, spontaneous emission (γ) reduces entanglement more than the cavity loss (κ). This is because a γ event (spontaneous emission) must come from the atom, while a κ event (cavity transmission) could come from either the drive or a photon emitted by the atoms into the cavity mode. A spontaneous emission event unambiguously leaves the atom in the ground state, and the system wavefunction factorizes into a product state. Figure 16.1 shows the entanglement of the system as a function of the three rates in the problem.

There is an optimal value for the coupling constant g given a set of dissipation rates κ, γ. For many interesting cavity QED effects, stronger coupling is generally better, such as the enhancement of the spontaneous emission by a factor of $1 + 2C_1 = 1 + 2g^2/\kappa\gamma$ (this formula strictly holds only in the bad cavity limit $\kappa \gg g, \gamma$). However, here increasing the coupling of the atom and the field mode eventually decreases the amount of entanglement. The central reason for this is the double-edged sword of coupling. Larger coupling should lead to more entanglement, but it also leads to larger decay rates. To explain this, it is instructive to recall that the concurrence $C = |2\alpha\beta(q - 1)|$, where α is the mean cavity field, and $\beta = -g\alpha/\gamma$ is the mean atomic dipole. As the coupling g increases for a fixed weak driving field ε, the intracavity field $\alpha = \varepsilon/(\kappa + 2g^2/\gamma)$ decreases. The intracavity field is the sum of the driving field in the cavity ε/κ, and the field radiated by the atom, $(-2C_1/(1 + 2C_1))\varepsilon/\kappa$, the minus sign resulting from the fact that the radiated field is π out of phase with the driving field on resonance. We see that as g and C_1 increase, the intracavity field decreases. This means that the steady-state wavefunction has a larger vacuum component and consequently less entanglement. Another way to view

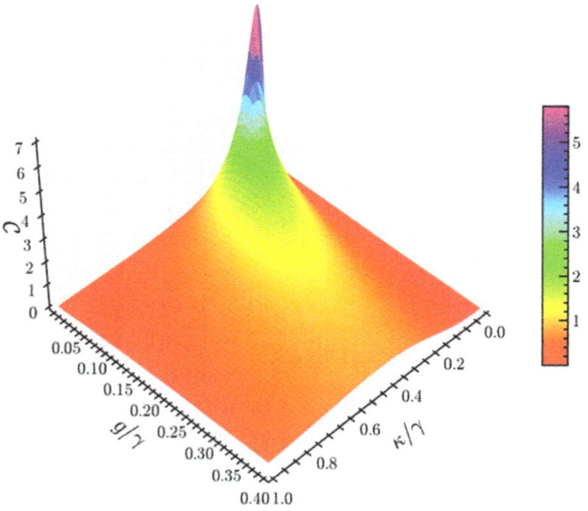

Figure 16.1. Concurrence as a function of cavity decay rate κ and atom–field coupling g.

this is that the cavity enhancement of the spontaneous emission rate means a larger loss rate for the system as the coupling increases, which is bad for entanglement. Detuning to a vacuum-Rabi peak ($\Delta = \pm g$), generates a steady-state wave function of the form

$$|\psi\rangle_{ss} = |0, g\rangle + \alpha\Gamma_1(g/\gamma)|1, \pm\rangle + \alpha^2\Gamma_2(g/\gamma)|2, \pm\rangle, \tag{16.195}$$

where $|n, \pm\rangle = (1/\sqrt{2})(|n, g\rangle \pm |n-1, e\rangle)$ is the n photon-dressed atom–field state one is tuned near and $\Gamma_1(g/\gamma)$ and $\Gamma_2(g/\gamma)$ are functions that are maximal when $g \simeq \gamma$. This is a state of mainly vacuum, plus a part that has entanglement between the atom and the cavity. Apparently, by continuing to tune to a vacuum-Rabi peak as g increases, it would be possible to maintain the entanglement, but figures 16.2 and 16.3 show that this is not the case.

Rather, as argued (for the on-resonance case) above, the crucial parameter for maximizing entanglement is $f = g/\gamma \propto 1/\sqrt{n_{\text{sat}}}$, where $n_{\text{sat}} = \gamma^2/8g^2$ is the

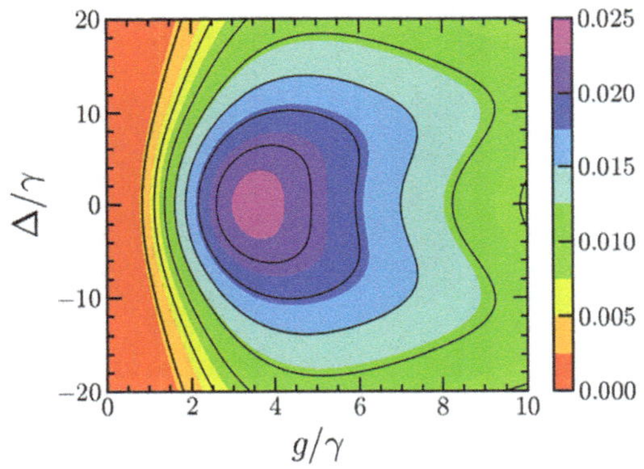

Figure 16.2. Concurrence as a function of cavity detuning and atom–field coupling in regime without vacuum-Rabi oscillations.

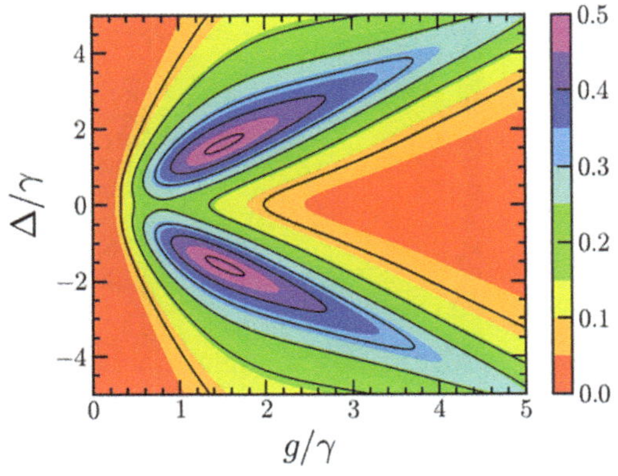

Figure 16.3. Concurrence as a function of cavity detuning and atom–field coupling in the vacuum-Rabi regime.

saturation photon number. This is the dependence on the nonlinearity of the atomic system. Recall that if these were two driven coupled harmonic oscillators, $q = 1$ and there would be no entanglement. A nonlinear interaction between the two harmonic oscillators is needed to entangle them, as in the signal and idler modes in the optical parametric oscillation. This nonlinear interaction would generate two-mode squeezing, which could be measured by the homodyne detection of mode $A(B)$ conditioned on the detection of a photon in mode B(A), just as squeezing in one mode can be detected via the conditioned homodyne detection of a mode based on a photo-detection from the other mode. The nonlinearity of the two-level atom is needed to generate two-mode squeezing and entanglement between the atom and the cavity field. Even though the driving field is weak and the atom never approaches saturation.

In figures 16.4 and 16.5 we see that the general behavior in which the log negativity vanishes as the atom saturates holds over different field-cavity detunings θ.

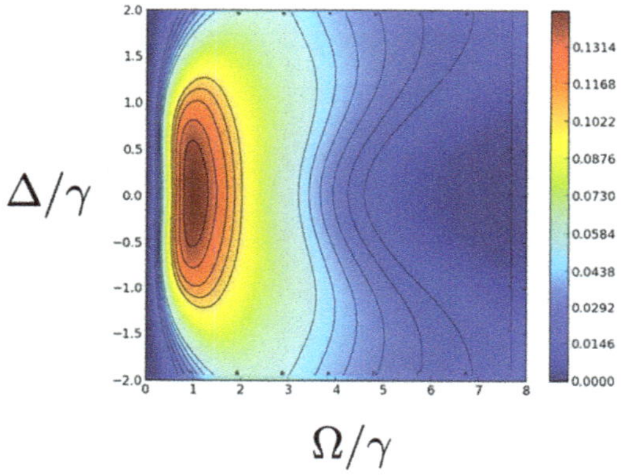

Figure 16.4. The log-negativity with $\theta = \Delta$ parameters used are κ and $g = 1$. 100 states of the cavity are kept.

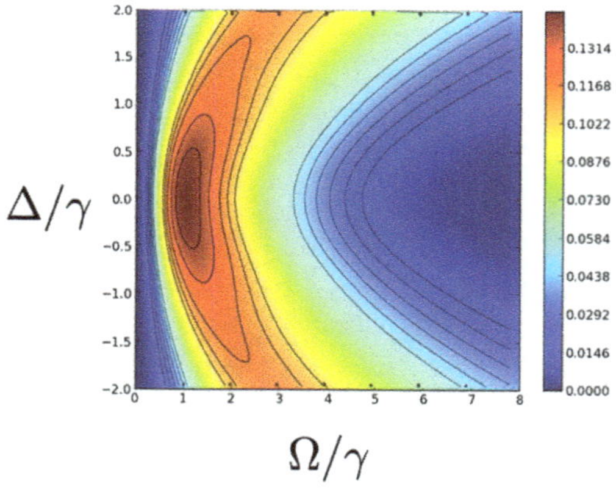

Figure 16.5. The log-negativity with $\theta = -\Delta$. The parameters used are κ and $g = 1$. 100 states of the cavity are kept.

Figure 16.5 shows that when $\theta = \pm\Delta$ this persistence is either enhanced or suppressed. Note that the relative signs of κ and γ matter a great deal. If $\theta = -\Delta$ the terms enhance one another, and hence we observe the behavior in figure 16.4, while if $\theta = \Delta$ we see behavior similar to to driving on resonance, producing the results in figure 16.4. The behavior in figure 16.5 is due to the persistence of the vacuum-Rabi oscillations when the detunings 'cancel' in this manner. One effectively tunes to a vacuum-Rabi sideband and sees two-level behavior. An explanation for this behavior can be found in the examination of the effect Δ and θ have on the vacuum-Rabi frequency. We note that including these detunings effectively makes κ and γ complex,

$$\kappa \rightarrow \kappa(1 + 1\theta) = \tilde{\kappa} \tag{16.196}$$

$$\gamma \rightarrow \gamma(1 + 1\Delta) = \tilde{\gamma}. \tag{16.197}$$

The vacuum-Rabi frequency becomes

$$\tilde{\Omega} = \sqrt{\left(\frac{1}{4}\right)\left(\tilde{\kappa} - \frac{\tilde{\gamma}}{2}\right)^2 - g^2}. \tag{16.198}$$

Although the results presented above seem to imply that entanglement vanishes for fields much larger than the field that saturates the atom, we instead find that when coupling is of sufficiently high magnitude, there is a regime where the entanglement reappears. This is due to the $g\sqrt{n}$ Rabi splitting via strong coupling and/or high driving fields. In general, these conditions produce multi-photon processes, as indicated in figure 16.6 [25].

As various multi-photon resonances are obtained, there are regimes in which the system effectively behaves like a two-level system, resulting in entanglement as in the resonant weak driving-field case. The results are shown in figure 16.7. Note that for a larger photon number, the accompanying entanglement is smaller.

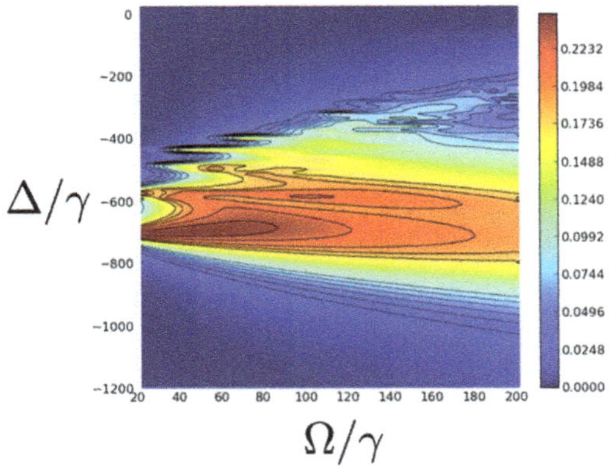

Figure 16.6. Log negativity function of atom detuning and drive strength, indicating entanglement at multi-photon resonances. The parameters used are $\kappa = 1$, $\gamma = 2$, and $g = 1000$. Twenty five states of the cavity kept in the basis used for calculations.

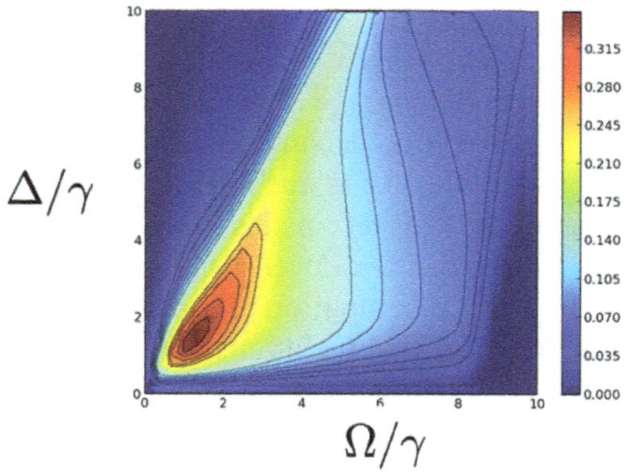

Figure 16.7. Log negativity as a function of the driving field and atom–field coupling for $\kappa = 1$ and $\gamma = 0.1$.

Another regime of interest is near the phase transition that occurs where $2\varepsilon/g = 1$. The atom decay rate must be the smallest rate for this to occur [26, 27]. The entanglement in this regime has been studied by Gea-Banacloche *et al* [28]. One finds that the atom and the field are entangled in the regime where $2\epsilon/g > 1$, with a sharp demarcation between this regime and the regime at fields below the phase transition. Furthermore, as for weak fields, when one drives the system to saturation, spontaneous emission kills the entanglement. This behavior is depicted in figure 16.6. One anticipates the presence of entanglement noting that Gea-Banacloche *et al* found that while is no steady state exists, for individual trajectories one finds

$$|\Psi\rangle_{\text{ss}} = \frac{1}{\sqrt{2}}\left[\left| \psi_+^0(|\alpha_{\text{ss}}|, -\Phi_{\text{ss}})\right\rangle + e^{-i\theta}\left| \psi_-^0(|\alpha_{\text{ss}}|, \Phi_{ss})\right\rangle\right] \tag{16.199}$$

with

$$\left| \psi_\pm^0(r, \theta_0)\right\rangle = \frac{1}{\sqrt{2}}[e^{-i\theta_0}|e\rangle \pm |g\rangle]|re^{-i\theta_0}\rangle, \tag{16.200}$$

which is clearly an entangled state.

References

[1] Jaeger G 2007 *Quantum Information Quantum Information: An Overview* (Berlin: Springer Nature)

[2] Wilde M 2017 *Quantum Information Theory* 2nd edn (Cambridge: Cambridge University Press)

[3] Vedral V 2007 *Introduction to Quantum Information Science* (Oxford: Oxford University Press)

[4] Nielsen M A and Chuang I L 2011 *Quantum Computation and Quantum Information: 10th Anniversary Edition* (Cambridge University Press)

[5] Bell J S 1964 On the Einstein Podolsky Rosen paradox *Phys. Physique Fizika* **1** 195–200

[6] Franson J D 1989 Bell inequality for position and time *Phys. Rev. Lett.* **62** 2205–8

[7] Tsirel'son B S 1987 Quantum analogues of the Bell inequalities. The case of two spatially separated domains *J. Sov. Math.* **36** 557–70
[8] Clauser J F, Horne M A, Shimony A and Holt R A 1969 Proposed experiment to test local hidden-variable theories *Phys. Rev. Lett.* **23** 880–4
[9] Mermin N D 1990 *Boojums All The Way Through: Communicating Science in a Prosaic Age* (Cambridge: Cambridge University Press)
[10] Mermin N D 1981 Bringing home the atomic world: quantum mysteries for anybody *Am. J. Phys.* **49** 940–3
[11] Stuckey W M, Silberstein M, McDevitt T and Le T D 2020 Answering Mermin's challenge with conservation per no preferred reference frame *Sci. Rep.* **10** 15771
[12] Greenberger D M, Horne M A, Shimony A and Zeilinger A 1990 Bellas theorem without inequalities *Am. J. Phys.* **58** 1131–43
[13] Schrödinger E 1935 Discussion of probability relations between separated systems *Math. Proc. Cambridge Philos. Soc.* **31** 555–63
[14] Einstein A, Podolsky B and Rosen N 1935 Can quantum-mechanical description of physical reality be considered complete? *Phys. Rev.* **47** 777–80
[15] Peres A 1996 Separability criterion for density matrices *Phys. Rev. Lett.* **77** 1413–5
[16] Horodecki M, Horodecki P and Horodecki R 1996 Separability of mixed states: necessary and sufficient conditions *Phys. Lett.* A **223** 1–8
[17] Bennett C H and Brassard G 2014 Quantum cryptography: public key distribution and coin tossing *Theor. Comput. Sci.* **560** 7–11
[18] Svetlichny G 1987 Distinguishing three-body from two-body nonseparability by a Bell-type inequality *Phys. Rev.* D **35** 3066–9
[19] Rungta P, Bužek V, Caves C M, Hillery M and Milburn G 2001 Universal state inversion and concurrence in arbitrary dimensions *Phys. Rev.* A **64** 042315
[20] Rungta P and Caves C M 2003 Concurrence-based entanglement measures for isotropic states *Phys. Rev.* A **67** 012307
[21] Nha H and Carmichael H J 2004 Entanglement within the quantum trajectory description of open quantum systems *Phys. Rev. Lett.* **93** 120408
[22] Rogers R, Cummings N, Pedrotti L M and Rice P 2017 Atom-field entanglement in cavity QED: nonlinearity and saturation *Phys. Rev.* A **96** 052311
[23] Carmichael H J, Brecha R J and Rice P R 1991 Quantum interference and collapse of the wavefunction in cavity QED *Opt. Commun.* **82** 73–9
[24] Rice P R, Gea-Banacloche J, Terraciano M L, Freimund D L and Orozco L A 2006 Steady state entanglement in cavity QED *Opt. Express* **14** 4514–24
[25] Shamailov S S, Parkins A S, Collett M J and Carmichael H J 2010 Multi-photon blockade and dressing of the dressed states *Opt. Commun.* **283** 766–72
[26] Carmichael H J 2015 Breakdown of photon blockade: a dissipative quantum phase transition in zero dimensions *Phys. Rev.* X **5** 031028
[27] Alsing P and Carmichael H J 1991 Spontaneous dressed-state polarization of a coupled atom and cavity mode *Quantum Opt.: J. Eur. Opt. Soc.* B **3** 13
[28] Gea-Banacloche J, Burt T C, Rice P R and Orozco L A 2005 Entangled and disentangled evolution for a single atom in a driven cavity *Phys. Rev. Lett.* **94** 053603

Chapter 17

Further!

17.1 For future study

Here I present areas of very active interest with a brief description, and a starting reference. These are sometimes a little bit dated, but provide exceptional starting points in my opinion.

1. **Circuit quantum electrodynamics.** This is a quite fertile area, where an atom in an optical cavity is replaced by (often) a superconducting qubit, or transmon, in a microwave or radio frequency resonator. It is one of the most popular platforms for quantum computers. See:

   Blais A, Grimso A, Girvin S M and Walraff A 2021 Circuit quantum electrodynamics *Rev. Mod. Phys.* **93** 025005.
2. **Cooling and trapping.** This area goes back to the 1980s and has seen a number of Nobel Prizes awarded. One can trap ions in a radio frequency trap, and use Doppler sideband methods to cool them to the vibrational ground state. A promising platform for quantum computers, you can use fields to move ions around in the trap to generate qubit interactions. Also one can use lasers to cool and trap neutral atoms and simple molecules, using a gradient in a dipole, an optical field, or their orientation. Another platform for quantum computing, consider a 2D or 3D array of optical tweezers with an average occupation of one atom, and use Rydberg blockade effects to accomplish qubit–qubit interactions, or trapping a radioactive isotope to do spectroscopic studies that have relevance to weak nuclear interactions. See:

   Haffner H, Roos C F and Blatt R 2008 Quantum computing with trapped ions *Phys. Rep.* **469** 155–203, and

   Metcalf H J and van der Straten P 2008 *Laser Cooling and Trapping* (Berlin: Springer).
3. **Feedback and control.** This is the study of measuring the output of a quantum system acting as a sensor or quantum logic circuit, and applying a

doi:10.1088/978-0-7503-6014-2ch17

feedback signal to control or steer the system in some way. Particularly useful for error correction. One must be careful of course, as the measurement process will disturb the system. See:

Wiseman H M and Milburn G J 2009 *Quantum Measurement and Control* Illustrated Edn (Cambridge: Cambridge University Press).

4. **Many-body physics.** Many-body systems are ubiquitous in condensed matter physics, as well as quantum optics. Bose–Einstein condensation is a great example. Quantum optics methods are used more often now, as the level of control and isolation of such systems is at ever increasing levels. See:

   Shuryak E 2018 *Quantum Many-Body Physics in a Nutshell* (Princeton, NJ: Princeton University Press).
5. **Non-Markovian systems.** We have focused exclusively on systems that are coupled to a bath of some type, which has a rapid response function. This of course is not always the case, the environment does have a memory, even if it is short-lived. Current methods for dealing with such system include enlarging the system to include slow varying fields at the cost of enlarging the Hilbert space, some form of perturbation theory, or numerical simulation. Finding even better ways to implement these and newer methods are a very active sphere of research. See:

   Milz S and Modi K 2021 Quantum stochastic processes and quantum non-Markovian phenomena *PRX Quantum* **2** 030201.
6. **PT-symmetric and non-Hermitian systems.** These are systems that entail both gain and loss as in a laser. The name refers to a system's invariance under combined parity (spatial reflection) and time-reversal transformation, where the operator PT commutes with the Hamiltonian H. One often explores systems whose Hamiltonians are not necessarily Hermitian. While Hermitian operators are commonly used to describe closed quantum systems, non-Hermitian systems can model open systems. Non-Hermitian Hamiltonians are generally complex, leading to complex eigenvalues. However, certain non-Hermitian systems, especially those with PT symmetry, can exhibit entirely real spectra. This opens the door to exploring unusual phenomena such as exceptional points (EPs), where eigenvalues coalesce, and systems with non-trivial topological and dynamical behavior despite non-Hermitian dynamics. See:

   Moiseyev N 2011 *Non-Hermitian Quantum Mechanics* Illustrated Edn (Cambridge: Cambridge University Press).
7. **Quantum communication.** With the advent of nonclassical states of the electromagnetic field, researchers are exploring limitations place on communications due to Heisenberg uncertainty and quantum noise in general. They are also investigating ways in which nonclassical entangled or squeezed states can yield a performance superior to classical communication systems. See:

   Djordjevic I B 2022 *Quantum Communication, Quantum Networks, and Quantum Sensing* (New York: Academic).

8. **Quantum computing and error correction**. Quantum computing references were given in chapter 16. One of the greatest obstacles to a useful quantum computer is a lack of scaling. That is, due to errors, a logical qubit must be comprised of a few to many physical qubits. Lowering the error rate of qubits requires error correction, and enhances the scalability of the system. See:
   Lidar D S and Brun T A 2013 *Quantum Error Correction* (Cambridge: Cambridge University Press).
9. **Quantum gravity.** This is a very interesting field with much to be done. Consider the quantum nature of two-photon processes in the dynamical Casimir effect, the Hawking effect, the Unruh effect, and the optical parametric oscillator; there is likely some stuff in there applicable to gravitational systems at the quantum/classical boundary. Also OTOC correlations can be used to quantify information scrambling in black holes, and it has been proposed that space-time itself is knitted together via entanglement. See:
   Mukhanov V and Winitzki S 2007 *Introduction to Quantum Effects in Gravity* (Cambridge: Cambridge University Press).
10. **Quantum machine learning.** Machine learning, neural networks, and artificial intelligence (the latter a misnomer in my opinion) have been used and abused in a widespread manner. It is only natural to ask if there is a quantum advantage to using quantum circuits or the like to perform analogous operations. See:
   Wittek P 2014 *Quantum Machine Learning: What Quantum Computing Means to Data Mining* (New Work: Academic) and
   Schuld M and Petruccione F 2021 *Machine Learning with Quantum Computers* 2nd edn (Berlin: Springer).
11. **Quantum opto-mechanics.** In this area, researchers study the coupling of optical systems to mechanical ones, such as a cavity mirror on a spring. Essentially the interaction of photons and photons, quite valuable for transducing optical frequencies to lower ones for signal processing with conventional techniques. See:
   Bowen W P 2020 *Quantum Optomechanics* (Boca Raton, FL: CRC Press).
12. **Quantum phase transitions.** What is a quantum phase transition?Is quantum physics not involved somewhere in everything? Yes, we refer here to phase transitions at $T = 0$, and the order parameter is a ratio of couplings in a system, such as the atom–field coupling and the cavity decay rate. The Dicke system is quite rich as well as cavity QED systems. The lab accessible knob is not a temperature, but some adjustable parameter. See:
   Sachdev S 2011 *Quantum Phase Transitions* 2nd edn (Cambridge: Cambridge University Press).
13. **Quantum sensing.** This refers to the use of quantum mechanical principles to measure physical quantities (such as time, position, magnetic fields, temperature, or acceleration) with extremely high precision, beyond the capabilities of classical sensors. Quantum sensors take advantage of phenomena such as quantum superposition, entanglement, and quantum interference to

achieve sensitivity and accuracy that would be impossible with classical techniques. There are even proposals for quantum radar. See:

Degen C L, Reinhart F and Cappellaro P 2017 Quantum sensing *Rev. Mod. Phys.* **89** 035002.

14. **Quantum spectroscopy.** Is there a quantum advantage to using squeezed or entangled light sources in lieu of classical laser sources? There seems to be, but the quantum sources do not produce a large photon flux. Of note is entangled pair spectroscopy of molecules. See:

    Schlawin F 2017 Entangled photon spectroscopy *J. Phys. B: At. Mol. Opt. Phys.* **50** 203001, and

    Landes T, Allgaier M, Merkouche S, Smith B J, Marcus A H and Raymer M G 2021 Experimental feasibility of molecular two-photon absorption with isolated time–frequency-entangled photon pairs *Phys. Rev. Res.* **3** 033154.

15. **Quantum thermodynamics.** This is an interdisciplinary field that combines quantum mechanics with thermodynamics to study the behavior of energy, entropy, and other thermodynamic quantities in systems governed by quantum laws. Unlike classical thermodynamics, which deals with macroscopic systems, quantum thermodynamics addresses how thermodynamic processes operate on microscopic scales, where quantum effects such as superposition, entanglement, and coherence play significant roles. How are heat transfer, energy, and entropy defined in terms of a density matrix. See:

    Deffner S and Campbell S 2019 *Quantum Thermodynamics: An Introduction to the Thermodynamics of Quantum Information* (Bristol: Institute of Physics Publishing).

16. **Ultrastrong coupling between light and matter.** This is a regime in quantum optics where the interaction strength between a light field (typically a photon) and a matter system (such as an atom, molecule, or quantum dot) is so strong that the traditional weak-coupling approximation no longer holds. In this regime, the coupling strength between the light and matter exceeds the typical energy scales of both the system and the light field. the interaction between the light and matter is intense enough that the Rabi frequency (which describes the oscillation frequency between the states of light and matter) becomes comparable to or larger than the frequency of the individual systems (the matter and photon). This means that the energy exchange between light and matter occurs on a fast timescale, influencing both systems simultaneously. One must go beyond the Jaynes–Cummings model, to include nonlinear effects due to a single photon. See:

    Forn-Díaz P, Lamata L, Rico E, Kono J and Solano E 2019 Ultrastrong coupling regimes of light–matter interaction *Rev. Mod. Phys.* **91** 025005.

17. **Tensor networks.** In the context of quantum systems, tensor networks are a powerful mathematical framework used to represent and compute the properties of quantum many-body systems, in particular when these systems are too large to be treated exactly using traditional methods. Tensor networks allow for efficient approximations of quantum states, and they have become particularly important for studying strongly correlated quantum systems,

such as those found in condensed matter physics, quantum information, and quantum computing. In particular, one has a systematic way of ignoring unlikely states, for example in a gas of 10^23 atoms, the odds that they are all in the third excited state is quite small, and that state can be ignored without losing track of the physics. All quantum gates can be represented by a tensor network, and also continuous variable quantum systems. See:

Bañuls M C 2023 Tensor network algorithms—a route map *Ann. Rev. Condens. Matter Phys.* **14** 173–91, and

Orus R 2019 Tensor networks for complex quantum systems *Nat. Rev. Phys.* **1** 538–50.

18. **Topological optics.** This is a field of optics that applies the principles of topology—a branch of mathematics dealing with properties of spaces that are preserved under continuous deformations (such as stretching or bending, but not tearing or gluing)—to the study of light and its interactions with materials. It explores how the topological properties of optical systems, such as their band structures and the behaviors of light at their boundaries, can lead to novel and robust phenomena in the propagation of light. Topological effects in optics are inspired by the field of topological condensed matter physics, where concepts such as topological insulators and topologically protected edge states are used to describe materials whose properties are dictated by their topological features rather than their detailed microscopic structure. See:

    Hashemi A, Zakeri M J, Jung P S and Blanco-Redondo A 2025 Topological quantum photonics *APL Photonics* **10** 010903.

19. **Waveguide quantum electrodynamics**. In this field, atoms are interacting with a semiconductor waveguide or fiber optic. The atoms are trapped near the waveguide and couple to field modes (and then to each other) via the evanescent field. The coupling between the atoms does not fall off as a dipole ($1/r^3$) but rather as $\sin(kr)$. Hence strong coupling and collective effects do not require the atoms to be in a volume on the order λ^3. This also happens in a single mode traveling wave optical cavity, and is responsible for the replacement of the single-atom coupling to the field g with Ng for mean field effects. The atoms are trapped, and can be manipulated to a position of choice. See:

    Sheremet A S, Petrov M I, Iorsh I V, Poshakinskiy A V and Poddubny A N 2023 Waveguide quantum electrodynamics: collective radiance and photon–photon correlations *Rev. Mod. Phys.* **95** 015002, and

    Masso S J and Asenjo-Garcia A 2020 Atomic-waveguide quantum electrodynamics *Phys. Rev. Res.* **2** 043213.

I am sure I have missed someone's favorite, and I apologize. The rate at which the field of quantum technology is growing is amazing. Hopefully these suggestions and this book will help point the way to your quantum calling. So for now, be excellent to one another and carry on!

Index

doi:10.1088/978-0-7503-6014-2ch18

www.ingramcontent.com/pod-product-compliance
Ingram Content Group UK Ltd.
Pitfield, Milton Keynes, MK11 3LW, UK
UKHW051832301026
9729IPUK00005B/56

9 780750 360128